COMPOUND SEMICONDUCTOR RADIATION DETECTORS

Series in Sensors

Series Editors: Barry Jones and Haiying Huang

Other recent books in the series:

Nanosensors: Physical, Chemical, and Biological
Vinod Kumar Khanna

Handbook of Magnetic Measurements
S. Tumanski

Structural Sensing, Health Monitoring, and Performance Evaluation
D. Huston

Chromatic Monitoring of Complex Conditions
Edited by G. R. Jones, A. G. Deakin, and J. W. Spencer

Principles of Electrical Measurement
S. Tumanski

Novel Sensors and Sensing
Roger G. Jackson

Hall Effect Devices, Second Edition
R. S. Popovic

Sensors and Their Applications XII
Edited by S. J. Prosser and E. Lewis

Sensors and Their Applications XI
Edited by K. T. V. Grattan and S. H. Khan

Thin Film Resistive Sensors
Edited by P. Ciureanu and S. Middelhoek

Electronic Noses and Olfaction 2000
Edited by J. W. Gardner and K. C. Persaud

Sensors and Their Applications X
Edited by N. M. White and A. T. Augousti

Sensor Materials
P. T. Moseley and J. Crocker

COMPOUND SEMICONDUCTOR RADIATION DETECTORS

Alan Owens
European Space Agency
Noordwijk, The Netherlands

CRC Press is an imprint of the
Taylor & Francis Group, an **informa** business
A TAYLOR & FRANCIS BOOK

Cover: A low-noise, Peltier-cooled, 6×6×1-mm^3 CdTe detector. (Image courtesy of Oxford Instruments Analytical Oy.)

CRC Press
Taylor & Francis Group
6000 Broken Sound Parkway NW, Suite 300
Boca Raton, FL 33487-2742

First issued in paperback 2016

Version Date: 20120202

ISBN 13: 978-1-138-19958-3 (pbk)
ISBN 13: 978-1-4398-7312-0 (hbk)

Library of Congress Cataloging-in-Publication Data

Owens, Alan.
Compound semiconductor radiation detectors / Alan Owens.
p. cm. -- (Series in sensors)
Includes bibliographical references and index.
ISBN 978-1-4398-7312-0 (hardback)
1. X-rays--Measurement--Materials. 2. Gamma ray detectors--Materials. 3. X-ray diffractometer--Materials. 4. Compound semiconductors. I. Title.

QC481.5.O94 2012
537.5'352--dc23 2012000808

Visit the Taylor & Francis Web site at
http://www.taylorandfrancis.com

and the CRC Press Web site at
http://www.crcpress.com

Dedication

For Cecilia, Katie, Andrea, and Thelma

Contents

List of Figures .... **xiii**
Preface .... **xxxiii**
About the Author .... **xlv**

Chapter 1: Semiconductors **1**
1.1 Metals, Semiconductors, and Insulators
1.2 Energy Band Formation
1.3 General Properties of the Bandgap
1.3.1 Carrier Generation and Recombination
1.3.2 Pressure Dependence of the Bandgap
1.3.3 Temperature Dependence of the Bandgap
1.3.4 Direct and Indirect Bandgaps
1.3.4.1 Electrons in Solids
1.3.4.2 Electrons in the Conduction Band
1.3.4.3 Band Morphology
1.4 Carrier Mobility
1.5 Effective Mass
1.6 Carrier Velocity
1.6.1 Saturated Carrier Velocities
1.7 Conduction in Semiconductors
1.7.1 Intrinsic Semiconductors
1.7.1.1 Intrinsic Carrier Concentration
1.7.2 Extrinsic Semiconductors
1.7.2.1 Donors and Acceptors
1.7.2.2 Extrinsic Carrier Concentration
1.7.2.3 Doping Dependence of the Energy Bandgap
1.7.2.4 Practical Considerations
1.7.3 Conductivity and Resistivity
References

Chapter 2: Growth Techniques **49**
2.1 Crystal Lattices
2.1.1 The Unit Cell
2.1.2 Bravais Lattice
2.1.3 The Pearson Notation
2.1.4 Space Groups
2.1.5 Miller Indices
2.2 Underlying Crystal Structure of Compound Semiconductors
2.2.1 Lattice Constant and Bandgap Energy of Alloy Semiconductors
2.2.2 Bonding

2.2.3 Common Semiconductor Structures
2.2.4 Polycrystalline and Amorphous Structures
2.3 Crystal Formation
2.4 Crystal Defects
2.4.1 Point Defects
2.4.2 Line Defects (Dislocations)
2.4.2.1 Edge Dislocations
2.4.3 Plane Defects
2.4.4 Bulk Defects
2.5 Crystal Growth
2.5.1 Material Purification
2.6 Bulk Growth Techniques
2.6.1 Czochralski (CZ)
2.6.2 Liquid Encapsulated Czochralski (LEC)
2.6.2.1 Limitations of the Czochralski Method
2.6.3 Vapor Pressure Controlled Czochralski (VCz)
2.6.4 Float-Zone Growth Technique (FZ)
2.6.5 Bridgman–Stockbarger (B-S)
2.6.6 High Pressure Bridgman (HPB)
2.6.7 Travelled Molten Zone (TMZ) or Heater Method (THM)
2.6.8 Vertical Gradient Freeze (VGF)
2.7 Discussion
2.8 Epitaxy
2.8.1 Substrates
2.8.2 Strain and Electronic Properties
2.8.3 Lattice Matching
2.8.4 Bandgap Engineering
2.9 Growth Techniques: VPE, LPE, MBE, and MOCVD
2.9.1 Liquid-Phase Epitaxy (LPE)
2.9.2 Chemical Vapor Deposition (CVD)/ Vapor-Phase Epitaxy (VPE)
2.9.2.1 Doping in Vapor Deposition Systems
2.9.3 The Multi-Tube Physical Vapor Transport (MTVPT) Technique
2.9.4 Molecular-Beam Epitaxy (MBE)
2.9.5 Metal Organic Chemical Vapor Deposition (MOCVD)
References

Chapter 3: Detector Fabrication 119
3.1 Mechanical Processing Overview
3.1.1 Thermal Annealing
3.1.2 Cutting
3.1.3 Lapping and Polishing
3.1.4 Etching
3.1.5 Cleaning
3.1.6 Contact Deposition
3.1.7 Lithography

3.2 Detector Characterization
3.2.1 Chemical Analysis
3.2.1.1 Inductively Coupled Plasma Spectroscopy (ICP-MS and ICP-OES)
3.2.1.2 Glow-Discharge Mass Spectrometry (GDMS)
3.2.2 Crystallographic Characterization
3.2.2.1 Single-Crystal X-Ray Diffraction
3.2.2.2 Powder Diffraction
3.2.2.3 Rocking Curve (RC) Measurements
3.2.2.4 XRD and Detector Performance
3.2.3 Electrical Characterization
3.2.3.1 Current-Voltage (I-V) Measurements
3.2.3.2 Contact Characterization
3.2.3.3 Measuring Contact Resistance
3.2.3.4 Capacitance-Voltage (C-V) Measurements
3.2.4 Electronic Characterization
3.2.4.1 Determining the Majority Carrier
3.2.4.2 Determining Effective Mass
3.2.4.3 The Hall Effect
3.2.5 Evaluating the Charge Transport Properties
3.2.5.1 Estimating the Mobilities
3.2.5.2 Estimating the Mu-Tau ($\mu\tau$) Products
3.2.5.3 Limitations of the Hecht Equation
3.2.5.4 Measuring the Charge Collection Efficiency
3.2.6 Defect Characterization
3.2.6.1 Thermally Stimulated Current (TSC) Spectroscopy
3.2.6.2 Deep Level Transient Spectroscopy
3.2.6.3 Photo-Induced Current Transient Spectroscopy (PICTS)
3.2.7 Photon Metrology
3.2.7.1 Synchrotron Radiation
3.2.7.2 Light Sources
3.2.7.3 Synchrotron Radiation Facilities
3.2.7.4 Properties of the Beam
3.2.7.5 Beamline Design
3.2.7.6 Installing the Detector
3.2.7.7 Harmonic Suppression
3.2.7.8 Extending the Energy Range
3.2.7.9 Detector Characterization
3.2.7.10 Probing Depth Dependences
3.2.7.11 Defect Metrology
3.2.7.12 Pump and Probe Techniques
3.2.7.13 X-Ray Absorption Fine Structure (XAFS) Metrology
3.2.7.14 Structural Studies
3.2.7.15 Topographical and Surface Studies
References

Chapter 4: Contacting Systems 207
4.1 *Metal Semiconductor Interfaces*
4.2 *Schottky Barriers*
4.2.1 *Image Force Reduction of the Schottky Barrier*
4.2.2 *Barrier Width*
4.2.3 *Measured Barrier Heights*
4.2.3.1 *Metal-Induced Gap States (MIGS)*
4.2.3.2 *Fermi Level Pinning*
4.3 *Current Transport across a Schottky Barrier*
4.3.1 *Thermionic Emission (TE)*
4.3.2 *Thermionic Assisted Field Emission (TFE)*
4.3.3 *Field Emission (FE)*
4.3.4 *Relative Contributions of TE, TFE, and FE*
4.3.5 *Estimated Contact Resistances for TE, FTE, and FE Current Modes*
4.3.6 *Other Current Components*
4.3.6.1 *Current Due to Image Force Lowering of the Potential Barrier*
4.3.6.2 *Generation–Recombination Effects*
4.3.6.3 *Surface Leakage Current*
4.4 *Ohmic Contacts*
4.4.1 *Practical Ohmic Contacts*
4.4.2 *Barrier Height Reduction*
4.4.2.1 *Choice of Metal*
4.4.2.2 *Doping Concentration*
4.4.2.3 *Annealing*
4.4.2.4 *Interface Doping*
4.4.3 *Barrier Width Reduction*
4.4.4 *Introducing Recombination Centers*
4.4.5 *Desirable Properties of Ohmic Contacts*
4.4.6 *Nonideal Effects in Metal–Semiconductor Junctions*
4.5 *Contactless (Proximity Effect) Readout*
References

Chapter 5: Radiation Detection and Measurement.... 247
5.1 *Interaction of Radiation with Matter*
5.2 *Charged Particles*
5.2.1 *Energy Loss of Secondary Electrons—Collisional and Bremsstrahlung*
5.3 *Neutron Detection*
5.4 *X- and Gamma Rays*
5.4.1 *Photoelectric Effect*
5.4.2 *Coherent Scattering—Thomson and Rayleigh Scattering*
5.4.3 *Incoherent Scattering—Compton Scattering*
5.4.4 *Pair Production*
5.5 *Attenuation and Absorption of Electromagnetic Radiation*

5.6 *Radiation Detection Using Compound Semiconductors*
5.6.1 *Photoconductors*
5.6.2 *The Solid-State Ionization Chamber*
5.6.2.1 *Spectral Broadening in Radiation Detection Systems*
References

Chapter 6: Present Detection Systems 287
6.1 *Compound Semiconductors and Radiation Detection*
6.2 *Group IV and IV–IV Materials*
6.2.1 *Silicon Carbide*
6.2.2 *Diamond*
6.3 *Group III–V Materials*
6.3.1 *Gallium Arsenide*
6.3.2 *Gallium Phosphide*
6.3.3 *Gallium Nitride*
6.3.4 *Indium Phosphide*
6.3.5 *Indium Iodide*
6.3.6 *Narrow-Gap Materials*
6.3.6.1 *Indium Arsenide*
6.3.6.2 *Indium Antimonide*
6.4 *Group II–VI Materials*
6.4.1 *Cadmium Telluride*
6.4.2 *Cadmium Zinc Telluride*
6.4.3 *Cadmium Manganese Telluride*
6.4.4 *Cadmium Selenide*
6.4.5 *Cadmium Zinc Selenide*
6.4.6 *Cadmium Telluride Selenide*
6.4.7 *Zinc Selenide*
6.5 *Group III–VI Materials*
6.5.1 *Gallium Selenide*
6.5.2 *Gallium Telluride*
6.6 *Group n–VII Materials*
6.6.1 *Mercuric Iodide*
6.6.2 *Mercuric Bromoiodide*
6.6.3 *Thallium Bromide*
6.6.4 *Thallium Bromoiodide*
6.6.5 *Lead Iodide*
6.6.6 *Bismuth Triiodide*
6.7 *Ternary Compounds*
6.7.1 *Thallium Lead Iodide*
6.7.2 *Thallium Chalcohalides*
6.7.2.1 *Thallium Gallium Selenide*
6.7.2.2 *Thallium Iodide Selenide*
6.8 *Other Inorganic Compounds*
6.9 *Organic Compounds*

6.10 Discussion
6.11 Neutron Detection
6.11.1 Indirect Neutron Detection
6.11.2 Direct Neutron Detection
6.11.3 Choice of Compound
References

Chapter 7

Chapter Improving Performance 369
7.1 Single Carrier Collection and Correction Techniques
7.1.1 Rise Time Discrimination
7.1.2 Bi-Parametric Techniques
7.1.3 Stack Geometries
7.1.4 Hemispherical Detectors
7.1.5 Coaxial Geometries
7.2 Electrode Design and the Near-Field Effect
7.2.1 Frisch Grid/Ring Detectors
7.2.2 Small-Pixel Effect Detectors
7.2.3 Drift-Strip Detectors
7.2.4 Coplanar Grid Detectors
7.2.5 Ring-Drift Detectors
7.2.6 Other Implementations
7.2.7 Combinations of Techniques
7.3 Discussion and Conclusions
7.4 The Future
7.4.1 General Requirements on Detector Material
7.4.2 The Longer Term
References

Appendix A: Table of Physical Constants 409

Appendix B: Units and Conversions............................413

Appendix C: Periodic Table of the Elements..............417

Appendix D: Properties of the Elements.....................419

Appendix E: General Properties of Semiconducting Materials...................... 427

Appendix F: Table of Radioactive Calibration Sources 483

Index 511

List of Figures

Figure 0.1 Left: the measured energy loss spectrum of a AgCl crystal counter to 400 keV beta particles (from Heerden and Milatz [16]). The applied bias was 200 V. The dotted curve is the theoretically predicted distribution. Right: the relation between the ionization and the beta particle energy (from [16]). The dotted line is the so-called "saturation curve" which in effect is the measured curve corrected for charge collection efficiency....................................xxxviii

Figure 0.2 A schematic diagram of the key components of the crystal conduction counter of Hofstadter (reproduced from [18], courtesy of Platts, a division of the McGraw-Hill Companies, Inc.)...xxxviii

Figure 1.1 Typical range of resistivities/conductivities for insulators, semiconductors, and conductors (adapted from [1]). Semiconductors exist in the shaded region. At resistivities above 10^8 Ω-cm, the distinction between insulators and semiconductors is blurred and ultimately depends on temperature, because an "insulating" material can only become semiconducting (in the sense it can pass a current) if a sufficient number of electrons can be excited into the conduction band. 2

Figure 1.2 Formation of energy bands as a diamond crystal is achieved by bringing together isolated carbon atoms—the energy bands essentially reflect the hybridization of the *s* and *p* levels. At the nominal lattice spacing, the forces of attraction and repulsion between atoms balance. The equilibrium position can be adjusted by pressure and temperature. The conventional one-dimensional representation of band structure at the nominal lattice spacing is given by the left-hand diagram..... 4

Figure 1.3 Schematic energy band diagrams for an insulator, a semiconductor, and a metal. Two cases are given for the latter: (1) semimetals with overlapping valence and conduction bands. Examples are Sn, Zn, Pb, or graphite and (2) classical metals with partially filled conduction bands (e.g., Cu, Au, and Ag). The Fermi level, E_F, is the energy level at which an average of 50% of the available quantum states are filled by an electron and relates the probable location of electrons within a band. For metals the Fermi level lies in the conduction band, while for insulators and semiconductors the Fermi level lies in the band gap. 6

Figure 1.4 Schematic of carrier recombination mechanisms in semiconductors illustrating, (a) radiative emission, (b) deep-level trap mediated, or (c) nonradiative band-to-band Auger recombination...... 8

Figure 1.5 Temperature dependence of the bandgap energy for common semiconductors from groups IV, III–V, and II–VI compounds. Bandgap energies decrease by ~0.4 meV per degree K for most semiconductors. 12

Figure 1.6 Energy band structures of (a) Si and (b) GaAs. Circles (o) indicate free holes in the valence bands, and filled circles (•) indicate free electrons in the conduction bands. Γ, X, and L refer to different conduction band minima along the main crystallographic directions in the crystal. Si is an indirect bandgap material, while GaAs is a direct bandgap material (adapted from [16], © 1976 American Physical Society). The important difference is that, for the direct band gap material, an electron can transit between the lowest potential in the conduction band to the highest potential in the valence band without a change in momentum, $\Delta \boldsymbol{k}$, whereas for an indirect band gap material it cannot do so without the mediation of a third body (e.g., phonon) to conserve momentum. Here m_{hh} and m_{lh} show the valence band maxima that contain heavy and light holes (see Section 1.5). 13

Figure 1.7 A simplified E-$\boldsymbol{k}$ diagram of GaAs showing the three valleys (L, Γ, and X) in the conduction band. For direct gap materials, the height of the Γ valley represents the normally quoted bandgap. The valence band is comprised of three bands with different curvatures. Thus, there are three effective hole masses—heavy holes, light holes, and split-off holes (see Section 1.5). 17

Figure 1.8 Temperature dependence of (a) electron mobilities and (b) hole mobilities for a number of group IV and III–V semiconductors [17–23]. The rollover in mobilities for some group III–V materials below ~100 K is due to impurity scattering. For comparison, the inset in Figure (b) illustrates the mobility temperature dependence on the two main scattering processes. 20

Figure 1.9 The electron velocity as a function of electric field for a number of semiconducting materials at 300 K [18,31,32]. At high fields the drift velocity saturates for the elemental semiconductors. However for compound semiconductors the velocity reaches a maximum at fields around 4 kV cm^{-1} and then begins to decrease due to the increasing influence of additional minima in the conduction bands. 27

Figure 1.10 The Fermi energy distribution function $f(E)$ as a function of electron energy, E. Here, E_v and E_c are the energy levels of the valence and conduction bands, respectively. The defining parameter of this distribution is the Fermi energy E_F, which is the energy at which the probability of occupation by an electron is exactly one-half at $T > 0$ K. Note the reduction of E_F with increasing temperature. However, at room temperature it is still very close to mid-bandgap. 30

Figure 1.11 Schematic of the various distributions discussed in the text leading up to the energy density of electrons and holes in the conduction and valence bands, respectively (from reference [36]). The example given is for an intrinsic semiconductor. (a) The energy band diagram. (b) The density of states (number of states per unit energy per unit volume. The total number of states in the valence band and conduction bands is equal to the number of valence electrons. However in metals the total number of states in the valence band is much larger, which is the reason why electrons in metals need no activation to become mobile. (c) The Fermi–Dirac probability function (probability of occupancy of a state) and (d) the product of $g(E)$ and $f(E)$, which gives the energy density of electrons in the conduction band. The area under $n_E(E)$ versus E is the electron concentration in the conduction band. 33

Figure 1.12 Intrinsic carrier concentrations in Ge, Si, GaAs, and GaP as a function of reciprocal temperature. The effect of increasing bandgap on the intrinsic carrier density is apparent.......... 37

Figure 1.13 Schematic illustrating the physical implementation of doping in silicon. Left: n-type silicon doped with phosphorus. The effect of the phosphorus atom is to introduce an extra electron into the lattice, and thus the phosphorus now acts as a donor for electrons. Right: p-type silicon doped with boron. In this case, the boron atom is missing one electron to fill its outer shell, and thus the boron behaves as an electron acceptor. 38

Figure 1.14 Current conduction in a uniformly doped semiconductor bar of length ℓ and cross-sectional area $A = t \times w$. 44

Figure 2.1 Schematic of the 7 basic crystal systems and 14 conventional Bravais lattices (for a review, see [3]). The lattice centering are: P = primitive centering, I = body centered, F = face centered, C = base centered, and R = rhombohedral (hexagonal class only). 52

Figure 2.2 The interrelationship between the basic crystal systems, the Bravais lattices, point groups, and the Pearson symbols and space groups. The macroscopic symmetry elements are those operations (e.g., reflection and translation) which take place over unit cell dimensions, whereas the microscopic symmetry elements add small translations (less than a unit cell vector) to the macroscopic symmetry operations. A point group is a representation of the ways that the macroscopic symmetry elements (operations) can be self-consistently arranged around a single geometric point. There are 32 unique ways in which this can be achieved.......... 55

Figure 2.3 Examples of different crystallographic planes within a cubic lattice. The Miller indices for each plane are indicated in parenthesis (from [8]). 56

Figure 2.4 (a) Schematic of an ordered substitutional cubic lattice in which an atom of one element replaces an atom of the host element in an alternating sequence. The ability to form a stable lattice depends on whether the two species can satisfy the Hume-Rothery rules. (b) Example of an interstitial lattice in which the atoms of one element fit interstitially into the spaces in the lattice of the host element.................... 58

Figure 2.5 Diagram illustrating the relationship of the elemental and compound semiconductors. Examples of the compound type are given and are listed by increasing bandgap energy or alternatively, decreasing wavelength, from the infrared to the ultraviolet. InSb and AlN delineate the extremes of the range in which compound semiconductors lie (0.17 eV–6.2 eV).. 60

Figure 2.6 Illustration showing how semiconductors bond to form closed valence shells. Examples are given for the four most common semiconductor groups; (a) group IV elemental semiconductors, such as Si and Ge, (b) group III–V semiconductors, such as GaAs and InP, (c) group II–VI compounds, such as ZnS and CdTe, and (d) group I–VII compounds, such AgCl and AgBr. 62

Figure 2.7 The diamond lattice structure. Each atom has four equidistant, tetrahedrally coordinated, nearest neighbors. The unit cell is outlined by the cube of dimension (lattice parameter), *a*. The primitive cell containing one lattice point is shown by the black bonds... 65

Figure 2.8 The zinc blende lattice structure which is most common structure for binary compound semiconductors. Here, the light and dark gray spheres denote the atoms of the binary elements. The unit cell is outlined by the cube of dimension (lattice parameter) *a*. The primitive cell is shown by the black bonds.................. 66

Figure 2.9 The wurtzite lattice structure which is the second most common structure for binary compound semiconductors. Here, the light and dark gray spheres denote the atoms of individual elements. 66

Figure 2.10 A hexagonal lattice structure. The light and dark gray spheres denote the atoms of individual elements.............................. 67

Figure 2.11 Illustration of macroscopic crystal structures in a semiconductor. While the majority of semiconductors solidify into regular periodic patterns shown in (a), they can also form polycrystals shown in (b); that is, a collection of individual grains of crystalline material separated by grain boundaries or (c) amorphous solid solutions in which there is little long-range order. 68

Figure 2.12 Schematic illustrating the different types of point defects in a crystalline material. These are: (a) self-interstitial atom, (b) substitutional impurity atom, (c) Schottky defect, (d) Frenkel defect, (e) interstitial impurity atom. 71

Figure 2.13 Transmission electron microscopy (TEM) image of a large number of dislocations in a GaN crystal (grown by MOCVD). 73

Figure 2.14 Illustrations of line defects. (a) Edge dislocation in a cubic primitive lattice (modified from [21]). (b) Screw dislocation (modified from [22]). 75

Figure 2.15 A screw dislocation growth spiral (from [24]). 76

Figure 2.16 (a) Optical images of two 50 × 50 mm^2, 3-mm-thick slices of a CdZnTe crystal grown by the high pressure Bridgman method (from [27]). Numerous grain boundaries and twins are apparent in the image. (b) The crystals count rate response, measured with a ^{57}Co radioactive source is shown in the lower images, illustrating poor charge collection at the grain boundaries. Interestingly, no correlation was found with the numerous twin boundaries observed inside the grains, indicating that twins have a negligible effect on the electric field and charge collection of semi-insulating CdZnTe devices. 78

Figure 2.17 (a) IR micrograph images of Te precipitates measured in a 1-mm-thick CZT crystal (from [28]). (b) The lower images for the corresponding X-ray response of the crystal when operated as a simple planar detector. The dark spots in this case correspond to a drop in the detector response, demonstrating the link between precipitates and poor device performance. The scans were performed by using a 10 × 10 μm^2, 85 keV X-ray beam. In some cases, the typical triangular shapes of precipitates are recognizable in the X-ray maps. 79

Figure 2.18 Steps required to produce single-crystal material for detector production. 80

Figure 2.19 Upper left: illustration of the principle of the zone refining process (from [30]). Lower left: the practical implementation. Right: the relative impurity distribution along a 10-zone-lengths-long ingot for various numbers (n) of zone passes for a distribution coefficient of 0.5 (adapted from [30]). Here l is the length of the zone, and x is the length of the ingot being refined. The distribution coefficient is the ratio of the solute concentration in the solid to that in the liquid in equilibrium and should be less than unity for purification to occur. 81

Figure 2.20 (a) Schematic of a Czochralski crystal growth system used to produce Si, GaAs, and InP substrate ingots (from [32]), image courtesy of the Center of Advanced European Studies and Research). (b) A typical boule (ingot). The top of the crystal is called the seed end, or alternatively, the first-to-solidify end. The bottom is known as the "tail" or "tang" end. 83

Figure 2.21 Growth sequence in a Czochralski furnace. (Images courtesy of Kinetics Systems, Inc.). (a) Seed crystal is lowered into the melt. (b) As crystallization begins the rod holding the seed crystal is slowly withdrawn. (c) By varying the pull rate the diameter of crystal can be controlled, forming the basis of the ingot. (d) A view into an actual crucible during the drawing of an ingot............ 84

Figure 2.22 Schematic of liquid encapsulated Czochralski furnace showing the position of the encapsulate (from [35]). 86

Figure 2.23 An implementation of the vapor controlled Czochralski technique used for the production of GaAs (from [37]). The essential difference over the standard Czochralski method is the inner pressure chamber and an arsenic source to provide an ambient overpressure of the most volatile component. 88

Figure 2.24 (a) Schematic of the float-zone growth technique (from [40]). (b) Photograph of the growth area. (Courtesy of Topsil Semiconductor Materials A/S.) 89

Figure 2.25 Schematic of the horizontal Bridgman method. The crystal is solidified by slowly withdrawing the charge from the heater..... 90

Figure 2.26 (a) Crystal growth by high pressure vertical and horizontal Bridgman showing the temperature profiles across the charge (modified from [43]). T_m is the melting temperature. (b) A sealed vertical Bridgman charge... 92

Figure 2.27 Schematic of the travelled heater method (from [46], © 2007 IEEE). 94

Figure 2.28 Essential elements of the vertical freeze growth (VGF) method. Left: the furnace temperature profile. Right: the material state along the charge. T_m is the melting temperature. 95

Figure 2.29 Examples of lattice mismatch between the substrate and the grown film showing the effects of bowing, which will occur if one or both crystals are inelastically strained. If the lattice parameter for the substrate a_s is less that that of the epi film case, a_e, (a) then the film will be in compression and the wafer will bow downward. If the converse is true (b), then the film is under tension and the film will bow upward. 98

Figure 2.30 Bandgap energy, E_g, versus lattice constant, a_o, for the most common III–V ternary alloys at room temperature and their relationship to the participating binary (from [49]). The solid interconnect lines represent direct bandgap compounds, while the dotted interconnect lines represent indirect bandgap compounds.......... 100

Figure 2.31 Bandgap energy, E_g, of II–VI compounds as a function of lattice constant a_o (from [49]). The dotted lines illustrate how altering the zinc fraction, x, in $Cd_{(1-x)}Zn_xTe$ alters the bandgap energy. Two cases are shown; $x = 0.1$, which provides optimum energy resolution at T = 243 K, and $x = 0.7$, which provides optimum energy resolution at room temperature (see text)........ 102

Figure 2.32 A schematic of LPE growth system suitable for growing heterostructures. The slider can be moved so that it is aligned with the different melts. 104

Figure 2.33 Reactors for VPE growth (adapted from [55]). (a) Horizontal reactor, (b) vertical reactor, and (c) multi-barrel reactor. In all cases, the substrate temperature must be maintained uniformly over its area. 105

Figure 2.34 Schematic of the essential components of the Multi-Tube Physical Vapor Transport (MTPVT) growth system used to grow CdZnTe. (Image courtesy of Kromek ®.)........ 107

Figure 2.35 Schematic diagram of MBE system (from [62])........ 109

Figure 2.36 Figure illustrating the principle behind the MOCVD growth of GaAs. 111

Figure 2.37 Schematic of the essential components of a typical MOCVD reactor system for the growth of a range of III–V compounds (modified from [66]). Here, the organometallic precursors are: TMA = trimethylaluminum, TMG = trimethylgallium, TMI = trimethylindium, DEZ = diethylzinc, AsH_4 = arsine, and SiH_4 = silane..... 111

Figure 3.1 The entire sequence for producing a detector can be separated into three phases or stages: the growth phase, the material processing stage, and the detector fabrication stage........ 120

Figure 3.2 Left: a diamond disk saw. Right: a wire saw used for cutting ingots into slices prior to detector preparation. (Images courtesy of the European Space Agency and Kromek®.)........ 123

Figure 3.3 Essential elements of a lapping and polishing machine (from [1], image courtesy Logitech Ltd.). The crystal being lapped and polished is held in the metal chuck between two rings. In operation, the rings stay in one location as the lapping plate rotates beneath them. The abrasive is applied onto the plate in the form of slurry, which also aids in lubrication. 124

Figure 3.4 Etching cabinet with automatic titration system for preparing different etch solutions. The cabinet also contains services such as a deionized water supply and compressed air system to clean and dry the crystals. (Image courtesy of the European Space Agency.)........ 126

Figure 3.5 Dislocation etch pits in Si, showing the effects of preferential etching along different crystallographic directions. In the [100] orientation, the etch pits appear (a) elliptical in shape, whereas along the [111] direction they can assume (b) triangular or (c) pyramidal shapes. 126

Figure 3.6 Left: a photolithographic mask used to produce a number of 3 × 3 InAs pixel arrays with guard rings (Image courtesy of Oxford Instruments Analytical Oy). Right: example of a typical sequence of lithographic processing steps for forming the contacts to an array, illustrated for a positive resist. 128

Figure 3.7 Schematic illustrating Bragg reflection. The diffracted X-rays exhibit constructive interference when the distances between paths AA′ and BB′ and CC′ differ by integer numbers of wavelengths (λ). 134

Figure 3.8 A single-crystal X-ray diffraction pattern from a ZnTe crystal. The peaks in the pattern are due to diffraction of X-rays off the various crystallographic planes of atoms within the crystal. In this case, the (200), (400), and (600) reflections are clearly evident. The angular position of each peak (2θ) can be used to calculate the spacing between planes within the structure using Equation (3.1). The derived lattice spacing was 6.18 ± 0.02 Å. The wavelength of the incident X-ray beam was 1.54 Å. 135

Figure 3.9 A powder XRF scan of a $Cd_{1-x}Zn_xTe$ crystal. The wavelength of the incident X-ray beam was 1.54 Å. Each of the crystal planes has been identified. The derived lattice spacing was 6.46 Å. By comparing this value with that expected from pure CdTe and using Vegard's law [6], the Zn fraction x, was determined to be 4.4%. 138

Figure 3.10 Examples of (a) double- and (b) triple-axis X-ray rocking curve (RC) scans of a $Cd_{1-x}Zn_xTe$ crystal, taken before and after various postprocessing treatments have been applied to remove surface damage. The wavelength of the incident X-ray beam was 1.54 Å. Triple-axis RC is generally only used on the highest quality crystals and can give quantitative information on mosaicity and strain in the crystal. 139

Figure 3.11 A Micromanipulator 6000 series probe station. The sample is held on the central 4-inch-diameter chuck by vacuum suction. It can be raised to contact the micromanipulator needles (two shown) with a positional accuracy of a few microns using the microscope. The chuck can also rotate through 360° and can move ±2 inches in X and Y and ±0.7 inches in Z (the vertical direction). 141

Figure 3.12 Current-voltage data for an epitaxial Al/GaAs Schottky diode, modified from Missous and Rhoderick [10]. (a) Conventional logarithmic plot of I versus V. (b) Logarithmic plot of $I/\{1 - \exp(-qV/kT)\}$ versus V. For comparison, we overplot data for a nonideal diode with an ideality factor of ~1.18. Note the calculated curve for this diode overlies the epitaxial diode calculated curve. 144

Figure 3.13 Typical detector construction, which consists of a sandwich of a metal contact, semiconductor, and further metal contact. R_c and R_{bulk} represent the resistances of the contact and the bulk semiconductor. The contact resistance region is that region over which the contact is formed and may have a finite extent depending on how the contact was formed... 145

Figure 3.14 Contact resistance test patterns (from [12]). (a) Measurement configurations for the transfer length method (TLM) and (b) circular transfer length method (CTLM). For the TLM measurement, the semiconductor has been etched away around the contacts to form a mesa in order to restrict current flow to adjacent contacts. For both TLM and CTLM, measurements are usually carried out using a four-probe technique as illustrated in the cross-sectional views. 147

Figure 3.15 Evaluation of the contact and sheet resistances using TLM and CTLM measurements. 149

Figure 3.16 Capacitance (left ordinate—solid line) and $1/C^2$ (right ordinate—dashed line) versus bias voltage of a p-n diode with N_a = 10^{16} cm^{-3}, $N_d = 10^{17}$ cm^{-3}. The area of the diode is 10^{-4} cm^2..................... 151

Figure 3.17 Illustration of the hot probe technique for determining the majority carrier type in semiconductors. Carriers diffuse more rapidly near the hot probe. This leads a flow of majority carriers away from the hot probe and a resultant electrical current toward (p-type) or away from (n-type) the hot probe............................... 152

Figure 3.18 Schematic illustrating the sign convention and terminology for the Hall effect.. 154

Figure 3.19 Hall measurement layouts and contact configurations (from [19], courtesy of Lake Shore Cryotronics Inc.). For measurements, the magnetic field is applied alternately into and out of the pads. 156

Figure 3.20 Van der Pauw arrangement [17] for (a) measuring resistivity and (b) the Hall coefficient in an arbitrarily shaped sample (from [19], courtesy of Lake Shore Cryotronics Inc.). 157

Figure 3.21 Common van der Pauw geometries (from [19], courtesy of Lake Shore Cryotronics Inc). The cross appears as a thin film pattern, and the others are bulk samples. Note: the contacts are shown in black, and the magnetic field is out of the page....................... 159

Figure 3.22 Experimental setup for determining carrier mobilities in a detector by the time-of-flight method. The apparatus is set up for electron measurements. For hole measurements, the source should be placed under the detector or the charge signal on the cathode monitored and the bias reversed. For materials suitable for detector applications, rise times will typically be in the range 0.1 to 10 μs........ 161

Figure 3.23 Measured drift time of electrons for a near cathode event in a 1-cm-thick HgI_2 pixelated detector (adapted from [23], © 2003 IEEE). Waveform *A* is the anode (pixel) signal, while waveform *B* is the cathode signal. Note that the pixel signal is generated only when the electrons are close to the pixel. From the figure, the electron drift time is estimated to be 7 μs. The inset shows pulse waveforms from an event near the anode. In this case, holes generate the signal. However, because they cannot travel the full thickness of the detector, the signal from the cathode is indistinguishable from noise (curve B), while a significant charge can be induced between the anode plane and pixel (curve A)........ 162

Figure 3.24 Charge collection efficiency (CCE) versus applied bias relationship for a 20-mm^2, 0.5-mm-thick TlBr planar detector. From a best fit of Equation (3.44), the mobility-lifetime product of electrons was estimated to be 2.8×10^{-3} cm^2/V (adapted from [25], © 2009 IEEE). The deviations from the fitted curve indicated by the arrows are most likely due to the assumptions implicit in the Hecht equation—no de-trapping or surface recombination. The dashed line shows the improvement in fit using a modified Hecht equation in which the fitted bias is offset by a delta amount........ 164

Figure 3.25 Anode and cathode pulse height distributions from a 4-mm-thick HgI_2 pixel detector (from [23], © 2003 IEEE). The effect of significant hole trapping is observed in the much reduced hole pulse height distribution........ 166

Figure 3.26 Charge collection efficiencies measured for two sides of a CdZnTe sample as a function of bias for the 59.54 keV photopeak of ^{241}Am (from [28]). The fitted parameters of a Hecht curve are given in Table 3.3........ 168

Figure 3.27 Schematic of the third generation 3-GeV Diamond Light Source [47] at Harwell, Oxford, UK, showing its major components. The ring is not truly circular, but is shaped as a twenty-four-sided polygon with a beamline at each vertex. Each beamline is optimized to support an experimental station which specializes in a specific area of science, including the life, physical, and environmental sciences. A station is generally comprised of an optics hutch, experimental hutch, and a control room. (Image courtesy of Diamond Light Source 2011.)........ 178

Figure 3.28 Spectral brilliance for several synchrotron radiation sources and conventional X-ray sources. The data for conventional X-ray tubes should be taken as rough estimates only, because brightness strongly depends on operating conditions (adapted from [48], courtesy LBNL). 180

Figure 3.29 The X1 hard X-ray beamline at the HASYLAB synchrotron radiation source at DESY [49]. 181

Figure 3.30 Photograph showing the installation of a detector ready for characterization on beamline X1 at the HASYLAB synchrotron research facility at DESY. A laser attached to the front of the detector is used to align its aperture with the center of the slits. A coaligned reference detector is then used to establish precisely the position of the beam with respect to the detector principal axis prior to scanning........ 184

Figure 3.31 Top: measured energy-loss spectra at 30 and 98 keV of a 250×250 μm^2 GaAs pixel detector. For the latter distribution, a Compton edge is apparent near 30 keV, below which is a Compton continuum. The measurements were carried out at a temperature of −40°C, a bias of 100 V, and a shaping time of 2 μs. Bottom: an expansion of the 30 keV data. 187

Figure 3.32 The energy-dependent FWHM energy resolution of a 250×250 μm^2 GaAs pixel detector under pencil beam illumination. The best-fit energy resolution function together with its key components is also shown........ 188

Figure 3.33 Left: photograph of a 4×4 GaAs pixel array and its associated front-end electronics. The inset shows a blow-up of the array before wire bonding. The pixel size is 350 μm × 350 μm, and the interpixel gap is 50 μm. Right: a surface plot of the spatial variation of the gain (the fitted centroid position) across the array measured at HASYLAB using a 15 keV, 20×20 μm^2 pencil beam (from [53]). The spatial sampling in X and Y was 10 μm........189

Figure 3.34 Defect diagnostics carried out on a $15 \times 15 \times 10$ mm^3 CdZnTe coplanar grid detector [50]. The figure shows spectrally resolved count rate maps obtained by raster scanning a 20×20 μm^2, 180 keV normally incident beam in 40-μm steps across the detector. Crystal defects are plainly evident in the detector count rate response and account for ~2% of the active area of the detector (light color corresponds to photopeak events, black to a lack of events). The lower right-hand image shows that the counts in the defects do not originate from the photopeak and, in fact, mainly emanate from energies 120 to 170 keV (see lower left-hand image). In this energy region there are virtually no events from the rest of the detector area. ... 191

Figure 3.35 Pulse height data from repeated 60 keV "probe" scans of a polarized volume following a "pump" pulse of 2.4×10^6 60 keV photons. The figures show the evolution of the 60 keV energy-loss peak: (a) peak channel position (initial value 740), (b) relative efficiency (initial value unity), (c) FWHM energy resolution (initial value 27), and (d) pulse height spectrum as a function of elapsed time (from [56])....... 194

Figure 3.36 The measured quantum efficiency across the Si K-edge of an X-ray CCD [64]. Individual edges and bonds are identified. For comparison, we show our calculated values based on new linear absorption coefficients abstracted from photocurrent measurements along with the classical predictions of Cromer and Liberman [65]....... 197

Figure 3.37 Detailed cross section through the overlying dead layers (electrodes, gate dielectrics, polysilicon gates, and passivation layers) above the active depletion region (from [64])....... 198

Figure 3.38 The derived linear attenuation coefficients across the Si K-edge. For Si, the letters *c* and *a* refer to crystalline and amorphous (from [66]). We also show the "classical" Si curve based on the calculation of Cromer and Liberman [65]....... 199

Figure 4.1 Examples of the current-voltage characteristics from ohmic and Schottky metal–semiconductor contacts to *n*-type GaAs doped to 10^{15} cm^{-3}....... 208

Figure 4.2 Energy band diagram of a metal and *n*-type semiconductor (a) before and (b) after contact is made....... 210

Figure 4.3 (a) Field lines and surface charges due to an electron in close proximity to a perfect conductor and (b) the field lines and image charge of an electron....... 213

Figure 4.4 The effect of the image force on the shape of the potential barrier at a metal–semiconductor interface for *n*-type GaAs with a band-bending energy of 1 *eV*; x_m is the maximum height of the barrier....... 217

Figure 4.5 (a) The location of the Fermi level relative to the conduction band (E_c–E_f) barrier heights for Au contacts on various covalent semiconductors, plotted as a function of energy gap, illustrating the two-thirds rule for barrier height pinning for *n*-type semiconductors at the interface (adapted from [20]). (b) Experimentally determined barrier heights for various metals on an ionic semiconductor (ZnS) showing a clear dependence of the barrier height on work function. For contrast, we show data for a covalent semiconductor (GaAs), which has a much weaker dependence....... 220

Figure 4.6 Conduction mechanisms through a metal–semiconductor interface with different semiconductor donor levels for an n-type semiconductor, corresponding to (a) a low residual doping level of $<10^{17}$ cm^{-3}, (b) an intermediate doping level of 10^{17}cm^{-3} to 10^{18}cm^{-3}, and (c) a high doping level of $>10^{18}$cm^{-3}. It can be seen that, as the doping concentration increases, the barrier width decreases as the conduction band falls below the Fermi level. To the right of the barrier we show an additional schematic of the barrier in which the shaded regions give a visual indication of the fraction of carriers that can cross the barrier. 223

Figure 4.7 Ranges of temperature and donor concentration over which n-type GaAs Schotkky diodes exhibit field and thermionic-field emission (adapted from [26], © 1982 IEEE.).................................... 227

Figure 4.8 (a) Illustration of the current-voltage characteristics for a Schottky barrier contact (Au on n-type GaAs) for progressively higher carrier concentrations (from [30]). The curves are: (a) $N_d \leq 10^{17}$ cm^{-3}, for which thermionic emission dominates, (b) $N_d \approx 10^{17}$ cm^{-3}- 10^{18} cm^{-3}, for which thermionic field tunneling dominates, and (c) $N_d \geq 10^{19}$ cm^{-3}, for which field emission tunneling dominates. (b) Current-voltage characteristics as a function of carrier concentration (from [30]]. As can be seen the current can vary by over six orders of magnitude depending on the carrier concentration.... 232

Figure 4.9 (a) The distribution of work function values for the metallic elements, from which it can be seen that they all cluster around a narrow range of values. (b) Measured barrier heights for Au/n-type and Au/p-type metal/semiconductor interfaces. 234

Figure 4.10 Noncontact readout of a detector using proximity electrodes (from [35], © 2009 IEEE)... 242

Figure 5.1 Compton scattering geometry. The electron is assumed to be initially at rest and is ejected at angle ϕ by the incident photon with a kinetic energy E_e. The initial photon in turn is scattered through angle θ but with a reduced energy E_γ', which by conservation of energy is equal to $E_\gamma - E_e$. 259

Figure 5.2 Differential angular distribution of Compton-scattered photons as a function of the angle of scattering, θ, for various incident photon energies, γ, expressed in units of the electron rest mass energy...... 261

Figure 5.3 The lines delineate the regions where photoelectric effect, the Compton effect, and pair production dominate (from Evans [21]). Here the left line marks the region of space where the photoelectric and Compton cross sections are equal as a function of energy and Z. Thus, to the left of the line, the photoelectric effect dominates, while to the right, Compton dominates. The right-hand line marks the boundary in energy-Z space where the Compton and Pair effects dominate.................. 263

Figure 5.4 Schematic of a photoconductor detector consisting of a highly conductive semiconductor bar. Radiation incident on the bar creates additional electron–hole pairs, which lead to an increase in conductivity and consequently current though the device....................... 265

Figure 5.5 Schematic of a simple planar detection system (from [23]). For completeness, we also show the signal chain and the evolution of currents and voltages. Ionizing radiation absorbed in the sensitive volume generates electron–hole pairs in direct proportion to the energy deposited. These are subsequently swept toward the appropriate electrode by the electric field induced by the bias voltage V_b. 268

Figure 5.6 The limiting energy resolution achievable at 5.9 keV for a range of compound semiconductors as a function of bandgap energy at 5.9 keV (from [34]). For completeness we also include the superconductors. Curves are given for average values of the Fano factor (i.e., 0.22 for superconductors and 0.14 for semiconductors). NBG and WBG show the regions in which the narrow bandgap and wide bandgap semiconductors lie. 274

Figure 5.7 (a) A typical detector front-end detector-preamplifier-shaper circuit. (b) Equivalent circuit diagram (adapted from [35]). 275

Figure 5.8 Equivalent noise charge versus shaping time. At low shaping times, series (voltage) noise dominates, whereas at high shaping times current (parallel) noise dominates. 279

Figure 5.9 The energy resolution ΔE of a 3.1-mm^2, 2.5-mm-thick CdZnTe detector measured at two temperatures (from [40]). The solid line shows the best-fit resolution function to –20°C data. The individual components to the FWHM are also shown for this curve. These are: the noise due to carrier generation or Fano noise ΔF, electronic noise due to leakage current and amplifier shot noise, Δe, and incomplete charge collection or trapping noise Δc............................ 282

Figure 6.1 Am X-ray and γ-ray spectra recorded by a 0.03-mm^2 70-μm-thick epitaxial 4H-SiC detector at +27°C and +100°C (from [12]). The spectrum at +100°C is displaced on the y-axis for clarity. At 26.3 keV, the FWHM energy resolution is 670 eV at 27°C and is essentially the same at 100°C operating temperature............................ 296

Figure 6.2 The response of a 1-mm-diameter, 40-micron-thick GaAs detector to ^{241}Am and ^{55}Fe radioactive sources (from [21]). The detector operating temperature was –40°C. The FWHM energy resolutions are 435 eV at 5.9 keV and 670 eV at 59.54 keV...................... 299

Figure 6.3 Fe spectrum acquired with a 200 × 200 μm^2 GaAs pixel detector at –31°C. At room temperature the same pixel shows 242 eV FWHM on the pulser line and 266 eV FWHM at 5.9 keV (from [22]). 300

Figure 6.4 X-ray image of a "swatch," taken by a 64 × 64 GaAs pixel array using a conventional X-ray set operating at 55 kV with a W target and a 2.5 mm Al filter (from [23]). 300

Figure 6.5 The measured response of a 3.1-mm^2, 200-μm-thick InP detector to ^{241}Am under full area illumination at –170°C. The inset shows its response to ^{55}Fe. The measured FWHM energy resolutions were 911 eV at 5.9 keV and 2.5 keV at 59.54 keV (from [21]). 305

Figure 6.6 Energy spectrum of 5.5-MeV alpha particles recorded at 5.5 K by a 1-mm-diameter InSb detector grown by LPE (data courtesy of I. Kanno). The dotted line shows a pulser spectrum. The inset shows the response of the detector to gamma rays from a ^{133}Ba radioactive source. The amplifier gain is 1.75 times that of the gain used to acquire the alpha particle spectrum. 310

Figure 6.7 Co and ^{137}Cs spectra taken with a 1-cm^2, 2.1-mm thick P–I–N CdTe detector cooled to –35°C (from [60]). The operating bias was –3000 V. The FWHM energy resolution at 662 keV was 3.5 keV. 312

Figure 6.8 The temperature dependence of the FWHM energy resolutions measured with a 3.1-mm^2, 2.5-mm-thick CdZnTe at 5.9 keV and 59.54 keV under full-area illumination (from [68]). The solid lines are best-fit polynomials. 315

Figure 6.9 Composite of ^{241}Am and ^{55}Fe spectra taken with a 3.1-mm^2, 2.5-mm-thick CdZnTe detector (from [21]). The detector temperature was –37°C, and the applied bias +320 V. The measured FWHM energy resolutions were 311 eV at 5.9 keV and 824 eV at 59.54 keV. The corresponding pulser widths were 260 keV and 370 eV, respectively. 316

Figure 6.10 Composite ^{241}Am and ^{55}Fe spectra taken with a 7-mm^2, 0.5-mm-thick HgI_2 detector (from [21]). The detector temperature was +24°C, and the applied bias +800 V. The measured FWHM energy resolutions are 600 eV at 5.9 keV and 2.4 keV at 59.54 keV. 327

Figure 6.11 X-ray fluorescence spectrum of a sample of the Murchison meteorite taken with a 5-mm^2, 200-μm-thick HgI_2 detector operated at room temperature. The sample was excited using an X-ray tube source (from [105], © 1991 IEEE). The energy resolution is ~200 eV FWHM. 328

Figure 6.12 ^{55}Fe and ^{241}Am spectra, measured at room temperature with $Hg(Br_{0.2}I_{0.8})_2$ detectors (from [108]). The measured FWHM energy resolutions at 5.9 keV and 59.54 keV are 0.9 keV and 6.5 keV, respectively, under full area illumination. 329

Figure 6.13 The response of an ~8-mm^2, 0.8-mm-thick TlBr detector to ^{241}Am and to ^{55}Fe (insert) using radioactive sources under full area illumination. The detector temperature was –22°C, and the pulser noise width 690 eV FWHM (from [21]). 332

Figure 6.14 Room-temperature spectra of 122 keV (left) and 662 keV (right) gamma rays taken with a large $10 \times 10 \times 10$ mm^3 pixel detector (from [123], © 2009 IEEE). The pixel size was 1.8×1.8 mm^2. The measured FWHM spectral resolutions at 122 keV and 662 keV were 5.5% and 2.5%, respectively. 333

Figure 6.15 Recorded pulse height spectrum from gamma radiation from a ^{57}Co-57 source using a $6 \times 4 \times 2$ mm^3 ⟨001⟩ Tl_6I_4Se wafer (dark solid line) (from [139], © 2011 American Chemical Society). For comparison, the measured pulse height spectrum from a commercial $5 \times 5 \times 5$ mm^3 CZT detector is also shown (dashed line; SPEAR detector manufactured by EI Detection & Imaging Systems). The measurements were carried out at 295 K. 339

Figure 6.16 Cross-section depiction of a self-biased GaAs neutron detector (from [150]). The internal potential formed at the p+/ν-type junction is sufficient to deplete the high-purity GaAs material. The device is coated with pure ^{10}B. Neutrons absorbed in the ^{10}B layer discharge energetic ions (^{7}Li ions and α-particles) in opposite directions. Charges excited within the depleted region are drifted by the internal potential to form measurable pulses. 345

Figure 6.17 Schematic of the principal interaction mechanism in a direct neutron detection device based on a boron compound. 349

Figure 6.18 Pulse height spectra from a p-BN detector exposed to a thermal neutron beam from a reactor demonstrating a clear response (from [168]) 351

Figure 6.19 (a) The response of a 12.6-mm^2, 1.7-μm-thick diode to 5.5 MeV alpha particles. The bias was 10 V. (b) The response to thermal neutrons from the Delft 2 MW nuclear reactor. Even under zero bias the diode shows a response 352

Figure 6.20 Alpha and photon response of a 16-mm^2, 2.8-μm-thick boron diode to an ^{241}Am radioactive source. The inset shows an expansion of the spectrum in the vicinity of 60 keV, showing that the 60 keV gamma-ray line is clearly resolved. The detector bias was 6 V, and shaping time 2 μs. 353

Figure 7.1 Distribution of preamplifier output pulse rise times from a 5-mm^2, 0.5-mm-thick TlBr planar detector illustrating how the wide disparity in carrier mobilities leads to their clear separation in the time domain (from [1]). For complete charge collection, a trade-off is required between an optimum shaping time to minimize electronic noise and a longer shaping time needed to collect the entire signal. 371

Figure 7.2 (a) Schematic of a simple planar detector showing a photoelectric interaction. Ionizing radiation absorbed in the sensitive volume generates electron–hole pairs in direct proportion to the energy deposited. These are subsequently swept toward the appropriate electrode by the electric field, E, induced by the bias voltage, V_b. (b) The time dependence of the induced signal for three different interaction sites in the detector in the absence of trapping. The fast-rising part of the signal is due to the electron component, while the slower-rising part is caused by the holes (from [4])................. 372

Figure 7.3 Two ^{241}Am spectra taken with a $3 \times 3 \times 2$ mm^3 CdZnTe detector, illustrating the effectiveness of rise time discrimination (RTD). From the figure we see that hole tailing is substantially reduced when RTD is employed, and in fact the FWHM energy resolution improves from 1 keV to 700 eV at 59.54 keV. However, an energy-dependent decrease in efficiency is also apparent (from [7])....... 374

Figure 7.4 Stack detector concept. The signal path of each element is essentially or-ed. The bias is common between adjacent planes and alternate between cathodes and anodes. 375

Figure 7.5 ^{137}Cs spectra obtained using the first generation stack detectors, with one, three, and five elements. The improvement in both the resolution and efficiency with the number of elements is clearly evident (from [11], © 2004 IEEE)... 376

Figure 7.6 Distribution of the equipotential field lines in (a) hemispherical and (b) pseudo-hemisphere detector geometries............. 377

Figure 7.7 Comparison between the spectroscopic capabilities of a simple planar detector and a quasi hemispheric CdZnTe detector of the same size. Left: A simulated ^{137}Cs spectrum for a $10 \times 10 \times 5$ mm^3 planar detector. Right: simulated ^{137}Cs spectrum for a $10 \times 10 \times 5$ mm^3 quasi-hemispherical detector (from [13]). 378

Figure 7.8 Two implementations of the Frisch grid concept. Left: a prototype semiconductor design (from [18]). Two parallel contact strips are fabricated on the sides of the detector between the anode and cathode planes and act as a pseudo-Frisch grid. Charge carriers are excited in the interaction region, and the electrons are drifted through the parallel grids by an applied electric field. Right: schematic of a capacitively coupled Frisch grid detector, consisting of (a) a bar-shaped detector placed inside an isolated but conductive ring (b) (from [23]). This implementation effectively eliminates leakage current between the grid and the anode, because the ring is not actually connected to the detector.. 383

Figure 7.9 Left: response of a 5 × 2 × 5 mm^3 CdZnTe patterned Frisch grid detector to a ^{137}Cs radioactive source under full area illumination (from [18]). Shown are spectra taken with the grid turned off and on. No full energy peak is apparent when the parallel grid is off; however a full energy peak of 6.2% FWHM at 662 keV becomes obvious when the grid is activated. Right: ^{137}Cs spectra from a 3 × 3 × 6 mm^3 CdZnTe device with a 5 mm insulated capacitively coupled Frisch ring (from [26]). The thin-lined blue spectrum shows the response of the device with the ring connected, and the thick-lined red spectrum with it disconnected. With the ring connected, the FWHM energy resolution is 2.3% at 662 keV. 384

Figure 7.10 Small-pixel geometry. The detector consists of a pixelated array with pixels of size w, located above a planar cathode. The detector thickness is L. 385

Figure 7.11 (a) and (b). Illustration of the weighting potentials $\Phi_w(x)$ for two ratios of pixel size to detector thickness, namely, $w/L = 2.0$ and 0.38 (from [28]). The contour lines in (a) and (b) are equivalent with the fraction of the total charge generated at $z = 1$ that are detected when the electrons are drawn toward the pixel at $z = 0$ mm and all the holes are lost. Thick black lines indicate the location of the pixel electrodes. In (c) the fraction of the generated charge measured as a function of photon absorption depth is shown for different pixel geometries, characterized by w/L. 386

Figure 7.12 (a) Demonstration of the small-pixel effect in TlBr, in which its spectral properties are greatly influenced by pixel geometry and specifically the ratio of the pixel dimension, w, to its thickness, L. We show two ^{241}Am spectra (from [28]), one recorded by a pixel detector with a ratio $w/L = 0.35$ (broken line) and the other by a 2.7 × 2.7 × 0.8 mm^3 planar detector with a ratio $w/L = 3.4$ (solid line). (b) Modeled 60 keV spectra for the two detectors for the same operating conditions. Both spectra have been convolved with a Gaussian of FWHM 690 eV to simulate the electronic noise of the system. 387

Figure 7.13 Drift-strip detector geometry. The drift-strip electrodes are biased in such a way that the electrons move to the anode strips. 389

Figure 7.14 Schematic illustrating the essential components of a coplanar grid detector (from [31]). 390

Figure 7.15 Comparison of pulse height spectra measured with a 2.25-cm^3 device (15 × 15 × 10 mm^3), when operated as a planar detector and as a coplanar grid detector [32]. 390

Figure 7.16 Schematic image of the prototype ring-drift detector. The crystal has dimensions 5 × 5 × 1 mm^3. The inner anode has a diameter of 80 μm, and the centers of the other two ring electrodes, R1 and R2, are 0.19 and 0.39 mm, away from the center of the anode. The guard ring extends beyond a radius of 0.59 mm. 392

Figure 7.17 The average energy to create an electron–hole pair as a function of bandgap energy for a selection of semiconductors (from [48]). Two main bands are evident—the main branch found by Klein [49] (solid line) and the *n*-VIIB branch (dashed-dotted line). The dotted line denotes the limiting case when $\varepsilon_p = E_g$. The difference between this curve and the measured curves is due to optical phonon losses and the residual kinetic energy left over from impact ionization thresholding effects. Note the solid lines through the two branches are best-fit Klein functions of the form 14/5 $E_g + a_1$, in which a_1 is a free parameter. In order to obtain good fits to both the main and secondary branches, both AlN and diamond were fit as part of the secondary branch because they are clearly displaced from the main branch. Note also that, while the parameter $a_1 = 0.6$ for the main branch is reasonable, in that it should lie in the range $0.5 \leq a_1 \leq 1.0$ [50], the fitted value for the secondary branch is unphysical (i.e., $a_1 = -4.8$). 398

Preface

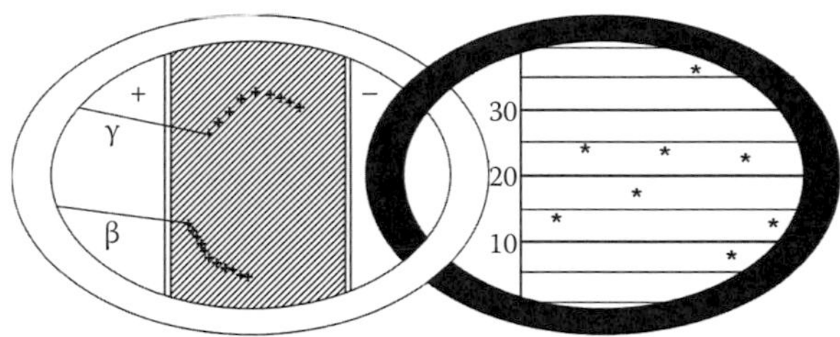

Frontispiece Reproduction of the illustration on the front cover of P.J. van Heerden's seminal doctoral thesis, entitled "*The Crystal Counter: A New Instrument in Nuclear Physics*" (Rijksuniversiteit Utrecht, July 1945). The left-hand side shows a stylistic depiction of the "crystal counter," which was essentially a solid state ionization chamber—the precursor of all modern semiconductor energy resolving radiation detectors. The right-hand side of the image shows the "deflections" (pulse heights) of individual events when the crystal was exposed to an external γ-ray source.

The term compound *semiconductor* encompasses a wide range of materials, most of which crystallize in the zinc blende, wurtzite, or rock salt crystal structures. They are extremely useful because of the sheer number of compounds available and the wide range of physical properties they encompass. In terms of radiation detection, compound semiconductors were among the first direct detection media to be investigated over 60 years ago. However, material problems caused by impurities, high defect densities, and stoichiometric imbalances limited their usefulness. It is only recently that substantial progress has been made in material growth and detector fabrication, allowing compound semiconductor radiation detectors to become serious competitors to laboratory standards such as Ge and Si.

Historical Perspective

The earliest radiation detectors used around the turn of the twentieth century were photographic emulsion and phosphorescent screens. The latter were the precursors of scintillation detectors. Emulsions and screens are integrating devices and, hence, provide no information on individual particles or events. This limitation was overcome in 1908 by Rutherford and Geiger [1], who developed a gas-filled radiation detector that could electronically register individual events in counting experiments. The design was later refined by Geiger and Müller [2] in 1928, culminating in the Geiger–Müller tube—a device still in extensive use today. Its main limitation is that it is essentially insensitive to radiation type or deposited energy. A significant advance in the state of art was achieved in 1930, with the invention of the Kubetsky tube [3], which later evolved into the photomultiplier tube [4]. When coupled to a scintillating material, not only could single-event scintillation signals now be efficiently discriminated and registered, but the energy they deposited could be simultaneously recorded.

Early work on semiconducting materials largely concentrated on the conduction properties of crystals and the rectification properties of crystal and metal contacts. However, during this time some surprising discoveries were made—although not recognized at the time. For example, in 1907 Round [5] recorded that during an investigation of the "unsymmetrical passage of current through a contact of carborundum, a curious phenomenon was noted." He went on to describe how the crystal gave out a yellow glow when a potential of 10 volts was applied between two points on the crystal. In other crystals a light green, orange, or blue glow was observed. Lossev [6] investigated these effects in more detail and correctly postulated that they were caused by a phenomenon that was essentially the inverse of the photoelectric effect that had recently been proposed by Einstein [7]. In actuality, what Round reported was the first observation of electroluminescence from SiC (carborundum), and what he had constructed was a light-emitting diode (based on a Schottky diode, rather than a p-n junction).

The observation of different colors from different crystals was presumably the influence of various impurities on the band-gap (i.e., unintentional band-gap engineering).

Photoconduction Detectors

In terms of radiation detection, early work was carried out by exploiting the photoconduction properties of semiconductors. The basic detection mechanism is as follows. A voltage is applied across a highly doped semiconducting material via two contacts. Light incident on the crystal generates electron–hole pairs, which increases the carrier concentration and, in turn, the conductivity of the material. By applying a small bias, a measurable current will flow through the crystal. Photoconductors differ from crystal counters* in several important ways; most notably they are integrating devices and because of their large intrinsic conductivity are not sensitive to individual particles or quanta.

Early work concentrated on diamond. For example, photoconductive diamond ultraviolet detectors were studied in the 1920s by Gudden and Pohl [8], who investigated their absorption characteristics. In diamond both the electrons with energies in the conduction band and those in the filled valence band are free to move under the influence of an applied electric field. The movement of electrons and holes creates a current which can be amplified. Unfortunately not all the electrons and holes are removed by the electric field because of trapping by crystal imperfections. This creates a space charge which generates an internal electric field opposed to the external field (the so-called polarization effect). The net result is a decreased effective electric field which reduces the amplitude of charge pulses to a level too small to be registered by the electronics. Hence, the count rate decreases with time. The use of diamond was restricted because of the high intrinsic value and the difficulty of obtaining large geological diamonds with predictable

* An early name for a semiconductor radiation detector operated as a "solid state ionization chamber."

properties. During this period a number of other materials were studied—most notably, the thallous halides. The "*photoelectric primary current phenomena*" in TlBr crystals was first described in 1933 by Lehfeldt [9], who later, in a study of the photoconductivies of thallium chloride, thallium bromide, silver chloride, and silver bromide, concluded that the photoconducting properties of the thallium salts and the silver salts showed many similarities [10]. Ionizing radiation was also observed to induce noticeable conductivity in crystals, although the effects were small and not reproducible. For example, Frerichs and Warminsky [11] demonstrated a response to alpha and beta radiation in crystal phosphors. A complete bibliography of this early work may be found in Jaffe [12].

Crystal Counters

In essence, a crystal counter is a particle detector in which the sensitive material is a dielectric (nonconducting) crystal mounted between two metallic electrodes. It is, in effect, a solid state ionization chamber, where the ionization released by the particle is a measure of its energy. The use of a solid as a counter was quickly recognized to provide several significant advantages over gaseous detectors; namely, high stopping power, small size (because the densities of solids are ~1000 times greater than gases), and high-speed operation. Most of the initial work was done with diamond and alkali halide crystals because of their insulating properties and relatively high purity. It was considered necessary to use an insulator so that an electric field could be applied for the purpose of charge collection without prohibitive electronic noise arising from leakage current through the device. The work was supported by the theoretical work of Hecht [13] who, in 1932, had already demonstrated a relationship between an ionization-induced current in a crystal and the electric field strength.

The first practical implementation of a crystal counter can be attributed to van Heerden [14], who in 1945 demonstrated that silver chloride crystals when cooled to low temperatures were capable of detecting individual gamma rays, alpha particles, and beta particles. This was not an integrating device, but was fully spectroscopic. Previous to this work, the energy spectra of charged particles had been determined *a priori* using a magnetic spectrograph—the detector then logged events either as a time averaged photocurrent or as individual nonspectroscopic pulses on an oscilloscope or galvanometer. Van Heerden reasoned that it must be possible to find substances in which a beta particle could be stopped over a short distance while releasing sufficient ionization to be measured. This led to the question: *what type of ionization is released in solids, liquids, and gases?* Long collection times coupled with high ion recombination rates eliminated liquids and gases, and van Heerden [15] concluded that the solids most likely to fit the detection criteria would be those crystals that show photoelectric conductivity. Such crystals show a response to light and therefore would presumably do the same to ionizing particles if made thick enough. Crystals of silver chloride were grown from the melt and diced into 1-cm-diameter cylinders of thickness 1.7 mm. Silver electrodes were deposited on each side of the crystal. At room temperature the crystal showed ionic conductivity, but when cooled to liquid air temperature the crystal became a perfect insulator, allowing a large bias of up to 2.5 kV to be applied. In Figure 0.1 we show the crystal counter response to magnetically selected 400 keV beta particles which is spectroscopic with a measured full width at half maximum (FWHM) of ~20% (from [16]). In addition, the response of the device was found to be linear over the energy range 0 to 1 MeV. Subsequent measurements showed that the device was also responsive to alpha particles and gamma rays.

Van Heerden's work was later extended to the thallous halides by Hofstadter [17] in 1947, who reasoned that, if the silver chloride behaved as a crystal counter when cooled, then it would be likely that the thallium salts would do the same, based on their similarities in photoconductivity observed by Lehfeldt [10]. Hofstadter [18] fabricated sample counters from a mixture of TlBr and TlI. The samples were 1 in. in diameter and 2 mm thick. Pt electrodes, half an inch in diameter were sputtered onto each side of the disk. A schematic diagram of the detector and preamplifier front-end is shown in Figure 0.2

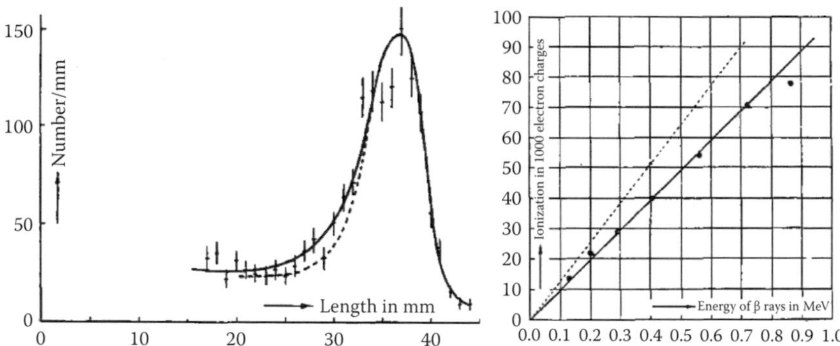

Figure 0.1 Left: the measured energy loss spectrum of a AgCl crystal counter to 400 keV beta particles (from Heerden and Milatz [16]). The applied bias was 200 V. The dotted curve is the theoretically predicted distribution. Right: the relation between the ionization and the beta particle energy (from [16]). The dotted line is the so-called "saturation curve" which in effect is the measured curve corrected for charge collection efficiency.

(from [18]). Like silver chloride, the sample was a conductor at room temperature. When connected to an amplifier and a power supply the detector showed no pulses on an oscilloscope when exposed to a radium gamma ray source, but it did, however,

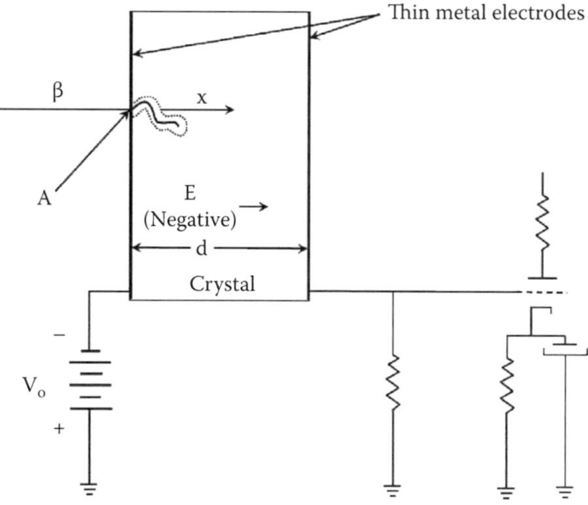

Figure 0.2 A schematic diagram of the key components of the crystal conduction counter of Hofstadter (reproduced from [18], courtesy of Platts, a division of the McGraw-Hill Companies, Inc.).

show photoconductivity and when cooled to liquid nitrogen temperature became insulating. When a 500-V bias was applied, it displayed a counting response to the gamma ray source with individual pulses observed on an oscilloscope. Experiments at different temperatures showed that the device became conductive at temperatures above –70°C. Hofstadter concludes, "*The question of what properties determine which crystal will count is not yet answered.*" Frerichs [19] addressed this question by showing that all crystals which give pulses belong to the class of crystal phosphors, at least under the temperature conditions at which they were studied. Specifically, they show the same optical phenomena as a crystal phosphor, and the pulses induced by nuclear particles are related to the primary and secondary photocurrents* characteristic of crystal phosphors.

By 1949, the number of materials found to exhibit "crystal counter" behavior had increased significantly [19] (CdS, ZnS, AgCl, AgBr, TlCl, TlBr, TlJ, type-2 diamonds, solid Ar). At this time, the elemental semiconductors were not anticipated to be responsive to radiation, because silicon and germanium are not insulators and the application of an electric field to sweep out the charge produced by an incident particle would result in unacceptably high leakage currents. However, several months after Hofstadter's work McKay [21] successfully measured the polonium alpha-ray spectrum using a Ge surface barrier detector. McKay circumvented charge trapping and low resistivity issues by constructing a very thin detector and using a surface barrier layer with a rectifying contact to allow the application of a high bias. McKay and McAfee [22] subsequently extended this work, producing both germanium and silicon p-n junction detectors. However, this work went unnoticed for almost a decade because of their very small counting efficiencies coupled with the disappointing performance of crystal counter detectors in general [23].

* The primary photocurrent is the direct result of the absorbed quanta. The secondary photocurrent is the result of an increase in conductivity arising from the passage of the primary photocurrent (for a review, see Hughes [20]).

Post-1960 Evolution

Renewed interest in "crystal counters" was aroused in the 1960s, sparked by two parallel events. The first was the rapid development of Ge and Si radiation detectors [24] as a consequence of the explosive rise of the semiconductor industry. Diffused junction and surface barrier detectors found widespread application in the detection of alpha particles [25]. The process of ion-drifting, first demonstrated by Pell in 1960 [26], provided a practical method by which large-volume semiconductor detectors could be made. The second major event was the successful synthesis of large crystals of CdTe by de Nobel in 1959 [27]. The latter led to the growth of a large number of other compound semiconductors, most notably a range of II–VI compounds. The potential of such materials was quickly recognized, especially their ability to operate without cryogenics [28]. Since the 1960s, progress has been incremental, but hardly spectacular, and the reader is referred to the reviews of McGregor and Harmon [29] and Owens and Peacock [30].

Currently, almost any compound can now be grown, but in general, detector development is still plagued by material problems caused by severe microcrystallinity, high defect and impurity densities, and stoichiometric imbalances. Recently, much effort has focused paradoxically on the qualities which degrade detector performance, specifically, the pivotal work of Luke [31] and others in developing single carrier sensing techniques. These techniques not only improve performance by mitigating the effects of the poorest carrier, but also point the way to future detector development—as for example, the controlled movement or channeling of charge, as dramatically demonstrated in silicon drift detectors [32]. Although these developments are still in their infancy, compound semiconductor radiation detectors have evolved sufficiently to fill certain niche applications and are slowly becoming viable competitors to laboratory standards like Si or Ge.

In this book, we shall examine the properties of compound semiconductors, their growth and characterization, and the fabrication of radiation sensors with emphasis on the X- and gamma-ray regimes. We explore their promise and limitations and discuss where the future may lie. The purpose of the book is twofold: (a) to serve as a text for those new to the field of compound semiconductor development and its application to radiation detection and measurement, and (b) as a general reference book for the established researcher. As with all enterprises, the production of this book has greatly benefited from the efforts and inputs of a large number of people, most notably, colleagues in the Science and Robotic Exploration Directorate of the European Space Agency. I am particularly indebted to Dr. Tone Peacock who, as Head of the Science Payload and Advanced Concepts Office, instigated and nurtured much of the Agency's research efforts in compound semiconductors. In addition, I would like to thank the following people: Dr. V. Gostilo of Baltic Scientific Instruments, Latvia, who elucidated much of the practical information presented, Professor J.K. Wigmore and Dr. A.G. Kozorezov of the Physics Department, Lancaster University, UK for theoretical insight and oversight.

Alan Owens
Noordwijk, The Netherlands
May 2011

References

1. E. Rutherford, H. Geiger, "An Electrical Method of Counting the Number of α-Particles from Radio-Active Substances," *Proc. R. Soc. A*, Vol. 81 (1908) pp. 141–162.
2. H. Geiger, W. Müller, "Elektronenzählrohr zur Messung schwächster Aktivitäten," *Naturwissenschaften*, Vol. 16, no. 31 (1928) pp. 617–618.
3. L.A. Kubetsky, "Multiple Amplifier," *Proc. Inst. Radio Engrs.*, Vol. 25 (1937) pp. 421–433.
4. V.K. Zworykin, G.A. Morton, L. Malter, "The Secondary Emission Multiplier—A New Electronic Device," *Proc. Inst. Radio Engrs.*, Vol. 24 (1936) pp. 351–375.

5. H.J. Round, "A Note on Carborundum," *Electrical World*, Vol. 49 (1907) p. 309.
6. O.V. Lossev, "Luminous Carborundum Detector and Detection Effect and Oscillations with Crystals," *Philosophical Magazine*, Vol. 5 (1928) pp. 1024–1044.
7. A. Einstein, "Über einen die Erzeugung und Verwandlung des Lichtes betreffenden heuristischen Gesichtspunkt," *Annalen der Physik*, Vol. 322, Issue 6 (1905) pp. 132–148.
8. B. Gudden, R. Pohl, "Das Quantenäquivalent bei der lichtelektrischen Leitung," *Z. Physik*, Vol. 17 (1923) pp. 331–346.
9. W. Lehfeldt, "Ober die elektrische. Leitfahigkeit von Einkristallen," *Z. Physik*, Vol. 85 (1933) pp. 717–726.
10. W. Lehfeldt, "Zur Elektronenleitung in Silber- und Thalliumhalogenid-kristallen", *Nachr. Ges. Wiss. Gottingen Math. Physik. Klasse, Fachgruppe II*, Vol. 1 (1935) p. 171.
11. R. Frerichs, R. Warminsky, "Die Messung von β- und γ-Strahlen durch inneren Photoeffekt in Kristallphosphoren," *Naturwissenschaften*, Vol. 33, no. 8 (1946) p. 251.
12. G. Jaffe, "Űber den Einfuß von α-Strahlen auf den Elektrizitätsdurchgang durch Kristalle," *Physikalische Zeitschrift*, Vol. 33 (1932) pp. 393–399.
13. K. Hecht, "Zum Mechanismus des lichtelektrischen Primärstromes in isolierenden Kristallen," *Z. Phys.*, Vol. 77 (1932) pp. 235–245.
14. P.J. Van Heerden, "The Crystal Counter: A New Instrument in Nuclear Physics," PhD dissertation, Rijksuniversiteit Utrecht, July (1945).
15. P.J. Van Heerden, "The Crystal Counter: A New Apparatus in Nuclear Physics for the Investigation of β and γ-Rays. Part I," *Physica*, Vol. 16, issue 6 (1950) pp. 505–516.
16. P.J. Van Heerden, J.M.W. Milatz, "The Crystal Counter: A New Apparatus in Nuclear Physics for the Investigation of β and γ-Rays. Part II," *Physica*, Vol. 16, Issue 6 (1950) pp. 517–527.
17. R. Hofstadter, "Thallium Halide Crystal Counter," *Phys. Rev.*, Vol. 72 (1947) pp. 1120–1121.
18. R. Hofstadter, "Crystal Counters-I," *Nucleonics*, Vol. 4, no. 2 (1949) pp. 20–27.
19. R. Frerichs, "On the Relations between Crystal Counters and Crystal Phosphors," *J. Opt. Soc. Am.*, Vol. 40 (1950) pp. 219–221.
20. A.L. Hughes, "Photoconductivity in Crystals," *Rev. of Mod. Phys.*, Vol. 8 (1936) pp. 294–315.
21. K.G. McKay, "A Germanium Counter," *Phys. Rev.*, Vol. 76 (1949) pp. 1537–1537.

22. K.G. McKay, K.B. McAfee, "Electron Multiplication in Silicon and Germanium," *Phys. Rev.*, Vol. 91 (1953) pp. 1079–1084.
23. A.G. Chynoweth, "Conductivity Crystal Counters," *Amer. J. Phys.*, Vol. 20 (1952) pp. 218–226.
24. G.L. Miller, W.M. Gibson, P.F. Donovan, "Semiconductor Particle Detectors," *Annual Review of Nuclear Science*, Vol. 12 (1962) pp. 189–220.
25. J.W. Mayer, "The Development of the Junction Detector," *IRE Trans. Nucl. Sci.*, Vol. NS-7 (1960) pp. 178–180.
26. E.M. Pell, "Ion Drift in an n-p Junction," *J. Appl. Phys.*, Vol. 31 (1960) pp. 291–302.
27. D. de Nobel, "Phase Equilibria and Semiconducting Properties of Cadmium Telluride," *Phillips Res. Rept.*, 14 (1959) pp. 361–399.
28. M.B. Prince, P. Polishuk, "Survey of Materials for Radiation Detection at Elevated Temperatures," *Trans. Nucl. Sci.*, Vol., NS-14 (1967) pp. 537–543.
29. D.S. McGregor, H. Harmon, "Room-Temperature Compound Semiconductor Radiation Detectors," *Nucl. Instr. and Meth.*, Vol. A395 (1997) pp. 101–124.
30. A. Owens, A. Peacock, "Compound Semiconductor Radiation Detectors," *Nucl. Instr. and Meth.*, Vol. A531 (2004) pp. 18–37.
31. P. Luke, "Unipolar Charge Sensing with Coplanar Electrodes—Applications to Semiconductor Devices," *IEEE Trans. Nucl. Sci.*, Vol. 42, no. 4 (1995) pp. 207–210.
32. E. Gatti, P. Rehak, "Semiconductor Drift Chamber—An Application of a Novel Charge Transport Scheme," *Nucl. Instr. and Meth.*, Vol. A225 (1984) pp. 608–661.

About the Author

Dr. Alan Owens holds an undergraduate honors degree in physics and physical electronics, and earned his doctorate in astrophysics from the University of Durham, United Kingdom. He spent over 30 years engaged in the design and construction of novel detection systems for X- and gamma-ray astronomy, and is currently a staff physicist at the European Space Agency's European Space Research and Technology Centre (ESTEC) in the Netherlands. He also holds an honorary senior lectureship in space science at the University of Leicester, United Kingdom.

Dr. Owens is currently involved in the development and exploitation of new technologies for space applications. Much of this work revolves around compound semiconductors for radiation detection and measurement, which by its very nature involves materials and systems at a low maturity level. Consequently, he has been involved in all aspects of a systematic and long-term program on material assessment, production, processing, detector fabrication, and characterization for a large number of compound semiconductors.

1 Semiconductors

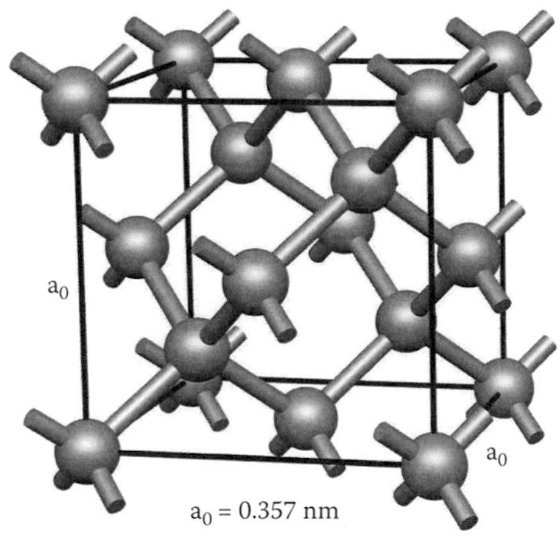

Frontispiece Ball-and-stick diagram of the conventional unit cell of diamond, where a_0 is the cubic lattice parameter. Source: Figure 1.1 of "*A First-Principles Study on Bulk Transfer Doping of Diamond*," S.J. Sque, PhD dissertation, The University of Exeter, UK (2007).

1.1 Metals, Semiconductors, and Insulators

Solids are usually divided into metals and nonmetals, with the nonmetals comprised of insulators and semiconductors. They may be grouped by their resistivity, ρ, which may range between 10^{-8} Ω-cm (metals) to 10^{18} Ω-cm (insulators) as illustrated in Figure 1.1. Semiconductors are intermediate

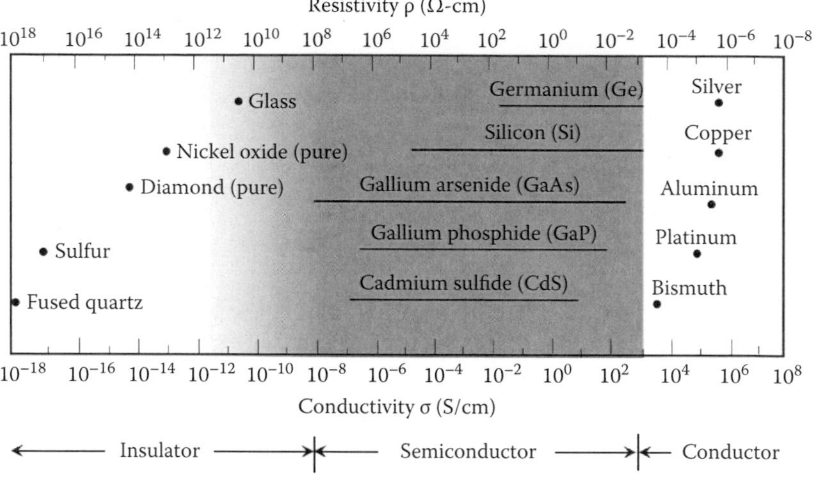

Figure 1.1 Typical range of resistivities/conductivities for insulators, semiconductors, and conductors (adapted from [1]). Semiconductors exist in the shaded region. At resistivities above 10^8 Ω-cm, the distinction between insulators and semiconductors is blurred and ultimately depends on temperature, because an "insulating" material can only become semiconducting (in the sense it can pass a current) if a sufficient number of electrons can be excited into the conduction band.

between the conductors and insulators and typically have resistivities between 10^{-2} Ω-cm and 10^8 Ω-cm. Virtually all inorganic and many organic materials belong to this group. The metals comprise those materials with very low resistivities (~10^{-6} Ω-cm). However, within this group, some have resistivities that are quite high—in fact ~10^2 to 10^3 times that of canonical metals, such as Cu. For this reason, metals such as As, Bi, and Sb are known as semi-metals. At the other end of the resistivity scale, the distinction between insulators and semiconductors can be somewhat vague, because some wide-gap materials that were previously regarded as insulators are now considered semiconductors (e.g., diamond). Because conductivity is a strong function of temperature, the difference between an insulator and a semiconductor really depends on the temperature at which the material's properties are being evaluated. We can qualitatively define a semiconductor as a material that displays a noticeable electrical conductivity at, or above, room temperature. A true insulator, on the other hand, will either melt or sublime before noticeable conductivity can be measured (e.g., quartz).

1.2 Energy Band Formation

Whether a material is classified as a metal, semiconductor, or insulator depends on its band structure, which is a natural consequence of its crystal structure, and specifically the periodic arrangement of its atoms. The potential energy associated with these atoms varies significantly over interatomic distances, and so the macroscopic properties of the crystal can only be adequately described using quantum mechanics. A comprehensive treatment of quantum mechanics is beyond the scope of this chapter, and so the reader is referred to standard texts, such as Kittel [2] and Wenckebach [3]. We will approach the formation of energy bands qualitatively using the Bohr theory of the atom as a starting point.

Bohr [4] combined a sun and planet model of the atom with quantum theory to explain why negatively charged electrons can form stable orbits around a small positively charged nucleus. Electrostatic attraction had long been thought to provide the force to hold the electron in orbit, but by postulating that the electrons could only occupy discrete "shells" or energy levels, Bohr was able to explain why the electrons did not lose energy continuously and spiral catastrophically into the nucleus. According to the Bohr model for a hydrogen atom, these energy levels are given by

$$E_H = -\frac{m_o q^4}{8\varepsilon_o^2 h^2 n^2} = -\frac{13.6}{n^2} \text{ eV}, \tag{1.1}$$

where m_o is the free electron mass, q is the electronic charge, ε_o is the permittivity of free space, h is Planck's constant, and n is the principal quantum number which defines each quantized energy level. For $n = 1$, that is the ground state, $E_H = -13.6$ eV, and for $n = 2$, the first excited state, $E_H = -3.4$ eV. In the limit, when $n \to \infty$, $E_H \to 0$ and the electron is no longer bound. The atom is ionized.

When two atoms approach one another, the electron wave functions begin to overlap and these energy levels split into two, as a consequence of Pauli's exclusion principle [5]. By the same argument, when N atoms come together to form a solid, each energy level now splits into N separate but closely spaced levels, thereby resulting in a continuous band of energy levels for each quantum number, n. It should be noted that the Bohr model is approximate and only applicable for the simplest atom—hydrogen. For crystalline solids, the detailed energy band structures can only be calculated using quantum mechanics. Figure 1.2 shows the effects on the energy levels when isolated C atoms are brought together to form a diamond lattice structure. As the atoms approach, the discrete *2s* and *2p* levels begin to split, into N closely spaced energy levels, where N is the number of atoms. As $N \to \infty$ these levels merge to form

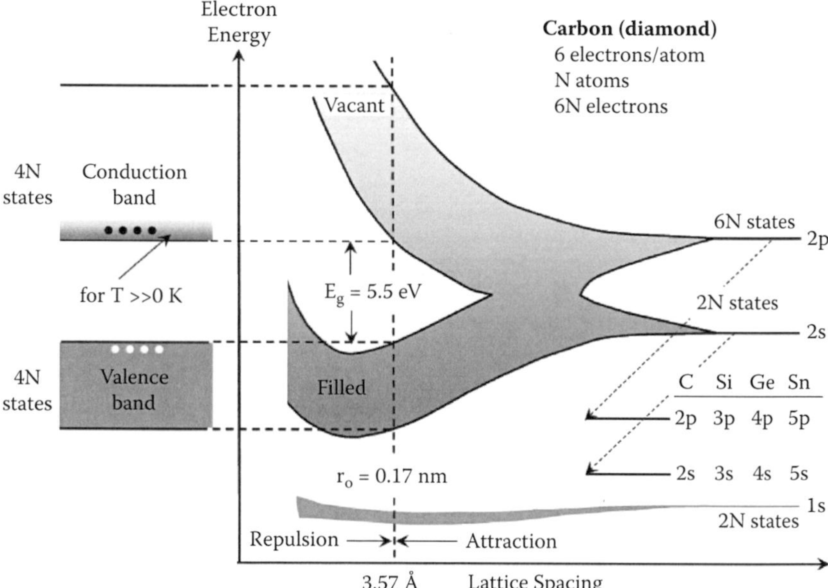

Figure 1.2 Formation of energy bands as a diamond crystal is achieved by bringing together isolated carbon atoms—the energy bands essentially reflect the hybridization of the *s* and *p* levels. At the nominal lattice spacing, the forces of attraction and repulsion between atoms balance. The equilibrium position can be adjusted by pressure and temperature. The conventional one-dimensional representation of band structure at the nominal lattice spacing is given by the left-hand diagram.

a continuous band—partly because the separation between energy levels becomes comparable with the lattice vibrational energy to which the electrons are strongly coupled, and partly because the separation in energy levels also becomes comparable with the time-averaged energy uncertainty due to the Heisenberg uncertainty principle [6]. As the atoms approach their equilibrium interatomic spacing, this band separates into two, becoming the conduction band and the valence* band. At the equilibrium position the total energy of the electrons and the lattice is minimized. The region separating the conduction and valence bands is termed the forbidden gap, or bandgap, and is characterized by a bandgap energy, E_g. It is this gap which uniquely determines the conduction properties of materials and represents the minimum energy to generate electron–hole pairs, which are the essential information carriers in a semiconductor. From Figure 1.2, we see that a reduction in lattice spacing from its equilibrium value will increase the size of the bandgap, while an increase in lattice spacing will decrease the bandgap. In reality, many other bands are formed out of the lower atomic levels, but only the valence and conduction bands play an active role in the conduction process, because all bands below the valence band are filled and all bands above the conduction band are empty. It should be noted that electrons must move between states to conduct electric current, and so full bands do not contribute to the electrical conductivity. It is only when electrons are excited into the conduction band that they are free to move between states and thus contribute to current flow.

1.3 General Properties of the Bandgap

Figure 1.3 shows energy band diagrams for the three classes of solids: insulators, semiconductors, and metals. These bands represent the outermost energy levels that an electron can occupy. Electrons in the valence band cannot move between atoms, whereas electrons in the conduction band can freely

* Named the valence band by analogy to the valence electrons of individual atoms.

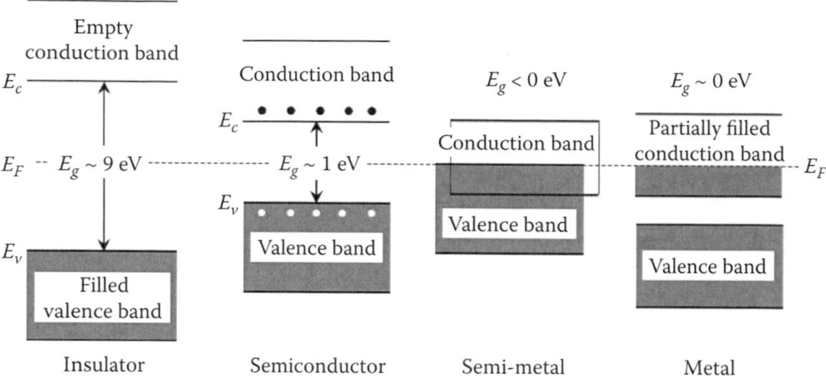

Figure 1.3 Schematic energy band diagrams for an insulator, a semiconductor, and a metal. Two cases are given for the latter: (1) semimetals with overlapping valence and conduction bands. Examples are Sn, Zn, Pb, or graphite and (2) classical metals with partially filled conduction bands (e.g., Cu, Au, and Ag). The Fermi level, E_F, is the energy level at which an average of 50% of the available quantum states are filled by an electron and relates the probable location of electrons within a band. For metals the Fermi level lies in the conduction band, while for insulators and semiconductors the Fermi level lies in the band gap.

move from atom to atom and act as mobile charge carriers with which to conduct current. The strength of valence bonds is 10 times that of the free electron bonding, so the movement of these free carriers will not compromise the integrity of the lattice structure. In insulators, the bandgap is relatively large, and thermal energy or an applied electric field cannot raise the uppermost electron in the valence band to the conduction band. In metals, the conduction band is either partially filled, as in the case of a classical metal, such as Ag and Cu, or overlaps the valence band, as in the case of a semimetal such as Sn or graphite. In both cases there is no bandgap, and current can readily flow in these materials. In semiconductors, the bandgaps are smaller than those of insulators to the extent that thermal energy can excite electrons to the conduction band. However, if the temperature (and therefore thermal energy) is lowered sufficiently, semiconductors become insulators. Metals, on the other hand, are conductors at any temperature. Table 1.1 classifies materials according to bandgap energy and,

TABLE 1.1
Classification of Materials According to Their Energy Gap, E_g, and Carrier Density, n, at Room Temperature

Material	E_g(eV)	n (cm^{-3})
Metal	no energy gap	10^{22}
Semimetal	$E_g \leq 0$	10^{17}–10^{21}
Semiconductor	$0 < E_g < 4$	$<10^{17}$
Insulator	$E_g \geq 4$	$<<1$

Note: Values are given for pure, that is, intrinsic, materials. We arbitrarily choose a bandgap of 4 eV as the dividing line between insulators and semiconductors, because the actual distinction between the two is somewhat blurred and depends on the temperature in practice, because that determines the number of free carriers in the conduction band.

Source: From ref. [7].

consequently, free carrier density, because its magnitude in pure materials is a direct consequence of the bandgap.

1.3.1 Carrier Generation and Recombination

Free electrons may be generated by exciting electrons from the valence band into the conduction band, leaving behind a free hole in the valence band. Thus, the carrier generation process simultaneously creates equal numbers of electrons and holes. There are several mechanisms by which free carriers can be generated, and these may be individually targeted for specific applications. For example, free carriers may be generated directly by thermal excitation (as in the case of thermal imaging), optical excitation (as in the case of optoelectronic devices), or direct ionization by charged particles (as in the case of radiation detectors). However, it should be appreciated that, even in the absence of any external energy source, electron–hole pairs are constantly being generated by the thermal energy stored in the crystal lattice. However, at any temperature, an equilibrium will be reached when the mean creation rate of electron–hole pairs is equal to the recombination rate.

The mean lifetime, τ_e, of a free electron is the average time that the electron exists in a free state before recombination. Similarly, the mean lifetime, τ_h, of a free hole is the average time that it exists in a free state before recombination. In an intrinsically pure semiconductor, $\tau_e = \tau_h$. In Si and Ge these lifetimes are typically of the order of 10^{-3} s at room temperature. However, the lifetimes in compound semiconductors are considerably shorter, by some three to five orders of magnitude. Carrier recombination can take place by several mechanisms, the most important of which are illustrated in Figure 1.4. These are (a) direct band-to-band radiative recombination, (b) deep level defect (trap) mediated, or (c) nonradiative band-to-band Auger recombination.

In the recombination process, electrons drop from the conduction band to the valence band—the energy difference being released as photons, phonons, photons and phonons, or the transmission of the energy to a third particle. As shown in Figure 1.4, band-to-band recombination occurs when an electron falls from its conduction band state into the empty valence band state associated with a hole, releasing a photon within a 1 to 10 ns time scale. The probability of band-to-band recombination is proportional to the product of the

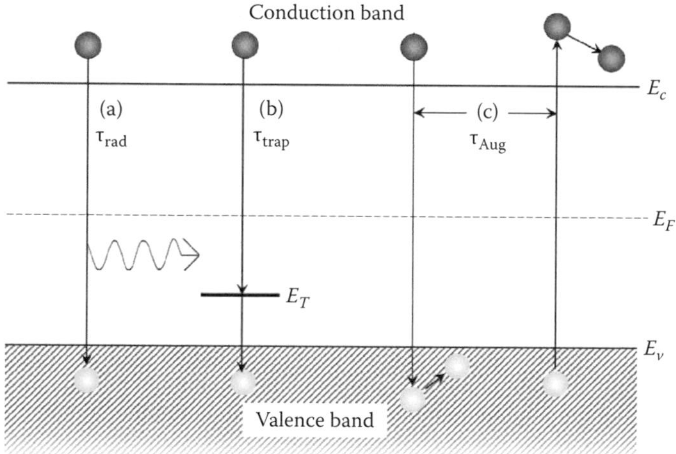

Figure 1.4 Schematic of carrier recombination mechanisms in semiconductors illustrating, (a) radiative emission, (b) deep-level trap mediated, or (c) nonradiative band-to-band Auger recombination.

free electron and vacant hole densities. For indirect bandgaps* this probability is small. Auger recombination involves three particles—an electron and a hole, which recombine in a band-to-band transition transferring the resulting energy to another electron or hole. The Auger recombination rate is proportional to the product of the concentrations of the respective reaction partners and is only relevant at high carrier concentrations. Carriers most effectively recombine via traps, which are energy states created in the forbidden energy gap by impurities or dislocations. These can hold an electron or hole long enough for a pair to be completed. The recombination process for deep-level traps is usually nonradiative, so the energy must be dissipated as phonons. The probability of recombination is dependent on the product of the free electron and hole densities and on the number of carrier traps and is described by the Shockley, Read, and Hall (SRH) model [8,9].

In most materials, one of the above processes will be much faster than the others and will largely determine the effective lifetime, τ, of the excited carriers. This is given by

$$\frac{1}{\tau} = \frac{1}{\tau_{rad}} + \frac{1}{\tau_{trap}} + \frac{1}{\tau_{Aug}} \tag{1.2}$$

where τ_{rad}, τ_{trap}, τ_{Aug}, are the radiative, trap, and Auger lifetimes, respectively. In high-quality direct bandgap semiconductors, radiative emission is likely to be the dominant mechanism. In heavily doped semiconductors (>10^{17} atoms cm^{-3}), Auger recombination is usually more rapid, because the Auger rate is proportional to the square of the doping density. For most semiconductors, however, especially those with indirect bandgaps, short lifetimes are mainly due to impurities and defects, especially for those which introduce traps near the middle of the gap.

1.3.2 Pressure Dependence of the Bandgap

The size of the bandgap is found to vary with both pressure and temperature, largely due to distortion of the lattice. Applying a high compressive stress reduces the lattice spacing, which

* See Section 1.3.4

increases the bandgap. An approximate expression relating pressure to bandgap energy is

$$E_g(P) = E_g(0) + aP + bP^2 \tag{1.3}$$

where $E_g(0)$ is the bandgap energy at zero pressure, P is the hydrostatic pressure, and a and b are empirical constants. To first order, it is found that the pressure dependences of many III–V direct gap materials are similar at ~50 meV per GPa (e.g., [10,11]). Because 1 GPa corresponds to 65 tons per square inch, it can be seen that the effect is very small and can be neglected for almost all applications. However, for optoelectronic applications, measuring the variation of photoluminescence with pressure can be very useful in understanding the underlying band structure of a material and its structural properties. At the highest hydrostatic pressures (usually, tens of GPa), some semiconductors can undergo a phase transition and lose their semiconducting properties completely, becoming metals.

1.3.3 Temperature Dependence of the Bandgap

As the temperature of a semiconductor increases, the amplitude of atomic vibrations increases, which in turn, increases interatomic spacings. The lattice therefore expands, as witnessed by the material's linear expansion coefficient. The motion of the atoms about their equilibrium lattice points affects the width of the energy levels—at higher temperatures they become broader. In addition, increasing the interatomic spacing decreases the potential seen by the electrons in the material, which in turn reduces the size of the energy bandgap. For example, the energy gap in silicon at 0 K is 1.17 eV. At room temperature it is 1.12 eV. The relationship between bandgap energy, E_g, and temperature, T, can be described by the semiempirical formulation of Varshni [12],

$$E_g(T) = E_g(0) - \frac{\alpha T^2}{T + \beta} \tag{1.4}$$

where $E_g(0)$ is the bandgap at 0 K, and α and β are material constants. In Table 1.2 we list these parameters for a

TABLE 1.2
Temperature Dependence of Bandgap Constants for a Number of Semiconductors from Groups IV, III–V, and II–VI

Parameter Material	$E_g(T = 0° K)$ (eV)	$\alpha \times 10^4$ (eV/K)	β (K)
	Group IV		
Ge	0.744	4.07	230
Si	1.170	4.18	406
4H-SiC	3.26	6.5	1300
C (diamond)	5.49	5.0	1067
	Group III–V		
InSb	0.234	2.50	136
InAs	0.420	2.50	75
InP	1.421	3.63	162
InN	1.994	2.1	453
GaSb	0.811	3.75	176
GaAs	1.519	5.41	204
GaP	2.338	5.771	372
GaN	3.47	5.99	504
AlSb	1.686	3.43	226
AlAs	2.239	6.0	408
AlP	2.49	3.5	130
AlN	6.20	8.3	575
	Group II–VI		
CdTe	1.606	3.10	108
CdSe	1.846	4.05	168
CdS	2.583	4.02	147
ZnTe	2.394	4.54	145
ZnSe	2.825	4.90	190
ZnS	3.841	5.32	240

Source: From [13].

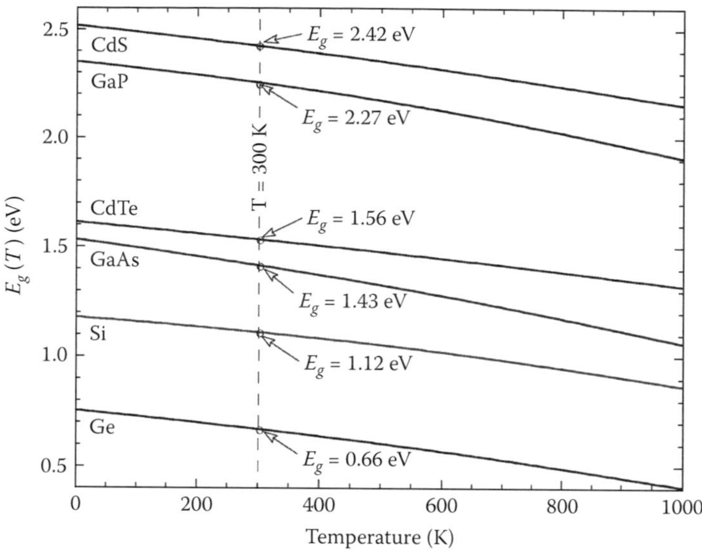

Figure 1.5 Temperature dependence of the bandgap energy for common semiconductors from groups IV, III–V, and II–VI compounds. Bandgap energies decrease by ~0.4 meV per degree K for most semiconductors.

range of materials from groups IV, III–V and II–VI semiconductors (taken from ref. [12]), and in Figure 1.5 we plot the functional relationship for some of the more common materials. For all semiconductors, like Si, GaAs, and CdS, the bandgap changes by ~0.3–0.5 meV per degree K, and so for the temperatures normally encountered by detection systems, this effect can be neglected. In reality, the dominant effect of temperature on semiconductor properties is on carrier densities which vary exponentially with temperature (see Section 1.7.1.1).

1.3.4 Direct and Indirect Bandgaps

The one-dimensional bandgap picture presented in Section 1.3 is oversimplified in that it does not adequately describe the true morphology of band structure and does not, for instance, distinguish between direct and indirect bandgap semiconductors. In order to do so, one has to consider the relationship between momentum and energy as a function of crystal

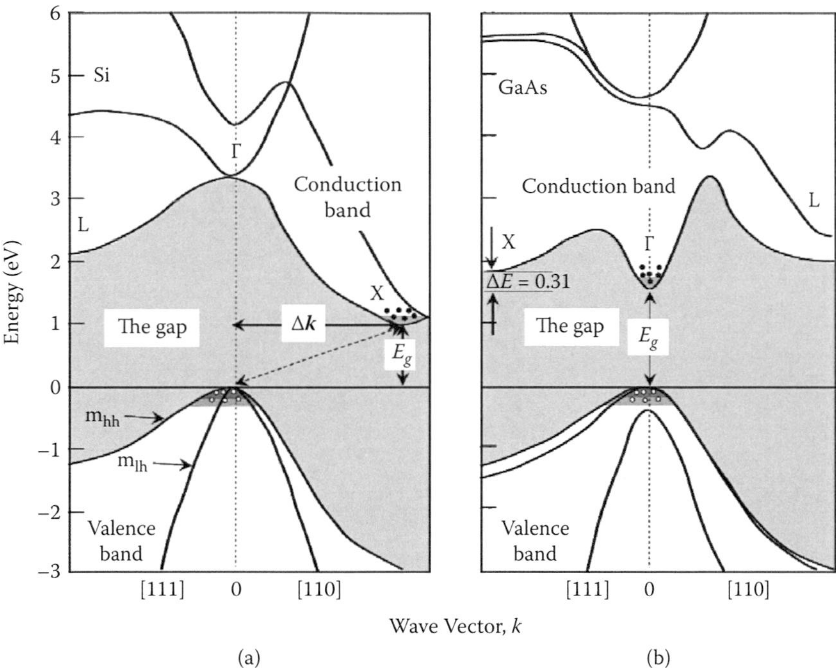

Figure 1.6 Energy band structures of (a) Si and (b) GaAs. Circles (o) indicate free holes in the valence bands, and filled circles (•) indicate free electrons in the conduction bands. Γ, X, and L refer to different conduction band minima along the main crystallographic directions in the crystal. Si is an indirect bandgap material, while GaAs is a direct bandgap material (adapted from [16], © 1976 American Physical Society). The important difference is that, for the direct band gap material, an electron can transit between the lowest potential in the conduction band to the highest potential in the valence band without a change in momentum, $\Delta \boldsymbol{k}$, whereas for an indirect band gap material it cannot do so without the mediation of a third body (e.g., phonon) to conserve momentum. Here m_{hh} and m_{lh} show the valence band maxima that contain heavy and light holes (see Section 1.5).

direction (e.g., see Figure 1.6). The reason that the energies of the states can be broadened into a band without violating any exclusion law is that the energy now also depends on the magnitude of the wave vector, $\boldsymbol{k}$, which in quantum mechanics is used to represent the momentum $h\boldsymbol{k}$ of a particle. This relationship is known as energy–momentum dispersion. In essence, the periodic potential imposed by the crystal lattice modulates the band structure at a basic scale length of the smallest unit cell. The E-$\boldsymbol{k}$ space encompassed by this cell is known as the first Brillouin zone and is a particularly important concept, because within this zone the bandstructure is unique—outside it is replicated.

1.3.4.1 Electrons in Solids

An electron in a vacuum can be described by a de Broglie plane wave, in which the electron momentum can be expressed as,

$$\boldsymbol{p} = \hbar \boldsymbol{k} \tag{1.5}$$

where $\hbar = h/2\pi$, and $\boldsymbol{k}$ is the wave vector of the de Broglie wave. The electrons velocity is then given by

$$\boldsymbol{v} = \hbar\, \boldsymbol{k}/m \tag{1.6}$$

and its energy by

$$E = E_o + \frac{\hbar^2(\boldsymbol{k})^2}{2m} \tag{1.7}$$

which is also known as the dispersion relationship.

The first term, E_o, represents the electron potential energy, and the second term its kinetic energy. In a similar fashion, a loosely bound electron in a crystalline solid can be represented by modified plane wave. The corresponding wave function is a solution of the Schrödinger equation [14] for a periodic potential and is called the Bloch function. It has the form

$$\psi_k(\boldsymbol{r}) = \exp(i\boldsymbol{k} \cdot \boldsymbol{r})\ \varphi_k(\boldsymbol{r}) \tag{1.8}$$

where $\boldsymbol{r}$ is the position vector, and $\varphi_k(\boldsymbol{r})$ is a periodic function with the period of the lattice potential field. Equation (1.8) may be regarded as a plane wave $\exp(i\boldsymbol{k}{\cdot}\boldsymbol{r})$ with an amplitude $\varphi_k(\boldsymbol{r})$ modulated by the period of the crystal lattice. The band structure of the electron energy spectrum in a crystalline solid arises as a consequence of the interference of the Bloch waves [15].

From Equation (1.7) we note that the electron energy is a function of $\boldsymbol{k}$, and also all parameters describing the motion of electrons are related to $\boldsymbol{k}$ and the function $E(\boldsymbol{k})$. In particular, whenever the Wulf–Bragg condition between the de

Broglie wavelength of an electron and the period of the crystal lattice is fulfilled, the electron wave will be totally reflected from the lattice. Owing to the interference of the incident and reflected waves, a standing wave arises. An electron with the corresponding wave vector $\boldsymbol{k}_o$ cannot propagate, and discontinuities in the electron energy spectrum occur corresponding to the energy levels at the lower and upper limits of both the conduction and valence bands, as shown in Figure 1.3.

1.3.4.2 Electrons in the Conduction Band

In k space a discontinuity in the electron energy spectrum corresponds to an extreme point of the function $E(\boldsymbol{k})$. At this point $dE/dk = 0$. Around this point the electron energy can be approximated by the quadratic function,

$$E(\boldsymbol{k}) = E(\boldsymbol{k}_o) + \frac{1}{2}\sum_{i=1}^{3} \frac{\partial^2 E}{\partial k_i^2}(k_i - k_{oi})^2 \tag{1.9}$$

where k_{oi} denotes the appropriate coordinate in $\boldsymbol{k}$ space, and $\boldsymbol{k}_o$ is the position vector of the extreme point. The term $E(\boldsymbol{k}_o)$ may be either a minimum or maximum in a permitted band. At the minimum of the conduction band, $E(\boldsymbol{k}_o) = E_c$, while at the maximum of the valence band $E(\boldsymbol{k}_o) = E_v$. This is illustrated in Figure 1.6. The electrons in the conduction band occupy the energy states slightly above the bottom of the conduction band, their distribution in $\boldsymbol{k}$ space being accurately described by Equation (1.9). Note that the number of electrons in the conduction band is fewer than the number of available states. Hence the electrons feel practically no restriction due to the exclusion principle to occupy any of the energy states. As such, within a crystal they can move and can be accelerated and in many respects behave like an electron in a vacuum.

In Figure 1.6 we show the detailed band structure for Si and GaAs in which the energy, E, is plotted against the crystal momentum, $\boldsymbol{k}$, for two crystal directions. For Si (see Figure 1.6a), we note that the momentum of holes at the top of the valence band is different from the momentum of electrons at the bottom of the conduction band. Therefore, in order to satisfy energy and momentum conservation laws, a transition between the two states requires a momentum transfer to the crystal lattice via its vibrational modes. Because a simple direct transition is not possible, Si is known as an indirect bandgap semiconductor. The equivalent diagram for a direct bandgap semiconductor (GaAs) is shown in Figure 1.6b. Here, we see that the minimum of the conduction band lies directly above the maximum of the valence band. Transitions can take place with the absorption or release of a photon of wavelength,

$$\lambda_o(\mu m) = \frac{1.239}{E_g(eV)} \tag{1.10}$$

without the mediation of a third body to conserve momentum. This is a highly efficient process and underlies the basic principle of operation behind the light emitting diode, semiconductor laser, and solar cell. In contrast, photon absorption or emission in indirect bandgap materials is very inefficient, because the process requires the simultaneous emission or absorption of a phonon. Such a correlated transition has a very low probability. Hence optoelectronic devices are nearly always fabricated from direct bandgap materials.

Note in Figure 1.6a that Si also has a direct bandgap and GaAs an indirect bandgap—albeit with larger bandgap energies. The classification into indirect and direct bandgap materials is based on which gap has the smallest energy and therefore will be the dominant source of transitions between the conduction and valence bands. The valence band maximum always occurs at $\boldsymbol{k} = 0$, so we can define a direct-band semiconductor as one where the conduction-band minimum also occurs at $\boldsymbol{k} = 0$ and an indirect-band semiconductor where it occurs at $\boldsymbol{k} \neq 0$. Semimetals are interesting variants of indirect materials in that the bottom of the conduction band is usually lower than the top of the valence band, and hence the bandgap energy can actually be negative.

1.3.4.3 Band Morphology

In Figure 1.7 we show an E-$\boldsymbol{k}$ diagram for GaAs in which we have simplified the band structure shown in Figure 1.6b to

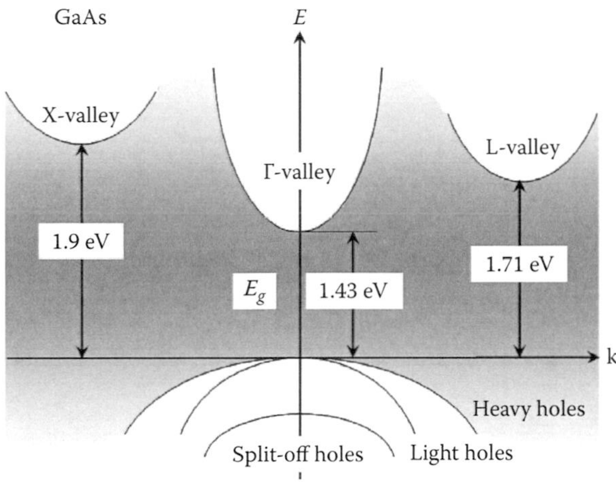

Figure 1.7 A simplified E-$\boldsymbol{k}$ diagram of GaAs showing the three valleys (L, Γ, and X) in the conduction band. For direct gap materials, the height of the Γ valley represents the normally quoted bandgap. The valence band is comprised of three bands with different curvatures. Thus, there are three effective hole masses—heavy holes, light holes, and split-off holes (see Section 1.5).

accentuate the conduction band minima. We see there are actually three valleys, which occur along different crystallographic directions in the crystal: the (000), (100), and (111) directions. These are known as the Γ (at $\boldsymbol{k}$ = 0), L, and X valleys (which both occur at nonzero values of $\boldsymbol{k}$). The Γ valley has the minimum energy and corresponds to the direct bandgap with E_g = 1.43 eV. Except under high field excitation, the L and X valleys contain very few electrons and therefore do not contribute to the conduction process. In common with group IV and other III–V semiconductors, GaAs has three valence bands with maxima at $\boldsymbol{k} = 0$. These are the *light-hole, heavy-hole* and *split-off bands.* The split-off band is due to spin–orbit interactions. Its maximum occurs at a lower energy than the other two valence bands and as a consequence contains virtually no holes. Because the heavy-hole band is wider than the light-hole band, holes tend to concentrate in it. The relative numbers of heavy holes and light holes (n_{hh} and n_{lh}) can be derived from a density of states calculation and is found to be a function only of their effective masses (see Sections 1.5 and 1.7.1.1). For Si, $n_{hh} = 6n_{lh}$, while for virtually all groups III–V and II–VI materials, $n_{hh} > 10n_{lh}$.

1.4 Carrier Mobility

Mobility is a fundamental parameter in semiconductor physics because it essentially describes the conduction process. Free electrons in a solid have a velocity which depends upon their energy. In the absence of an electric or magnetic field the electron energy approaches that of the Maxwell thermal agitation energy given by

$$\frac{1}{2}m_e^* v_{th}^2 = \frac{3}{2}kT \tag{1.11}$$

where m^*_e is the electron effective mass, and v_{th} is the average thermal velocity. Simply stated, a particle's effective mass is the mass it appears to have when calculated using a classical transport model. The thermal motion of an individual electron can be visualized as a succession of random scattering from collisions with lattice atoms, impurity atoms, and other scattering centers with no net direction. The average distance between collisions is called the mean free path, and the average time between collisions is termed the mean free time, τ_c. Upon applying an electric field, ξ, the electrons will be accelerated by a force equal to $-q\xi$ in an opposite direction to the electric field with a drift velocity, v_n. The net result is that the electrons are transported through the crystal with an average drift velocity. The momentum of the electron is equal to $-q\xi\tau_c$. Therefore, $m^*_e v_e = -q\xi\tau_c$, or

$$v_n = -\left(\frac{q\tau_c}{m_e^*}\right)\xi \tag{1.12}$$

Equation (1.12) shows that the drift velocity is proportional to the applied electric field and inversely proportional to the effective mass. The constant of proportionality is known as the electron mobility, μ_e. Hence,

$$v_n = -\mu_e \xi \tag{1.13}$$

where $\mu_e = q\tau_c/m^*_e$. A similar expression can be written for holes

$$v_p = \mu_p \xi \tag{1.14}$$

The difference in signs between Equations (1.13) and (1.14) arises because electrons drift in the opposite direction to the electric field while holes drift in the same direction. It should be noted that electrons and holes will have different drift velocities for the same electric field and therefore different mobilities.

Mobilities tend to be larger in direct bandgap materials than in indirect bandgap materials and are influenced by a number of factors. For example, because of differences in the properties of the valence and conduction bands, there is a significant difference in the mobilities of n and p type materials. Mobility is also strongly affected by temperature (see Figure 1.8). It is lower at very low and high temperatures and has its maximum value in the middle of the temperature range around 100 K. Numerically, electron mobilities are generally in the thousands of $cm^2V^{-1}s^{-1}$ range, while hole mobilities are in the hundreds of $cm^2V^{-1}s^{-1}$ range. Mobility also depends on lattice scattering and impurity scattering. Lattice scattering results from thermal vibrations of the crystal atoms. As lattice vibrations become more significant with increasing temperature, mobility decreases. It is particularly acute in crystals with high ionicity, because as the lattice deforms, dipoles are formed which scatter the carriers in the resulting electric field. In fact, at high temperature, lattice vibrations dominate the mobility function and its contribution to the overall mobility begins to decrease as $\mu_{\text{lattice}} \propto T^{-3/2}$. Impurity scattering results when a charge carrier travels past an ionized donor or acceptor and is perturbed by the Coulomb potential field. Unlike lattice scattering, impurity scattering becomes less significant at high temperatures because the carriers move faster and are less effectively scattered, and its contribution to the overall mobility begins to increase as $\mu_{\text{impurities}} \propto T^{3/2}$. An approximation of the overall mobility function can be written as

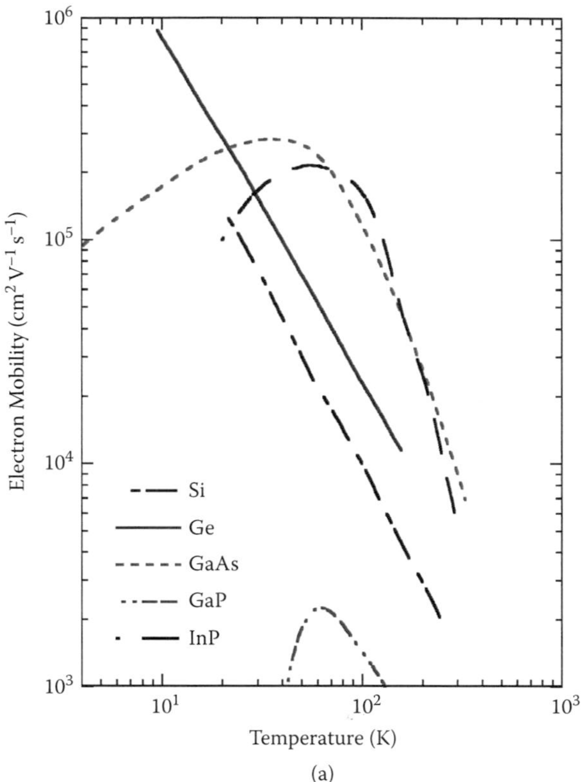

Figure 1.8 Temperature dependence of (a) electron mobilities and (b) hole mobilities for a number of group IV and III–V semiconductors [17–23]. The rollover in mobilities for some group III–V materials below ~100 K is due to impurity scattering. For comparison, the inset in Figure (b) illustrates the mobility temperature dependence on the two main scattering processes.

a combination of the contributions from lattice vibrations (phonons) and from impurities. Thus,

$$\frac{1}{\mu} = \frac{1}{\mu_{lattice}} + \frac{1}{\mu_{impurities}} \tag{1.15}$$

Hence the smallest mobility component dominates. Figure 1.8 shows the temperature dependence of electron and hole mobilities for a number of group III–V materials along with the elemental semiconductors Si and Ge. For comparison, the

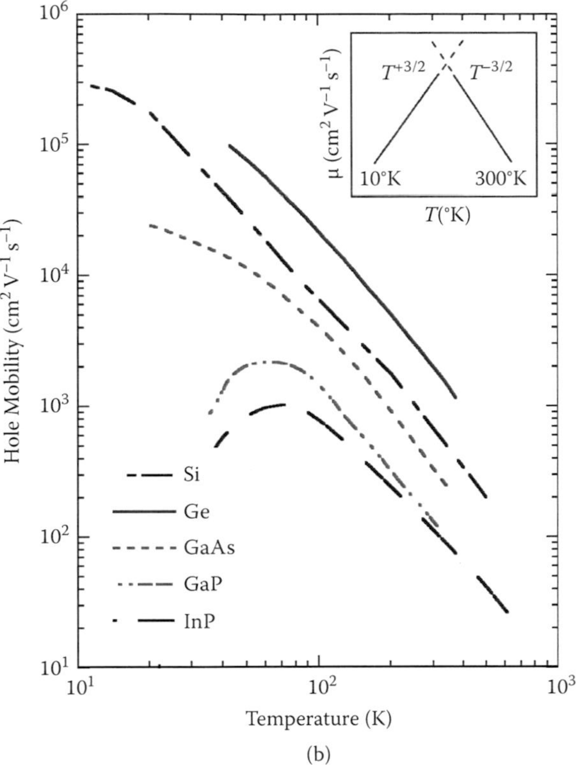

Figure 1.8 (Continued)

inset of Figure 1.8b also shows the mobility temperature dependence of the two main scattering processes. Because both Si and Ge can be grown with impurity levels less than one part in 10^{12}, there is no impurity scattering, even at low temperatures. Consequently, their electron mobilities continue to rise with decreasing temperature. Group III–V materials, on the other hand, cannot be grown to the same purity levels and show a peak in mobility around 100 K. At lower temperatures, impurity and defect scattering begins to dominate Equation (1.9), and electron mobilities fall. The same trends are seen in the hole mobilities (see Figure 1.8b). As expected, for extrinsic semiconductors, the mobility due to lattice scattering shows a strong dependence on doping levels at temperatures for which scattering is dominant (see Section 1.7.2.1). That is, as the impurities become ionized, scattering processes increase and begin to dominate the overall mobility function. For example, the electron and hole mobilities in Si at room temperature are μ_e = 1400 cm^{-2}V^{-1}s^{-1} and μ_h = 400 cm^{-2}V^{-1}s^{-1}, respectively, for dopant concentrations below 10^{15} cm^{-3}. These fall to μ_e ~ 300 cm^{-2}V^{-1}s^{-1} and μ_h ~ 100 cm^{-2}V^{-1}s^{-1} for concentrations above 10^{18} cm^{-3}.

1.5 Effective Mass

Effective mass is a useful concept when describing the transport of electrons through a crystal and is essentially a heuristic approach that relates the quantum behavior of electrons moving in the periodic potential of a crystal lattice to a semi-classical model. The effective mass is an important concept because it affects many of the electrical properties of the semiconductor, such as electron and hole mobilities, which in turn determine conductivity. From Figure 1.6 we see that a particle's energy is a function of its crystal momentum. For an electron close to the conduction band minimum, the energy versus momentum relationship given by Equation (1.7) can be approximated by a parabolic distribution

$$E(\boldsymbol{k}) \approx E_o + a_2(\boldsymbol{k} - \boldsymbol{k}_o)^2 \tag{1.16}$$

where a_2 is a constant which describes the curvature of the relationship. Here, $E_o = E_c$ is the "ground state energy" corresponding to a free electron at rest, and $\boldsymbol{k}_o$ is the wave vector at which the conduction band minimum occurs. The above approximation holds for sufficiently small increases in E above E_c. Combining Equations (1.7) and (1.16) and taking $\boldsymbol{k}_o = 0$, we can define the curvature coefficient a_2 associated with the parabolic minimum of the conduction band in terms of an *effective mass* m_e^*, thus

$$a_2 = \frac{\hbar^2}{2m_e^*} \tag{1.17}$$

Substituting into Equation (1.16) and taking the second derivative of the functional form of $E(\boldsymbol{k})$ with respect to $\boldsymbol{k}$ at the band minimum, $\boldsymbol{k}_o$, we can derive an analytical expression for the effective mass

$$m_e^* = \frac{\hbar^2}{\left(\dfrac{\partial^2 E_c(\boldsymbol{k})}{\partial k^2}\right)} \tag{1.18}$$

Note: this expression can be easily derived from Newton's second law of motion, taking into account the electron's interaction with the lattice and the effect on its inertia. The quantity $(\partial^2 E_c/\partial k^2)$ is in fact the curvature of the band. Thus the effective mass is an inverse function of the curvature of the E_c-$\boldsymbol{k}$ diagram: weak curvature gives a large effective mass and strong curvature a small effective mass. Depending on the nature of the periodic lattice potential, effective masses may be lighter or heavier than the free electron mass, m_o. From Figure 1.6 we see that the conduction band actually has three minima or valleys and thus three effective masses. In GaAs, these are 0.067 m_o, 0.85 m_o, and 0.85 m_o for the Γ, L, and X valleys, respectively. In this case, carrier dynamics are dominated by the effective mass of the Γ valley since the other two valleys are essentially unpopulated under normal conditions.

For holes we have a dispersive relationship similar to that of electrons, with the effective mass m_h^* now determined from the curvature of the valence band maximum at $\boldsymbol{k} = 0$, i.e.,

$$m_h^* = \frac{\hbar^2}{\left(\dfrac{\partial^2 E_v(\boldsymbol{k})}{\partial k^2}\right)} \tag{1.19}$$

However, the situation is somewhat more complicated because there are three valence bands centered at $\boldsymbol{k} = 0$. Thus, there are three effective masses, corresponding to heavy holes, light holes, and split-off holes. Because holes typically occupy the light-hole and heavy-hole bands only, the effective scalar mass of the valence band can be expressed as

$$m_h^* = \left(m_{hh}^{*\,\frac{3}{2}} + m_{lh}^{*\,\frac{3}{2}}\right)^{\frac{2}{3}} \tag{1.20}$$

where m^*_{hh} is the effective mass of heavy holes, and m^*_{lh} is the effective mass of light holes. Theoretically [24], it is found that the electron and light hole effective masses are strongly related to the bandgap energy and, for the Γ conduction band, are approximately equal to each other for most cubic group VI and groups III–V and II–VI materials (i.e., $m_e^* \sim m_{lh}^*$). The mass of the heavy hole, on the other hand, shows little dependence, being largely determined by interactions with other bands more distant in energy. As a result, it changes little between materials and typically has a value of $m_{hh}^* \sim 0.4m_o$. Holes in the split-off band have roughly the same effective mass as those in the light-hole band, owing to their similar band curvatures.

Equations (1.18) and (1.19) also show that for a parabolic dispersion relationship, the second derivative of E with respect to $\boldsymbol{k}$ is a constant. As a consequence, the effective mass is a constant as well, and is independent of energy. Since the energy bands depend on temperature and pressure, the effective masses can also be expected to have such dependences. (Note that the above discussions assume a simplified single dimension E-$\boldsymbol{k}$ space for the conduction band as depicted in Figs. 1.6 and 1.7.) However, the wave vector $\boldsymbol{k}$ is a tensor. Consequently, there are separate masses for each vector component. For a direct gap cubic material, like GaAs, the Γ valley has a spherical constant energy surface near its center and the effective masses are essentially the same in each crystallographic direction. Thus, we need only specify a single value for the effective mass. For indirect gap materials (e.g., Si and Ge), this is not the case; the relevant constant energy surfaces form six ellipsoids in $\boldsymbol{k}$-space in the case of Si (X valley) and eight ellipsoids in the case of Ge (L valley). Fortunately, the ellipsoids frequently have longitudinal and rotational symmetry and, thus, we need only specify two components for the effective mass: (1) a longitudinal component m^*_l (defined along the <100> axis for the X-valley and the <111> axis for the L valley) and (2) a transverse component m^*_t, taken as the minor or transverse axis of the ellipsoid. Generally, m^*_l is much larger than m^*_t.

Table 1.3 lists the average weighted effective electron and hole masses for a range of semiconducting materials. Values are given in units of the electron rest mass, m_o. As the table shows, group III–V compounds, such as GaAs and InSb, tend to

TABLE 1.3
Effective Electron and Hole Masses for a Number of Compound Semiconductors

Material	Electron Effective Mass $m_e^* = m/m_o$	Hole Effective Mass $m_h^* = m/m_o$
	Group IV	
Ge	0.55	0.37
Si	1.08	0.56
	Group III–V	
InSb	0.013	0.68
InAs	0.023	0.40
InP	0.077	0.64
InN	0.11	0.27–1.63 (direction dependent)
GaSb	0.042	0.39
GaAs	0.067	0.45
GaP	0.82	0.60
GaN	0.19	0.6
AlSb	0.12	0.98
AlAs	0.15	0.51
AlP	0.13	0.64
	Group II–VI	
CdS	0.21	0.80
CdSe	0.13	0.45
CdTe	0.11	0.35
ZnTe	0.12	0.6
ZnSe	0.17	0.57
ZnS	0.30	0.57
ZnO	0.19	1.21
	Group IV–VI	
PbTe	0.17	0.20
PbSe	0.30	0.34
PbS	0.25	0.25

Note: Depending on the nature of the periodic lattice potential, the effective masses may be lighter or heavier than the free electron mass, m_o. Note that the effective masses for the elemental semiconductors are larger than for other groups.
Source: From [25–30].

have far smaller electron effective masses than tetrahedrally coordinated group IV elements, such as Si and Ge. Group II–VI materials tend to have effective electron masses intermediate between groups IV and III–V. For all three groups, the hole effective masses are similar.

1.6 Carrier Velocity

The ultimate speed of semiconductor materials depends on carrier velocity, which in turn can be related to the effective mass through the carrier's mobility. Assuming a simple kinetic model of electronic transport, in which the electrons undergo elastic billiard ball–like collisions with fixed and immobile crystal ions, the maximum obtainable charge carrier velocity is given by

$$v_e = \mu_e \xi = (e\tau/m_e^*)\xi \tag{1.21}$$

where ξ is the applied external electric field, e is the electronic charge, and τ is the carrier collision lifetime. A similar expression exists for holes. Thus, a low effective mass is a necessary property for high-bandwidth applications like communications and explains why power amplifiers in cell phones are increasingly based on GaAs rather than Si technology. For radiation detectors, a low effective mass means the detector can operate effectively in high count rate environments, such as those encountered at synchrotron radiation sources.

1.6.1 Saturated Carrier Velocities

The drift velocity derived above is based on the so-called low-field mobility and applies only for sufficiently small electric fields. The low-field mobility is driven by scattering from distortions of the perfect lattice, with the electron gaining energy from the electric field between collisions at such distortion sites. At sufficiently high electric fields (> ~3 kV cm^{-1}) the electron gains sufficient energy to encounter inelastic collisions with the elemental atoms of the perfect crystal. Since the density of such atoms is high, compared to the density of scattering sites, this new mechanism determines carrier velocities at sufficiently high fields, causing them to become independent of the field. In this case, the semiconductor is said to be in a state of velocity saturation. For semiconductors that show saturation, such as group IV semiconductors, it can be approximated by

$$v(\xi) = \frac{\mu\xi}{1 + \mu\xi/v_{sat}} \tag{1.22}$$

where ξ is the electric field, and v_{sat} is the saturation velocity. The electric field at which velocities become saturated is referred to as the critical field, ξ_{cr}, and ranges from 0.1 MV cm^{-1} for Ge to 10 MV cm^{-1} for diamond.

In Figure 1.9 we show the electron drift velocity as a function of electric field for a number of semiconductors. The linear dependence at small fields illustrates the low-field mobility, with GaAs having a larger low-field mobility than silicon and

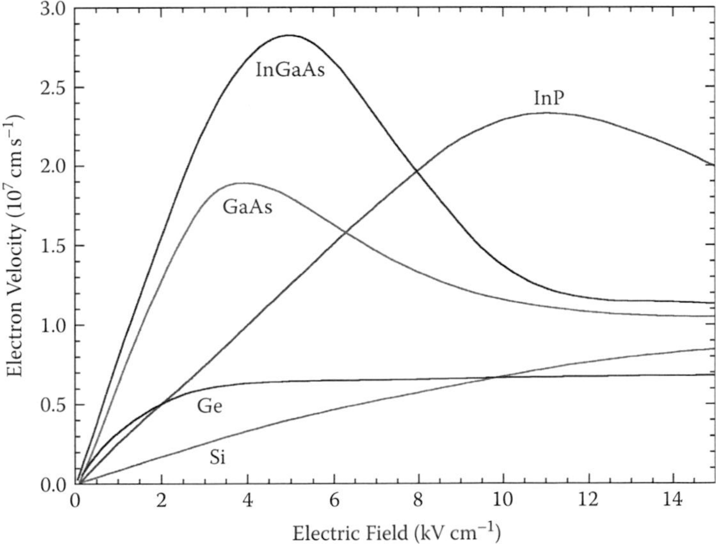

Figure 1.9 The electron velocity as a function of electric field for a number of semiconducting materials at 300 K [18,31,32]. At high fields the drift velocity saturates for the elemental semiconductors. However for compound semiconductors the velocity reaches a maximum at fields around 4 kV cm^{-1} and then begins to decrease due to the increasing influence of additional minima in the conduction bands.

hence a proportionally larger drift velocity. We also note that the function does indeed saturate for Si and Ge, but with group III–V materials, the high-field behavior of the drift velocity does not follow this pattern. Instead, the drift velocity reaches a peak and then starts to decrease. This is because the band structure in these materials contains more than one local minima in the conduction band (see Section 1.3.4). If we take GaAs as an example, under normal low-field conditions, carrier dynamics are dominated by transitions to and from the Γ valley (see Figure 1.7). This minimum has a lower effective mass than the other two valleys and therefore a higher mobility. However as the electric field increases beyond ~3 kV cm^{-1}, some of the conduction electrons that normally occupy the Γ band will have enough energy to occupy energies in the second minima (L band) with its higher effective mass and lower mobility. As the field is increased further, the proportion of electrons with the lower mobility increases, and the drift velocity will continue to decrease until all the electrons share the lower mobility and the drift velocity levels off. This effect was first suggested by Ridley and Watkins [33] and forms the operational basis of the Gunn diode [34] used extensively for microwave applications.

The saturation velocity and critical field are important parameters for several reasons. In communication applications, the speed advantages offered by high-mobility materials may be lost under high-field conditions. Good examples are GaAs and Si for which the saturated drift velocities are both ~10^7 cm s^{-1}, although under low-field conditions the drift velocity of GaAs is three times that of Si. Similarly in micropower electronics, under saturated velocity conditions, currents may no longer be proportional to the applied voltage, as in the low-field case.

1.7 Conduction in Semiconductors

The conduction process in semiconductors is strongly temperature dependent and relies on the fact that, above a particular temperature, electrons may be thermally excited into the conduction band where they are available for the conduction process. Once in the conduction band they are free to move around the crystal under the influence of an external magnetic or electric field.

1.7.1 Intrinsic Semiconductors

Intrinsic semiconductors are intrinsically pure, meaning that they do not intentionally contain impurities. At room temperature, the thermal energy of the atoms may allow a small number of the electrons to participate in the conduction process depending on the bandgap energy. The electron and hole carrier densities will be equal because the thermal activation of an electron from the valence band to the conduction band yields both a free electron in the conduction band and a hole in the valence band. In intrinsic silicon, the energy available at room temperature generates 1.45×10^{10} carriers per cubic meter of each carrier type (electrons and holes). This number doubles for every 11°C rise in temperature, but is a constant for a given temperature. For GaAs, the number of thermally generated carriers is 1.79×10^{6} cm^{-3} (GaAs has a lower intrinsic carrier density on account of its larger bandgap energy).

1.7.1.1 Intrinsic Carrier Concentration

The conductivity of a semiconductor depends on the number of free electrons in the conduction band, which in turn depends on two functions, namely, the density of states $g(E)$ in which the electrons exist and the energy distribution function $f_e(E)$ for free electrons. Although the number of states in all bands is effectively infinite, the number of electrons is not, since it must be equal to the number of protons in the neutral material. Therefore, not all of the states can be occupied at any given time. The likelihood of any particular state being filled is given by the Fermi–Dirac probability function, $f_e(E)$, which gives the probability that, at a given temperature T, an electronic state with energy E is occupied by an electron.

$$f_e(E) = \frac{1}{\exp\left(\dfrac{E - E_F}{kT}\right) + 1} \tag{1.23}$$

Here E_F is the Fermi level which is the energy at which the probability of occupation by an electron is exactly one-half for temperatures above absolute zero. The corresponding hole distribution function is

$$f_h(E) = 1 - f_e(E) = \frac{1}{\exp\left(\dfrac{E_F - E}{kT}\right) + 1} \quad (1.24)$$

which is the probability that a state at energy E is not occupied by an electron and, by default, is occupied by a hole. The functional form of f_e(E) is displayed in Figure 1.10 for three cases, $T = 0$, $T > 0$ and $T \gg 0$. At $T = 0$, the distribution has the shape of a Heavyside function. Above E_F the probability of occupancy is zero, and below it is unity, meaning that all states are filled up to the Fermi energy and empty above it. Its maximum value of unity merely reflects the fact that only one electron can occupy a particular state—a natural consequence of the Pauli exclusion principle [5]. At higher temperatures, the function begins to harden and smear out toward the high energies. The probability that an electron will occupy

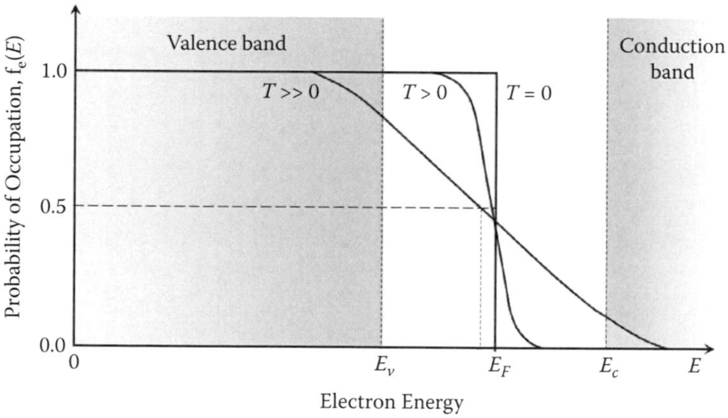

Figure 1.10 The Fermi energy distribution function $f(E)$ as a function of electron energy, E. Here, E_v and E_c are the energy levels of the valence and conduction bands, respectively. The defining parameter of this distribution is the Fermi energy E_F, which is the energy at which the probability of occupation by an electron is exactly one-half at $T > 0$ K. Note the reduction of E_F with increasing temperature. However, at room temperature it is still very close to mid-bandgap.

a state in the conduction band correspondingly increases. The distribution can therefore be characterized by the Fermi energy which, depending on temperature, determines the concentration of electrons in the conduction band. At room temperature, the intrinsic Fermi level lies very close to the middle of the bandgap. In most cases $f_e(E)$ can be approximated by a Maxwell–Boltzmann distribution and in general does not depend on the semiconductor material.

The density of states function, g(E), on the other hand, is intimately dependent on the material. For electrons, it can be approximated by a parabola about the band edge for energies close to the edge:

$$g_e(E) = M_c \frac{\sqrt{2}}{\pi^2} \frac{(E - E_c)^{\frac{1}{2}}}{\hbar^3} (m_{de}^*)^{\frac{3}{2}} \tag{1.25}$$

where, $\hbar = h/2\pi$. Here, M_c is the number of equivalent conduction band minima contained in the first Brillouin zone (smallest unit cell) and m_{de}^* is the density of states' effective mass. For indirect bandgap semiconductors, such as Si and Ge, M_c equals 6 and 4, respectively. For direct bandgap semiconductors, such as GaAs, $M_c = 1$. The density of states effective mass is given by

$$m_{de}^* = (m_1^* m_2^* m_3^*)^{\frac{1}{3}} \tag{1.26}$$

which is the geometric mean of the components of electron effective mass along each crystallographic direction. The magnitudes of these components depend on the shapes of the conduction band minima and can be conveniently visualized by considering the constant energy surfaces in $\boldsymbol{k}$-space around $\boldsymbol{k} = 0$. In general, this is a sphere for direct gap materials and an ellipsoid for indirect gap materials. For elliptical surfaces (for example, Si and Ge), the two transverse components are equal, and $m^*_{de} = (m^*_1 m_2^{*2})^{1/3}$. For spherical surfaces (for example, GaAs), the longitudinal and transverse components are equal, and $m^*_{de} = (m^*{}_1^3)^{1/3} = m^*_e$. (Note that the overall effective mass, m^*_e, is simply related to m^*_{de} by weighting Equation (1.26) by the number of equivalent conduction band minima, for example $M_c^{2/3}$.) The effective density of states for an arbitrary band structure may be calculated following the approach of van Vliet and Marshak [35]. The density of states for holes is given by a similar expression,

$$g_h(E) = M_v \frac{\sqrt{2}}{\pi^2} \frac{(E - E_c)^{\frac{1}{2}}}{\hbar^3} (m_h^*)^{\frac{3}{2}} \qquad (1.27)$$

However, in this case, $M_v = 1$, since there is no degeneracy factor due to crystal symmetries, because the top of the valence band is unique and always occurs at the center of the first Brillouin zone. The effective mass used in Equation (1.27) should be calculated from Equation (1.20) to take into account the two subbands: the heavy-hole and light-hole bands. The split-off band can be ignored as energetically much less favorable.

Note that E must be greater than E_c for free electrons and less than E_v for free holes due to the forbidden gap between E_c and E_v. The density of free electrons in the conduction band can then be calculated by integrating the product of f_e(E) and $g_e(E)$, over energy from the bottom of the conduction band to the top of the conduction band. The difficulty involved in evaluating the integral in such a case is due to the fact that it would require an additional parameter related to the width of the conduction band. However, because the Fermi level is significantly below the top of the conduction band, the upper limit for the integral can be replaced by infinity, and thus the total number of electrons in the conduction band is

$$n = \int_{E_c}^{\infty} f_e(E) g_e(E)\, dE \qquad (1.28)$$

A similar expression exists for holes. For Fermi levels significantly below E_c and above E_v, and a nondegenerate semiconductor (which intrinsic materials nearly always are), this integration reduces to

$$n = N_c \exp\left(-\frac{(E_c - E_F)}{kT} \right) \qquad (1.29)$$

and for holes,

$$p = N_v \exp\left(-\frac{(E_F - E_v)}{kT}\right) \tag{1.30}$$

Here n and p are the densities of free electrons in the conduction band and of holes in the valence band, respectively, and N_c and N_v are the effective densities of states which vary with temperature and effective mass. Figure 1.11 illustrates the various distributions described by Equations (1.23) to (1.28),

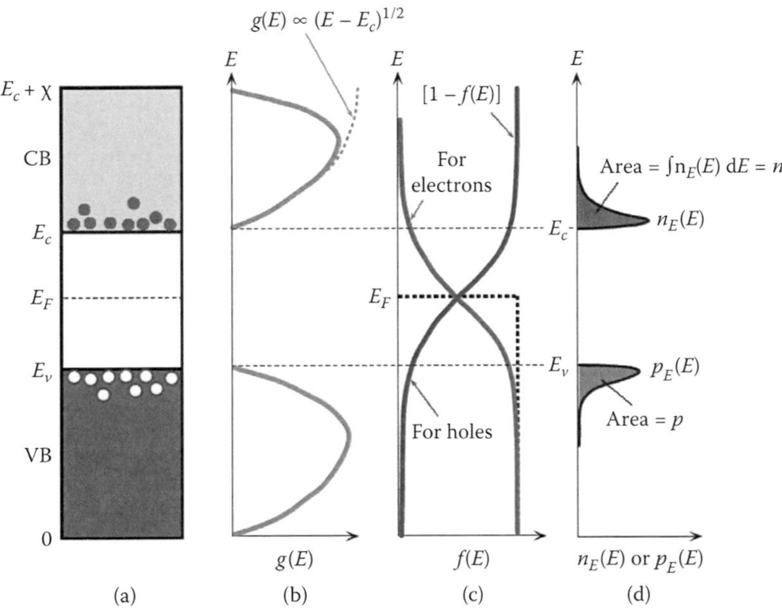

Figure 1.11 **(See color insert.)** Schematic of the various distributions discussed in the text leading up to the energy density of electrons and holes in the conduction and valence bands, respectively (from reference [36]). The example given is for an intrinsic semiconductor. (a) The energy band diagram. (b) The density of states (number of states per unit energy per unit volume. The total number of states in the valence band and conduction bands is equal to the number of valence electrons. However in metals the total number of states in the valence band is much larger, which is the reason why electrons in metals need no activation to become mobile. (c) The Fermi–Dirac probability function (probability of occupancy of a state) and (d) the product of $g(E)$ and $f(E)$, which gives the energy density of electrons in the conduction band. The area under $n_E(E)$ versus E is the electron concentration in the conduction band.

showing how they lead up to the electron and hole concentrations described by Equations (1.29) and (1.30) in the conduction and valence bands, respectively. The curves are given for an intrinsic semiconductor (that is, without any intentional impurities). The corresponding curves for an extrinsic semiconductor (that is, doped with intentional impurities) show an increase in either electron or hole concentrations, depending on whether the semiconductor is *n*-type or *p*-type.

To calculate the effective density of states, N_c and N_v, we must compute the effective pair creation rate. To do this, we recognize that an electron can make a transition from any state in the valence band to any state in the conduction band. Therefore we integrate over all possible transitions with a weighting factor to account for the fact that the most likely transitions are from the top of the valence band to states near the bottom of the conduction band. An approximate expression for the effective density of states in the conduction band, N_c, is

$$N_c = 2M_c\left(\frac{m_e^* kT}{2\pi\hbar^2}\right)^{\frac{3}{2}} = 2.51\times10^{19}M_c\left(\frac{m_e^* T}{300}\right)^{\frac{3}{2}} \text{ cm}^{-3} \quad (1.31)$$

At room temperature, N_c is of the order of ~10^{18} to 10^{19} atoms cm^{-3} for most materials. Similarly, the effective density of states in the valence band, N_v, is given by

$$N_v = 2M_v\left(\frac{m_h^* kT}{2\pi\hbar^2}\right)^{\frac{3}{2}} = 2.51\times10^{19}M_v\left(\frac{m_h^* T}{300}\right)^{\frac{3}{2}} \text{ cm}^{-3} \quad (1.32)$$

From Equations (1.31) and (1.32) we see that density of states, N, increases with temperature because higher states in the conduction band and deeper states in the valence band become more accessible as the thermal energy increases. Furthermore, we see that, apart from a weak material dependence through the effective masses, N depends only on T. By combining Equations (1.29) and (1.30), we get

$$E_F = \frac{1}{2}(E_c + E_v) + \frac{3}{4}kT\,\text{In}\left(\frac{m_h^*}{m_e^*}\right) \quad (1.33)$$

indicating that the Fermi energy in intrinsic semiconductors lies very close to the center of the bandgap (typically within ~0.05 eV at 300 K).

To preserve charge neutrality, the number of electrons per unit volume in the conduction band must equal the number of holes per unit volume in the valence band. Thus,

$$n = N_c \exp\left(-\frac{(E_c - E_F)}{kT}\right) = p = N_v \exp\left(-\frac{(E_F - E_v)}{kT}\right) \quad (1.34)$$

where n is the electron density, and p is the hole density. Note that, in the evaluation of Equation (1.34), N_c and N_v can be considered constants near room temperature, because the dominant effect on carrier densities is through the exponential term. We define n_i to be the intrinsic carrier density which must equal to n and also p. Therefore,

$$n_i^2 = np \quad (1.35)$$

This is known as the mass-action law* and is a powerful relationship in that it allows the hole density to be determined if the electron density is known and vice versa. Substituting,

$$n_i^2 = N_c N_v \exp\left(-\frac{(E_v - E_c)}{kT}\right) = N_c N_v \exp\left(-\frac{E_g}{kT}\right) \quad (1.36)$$

and therefore,

$$n_i = \sqrt{N_c N_v} \exp\left(-\frac{E_g}{2kT}\right) \quad (1.37)$$

Substituting for N_c and N_v, we obtain

$$n_i = 2(m_e^* m_h^*)^{\frac{3}{4}} \left(\frac{kT}{2\pi\hbar^2}\right)^{\frac{3}{2}} \sqrt{M_c M_v} \exp\left(-\frac{E_g}{2kT}\right) \quad (1.38)$$

* The term is borrowed from chemistry, where it is used to describe systems in dynamic equilibrium.

which, for direct bandgap semiconductors, reduces to

$$n_i = 2.51 \times 10^{19} (m_e^* m_h^*)^{\frac{3}{4}} \left(\frac{T}{300}\right)^{\frac{3}{2}} \exp\left(-19.35 E_g \left(\frac{300}{T}\right)\right) \quad (1.39)$$

Values of intrinsic carrier concentrations vary considerably between semiconductor materials, mainly through the exponential dependence on the bandgap energy. For example, at room temperature $n_i = 1.3 \times 10^{13}$ cm^{-3} for InSb ($E_g = 0.18$ eV) and 1.0×10^6 cm^{-3} for CdSe ($E_g = 1.8$). This represents roughly seven orders of magnitude change in intrinsic carrier concentration for an order of magnitude change in bandgap energy. Tabulated values of n_i for various semiconductors are given in Table 1.4. In Figure 1.12 we show the intrinsic carrier concentrations for the common semiconductors, Ge, Si, GaAs, and GaP as a function of temperature [37]. The effect of increasing bandgap energy is apparent, leading to a substantial reduction in free carriers at low temperatures for wide gap materials.

TABLE 1.4
The Intrinsic Carrier Concentration, (n_i) at 300 K in Various Semiconducting Materials

Semiconductor	Bandgap Energy (eV)	Intrinsic Carrier Concentration, n_i, at 300 K (cm^{-3})
InSb	0.18	1.3×10^{16}
InAs	0.35	1.0×10^{15}
Ge	0.66	2.0×10^{13}
Si	1.12	1.0×10^{10}
InP	1.34	1.3×10^{7}
GaAs	1.43	2.1×10^{6}
GaP	2.27	2.0×10^{0}
SiC	2.39–3.26	$0.12–2 \times 10^{-8}$

Note: For comparison, n_i in metals is ~10^{22} cm^{-3}. Note: the strong dependence on bandgap energy.
Source: From [38].

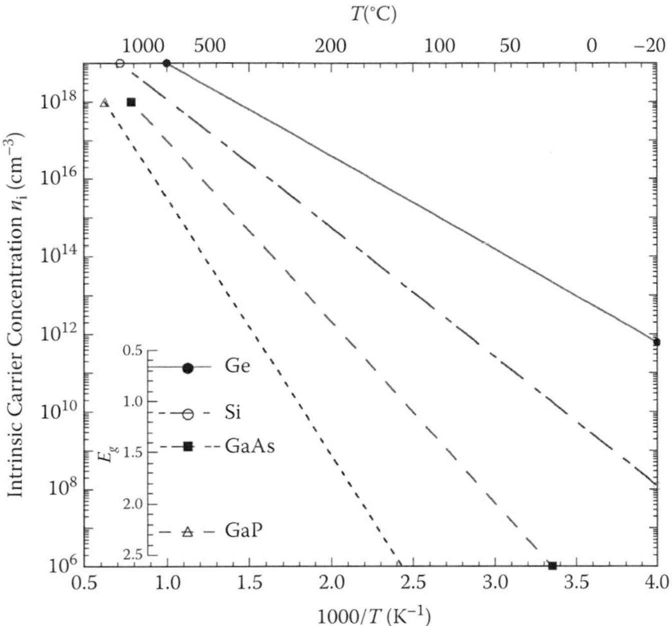

Figure 1.12 Intrinsic carrier concentrations in Ge, Si, GaAs, and GaP as a function of reciprocal temperature. The effect of increasing bandgap on the intrinsic carrier density is apparent.

1.7.2 Extrinsic Semiconductors

The physical properties of semiconductors can be considerably modified by the introduction of small quantities of foreign atoms. The resistivity is proportional to the free carrier density, which can be changed over a wide range by replacing a small fraction of the host crystal atoms with a different atomic species (doping atoms). For example, the resistivity of silicon can be altered by seven orders of magnitude when only one millionth of the host atoms are replaced by a dopant. Note the impurity substitution does not affect the bandgap energy or other basic characteristics of the host material, only the conduction of current. Depending on the impurity, whether it is a donor or an acceptor, this current may consist primarily of holes or electrons. An increase in the number of one type of carrier leads to the reduction in the number of the other type. In fact, it is found that the mass-action law given by Equation (1.35) still applies, and the product of the electron and hole number densities remains constant at a given temperature (as in the case of intrinsic semiconductors).

1.7.2.1 Donors and Acceptors

Figure 1.13 (left) shows schematically the doping of a silicon crystal with a phosphorus atom. Phosphorus has five electrons in its outer shell and forms covalent bonds with its four adjacent silicon atoms. The fifth electron is unpaired and becomes a conduction electron, leaving behind a positively charged phosphorus atom. As a consequence, the silicon crystal becomes n-type, and phosphorus is said to be a donor, because it donates weakly bound valence electrons to the material, creating an excess of negative charge carriers. These weakly bound valence electrons can move about in the crystal lattice relatively freely and can facilitate conduction in the presence of an electric field. Typical donor impurities are arsenic, phosphorus, and antimony in

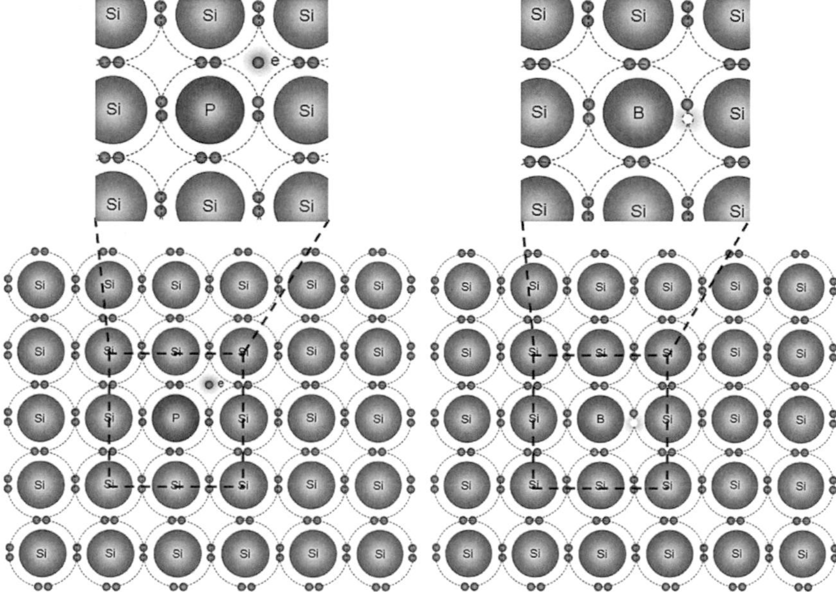

Figure 1.13 Schematic illustrating the physical implementation of doping in silicon. Left: n-type silicon doped with phosphorus. The effect of the phosphorus atom is to introduce an extra electron into the lattice, and thus the phosphorus now acts as a donor for electrons. Right: p-type silicon doped with boron. In this case, the boron atom is missing one electron to fill its outer shell, and thus the boron behaves as an electron acceptor.

Si and sulfur, selenium, and tin in GaAs. Boron, on the other hand, has three outer shell electrons and hence can covalently bond to only three silicon atoms, leaving a missing electron or hole in its outer shell, so that the material has an excess of positive charge. In this case the Si is *p*-type, and boron is said to be an acceptor (see Figure 1.13 right). Typical acceptor impurities are gallium, boron, and aluminum in Si and beryllium, magnesium, and zinc in GaAs. The *n* and *p* designations denote which particle, electron or hole, will be the majority carrier, the other being known as the minority carrier.

Common impurities such as phosphorus and boron have energy levels very close to the conduction band and valence bands, respectively. Shallow donors or acceptors such as these exist in ionized form at room temperature because thermal energy is sufficient to ionize them. This condition is called complete ionization, that is, $n = N_A$ or N_D. For *n*-type material, $n = N_D$ and

$$n = N_c \exp\left(-\frac{(E_c - E_F)}{kT}\right) = N_D \tag{1.40}$$

Rearranging,

$$E_c - E_F = kT \,\mathrm{In}\left(\frac{N_c}{N_D}\right) \tag{1.41}$$

One of the implications of Equation (1.41) is that, as N_D increases, $(E_c - E_F)$ becomes smaller, so that the Fermi level moves closer to the bottom of the conduction band. Similarly, for *p*-type semiconductors, the Fermi level moves toward the top of the valence band with increasing acceptor concentration.

For doping, impurity concentrations are typically one atom per 10^8 semiconductor atoms. When both donors and acceptors are present simultaneously, the impurity present at a higher concentration determines the type of conductivity in the semiconductor. Electrons in an *n*-type semiconductor are the majority carriers, whereas the holes are the minority carriers. Conversely, in a *p*-type semiconductor, holes are the majority carriers, and electrons the minority carriers. In reality, dopants form energy levels in the forbidden energy gap. Donor atoms introduce states below, but very close to the conduction band, while acceptor atoms introduce states above, but very close to the valence band. Deep energy levels close to the center of the gap are usually introduced by unintentional impurities and act as recombination sites. They are the dominant recombination mechanism in indirect bandgap materials. At $T = 0$ K, any free carriers from donors and acceptors are bound to their atoms, and no conduction occurs. For nonzero temperatures, the sites can be thermally ionized, releasing carriers into the bands so conduction can occur. For donor atoms, an electron orbits a lattice site, while for acceptors, a hole orbits around a lattice site with residual negative charge. The energy required to ionize these carriers is much less than the binding energy of the hydrogen atom,

$$E_H = -\frac{m_o e^4}{32\pi^2 \varepsilon_o^2 \hbar^2} = -13.6 \text{ eV} \tag{1.42}$$

because (1) the effective mass is smaller, (2) the radius of the carrier orbit larger, and (3) the dielectric constant of a semiconductor, ε_s, much greater (i.e., $\varepsilon_s > 10\ \varepsilon_o$). The last has the largest effect because it occurs in Equation (1.42) as an inverse square. Typically, ionization energies are ~0.015 eV in Ge and somewhat larger in Si (~0.04 eV) due to its lower dielectric constant.

For dopants to be effective, they must be ionized at the temperature of operation, and this depends on their activation energy, E_a. At room temperatures E_a should be $\leq kT$ or ~1/40 eV. For low enough temperatures, excitation from donors and acceptors is the only source of carriers, and the conductivity can be considered solely extrinsic. In this regime, the doping of the semiconductor determines whether the semiconductor, ε is *n*-type or *p*-type. At increasingly higher temperatures, direct thermal excitation from the valence band to the conduction band begins to dominate the extrinsic density. At some point, there will be essentially equal numbers of electrons and holes, and the conductivity can now be considered intrinsic. Impurities which can serve either as an acceptor or donor are called *amphoteric* impurities. Semiconductors which can be doped either *p*-type or *n*-type are called *ambipolar* semiconductors. Those which can only be made n or p type (such as ZnTe) are called *unipolar* semiconductors.

1.7.2.2 Extrinsic Carrier Concentration

To calculate the carrier concentration in extrinsic semiconductors, we make use of the fact that in equilibrium the material must be charge neutral. For fully ionized impurities this means that

$$p_n + N_D = n_n + N_A \tag{1.43}$$

and

$$p_p + N_D = n_p + N_A \tag{1.44}$$

where n_n and p_n are the electron and hole concentrations in an n-type semiconductor, and n_p and p_p are the corresponding concentrations in a p-type semiconductor. N_D and N_A are the donor and acceptor concentrations, respectively. Using Equations (1.43) and (1.44) and assuming $n_i \ll |N_D - N_A|$, the carrier concentrations in n-type semiconductors are,

$$n_n = N_D - N_A \quad \text{and} \quad p_n = \frac{n_i^2}{N_D - N_A} \tag{1.45}$$

where we have made use of the mass-action law (Equation 1.35) to derive p_n. Similarly for p-type material,

$$p_p = N_A — N_D \quad \text{and} \quad p_p = \frac{n_i^2}{N_D - N_A} \tag{1.46}$$

From Equations (1.45) and (1.46), we see that a reduction in charge carriers and therefore conductivity can be achieved in materials with a low $N_D - N_A$. This lowering can be achieved by intentionally adding donors into a p-type material or acceptors to an n-type material. Such a material is said to be compensated, and if N_D and N_A are approximately equal and less than n_i, the semiconductor behaves almost intrinsically. However, it should be appreciated that high impurity concentrations will alter mobilities and effective masses. Compensation is a commonly used method to increase the resistivity in extrinsic materials. This is usually highly desirable for a radiation detector, because higher resistivity material can support larger biases, resulting in better charge collection. However, high doping levels will also affect carrier mobilities.

1.7.2.3 Doping Dependence of the Energy Bandgap

High doping densities also causes the bandgap to shrink. This effect is explained by the fact that the wave functions of the electrons bound to the impurity atoms start to overlap as the density of impurities increases. For instance, at a doping density of 10^{18} cm^{-3}, the average distance between two impurities is only 10 nm. This results in an energy band rather than discrete levels. If the impurity level is shallow, this impurity band reduces the energy band of the host material by

$$\Delta E_g = \frac{3q^2}{16\pi\varepsilon_s}\sqrt{\frac{q^2 N}{\varepsilon_s kT}} \tag{1.47}$$

where N is the doping density, q is the electronic charge, ε_s is the dielectric constant of the semiconductor, k is Boltzmann's constant, and T is the temperature in kelvin. For silicon ($\varepsilon_r = 11.7$), this expression further reduces to,

$$\Delta E_g = -22.5\sqrt{\frac{N}{10^{18}\,\text{cm}^{-3}}}\ \text{meV} \tag{1.48}$$

From this expression we find that the bandgap shrinkage can typically be ignored for doping densities less than 10^{18} cm^{-3} and only becomes significant (~20%) for doping densities in excess of 10^{20} cm^{-3}, which is close to, or above, the solubility limits for known semiconductors. In fact, very highly (degenerately) doped semiconductors (>10^{18} cm^{-3}) have conductivity levels comparable to metals and are often used as metal replacements in modern integrated circuit fabrication.

1.7.2.4 Practical Considerations

Because the resistivity of intrinsic materials is high, doping is often used in the semiconductor industry to create lower resistivity material in a highly controlled manner, as well to establish both p-type and n-type conductivity, which is needed for many applications. In practice, doping a semiconductor requires that two conditions are met. The first is that dopants must introduce electronic energy levels E_{dop} in the bandgap that are very close to the band edges. In addition, the difference in the energy levels, E_C–E_{dop} and E_{dop}–E_V, must be comparable to kT (about 25 meV at room temperature); otherwise the transfer of electrons between the bands and the dopant level will not occur at room temperature. The second consideration is that the Fermi energy must be able to move freely with dopant level and must not be "pinned."* Specifically, the energy states introduced by intentionally introduced dopants should dominate conduction properties. If there are too many energy states stemming from defects such as dislocations, precipitates and interfaces, the Fermi level will be tied or "pinned" to these states and doping will be limited or totally ineffective. This problem is particularly acute in group II–VI materials. One final consideration is that, if doping levels are so high that the impurity atom wave functions begin to overlap, undesirable conductivity modes may occur, such as "hopping." In this case, conduction occurs directly between impurity atoms without having to raise electrons into the conduction band.

1.7.3 Conductivity and Resistivity

A perfect crystal has perfect periodicity so that the potential seen by a carrier is completely periodic. Such a potential does not impede the movement of the charge carriers, so the crystal has no resistance to current flow and behaves as a superconductor. However, in real materials, the presence of any impurities or crystalline defects upsets the periodicity of the potential seen by the carrier, creating an impedance to current flow. One of the most useful semiconductor material properties is conductivity, because it is intimately dependent, not only on imperfections in the crystal, but also on basic semiconductor parameters such as dopant concentration and mobility of charge carriers. It requires a relatively simple measurement

* A comprehensive discussion of Fermi level pinning is given in Chapter 4.

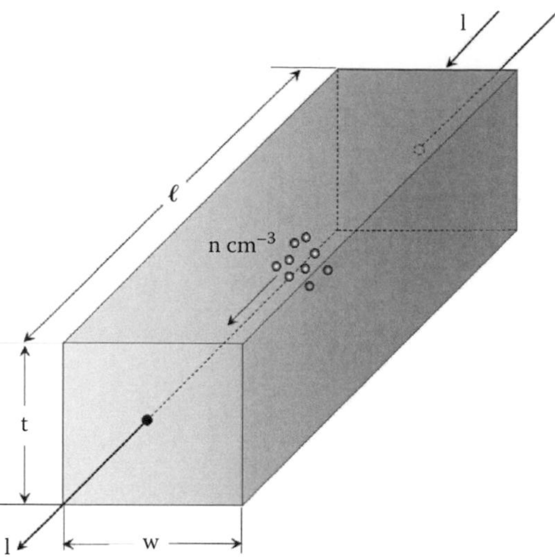

Figure 1.14 Current conduction in a uniformly doped semiconductor bar of length ℓ and cross-sectional area $A = t \times w$.

(see Chapter 3, Section 3.2.4) and therefore can be used as the basis on which many other parameters can be derived. Consider a rectangular bar of semiconducting material as shown in Figure 1.14. The current density, J, flowing through the sample is given by $J = I/A$, where A is the cross-sectional area and I is the current flowing through it. The electron current density is determined by summing the product of the charge on each electron times the electron's velocity over all electrons per unit volume n,

$$Jn = In/A = \sum -qvi = -qnvn \tag{1.49}$$

where q is the electronic charge, and v is electron drift velocity, Now since the drift velocity and mobility are related through $v = \mu\xi$, where ξ is the electric field, then

$$Jn = qn\mu_e\xi. \tag{1.50}$$

Mobility is related to the conductivity, σ, by

$$\sigma_e = qn\mu_e \tag{1.51}$$

Similarly for holes,

$$\sigma_p = qp\mu_p \tag{1.52}$$

Therefore, the total current density (electrons plus holes), J, is given by

$$J = q(n\mu_e + p\mu_p)\xi = \sigma\xi \tag{1.53}$$

The resistivity is inversely related to the conductivity ($\rho = 1/\sigma$). Thus,

$$\rho = [q(n\mu_e + p\mu_p)]^{-1} \tag{1.54}$$

For an extrinsic semiconductor, one of the carriers will be dominant. For an n-type semiconductor

$$\rho = (qn\mu_e)^{-1} \tag{1.55}$$

and for a p-type

$$\rho = (qp\mu_p)^{-1} \tag{1.56}$$

For a sample of material the resistance to current flow is then given by

$$R = \rho\ell/wt \tag{1.57}$$

where ℓ is the length of the sample, and w and t are the cross-sectional dimensions such that $w \cdot t = A$, the cross-sectional area of the sample. Thus by measuring the resistance of a sample, it is possible to derive the mobilities of extrinsic materials directly.

References

1. S.M. Sze, *Semiconductor Devices: Physics and Technology*, John Wiley & Sons, Hoboken, N.J., 2nd edition, (2002). ISBN: 978-0-471-33372-2.
2. C. Kittel, *Introduction to Solid State Physics*, John Wiley & Sons, Wiley, 8th edition, (2005) ISBN 978-0-471-41526-8.

3. W.T. Wenckebach, *Essentials of Semiconductor Physics*, John Wiley & Sons, Hoboken N.J. (1999) ISBN: 978-0-471-96539-8.
4. N. Bohr, "On the Constitution of Atoms and Molecules," *Philosophical Magazine*, Series 6, Vol. 26 (1913) pp. 1–25.
5. W. Pauli, "Über den Zusammenhang des Abschlusses der Elektronengruppen im Atom mit der Komplexstruktur der Spektren," *Z. Phys.*, Vol. 31 (1925) pp. 765–783.
6. W. Heisenberg, "Über den anschaulichen Inhalt der quantentheoretischen Kinematik und Mechanik," *Zeitschrift für Physik*, Vol. 43 (1927) pp. 172–198.
7. H.T. Grahn, *Introduction to Semiconductor Physics*, World Scientific Publishing Co. (1999) ISBN-10: 9810233027: ISBN-13: 978-9810233020.
8. W. Shockley, W.T. Read, Jr., "Statistics of the Recombinations of Holes and Electrons," *Phys. Rev.*, Vol. 87, no. 5 (1952) pp. 835–842.
9. R.N. Hall, "Electron-Hole Recombination in Germanium," *Phys. Rev.*, Vol. 87 (1952) p. 387.
10. S. Berrah, H. Abid, A. Boukortt, M. Sehil, "Band Gap of Cubic AlN, GaN and InN Compounds," *Turk. J. Phys.*, Vol. 30 (2006) pp. 513–518.
11. G.G.N. Angilella, N.H. March, I.A. Howard, R. Pucci, "Pressure Dependence of the Energy Gaps in Diamond-Type Semiconductors, and Their III–V Analogues Such as InSb," *Journal of Physics: Conference Series*, Vol. 121, Issue 3 (2008) 032006 (5 pp).
12. Y.P. Varshni, "Temperature Dependence of the Energy Gap in Semiconductors," *Physica*, Vol. 34 (1967) pp. 149–154.
13. R. Passler, "Parameter Sets Due to Fittings of the Temperature Dependencies of Fundamental Bandgaps in Semiconductors," *Phys. Stat. Sol. (b)*, Vol. 216 (1999) pp. 975–1007.
14. E. Schrödinger, "Quantisierung als Eigenwertproblem III," *Annalen der Physik (Leipzig)*, Vol. 80 (1926) pp. 734–756.
15. F. Bloch, "Über die Quantenmechanik der Elektronen in Kristallgittern," *Z. Physik*, Vol. 52 (1928) pp. 555–600.
16. J.R. Chelikowsky, M.L. Cohen, "Non-Local Pseudopotential Calculations for the Electronic Structure of Eleven Diamond and Zinc-Blende Semiconductors," *Phys. Rev.*, Vol. B14 (1976) pp. 556–582.
17. J.D. Wiley, *Semiconductor and Semimetals,* R.K. Willardson, A.C. Beer, eds., Academic Press, New York, Vol. 10 (1975) p. 91.
18. C. Jacoboni, C. Canali, G. Ottaviani, A.A. Quaranta, "A Review of Some Charge Transport Properties of Silicon," *Solid State Electron.*, Vol. 20, no. 2 (1977) pp. 77–89.

19. H. Casey Jr., C.F. Ermanis, K.B. Wolfstirn, "Variation of Electrical Properties with Zn Concentration in GaP," *J. Appl. Phys.*, Vol. 40, 7, (1969) pp. 2945–2958.
20. G. Ottaviani, L. Reggiani, C. Canali, F. Nava, A.A. Quaranta, "Hole Drift Velocity in Silicon," *Phys. Rev.*, Vol. B12, no. 8 (1975) pp. 3318–3329.
21. M. Razeghi, Ph. Maurel, M. Defour, F. Omnes, G. Neu, A. Kozacki, "Very High Purity InP Epilayer Grown by Metal organic Chemical Vapor Deposition," *Appl. Phys. Lett.*, Vol. 52, no. 2 (1988) pp. 117–119.
22. W. Walukiewicz, J. Lagowskii, L. Jastrzebski, P. Rava, M. Lichtensteiger, C.H. Gatos, H.C. Gatos, "Electron Mobility and Free Carrier Absorption in InP; Determination of Compensation Ratio," *J. Appl. Phys.*, Vol. 51, no. 5 (1980) pp. 2659–2668.
23. D.N. Nasledov, Y.G. Popov, N.V. Synkaev, S.P. Starosel'tseva, "Electrical Properties of *p*-Type InP at Low Temperatures," *Sov. Phys. Semicond.*, Vol. 3 (1969) 387–389.
24. E.O. Kane, "Energy Band Structure in *p*-Type Germanium and Silicon," *J. Phys. Chem. Solids*, Vol. 1 (1956) pp. 82–89.
25. C.M. Wolfe, N. Holonyak, G.E. Stillman, *Physical Properties of Semiconductors*, Prentice-Hall, Inc., Upper Saddle River, N.J. (1989).
26. S.M. Sze, *Physics of Semiconductor Devices,* John Wiley & Sons, 2nd edition (1981) ISBN-13: 9780471056614.
27. W.A. Harrison, *Electronic Structure and the Properties of Solids,* Dover Pub., New York (1989) ISBN 0-486-66021-4.
28. H. Föll, "Effective Masses," http://www.tf.uni-kiel.de/matwis/amat/semi_en/kap_2/backbone/r2_3_1.html.
29. O. Madelung, U. Rössler, M. Schulz, "Landolt-Börnstein—Group III Condensed Matter Numerical Data and Functional Relationships in Science and Technology." In *II–VI and I–VII Compounds; Semimagnetic Compounds*, Vol. 41B (1999) pp. 1–2, ISBN 978-3-540-64964-9.
30. I. Vurgaftman, J.R. Meyer, L.R. Ram-Mohan, "Band Parameters for III–V Compound Semiconductors and Their Alloys," *J. Appl. Phys.*, Vol. 89, (2001) pp. 5815–5875.
31. P. Smith, M. Inoue, J. Frey, "Electron Velocity in Si and GaAs at Very High Electric Fields," *Appl. Phys. Letts.*, Vol. 37 (1980) pp. 797–798.
32. J.G. Ruch, G.S. Kino, "Measurement of the Velocity Field Characteristics of Gallium Arsenide," *Appl. Phys. Lett.*, Vol. 10 (1967) pp. 40–42.

33. B.K. Ridley, T.B. Watkins, "The Possibility of Negative Resistance Effects in Semiconductors," *Proc. Phys. Soc. London*, Vol. 78 (1961) pp. 293–304.
34. J.B. Gunn, "Microwave Oscillation of Current in III–V Semiconductors," *Solid State Commun.*, Vol. 1 (1963) p. 88.
35. K.M. van Vliet, A.H. Marshak, "The Effective Density of States in the Conduction and Valence Bands for Arbitrary Band Structure," *Phys. Stat. Sol. (b)*, Vol. 101 (1980) pp. 525–530.
36. S.O. Kasap, *Principles of Electronic Materials and Devices*, McGraw Hill, 2nd edition (2002) ISBN 0-07-112237-0.
37. C.D. Thurmond, "The Standard Thermodynamic Functions for the Formation of Electrons and Holes in Ge, Si, GaAs, and GaP," *J. Electrochem. Soc.*, Vol. 122, no. 8 (1975) pp. 1133–1141.
38. Ioffe Institute, "Semiconductors on NSM," http://www.ioffe.rssi.ru/SVA/NSM/Semicond/.

2
Growth Techniques

Frontispiece A 2-inch-diameter, single CdZnTe crystal grown by the Multi-Tube Physical Vapor Transport (MTPV) technique. (Photo courtesy of Kromek®.)

To make a radiation detector, we first need to grow a crystal of the particular semiconductor material. To do this we need to understand a little of crystal structure.

2.1 Crystal Lattices

The term "crystal" is derived from the Greek word κρύσταλλος (*krystallos*), meaning "clear ice," and was originally used to refer to materials which looked similar to ice (such as "rock crystal" the colorless form of quartz)—because ice was the most obvious manifestation of crystal structure in ancient times. The structure of a crystal is usually defined in terms of lattice points, which mark the positions of the atoms forming the basic unit cell of the crystal. Cullity [1] defines a lattice point as "*an array of points in space so arranged that each point has (statistically) identical surroundings.*" The word "statistically" is introduced to allow for solid solutions, where fractional atoms would otherwise be required. The defining property of a crystal is thus its inherent symmetry, by which we mean that under certain operations the crystal remains unchanged. For example, rotating the crystal 180 degrees about a certain axis may result in an atomic configuration that looks identical to the original configuration. The crystal is then said to have a twofold rotational symmetry about this axis. In addition to rotational symmetries like this, a crystal may have symmetries in the form of mirror planes and translational symmetries.

2.1.1 The Unit Cell

In chemistry and mineralogy, a crystal is a solid in which the constituent atoms, molecules, or ions are packed in a regularly ordered, repeating pattern of unit cells extending in all three spatial dimensions. The unit cell is the smallest building block of a crystal, consisting of atoms, ions, or molecules whose geometric arrangement defines a crystal's characteristic symmetry and whose repetition in space produces a crystal lattice. The unit cell is defined by its lattice parameters, which are the lengths of the cell edges and the angles between them. The positions of the atoms inside the unit cell are described by a set of atomic positions (x_i,y_i,z_i) measured from a lattice point. The extent or limit of building the unit cells is the crystal face, and thus, the shape of a crystal face is, in part, related to the shape of the unit cell. Although there are an infinite number of ways to specify a unit cell, for each crystal structure there is a conventional unit cell, which is the smallest which can be chosen to display the full symmetry of the crystal.

2.1.2 Bravais Lattice

Crystal systems are groupings of crystal structures according to an axial system used to describe their lattice. Each system consists of a set of three axes in a particular geometrical arrangement. There are seven unique crystal systems. The simplest and most symmetric is the cubic (or isometric) system, having the symmetry of a cube, It is defined as having four threefold rotational axes oriented at 109.5 degrees (the tetrahedral angle) with respect to each other, forming the diagonals of the cube. The other six systems, in order of decreasing symmetry, are hexagonal, tetragonal, rhombohedral (also known as trigonal), orthorhombic, monoclinic, and triclinic.

Combining the central crystal systems with the various possible lattice centerings leads to the Bravais lattices (sometimes known as space lattices). Bravais [2] showed that crystals could be divided into 14 unit cells for which: (a) the unit cell is the simplest repeating unit in the crystal; (b) opposite faces of a unit cell are parallel; and (c) the edge of the unit cell connects equivalent points. There are 14 unique Bravais lattices, based on the seven basic crystal systems, which are distinct from one another in the translational symmetry they contain. All crystalline materials fit into one of these arrangements. The fourteen three-dimensional lattices, classified by crystal system, are shown in Figure 2.1.

2.1.3 The Pearson Notation

While the Bravias lattice designations identify crystal types, they cannot uniquely identify particular crystals. There are several systems for classifying crystal structure, most of which are based on assigning a specific letter to each of the Bravais lattices. However, with the exception of the Pearson classification, few are self-defining. The Pearson notation [4]

Bravias Lattice	Primitive (P)	Body Centered (I)	Side Face Centered (C)	All Face Centered (F)	Axes / Interaxial Angles	Examples
Cubic (Isomeric)					$a_1 = a_2 = a_3$ $\alpha_{12} = \alpha_{23} = \alpha_{31} = 90°$	Si, Ge, Au, Zinc Blende structures (e.g., GaAs, InAs InP, AlSb, CdTe, ZnTe, CZT)
Tetragonal	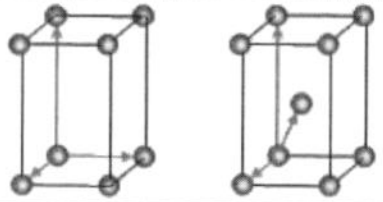				$a_1 = a_2 \neq a_3$ $\alpha_{12} = \alpha_{23} = \alpha_{31} = 90°$	β-Sn, In, HgI_2, $TiO2_2$, PbO Chalcopyrite ($CuFeS_2$)
Othorhombic	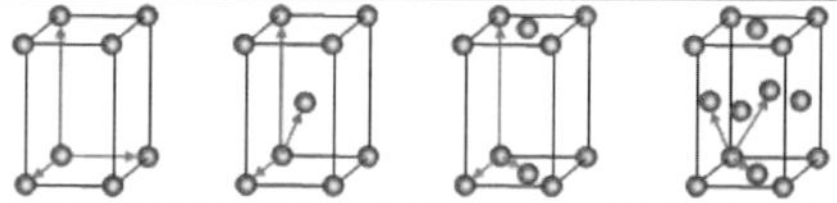				$a_1 \neq a_2 \neq a_3$ $\alpha_{12} = \alpha_{23} = \alpha_{31} = 90°$	Ga, Cl, I, perovskite ($CaTiO_3$), Bi_2S_3, SnS, $TlPbI_3$
Hexagonal	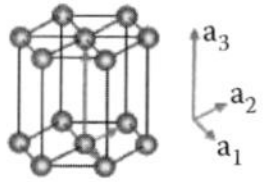 a_3 a_2 a_1				$a_1 = a_2 \neq a_3$ $\alpha_{12} = 120°$ $\alpha_{23} = \alpha_{31} = 90°$	Be, Mg, β-Quartz, BN, SiC Wurtzite sturctures (e.g., AlN, a-GaN, InN, ZnO)
Trigonal (R)	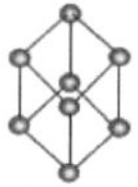				$a_1 = a_2 = a_3$ $\alpha_{12} = \alpha_{23} = \alpha_{31} < 120°$	Bi, As, Te, Al_2O_3, Bi_2Te_3 α-Quartz, cinnabar (HgS)
Monoclinic	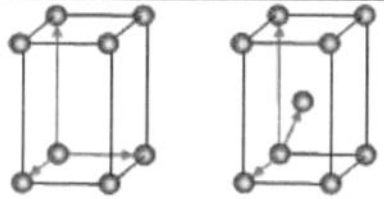				$a_1 \neq a_2 \neq a_3$ $\alpha_{23} = \alpha_{31} = 90°$ $\alpha_{31} \neq 90°$	Gypsum, As_4S_4, KNO_2
Triclinic	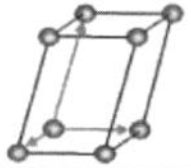				$a_1 \neq a_2 \neq a_3$ $\alpha_{12} \neq \alpha_{23} \neq \alpha_{31}$	B, Cd, Zn, B_4C, BiI_3, PbI_2

Figure 2.1 Schematic of the 7 basic crystal systems and 14 conventional Bravais lattices (for a review, see [3]). The lattice centering are: P = primitive centering, I = body centered, F = face centered, C = base centered, and R = rhombohedral (hexagonal class only).

is a simple and convenient scheme and is based on the so-called Pearson symbols, of which there are three. The first symbol is a lowercase letter designating the crystal type (i.e., a = triclinic, m = monoclinic, o = orthorhombic, t = tetragonal, h = hexagonal and rhombohedral, c = cubic). The second symbol is a capital letter which designates the lattice centering (i.e., P = primitive, C = side face centered, F = all face centered, I = body centered, R = rhombohedral). Thus, the 14 unique Bravais lattices can be characterized by a two-letter mnemonic as summarized in Table 2.1. The third Pearson symbol is a number which designates the number of atoms in the conventional unit cell. Therefore, a diamond structure which is cubic, face centered and has 8 atoms in its unit cell is represented by *cF8*. To use the Pearson system effectively, however, we need to know a structure, or prototype, which is

TABLE 2.1
The 14 Space (Bravais) Lattices and Their Pearson Symbols

Crystal System	Bravias Lattice	Pearson Symbol
Cubic	Primitive	cP
	Face centered	cF
	Body centered	cI
Tetragonal	Primitive	tP
	Body centered	tI
Orthorhombic	Primitive	oP
	Base centered	oS
	Face centered	oF
	Body centered	oI
Monoclinic	Primitive	mP
	Base centered	mS
Triclinic (anorthic)	Primitive	aP
Rhombohedral	Primitive	hR
Hexagonal	Primitive	hP

the classic example of that particular structure. For example, while both GaAs and MgSe have a Pearson designation cF8, GaAs has a classic zinc blende i.e., "ZnS" structure, while MgSe has a "NaCl" type structure. While both structures are formed by two interpenetrating face centered cubic lattices, they differ in how the two lattices are positioned relative to one another.

2.1.4 Space Groups

While Pearson symbols categorize crystal structure into particular patterns and are easy to use and conceptually simple, not every crystal structure is uniquely defined. The space group designation, also known as the International or Hermann–Mauguin [5] system, is a mathematical description of the symmetry inherent in a crystal's structure and is also represented by a set of numbers and symbols. The space groups in three dimensions are made from combinations of the 32 crystallographic point groups with the 14 Bravais lattices which belong to one of the 7 basic crystal systems. This results in a space group being some combination of the translational symmetry of a unit cell including lattice centering and the point group symmetry operations of reflection and rotation improper rotation. The combination of all these symmetry operations results in a total of 230 unique space groups describing all possible crystal symmetries. The International Union of Crystallography publishes comprehensive tables [6] of all space groups and assigns each a unique number. For example, rock salt is given the number "225" (in Hermann–Mauguin notation [5] it is designated "Fm$\bar{3}$m").The relationships between the basic crystal systems, the Bravais lattices and the point and space groups are shown in Figure 2.2.

2.1.5 Miller Indices

The classification of crystallographic directions and planes is important, because the physical and electrical properties of semiconducting materials can vary significantly, depending on the plane. This can have important consequences, particularly when processing semiconductor materials. Because

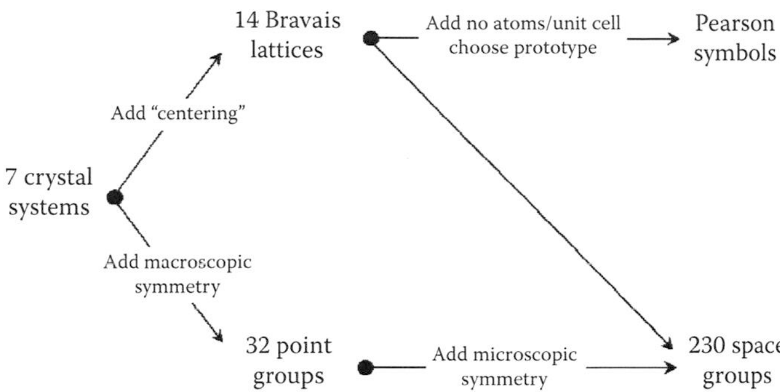

Figure 2.2 The interrelationship between the basic crystal systems, the Bravais lattices, point groups, and the Pearson symbols and space groups. The macroscopic symmetry elements are those operations (e.g., reflection and translation) which take place over unit cell dimensions, whereas the microscopic symmetry elements add small translations (less than a unit cell vector) to the macroscopic symmetry operations. A point group is a representation of the ways that the macroscopic symmetry elements (operations) can be self-consistently arranged around a single geometric point. There are 32 unique ways in which this can be achieved.

crystalline structures are repetitive, there exist families of equivalent directions and lattice planes centered on the unit cell. The orientation of a crystal plane may be defined by considering how the plane intersects the main crystallographic axes of the solid. The Miller Indices* [7] are a set of numbers which quantify the reciprocals of the intercepts and thus may be used to uniquely identify the plane. A direction is expressed in terms of its ratio of unit vectors in the form $[u,v,w]$, where u, v, and w are integers. A family of crystallographically equivalent directions is expressed as $<uvw>$. A plane is expressed as (lmn), where l, m, and n are integers, and a family of crystallographically equivalent planes is expressed as $\{lmn\}$. As an example, Figure 2.3 illustrates how Miller indices are used to define different planes in a cubic crystal. The (100), (010), and (001) planes correspond to the faces of the cube. The (111) plane is tilted with respect to the cube faces, intercepting the x, y, and z axes at 1, 1, and 1, respectively. In the case of a negative axis intercept, the corresponding Miller index is given as

* Named after the British mineralogist W.H. Miller (b. 1801; d. 1880).

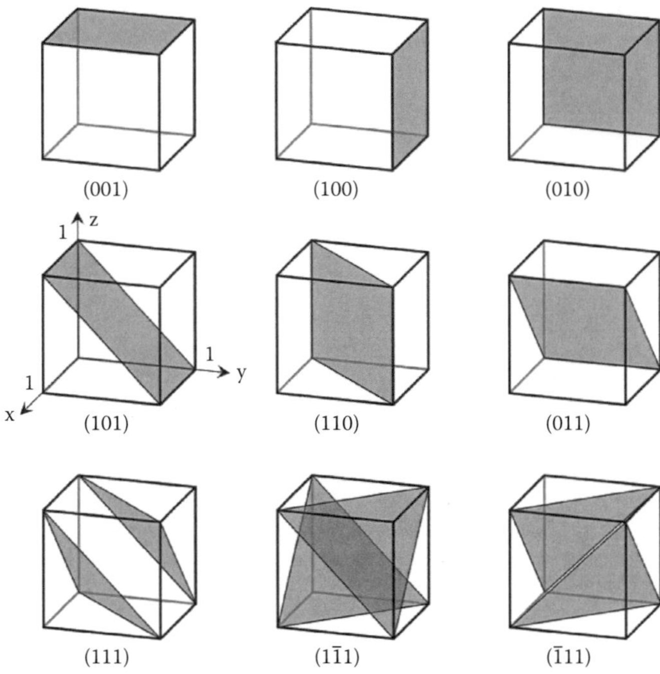

Figure 2.3 Examples of different crystallographic planes within a cubic lattice. The Miller indices for each plane are indicated in parenthesis (from [8]).

an integer with a bar over it, thus $(\bar{1}00)$, which is similar to the (100) plane but intersecting the *x*-axis at –1.

2.2 Underlying Crystal Structure of Compound Semiconductors

Compound semiconductors are derived from elements in groups I to VII of the periodic table. They are so useful because of the sheer range of compounds available, compared to the elemental semiconductors, Sn-α, C, Si, and Ge. Most elements in these groups are soluble within each other, forming homogeneous solid solutions. By solid solutions we mean a mixture of two crystalline solids that coexist as a new crystalline solid that remains in a single homogeneous phase. The mixing is usually accomplished by melting the two solids together at high temperatures and cooling the combination to form the new solid. Solutions occur when atoms of a particular element are able to be incorporated into the lattice of another element without altering its crystal structure. The solute may incorporate into the solvent crystal lattice *substitutionally*, by replacing a solvent particle in the lattice, or *interstitially*, by fitting into the spaces between solvent particles as illustrated in Figure 2.4. In general, compound semiconductors fall into the *substitutional* category, while ceramics and clays fall into the interstitial category. In both substitutional and interstitial solid solutions, the overall atomic structure is preserved, although there will be a slight distortion of the lattice. This distortion can disrupt the physical and electrical homogeneity of the solvent material.

In order that atoms can form solid solutions over large ranges of miscibility, they should satisfy the Hume-Rothery* rules [9], namely that,

- The two species have similar valences.
- They have comparable atomic radii,† allowing substitution without large mechanical distortion.
- Their electronegativities are similar to avoid the creation of intermetallic compounds.
- Individually, their crystal structures are the same.

The rationale behind the Hume-Rothery rules is as follows. The atomic radii must be similar, because if the substituted atom is too large, considerable strain will develop in the crystal lattice, and the components must have similar crystal structure if solubility is to occur over all proportions. However, this is less important if small proportions of solute are being added—such as in the doping of semiconductors. Similar valences and electronegativities indicate that components have similar bonding properties. In fact, components should have the same valence and an electronegativity difference close to zero to achieve maximum solubility. The more electropositive one element and electronegative the other, the greater

* After the British metallurgist William Hume-Rothery (b. 1899; d. 1968).
† Experimentally it is found they should differ by no more than 15%.

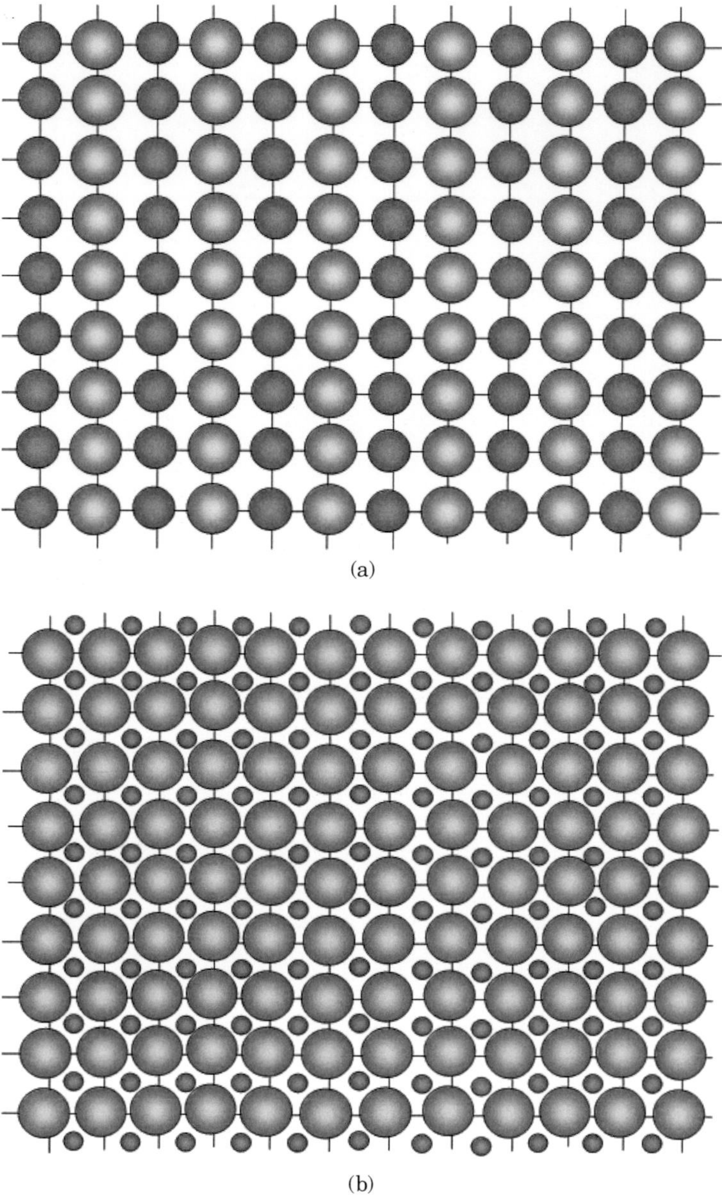

Figure 2.4 **(See color insert.)** (a) Schematic of an ordered substitutional cubic lattice in which an atom of one element replaces an atom of the host element in an alternating sequence. The ability to form a stable lattice depends on whether the two species can satisfy the Hume-Rothery rules. (b) Example of an interstitial lattice in which the atoms of one element fit interstitially into the spaces in the lattice of the host element.

the likelihood of forming an intermetallic compound instead of a substitutional solid solution. It should be appreciated that the Hume-Rothery rules are a semiempirical guide and by themselves not sufficient to ensure extensive solid solutions for all qualifying systems. For example, binary alloys of silver or copper satisfy all four rules but do not form a continuous series of solid solutions; instead they form a eutectic system with limited primary solubility.

Mooser and Pearson [10,11] expanded on the work of Hume-Rothery [9] and elaborated a set of simple rules to test whether an element or compound will show semiconducting properties. For materials which contain at least one atom per molecule from groups IV to VII of the periodic table, these are:

1. The bonds are predominantly covalent.
2. The bonds are formed by a process of electron sharing which leads to completely filled *s* and *p* subshells on all atoms in elemental semiconductors. For compound semiconductors, the condition has to apply for each constituent element.
3. The presence of vacant "metallic" orbitals on some atoms in a compound will not prevent semiconductivity *provided* that these atoms are not bonded together.
4. The bonds form a continuous array in one, two, or three dimensions throughout the crystal.

Elements from group IV to group VII will satisfy conditions 1 and 2.* Compounds which follow the second condition can be represented by the following equation [11].

$$n_e/n_a + b = 8 \tag{2.1}$$

Here, n_e is the number of valence electrons *per molecule*, n_a is the number of group IV to VII atoms *per molecule*, and b is the average number of covalent bonds formed by one of these atoms with other atoms of groups IV to VII. Thus for

* Completed octets, that is closed s and p subshells, can only occur in atoms from groups IV to VII of the periodic table (excluding the transition metals).

GaAs, $n_e = 8$, $n_a = 1$, and $b = 0$—which satisfies the above condition. Note, for the elemental semiconductors, Equation (2.1) is equivalent to the 8–*N* rule, which states that each atom in a covalent crystal should have 8–*N* nearest neighbors—*N* being the ordinal number of the atomic group to which the atom belongs.

In addition to binary materials (such as GaAs or InP), most binary systems are also soluble within each other, making it possible to synthesize *ternary* (e.g., AlGaAs, HgCdTe) *quaternary* (e.g., InGaAsP, InGaAlP) and higher-order solutions, simply by alloying binary compounds together. Thus semiconductor compounds can be built up hierarchically as illustrated in Figure 2.5.

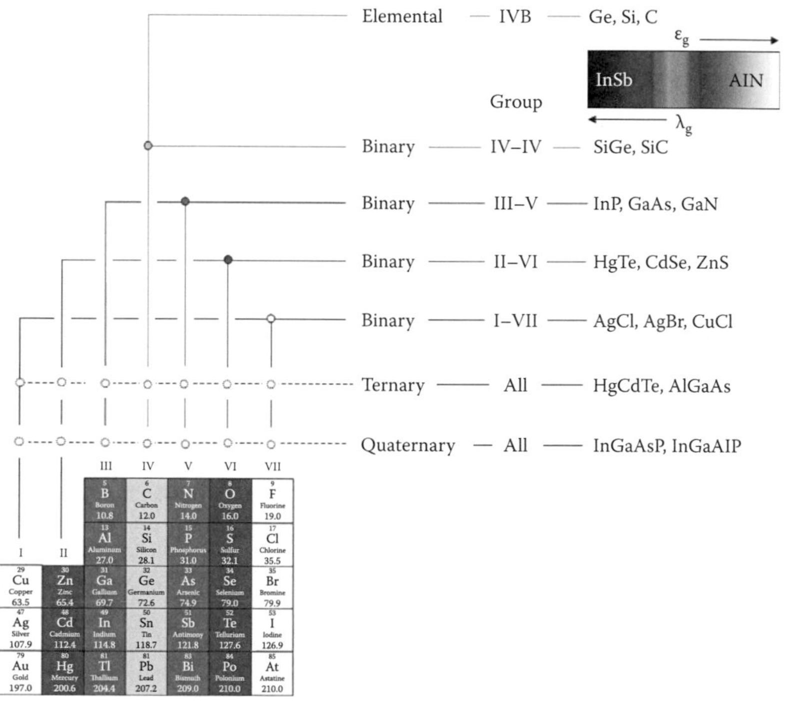

Figure 2.5 **(See color insert.)** Diagram illustrating the relationship of the elemental and compound semiconductors. Examples of the compound type are given and are listed by increasing bandgap energy or alternatively, decreasing wavelength, from the infrared to the ultraviolet. InSb and AlN delineate the extremes of the range in which compound semiconductors lie (0.17 eV–6.2 eV).

2.2.1 Lattice Constant and Bandgap Energy of Alloy Semiconductors

Consider two semiconductors A and B and the semiconductor alloy A_xB_{1-x}, where x is the alloy composition or alloy mole fraction. If we assume that A has a lattice constant a^A, and B has a lattice constant a^B, then the lattice constant of the alloy, $a^{AB}(x)$, is given by Vegard's law [12], which at constant temperature, empirically relates the crystal lattice constant of an alloy to the concentrations of its constituent elements, thus,

$$a^{AB}(x) = xa^A + (1-x)a^B. \tag{2.2}$$

Similarly, the effective bandgap energy $E_g(x)$ is given by

$$E_g(x) = xE_g^A + (1-x)E_g^B - x(1-x)E_b \tag{2.3}$$

where $E_g{}^A$ and $E_g{}^B$ are the corresponding bandgap energies of elements A and B, and E_b is a constant known as the bowing parameter, which is a consequence of the microscopic bonding and structural properties of the alloys [13]. Tabulated values of E_b for an extensive range of binary, ternary, and quaternary alloys may be found in [14,15]. In effect, Equations (2.2) and (2.3) mean that most material properties (such as lattice constant, bandgap, refractive index, thermal properties and mechanical constants) can be adjusted smoothly by the composition of these alloys. This dramatically increases the freedom of choosing the best compound for a particular application. However, it should be noted that for higher-order alloys (quaternaries and higher), Equation (2.3) does not hold, due to the presence of multiple minima in the conduction band which, in theory, should be taken into account. In fact, for the higher-order alloys, the bandgap can change type (direct to indirect and vice versa), depending on composition and whether the semiconductor is strained or not. Generally, however, a so-called "one valley" bandgap model can give reasonable agreement over a limited range of compositions.

2.2.2 Bonding

Structurally, compound semiconductors are crystals formed by combinations of elements in which each element tries to attain a closed valence shell of eight electrons. It achieves this by sharing (donating or accepting) electrons with its nearest neighbors. The simplest semiconducting materials are composed of a single atomic element, with the basic atom having four electrons in its valence band, supplemented by covalent bonds to four neighboring atoms to complete the valence shell (see Figure 2.6a). These elemental semiconductors use atoms from group IV of the periodic table. Binary semiconductors are composed of two

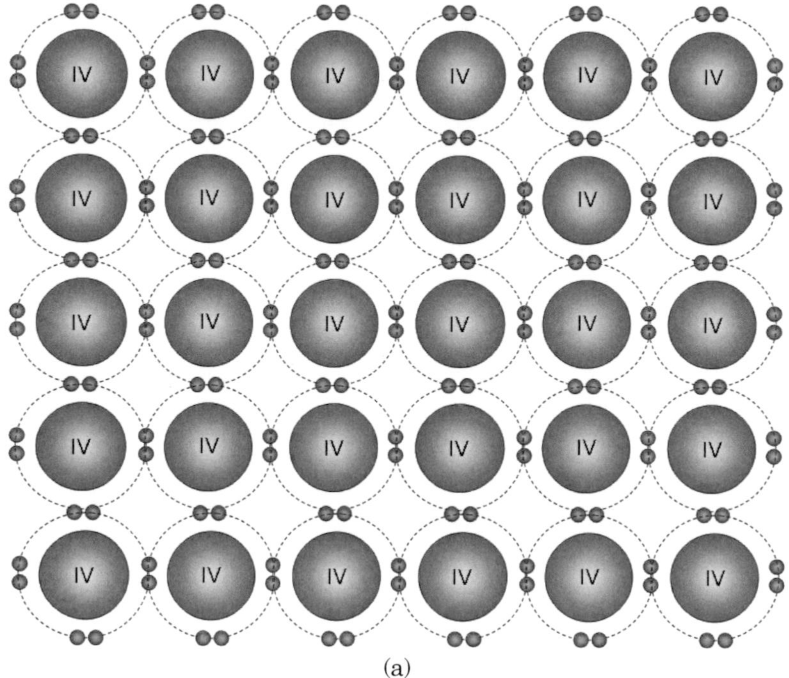

(a)

Figure 2.6 Illustration showing how semiconductors bond to form closed valence shells. Examples are given for the four most common semiconductor groups; (a) group IV elemental semiconductors, such as Si and Ge, (b) group III–V semiconductors, such as GaAs and InP, (c) group II–VI compounds, such as ZnS and CdTe, and (d) group I–VII compounds, such AgCl and AgBr.

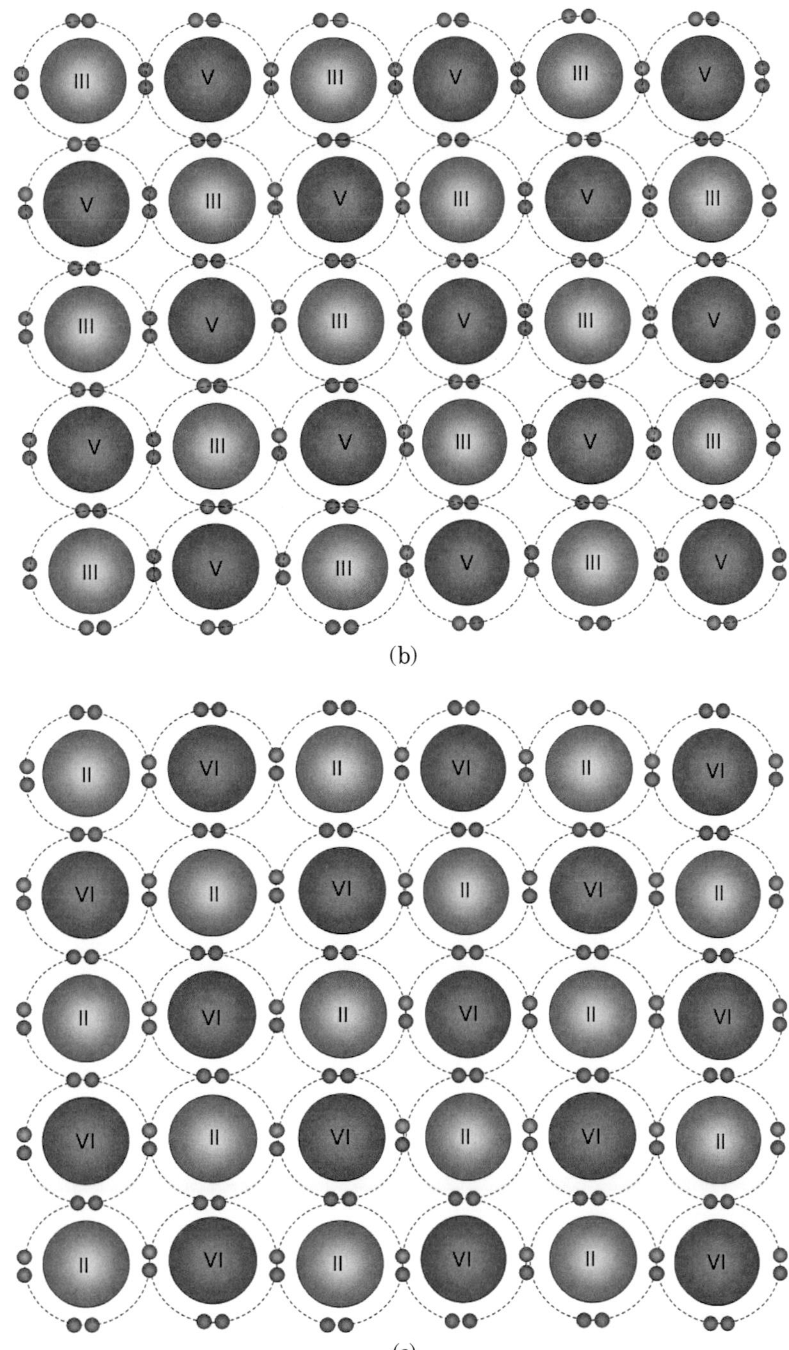

Figure 2.6 (Continued)

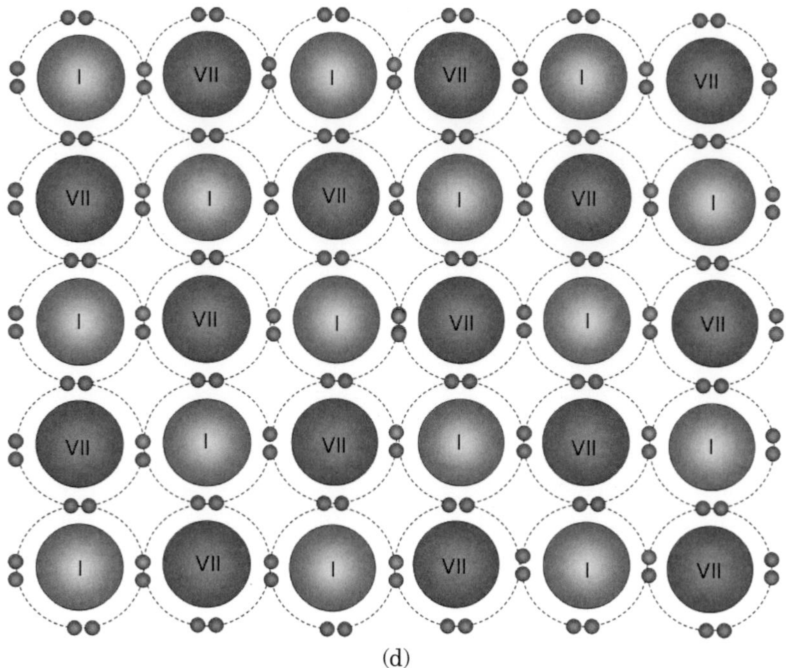

Figure 2.6 (Continued)

atoms—one from group N (where $N < 4$) and the other from group M (where $M > 4$) such that $N + M = 8$, thus filling the valence band. Figure 2.5 shows the relevant section of the periodic table from which the bulk of the semiconductor groups are derived.

2.2.3 Common Semiconductor Structures

Most semiconductors used in radiation detection assume a cubic lattice structure and fall into basic structures which are described below. Note that some compounds can also crystallize into both cubic and hexagonal forms. For historical reasons, actual crystal structures are generally named after the minerals that most commonly assume that shape.

Diamond. The "diamond structure" is the arrangement of the Group IV atoms in diamond, silicon, and germanium.

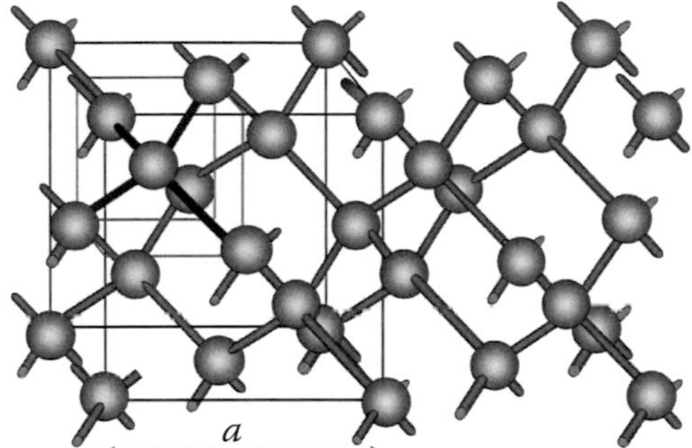

Figure 2.7 The diamond lattice structure. Each atom has four equidistant, tetrahedrally coordinated, nearest neighbors. The unit cell is outlined by the cube of dimension (lattice parameter), *a*. The primitive cell containing one lattice point is shown by the black bonds.

The structure is illustrated in Figure 2.7. The configuration belongs to cubic crystal family and can be envisioned as two interpenetrating face-centered cubic sublattices with one sublattice staggered from the other by one quarter of the distance along a diagonal of the cube. Each atom has four equidistant, tetrahedrally coordinated, nearest neighbors. The atom–atom bond direction is <111>, and the interbond angle is 109 degrees 28 min.

Zinc blende. Named after the mineral zinc blende or "sphalerite," which is the principal ore of zinc. In fact it consists largely of crystalline ZnS. The zinc blende structure is the stable form of many III–V and II–VI compounds and is structurally identical to the diamond described previously, except that alternate atoms are respectively from Group III and Group V for III–V materials and groups II and VI for II–VI materials (see Figure 2.8). Among the materials that classify into this structure are virtually all of the group III–V compounds (e.g., GaAs, InAs, InSb, and InP).

Wurtzite is the less frequently encountered mineral form (a polymorph) of zinc sulfide and is named after French chemist Charles-Adolphe Wurtz.* In some semiconductors, such as

* b. 1817; d. 1884.

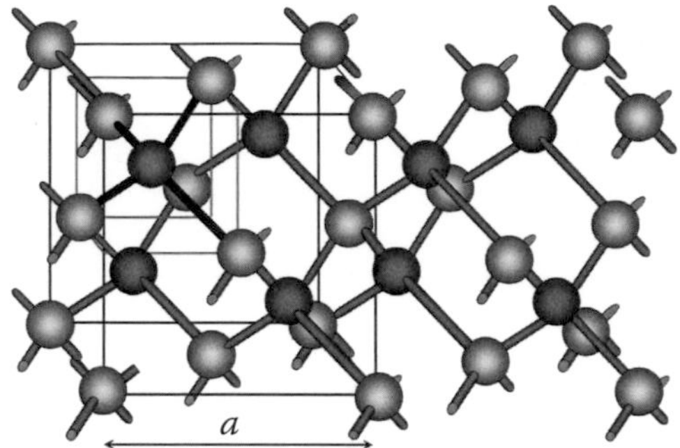

Figure 2.8 The zinc blende lattice structure which is most common structure for binary compound semiconductors. Here, the light and dark gray spheres denote the atoms of the binary elements. The unit cell is outlined by the cube of dimension (lattice parameter) *a*. The primitive cell is shown by the black bonds.

GaN, although a cubic structure is metastable, the stable form is an atomic arrangement (the "wurzite" structure) in which the atoms are slightly displaced from their cubic-structure positions and form a crystal of hexagonal symmetry. Thus, this structure is a member of the hexagonal crystal system. Specifically in the case of ZnS, the structure consists of tetrahedrally coordinated zinc and sulfur atoms that are stacked in an alternating pattern (see Figure 2.9). The principal

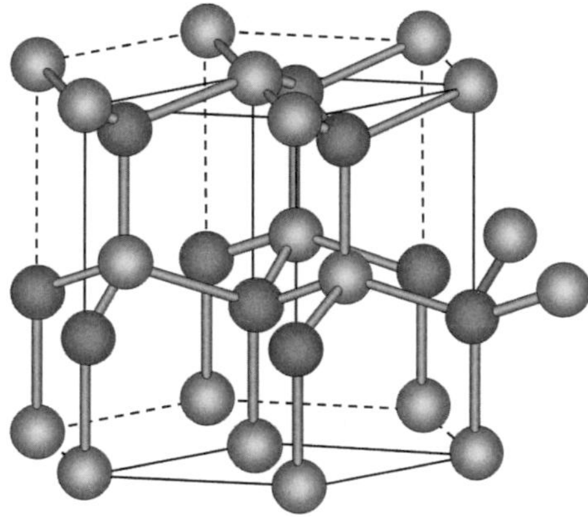

Figure 2.9 The wurtzite lattice structure which is the second most common structure for binary compound semiconductors. Here, the light and dark gray spheres denote the atoms of individual elements.

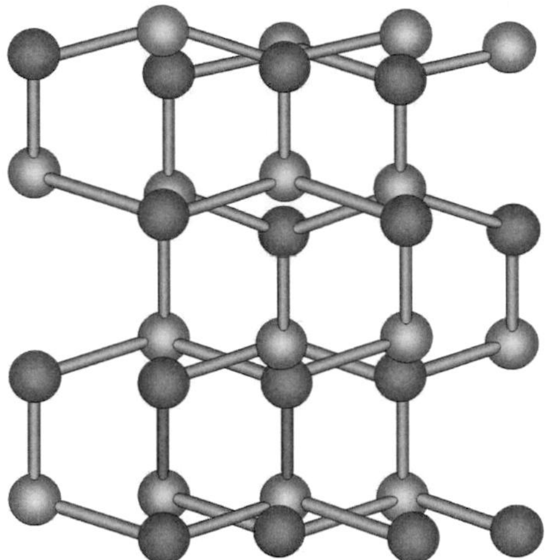

Figure 2.10 A hexagonal lattice structure. The light and dark gray spheres denote the atoms of individual elements.

difference with the zinc blende lattice structure is that wurtzite contains four atoms per primitive unit cell instead of two. Among the compounds that can assume a wurtzite configuration are AgI, ZnO, CdS, CdSe, α-SiC, GaN, and AlN. In most of these compounds, wurtzite is not the favored form of the bulk crystal, but the structure can be observed in some nanocrystal forms.

Hexagonal. The hexagonal structure is depicted in Figure 2.10. It has the same symmetry as a right prism with a hexagonal base. The hexagonal system is uniaxial, meaning it is based on one major axis, in this case a sixfold rotational axis that is perpendicular to the other axes. Examples of materials that crystallize into a hexagonal structure include silicon carbide and boron nitride and of course all wurtzite forms.

2.2.4 Polycrystalline and Amorphous Structures

The majority of semiconductors solidify into regular periodic crystallographic patterns (Figure 2.11a). However, they can also form amorphous solid solutions in which the

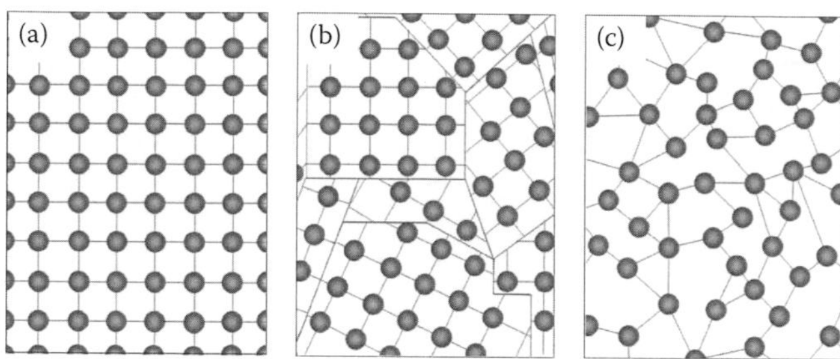

Figure 2.11 Illustration of macroscopic crystal structures in a semiconductor. While the majority of semiconductors solidify into regular periodic patterns shown in (a), they can also form polycrystals shown in (b); that is, a collection of individual grains of crystalline material separated by grain boundaries or (c) amorphous solid solutions in which there is little long-range order.

arrangement of the atoms exhibits no periodicity or long-range order at all (Figure 2.11c). It is the presence or absence of this long-range order that has a profound influence on the electronic and photonic properties of a material, and in fact, carrier mobilities and diffusion lengths degrade as the order is reduced. Even among crystalline materials, we can distinguish between single-crystal and polycrystalline materials. Polycrystalline materials are solids that are comprised of a collection of crystallites (grains), connected to each other with random orientations and separated from one another by areas of relative disorder known as grain boundaries (see Figure 2.11b). In this case, order can only be maintained within the grains.

2.3 Crystal Formation

Crystal growth can occur naturally or artificially. The crystallization process consists of two major events, nucleation and crystal growth. Crystal growth occurs because it is energetically favorable, in that the potential energy of the atoms or molecules is lowered when they form bonds to each other.

The process starts with the nucleation stage. Several atoms or molecules in a supersaturated vapor or liquid start forming clusters; the bulk free energy of the cluster is less than that of the vapor or liquid. Nucleation and crystal growth are dependent upon temperature, pressure, and the composition of the surrounding fluid/vapor (including ionic concentrations necessary for crystal growth and the effects of other ions that may act as poisons to inhibit crystal growth). They are also dependent on the availability of surface area (in the case of a crystal that has already nucleated). To effect growth, the ions must be transported to the surface of the crystal and react with the surface. The reaction products, in turn, must be removed to allow the nucleation process to continue. Crystal morphology is also dependent upon the manner (that is, rate and direction) in which the crystal grows. Many compounds have the ability to crystallize with different crystal structures, a phenomenon called polymorphism. Each polymorph is a different thermodynamic solid state, and crystal polymorphs of the same compound exhibit different physical properties, such as dissolution rate, shape (angles between facets and facet growth rates), and melting point.

2.4 Crystal Defects

During the growth process a number of defects are invariably introduced into the crystal which can adversely affect semiconducting properties. These can be classified into three broad groups: point defects, plane defects, and bulk defects which are described below and summarized in Table 2.2. It is important to note that, unlike the elemental semiconductors, any deviation from perfect stoichiometry in a compound semiconductor must necessarily result in defects—a surplus of atoms of whatever kind will be incorporated into the lattice as point defects, agglomerates, or precipitates. The large thermal gradients encountered during the growth process

TABLE 2.2
List of Crystal Imperfections

Imperfection		Description
Class	**Subclass**	
Point defects		
	Interstitial	Extra atom in an interstitial site
	Schottky	Atom missing from correct site
	Frenkel	Atom displaced to an interstitial site, creating nearby vacancy
	Antisite	One component of a binary component occupies a lattice site of the other component
Line defects		
	Edge dislocation	Row of atoms marking the edge of a crystallographic plane extending only part way into the crystal
	Screw dislocation	Row of atoms, about which a normal crystallographic plane appears to spiral
Plane defects		
	Lineage boundary	Boundary between two adjacent perfect regions in the same crystal that are slightly tilted with respect to each other
	Grain boundary	Boundary between two crystals in a polycrystalline solid
	Stacking fault	Boundary between two parts of a closed packing region having alternate stacking sequences
Bulk defects		
	Precipitates	Conglomeration of impurity atoms
	Voids	Macroscopic holes in the lattice

Source: Adapted from [16].

invariably lead to mechanical stress, which gives rise to dislocations. In addition, most compounds are mechanically soft, which means that relatively small stresses are sufficient to cause plastic deformation, even at low temperatures—again giving rise to dislocations.

2.4.1 Point Defects

Point defects are localized defects of atomic dimensions and are the smallest structural elements, or imperfections to cause a departure from a perfect lattice structure. They include self-interstitial atoms, interstitial impurity atoms, substitutional atoms, and vacancies, as illustrated in Figure 2.12.

A self-interstitial atom is an extra atom that has crowded its way into an interstitial void in the crystal structure. A substitutional impurity atom is an atom of a different type from the bulk atoms that has replaced a host atom in the lattice. Substitutional impurity atoms are usually close in size (within approximately 15%) to the bulk atom; otherwise they will mechanically disrupt the lattice. Interstitial impurity atoms, on the other hand, are much smaller than the bulk matrix atoms and fit into the open spaces between the host atoms of the lattice structure. In general, substitutional impurities are electronically active and will alter the electronic properties of the material, whereas interstitial contaminants are generally electronically inactive. Vacancies are empty spaces where an atom should be, but is missing. They are particularly common at elevated temperatures when atoms frequently and randomly change their positions, leaving behind empty lattice sites. If the missing atom is no longer in the vicinity of the hole or has migrated to the surface, it is known as a Schottky defect.

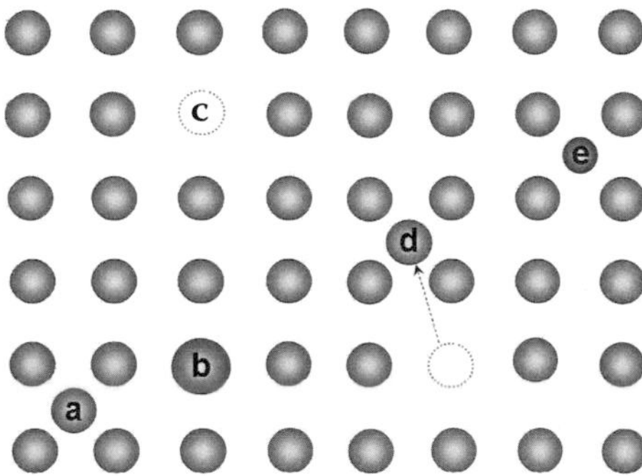

Figure 2.12 Schematic illustrating the different types of point defects in a crystalline material. These are: (a) self-interstitial atom, (b) substitutional impurity atom, (c) Schottky defect, (d) Frenkel defect, (e) interstitial impurity atom.

Those that migrate to the surface generally become incorporated into the lattice at the surface. If the atom has moved into an adjacent interstitial site, it is known as a Frenkel defect. Thus, a Frenkel defect is in reality a pair of defects—an empty lattice site and an extra interstitially positioned atom. Schottky and Frenkel defects tend to form during the growth process and are frozen into the lattice as the crystal crystallizes. Although in both cases, the crystal remains neutral (since the total number of positive and negative ions is the same) these defects affect semiconductor properties because they allow ionic conduction. However, because covalent bonding is much stronger than ionic bonding in tetrahedrally coordinated materials, activation energies are much higher in covalently bonded materials, and as such, their effects tend to be less important in groups IV, or III–V semiconductors.

Another form of point defect is the antisite defect, which is neither a vacancy nor an interstitial, nor an impurity. It occurs when an anion* (cation†) replaces a cation (anion) on a regular cation (anion) lattice site. A defect is known as "native" if it is a vacancy, a self-interstitial, an antisite, or a complex formed of these.

2.4.2 Line Defects (Dislocations)

A line defect is formed in a crystal when an atomic plane is misaligned and terminates within the crystal instead of passing all the way through it. The resulting irregularity in spacing is most severe along a perpendicular line called the line of dislocation. The concept of the dislocation was first proposed by Orowan [17], Polanyi [18], and Taylor [19] in 1934 to explain two key observations about the plastic deformation of crystalline materials. The first was that the stress required to plastically deform a crystal is much less (by a factor of 10^2 to 10^3) than the stress calculated for a defect-free crystal structure. The second observation was that a material *work hardens* when it has been plastically deformed and subsequently requires a greater stress to deform it further. However, it was not until 1947 that the existence of dislocations was experimentally verified [20]. Later advances in electron microscopy techniques showed that

* The electronegative component in a compound (e.g., As in GaAs, Te in CdTe).
† The electropositive component in a compound (e.g., Ga in GaAs, Cd in CdTe).

dislocations were mobile in a material and caused its plastic (ductile versus brittle) behavior. Only in the last few years have electron microscopy techniques advanced sufficiently to allow the atomic structure around a dislocation to be resolved.

A line defect can be thought of as a one-dimensional array of point defects which are out of position in the crystal structure and occur when the crystal is subjected to stresses beyond the elastic limit of the material. These stresses generally arise from thermal and mechanical processing during the growth and detector fabrication process. The energy to create a dislocation is of the order of ~100 eV per mm of dislocation line, whereas it takes only a few eV to form a point defect which is a few nanometers in length. Thus, forming a number of point defects is energetically more favorable than forming a dislocation. Dislocations interact with chemical and other point defects. The presence of a dislocation is usually associated with an enhanced rate of impurity diffusion leading to the formation of diffusion pipes. This effect gives rise to the introduction of trapping states in the bandgap, altering the electrical properties of the devices. In addition, dislocations move when a stress is applied, leading to slip and plastic deformation of the lattice. Note that a line of dislocation cannot end within a crystal unless it forms a complete loop. Generally, it ends at the sample surface, in which case it can be visualized by etching (acids preferentially etch the intersections of dislocations with the crystal surface), infrared, X-ray, or electron transmission techniques (see Figure 2.13). In fact,

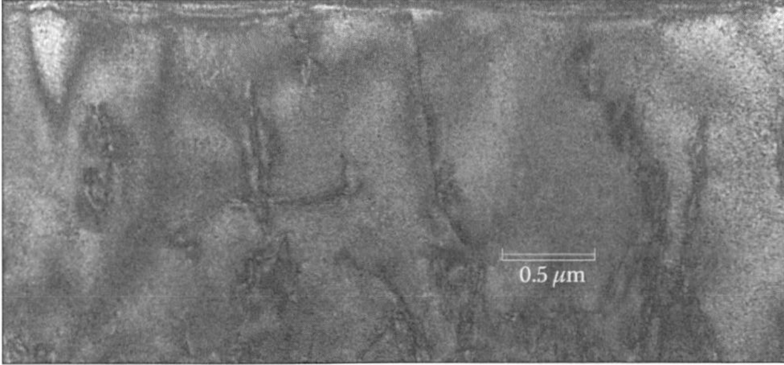

Figure 2.13 Transmission electron microscopy (TEM) image of a large number of dislocations in a GaN crystal grown by metal organic chemical vapor deposition (MOCVD).

etching is routinely used to quickly assess material quality in terms of number of dislocations at a surface—the result is usually expressed in terms of the etch pit density (EPD). Dislocations can also be visualized by decoration (trace elements will precipitate on dislocations, usually known as decorated boundaries).

2.4.2.1 Edge Dislocations

An edge dislocation can easily be visualized as an extra half-plane of atoms inserted into the lattice orthogonal to the growth direction as illustrated in Figure 2.14 (a). The dislocation is called a line defect because the locus of defective points produced in the lattice by the dislocation lie along a line. This line runs along the top of the extra half-plane. The interatomic bonds are significantly distorted only in the immediate vicinity of the dislocation line (denoted by the symbol, ⊥, which runs into the paper) as can be seen in Figure 2.14 (b), forming a strain field on either side of the line with associated impacts on the mechanical strength of the crystal. In addition, the disruption of the lattice in this region leaves behind a string of dangling bonds* that act as trapping centers with subsequent impacts on electronic performance.

If the dislocation is such that a step or ramp is formed by the displacement of atoms in a plane in the crystal, then it is referred to as a screw dislocation and is essentially a shearing of one portion of the crystal with respect to another by one atomic distance. The displacement occurs on either side of a "screw dislocation line" (see Figure 2.14b) which forms the boundary between the slipped and unslipped atoms in the crystal and forms a spiral ramp winding around the line of the dislocation. In some materials, such as SiC, growth can commonly proceed via spiral growth. It was suggested by Frank [23] that a a screw dislocation in a crystal provides a step onto which atoms can be adsorbed under vapor growth. The whole spiral can then rotate around its emergence point with uniform angular velocity and stationary

* Unsatisfied valences on an immobilized atom.

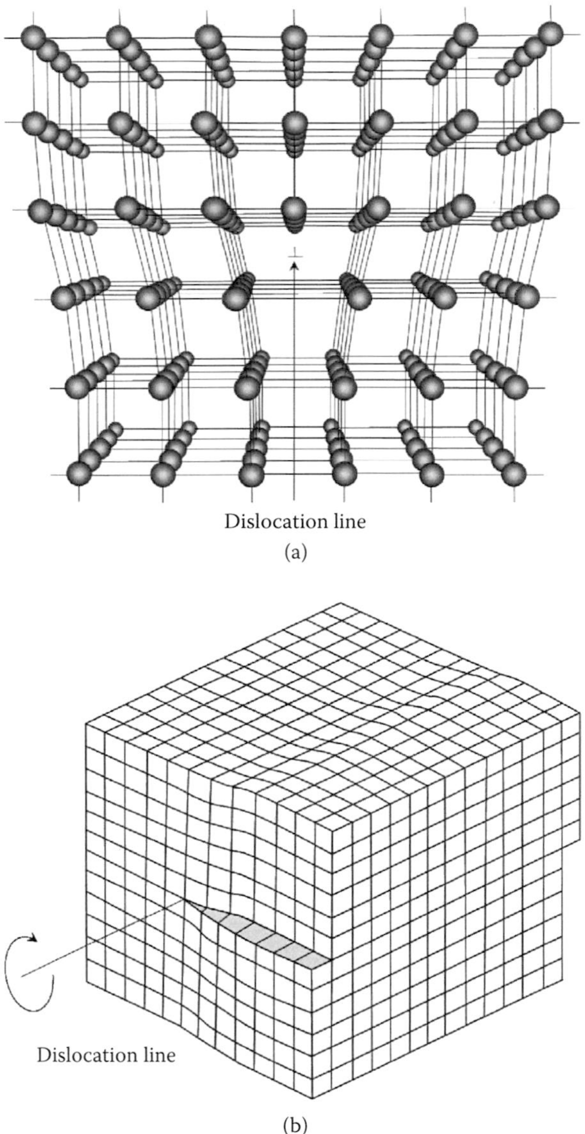

Figure 2.14 Illustrations of line defects. (a) Edge dislocation in a cubic primitive lattice (modified from [21]). (b) Screw dislocation (modified from [22]).

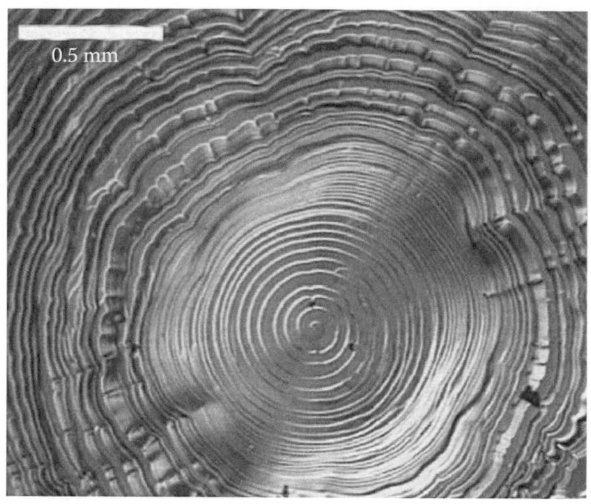

Figure 2.15 A screw dislocation growth spiral (from [24]).

shape, giving rise to the growth morphology illustrated in Figure 2.15.

In general, dislocations cannot end within a crystal. Instead, they form a closed loop within the crystal or a half loop extending in from a face and then back out of the same face.

2.4.3 Plane Defects

A plane defect is a disruption of the long-range stacking sequences of atomic planes. Plane defects are generally manifested as a stacking faults or twin regions. A stacking fault is a change in the stacking sequence over a few atomic spacings. If a stacking fault continues over a large number of atomic spacings, it will produce a second stacking fault that is the twin of the first one—thus forming a twin region. Another type of plane defect is a grain boundary, which is the boundary between crystallites or grains in polycrystalline material. These are structural defects that decrease the electrical and thermal conductivity of the material. The high interfacial energy and relatively weak boundaries often make them preferred sites for the onset of corrosion or the precipitation of new phases from the solid during growth. Grains can range in size from nanometers to millimeters, and their orientations are usually rotated with respect to neighboring grains. The interface between grains is typically two to five atomic diameters wide and is known as a grain boundary. The manipulation of grain sizes can be used to limit the lengths and motions of dislocations, because they essentially act as obstacles to dislocation motion. This is a well-known strengthening technique in metallurgy, known as "grain-boundary" or "Hall–Petch" strengthening (named after E.O. Hall [25] and N.S. Petch [26], who showed that the strength of mild steel varies reciprocally as its grain size).

Twins and grain boundaries can be considered as arrays of linear dislocations that can affect charge transport in several ways. For example, they can form potential barriers that guide the carriers, they can accumulate secondary phases that block carrier transport, they can attract impurities that increase charge trapping, and they can form conduction bands that make the boundaries highly conductive. This is illustrated in Figure 2.16.

2.4.4 Bulk Defects

Bulk defects occur on a macroscopic scale compared to point defects. Perhaps, the most common form of bulk defect is the precipitate, which is a cluster of impurity atoms that form small regions of a different phase—meaning a region of space occupied by a physically homogeneous material. Precipitates tend to be more common in bulk growth materials, such as CdTe, and have sizes ranging up to tens of microns. Examples of precipitates in CdZnTe are shown in Figure 2.17. A gross type of bulk defect is a void, which is a region where there are a large number of missing atoms (that is holes) in the lattice.

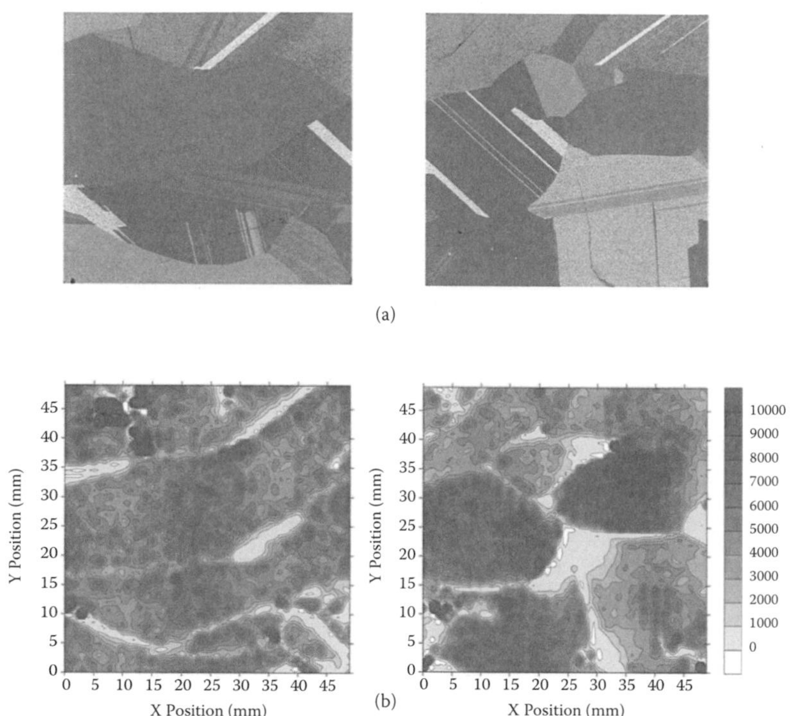

Figure 2.16 **(See color insert.)** (a) Optical images of two 50 × 50 mm^2, 3-mm-thick slices of a CdZnTe crystal grown by the high pressure Bridgman method (from [27]). Numerous grain boundaries and twins are apparent in the image. (b) The crystals count rate response, measured with a ^{57}Co radioactive source is shown in the lower images, illustrating poor charge collection at the grain boundaries. Interestingly, no correlation was found with the numerous twin boundaries observed inside the grains, indicating that twins have a negligible effect on the electric field and charge collection of semi-insulating CdZnTe devices.

2.5 Crystal Growth

To produce radiation detectors with good charge collection properties, high-quality single-crystal material must be grown which is both crystallographically and stoichiometrically as perfect as possible. The overall process is illustrated in Figure 2.18 and begins with the production of the starting materials or precursors. Typically, these are synthesized from pure elements or purchased from commercial vendors with

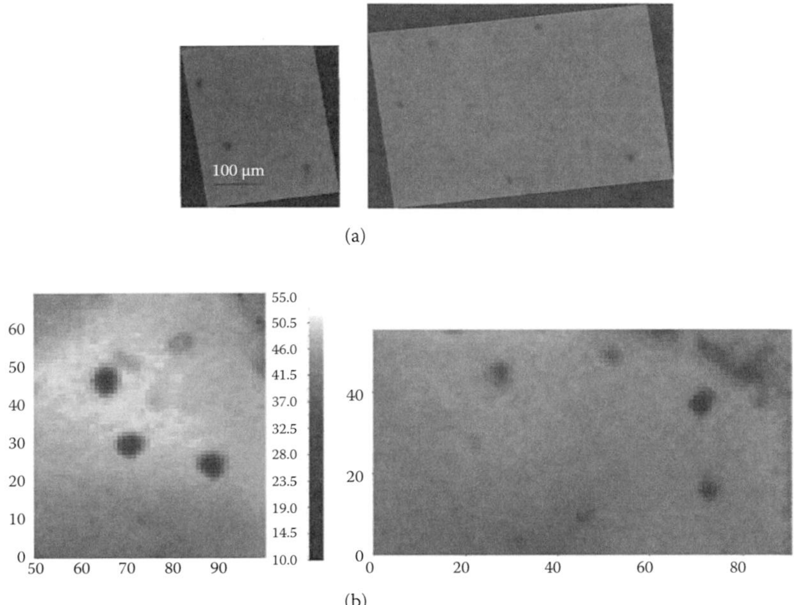

Figure 2.17 (a) IR micrograph images of Te precipitates measured in a 1-mm-thick CZT crystal (from [28]). (b) The lower images for the corresponding X-ray response of the crystal when operated as a simple planar detector. The dark spots in this case correspond to a drop in the detector response, demonstrating the link between precipitates and poor device performance. The scans were performed by using a 10×10 μm², 85 keV X-ray beam. In some cases, the typical triangular shapes of precipitates are recognizable in the X-ray maps.

nominal purity of 99.9999%.* The precursors are then grown into the desired semiconductor, usually in a polycrystalline form, using a melt technique. This material is then purified using zone refining or distillation techniques before being used to produce thin seeds or substrates with as near perfect single crystal properties as possible. These will be used to grow ingots or wafers of similar or higher quality crystalline material. For some materials, these ingots can be grown directly during the last zone refining stage. Wafers will subsequently be divided into detector crystals and ingots, sliced into wafers, and then diced into detectors. Substrates are usually grown by bulk growth techniques because of cost consid-

* Commonly known as 6 N, where "6" refers to the number of "9s."

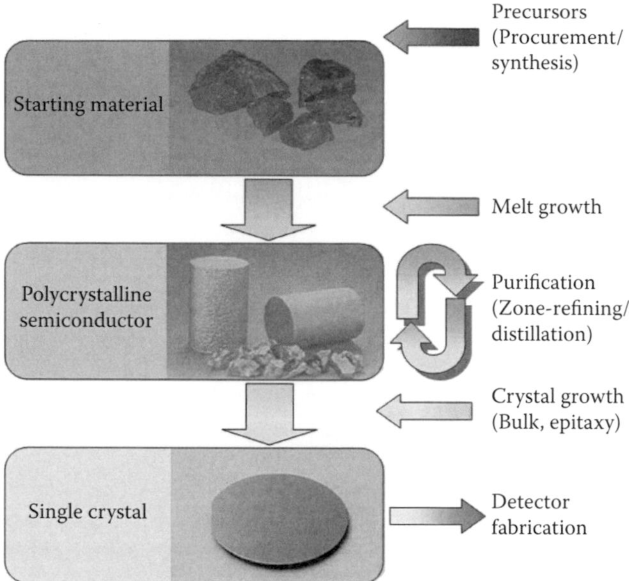

Figure 2.18 **(See color insert.)** Steps required to produce single-crystal material for detector production.

erations, starting with the purified forms of the elements that make up the compound.

2.5.1 Material Purification

The preparation of semiconductor material usually begins with the purification of the base material down to impurity levels below 1 ppm (parts per million) and in some cases ppt (parts per trillion). Zone refining and vacuum sublimation are two standard processes to achieve this goal. The zone-refining technique was first developed in 1952 by Pfann [29] to purify germanium and has since been applied to many elements and compounds—in fact to any solute–solvent system having an appreciable concentration difference between the solid and liquid phases at equilibrium. The technique works as follows (see Figure 2.19). A moving heater traverses a long ingot of raw material and melts a small zone as it moves. The impurities will congregate in this zone if there is a difference in the solubility of the impurities between the liquid and solid

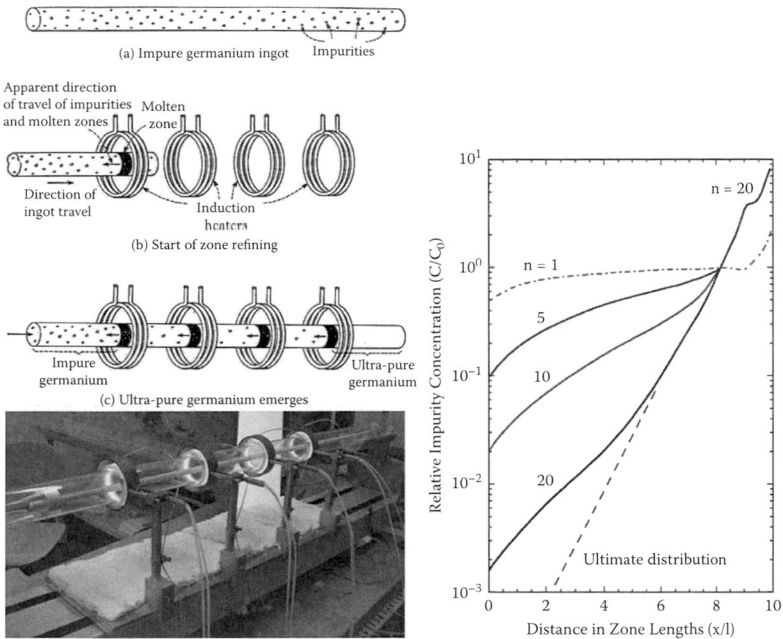

Figure 2.19 Upper left: illustration of the principle of the zone refining process (from [30]). Lower left: the practical implementation. Right: the relative impurity distribution along a 10-zone-lengths-long ingot for various numbers (n) of zone passes for a distribution coefficient of 0.5 (adapted from [30]). Here l is the length of the zone, and x is the length of the ingot being refined. The distribution coefficient is the ratio of the solute concentration in the solid to that in the liquid in equilibrium and should be less than unity for purification to occur.

phases of the raw material. The net effect is that the impurities accumulate at one end of an ingot (usually in the last-to-freeze region) leaving pure material in the central section. The effectiveness of the process can be quantified by the distribution coefficient, k, defined to be the ratio of the solute concentration in the solid to that in the liquid in equilibrium, and should be less than unity. The process is usually repeated a number times.

Vacuum sublimation is another method routinely applied to the purification of materials in which the sample is placed under reduced pressure to permit sublimation at lower temperatures. Sublimation is defined as a direct change of state from the solid to the gas phase without going through the liquid state. Since elements and compounds condense at unique temperatures, the use of a controlled cold surface will recover the required solid, leaving the nonvolatile residue impurities behind. There are numerous variations of this method using either static or dynamic vacuum. Dynamic vacuum works better for less volatile samples.

2.6 Bulk Growth Techniques

Bulk single crystals are usually grown from a liquid phase. The liquid may have approximately the same composition as the solid. It may be a solution consisting primarily of one component of the crystal, or it may be a solution whose solvent constitutes a minor fraction of the crystal's composition. Generally melt growth techniques are used for the elemental semiconductors and group II–VI materials. They are usually not used for III–V materials (e.g., GaAs, GaP, GaN), because the large difference in vapor pressures of the constituent components requires a high-pressure environment to keep the group V element in the system.

2.6.1 Czochralski (CZ)

Perhaps the most widely used bulk growth technique is the crystal-pulling or Czochralski (CZ) method* [31], in which the melt of the charge, usually high-quality polycrystalline material, is held in a vertical crucible, as illustrated in Figure 2.20a. The top surface of the melt is held just above the melting temperature. A seed crystal is then lowered into the melt and slowly withdrawn (see Figure 2.21a). Both the seed crystal and the crucible are rotated during the pulling process, in opposite directions. As the heat from the melt flows up the seed, the melt surface cools and the crystal begins to grow. The surface tension of the melt at the interface allows molten material to be pulled out of the crucible, where it cools

* Named after the Polish chemist Jan Czochralski (b.1885; d.1953).

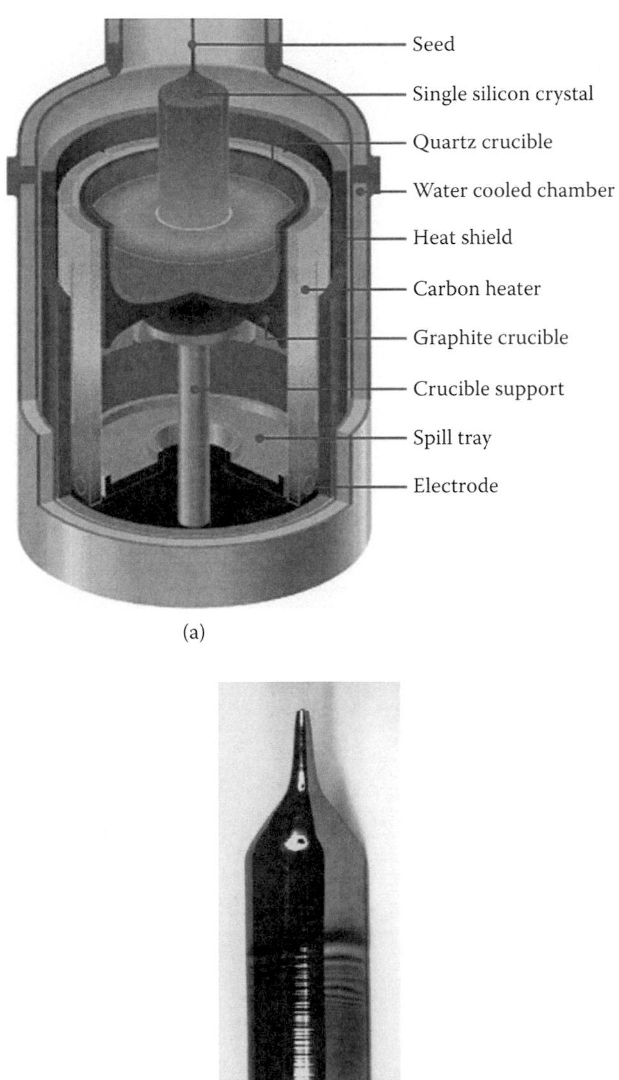

Figure 2.20 (a) Schematic of a Czochralski crystal growth system used to produce Si, GaAs, and InP substrate ingots (from [32]), image courtesy of the Center of Advanced European Studies and Research). (b) A typical boule (ingot). The top of the crystal is called the seed end, or alternatively, the first-to-solidify end. The bottom is known as the "tail" or "tang" end.

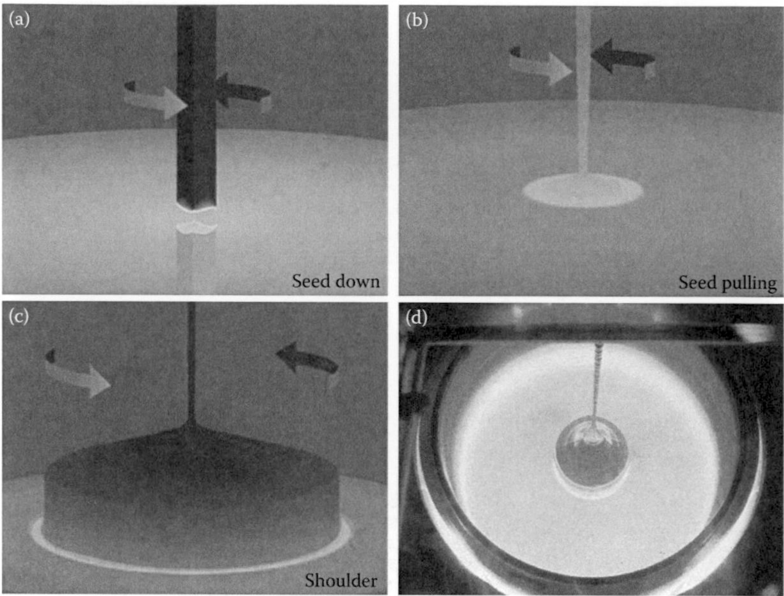

Figure 2.21 Growth sequence in a Czochralski furnace. (Images courtesy of Kinetics Systems, Inc.). (a) Seed crystal is lowered into the melt. (b) As crystallization begins the rod holding the seed crystal is slowly withdrawn. (c) By varying the pull rate the diameter of crystal can be controlled, forming the basis of the ingot. (d) A view into an actual crucible during the drawing of an ingot.

and solidifies on the seed, thus forming a single crystal. The rotation of the seed about its axis ensures a roughly circular cross section crystal and reduces radial temperature gradients. The rotation also has the added advantage that it inhibits the natural tendency of the crystal to nucleate along natural orientations to produce a faceted crystal.

The initial pull rate is relatively rapid so that a thin neck is produced (Figure 2.21b), because if the cross section of the neck is less than that of the seed, thermal stress is reduced, which helps prevent the generation of dislocations in the pulled crystal. The melt temperature is now reduced and stabilized so that the desired ingot diameter can be formed (Figure 2.21c). This diameter is generally maintained by controlling the pull rate. The pulling continues until the melt is nearly exhausted, at which time a tail is formed. The crystal diameter and length depend upon the temperature, pulling rate, and the dimensions of the melt container, and the crystal quality depends critically upon minimization of temperature gradients, which enhance the formation of dislocations.

Pellets of dopant material can be added to the melt if extrinsic semiconductor material is required. The impurities are usually added and dissolved in the melt using solid impurities and are primarily used to increase the concentration of mobile carriers and subsequently the conductivity of the device.

The CZ technique is commonly used in Si, GaAs, and InP production to produce substrate material and is not suited to detector-grade material because too many impurities, originating mostly from the crucible, are left in the crystal. This reduces minority carrier lifetimes to unacceptably low levels, making the production of high-resistivity material impossible. The technique yields long ingots (boules) with good circular cross sections (see Figure 2.20b) [33]. The largest silicon ingots produced today are around 400 mm in diameter, 1 to 2 m in length, and weigh up to 100 kg.

2.6.2 Liquid Encapsulated Czochralski (LEC)

Synthesizing compound semiconductor crystals is much more difficult than elemental semiconductors, because the vapor pressures of the constituent materials are generally quite different. At the temperature required to melt the higher-temperature material, the lower-melting-point material has evaporated. For example, in the case of the compounds GaAs and InP, As and P have the lower melting points and tend leave the melt and condense on the sidewalls. Metz et al. [34] circumvented this problem by using a liquid lid or encapsulate to seal the melt and thus prevent evaporation. The technique is commonly referred to as liquid encapsulated Czochralski, or LEC. The encapsulate is made of an inert material that is (a) less dense than the melt—so it floats, forming the "lid" and (b) does not interact with or contaminate the melt (see Figure 2.22). For most compounds, a molten layer of boron oxide (B_2O_3) is commonly used to prevent the evaporation of the volatile component as well as block oxygen and carbon contamination of the melt. The growth system is contained in a

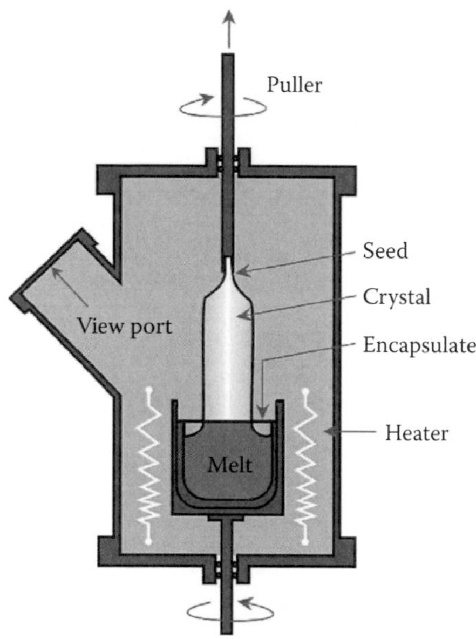

Figure 2.22 Schematic of liquid encapsulated Czochralski furnace showing the position of the encapsulate (from [35]).

high-pressure atmosphere (up to 100 atm) of argon or nitrogen to prevent the volatile constituent from bubbling through the B_2O_3 and escaping. Metz et al. [34] originally developed the method to grow IV–VI Pb compounds. It was further developed by Mullin et al. [36] for III–V compounds and is now the industry standard for the production of most semi-insulating III–V compound materials, such as GaAs, GaP, InP, and InAs.

2.6.2.1 Limitations of the Czochralski Method

There are a number of limitations of the CZ and LEC techniques, which mainly relate to contamination issues and dislocation defect densities (usually expressed in terms of the etch pit density, EPD*). Contamination usually arises from the encapsulate (usually boron when using B_2O_3) and/or the crucible

* The EPD is usually determined by etching the crystal surface with a molten alkali, such as KOH, and measuring the number of "pits" in the surface using an optical microscope. Each pit represents a single dislocation ending at the surface.

(which can be fabricated from any of a number of materials, typically quartz, graphite, glassy carbon, BN, AlN). Crystal defects arise largely from temperature gradients across the melt. While these can be minimized using multi-zone heaters or careful heat-shield design; a reduction in gradients raises the temperature at the crystal surface, which can diminish the compositional stability from the dissociation of volatile elements such as phosphorus or arsenic. Dissociation can be reduced by the introduction of an ambient of the volatile into the space above the melt. For LEC, growth dissociation can also occur from the crystal surface left exposed above the B_2O_3 encapsulate layer leading to twining production. Crystal twinning emanating from the neck of the pulled crystal (see Figure 2.21c) or the cone of the crucible can also be a problem. Twinning occurs when two separate crystals share some of the same crystal lattice points, resulting in an intergrowth of two different crystals joined by a so-called twin boundary. Effective methods to reduce the formation of twinning are an optimization of the cone angle at the crystal shoulder and the application of a magnetic field to suppress temperature variations in the melt during growth.

2.6.3 Vapor Pressure Controlled Czochralski (VCz)

In LEC, the main drawback of using low temperature gradients (<30 K cm^{-1}) to reduce the defect density is the high surface temperature of the crystal as it emerges from the encapsulate. For GaAs and InP, the resulting dissociation of the recently solidified material due to the high partial pressures of As and P adversely affects the growth kinetics and crystal perfection. The principle of vapor pressure controlled Czochralski (VCz) is to seal the crystal-growing system inside a pressurized container, which establishes an ambient of the most volatile component to suppress the dissociation of the crystals, resulting in mirrored rather than dull, pitted surfaces. The essential components of VCz are shown in Figure 2.23. In a variation on this theme, Rudolph and Kiessling [38] have proposed growing GaAs crystals by the VCz method without a boron oxide encapsulate. Material thus grown shows a markedly reduced boron content.

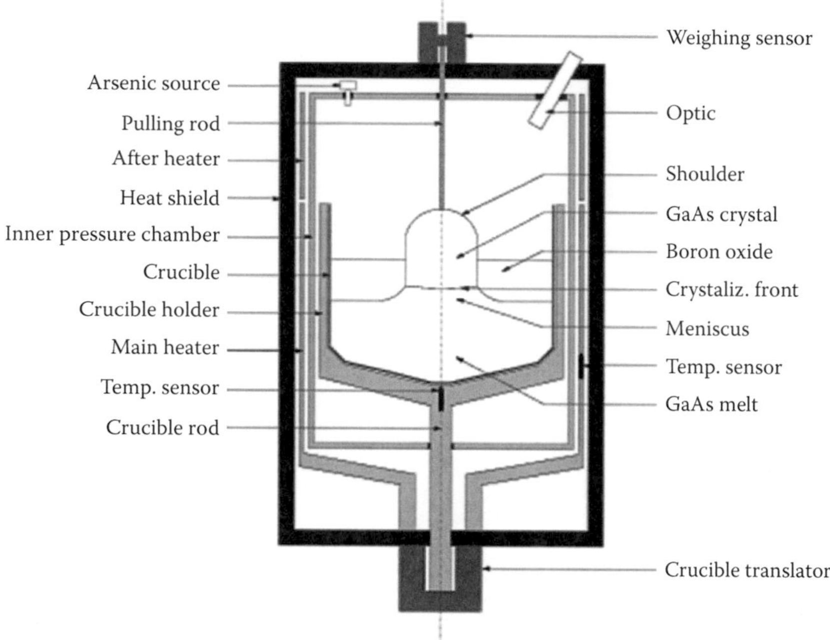

Figure 2.23 An implementation of the vapor controlled Czochralski technique used for the production of GaAs (from [37]). The essential difference over the standard Czochralski method is the inner pressure chamber and an arsenic source to provide an ambient overpressure of the most volatile component.

2.6.4 Float-Zone Growth Technique (FZ)

Because one of the major sources of crystal defects and contamination in conventional growth techniques (e.g., CZ, LEC) is the contact of the melt with the crucible, a major improvement in crystal quality (impurity introduction, formation of dislocations, and residual stress) can be achieved by dispensing with the crucible altogether. The float-zone (FZ) technique is one such method and is widely used for the production of high-purity silicon [39]. The technique is illustrated in Figure 2.24a. A polysilicon rod with a seed crystal at the bottom is held in a vertical position. The rod is then lowered through an electromagnetic coil. An RF (radio frequency) field is used to produce a local melted zone in the polycrystalline rod, which is dragged from the seed end to the holder, solidifying into crystalline material in its wake—as illustrated in Figure 2.24b. This molten zone has

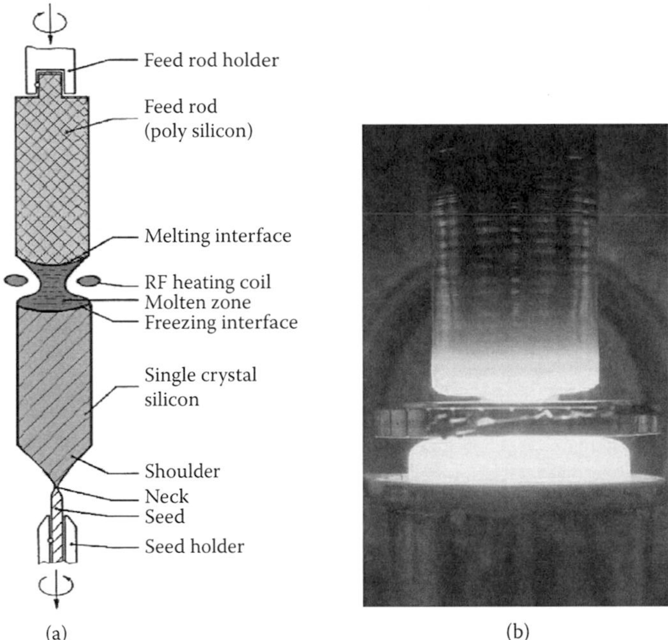

Figure 2.24 (a) Schematic of the float-zone growth technique (from [40]). (b) Photograph of the growth area. (Courtesy of Topsil Semiconductor Materials A/S.)

the added advantage of carrying impurities away with it, taking advantage of the low segregation coefficients of many impurities (impurities are more soluble in the melt than the crystal). The relative advantages and disadvantages of this technique are listed in Table 2.3.

TABLE 2.3
Advantages and Disadvantages of the Float-Zone Crystal Growth Technique

Advantages	Disadvantages
No crucible contamination	Expensive
Shape control	Incongruent melting
In-built impurity control	Sensitivity of control parameters
Vapor pressure control	Relatively poor radial doping profiles
High growth rates (0.5–50 mm/min)	Boule diameter limited to ~15 cm

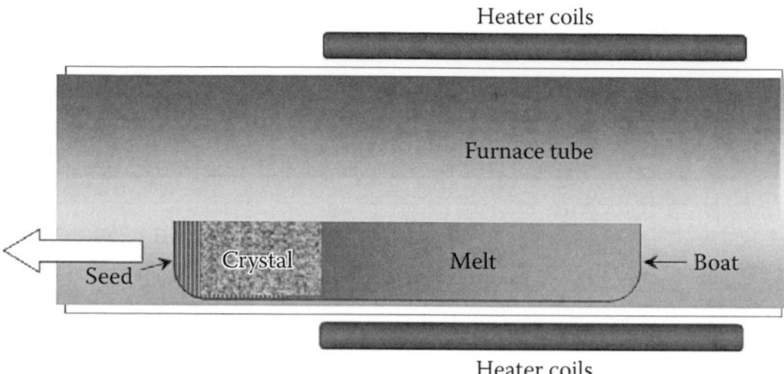

Figure 2.25 Schematic of the horizontal Bridgman method. The crystal is solidified by slowly withdrawing the charge from the heater.

The method was initially used for purification (zone refining), but is now widely used to produce high-quality silicon for power devices and detector applications. The dimensions of ingots produced are generally less than 15 cm due to surface tension limitations during growth.

2.6.5 Bridgman–Stockbarger (B-S)

The second major bulk crystal growth technique is the Bridgman–Stockbarger* (B-S) method, which is essentially a controlled freezing process taking place under liquid–solid equilibrium conditions. The technique is an amalgam of two earlier methods. The Bridgman method [41] is characterized by the translation of a crucible containing a melt along a single axial temperature gradient in a furnace as illustrated in Figure 2.25. The Stockbarger method [42] is a modification of Bridgman and employs a single heat insulation buffer separating a vertical furnace into two zones, a high-temperature zone and an upper low-temperature zone. A B-S system consists of three major components: a growth chamber, temperature controllers, and "boat" translation assembly. The "boat" is a crucible into which the charge is loaded with a seed crystal at one end. The boat is then heated until the charge melts and wets the seed crystal.

* Named after the American physicists P.W. Bridgman (b.1882; d.1961) and D.C. Stockbarger (b.1895; d.1952).

The seed is used to crystallize the melt by slowly lowering the boat temperature starting from the seed end. The charge may be composed of either high-quality polycrystalline material or carefully measured quantities of elements which make up the desired compound. The boat is kept stationary during this process while the longitudinal furnace temperature is varied to form the crystal. Although the process can be carried out in a horizontal or vertical geometry, the easiest approach is to use a horizontal boat. However, in this case, the boule produced has a D-shaped cross section. To produce circular cross sections, vertical configurations have been developed for GaAs and InP.

2.6.6 High Pressure Bridgman (HPB)

High pressure Bridgman (HPB) is commonly used for compounds with high melting points and disparate vapor pressures and thus is ideal for II–VI compounds such as CdTe and CdZnTe. In this implementation, the charge and seed are contained in a container, either a graphite crucible with tight lid or a sealed quartz ampoule. The container is also backfilled (as in the case of CdZnTe) with a high pressure inert gas (~100 atm) such as Ar, to suppress the loss of the most volatile component. Figure 2.26 shows the horizontal and vertical versions of this technique. In both cases the crystal grows from the melt by moving it along a region with a temperature gradient that extends from above to below the melting point. The growth may proceed by mechanically either moving the ampoule or the heating furnace. In recent systems the furnace consists of many heating zones with computer control of the temperature profile. The computer shifts the profile electronically so there are no moving parts within the furnace. However the crucible, which is generally made of carbon-layered silica glass, gives rise to a number of problems, solid–liquid interface curvature, spurious nucleation of grains and twins, and large thermal stresses during cooling of the crystal. Crystals grown by HPB growth are polycrystalline with large grains and twins. Nevertheless, the grains are large enough to obtain single crystals of volume of several cubic meters. In addition, HPB material tends to suffer from macroscopic cracking, the formation of pipe defects

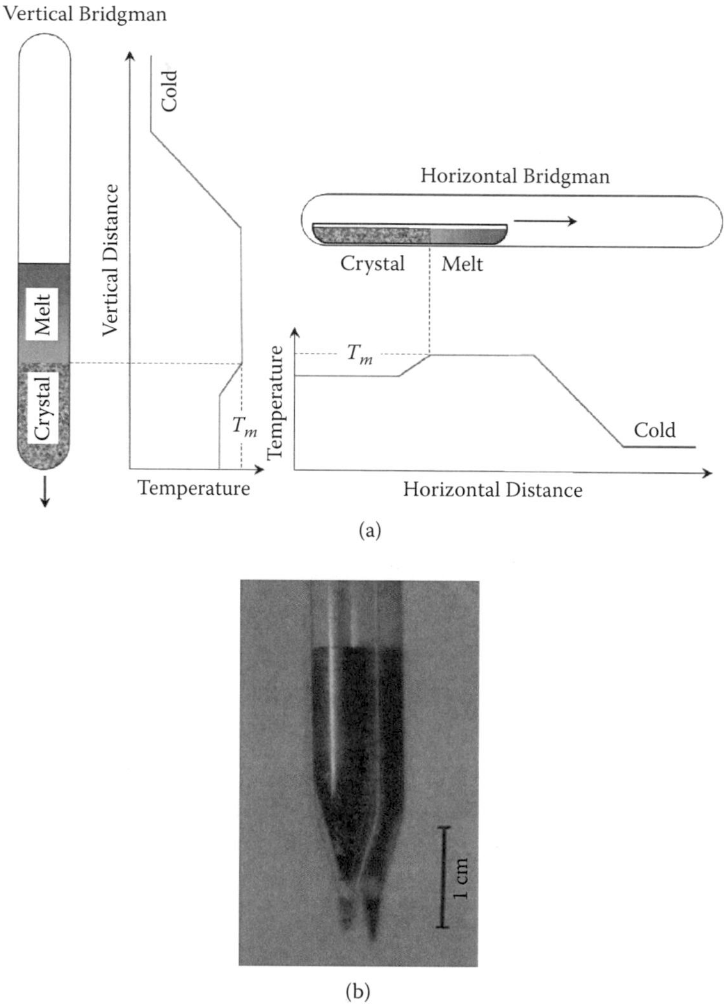

Figure 2.26 (a) Crystal growth by high pressure vertical and horizontal Bridgman showing the temperature profiles across the charge (modified from [43]). T_m is the melting temperature. (b) A sealed vertical Bridgman charge.

and, in the case of CdZnTe, the formation of Te inclusions. For CdZnTe growth, precise control of stoichiometry can be achieved by controlling the partial pressure of its constituent elements. This is usually effected through Cd because it is has the highest vapor pressure and can thus be expected to have the greatest influence on the composition of the melt. Without such control the melt will lose Cd and become Te rich. For commercially produced CdZnTe, a typical growth cycle lasts around 4 weeks, producing ingots up to 10 kg in weight [44].

2.6.7 Travelled Molten Zone (TMZ) or Heater Method (THM)

The conventional technique used for compounding II–VI semiconductors such as CdTe consists of unidirectional freezing of a molten charge within a Bridgman or modified Bridgman reactor. The Bridgman process, being a melt technique, has two inherent drawbacks: (a) the high compounding temperature can lead to material contamination from the crucible walls, and (b) the potential exists for explosions associated with the highly exothermic synthesis reaction and the possibility of trapping uncompounded Cd within a "shell" of compounded material. This limits the size of the charges that can be safely processed in a single batch and thus increases production costs. The travelled heater method (THM) is a low temperature/pressure alternative technique based on solution growth and is the method of choice for the growth of CdTe [45]. In the THM synthesis of cadmium telluride, a saturated solution of cadmium in tellurium is held molten by a narrow heater and swept vertically through the length of the charge consisting of the precursor elements. The technique is illustrated in Figure 2.27. The movement of the molten zone through the charge leads to progressive dissolution of the charge at the top liquid-to-solid interface, while simultaneously leading to synthesis and growth of the compound at the bottom interface. With this process, a constant, controlled, but slow growth is achieved. The quality of grown crystals is very sensitive to the relative movement of the temperature profile that determines the growth rate. THM has a number of advantages over melt techniques, namely easy growth of binary and ternary alloys, fewer thermal stresses, uniform crystal composition [46], and a low temperature and pressure environment compared to HPB. In addition, whereas the CdTe synthesis with conventional Bridgman is typically limited to around 1.5 kg, mainly because of crucible limitations, CdTe ingots produced by THM can easily exceed 3.5 kg. This offers significant cost advantages, as well as benefits in terms of reduced material

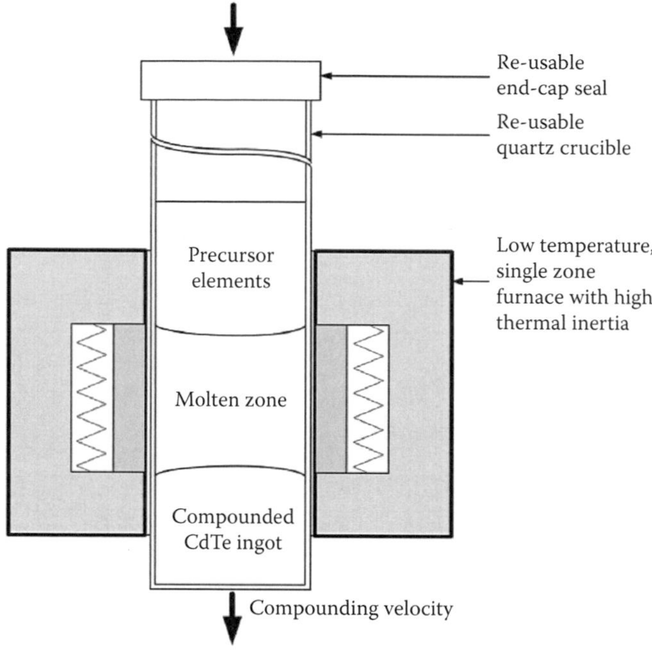

Figure 2.27 Schematic of the travelled heater method (from [46], © 2007 IEEE).

contamination (large ingots have a smaller surface to volume ratio, and so less material is in contact with the crucible).

2.6.8 Vertical Gradient Freeze (VGF)

The vertical gradient freeze (VGF) method of Ramsperger and Melvin [47] is a derivative of vertical Bridgman. Raw polycrystalline material is melted in a preshaped crucible and is directionally solidified from a single-crystal seed at the bottom of the crucible (Figure 2.28). This is achieved by lowering the temperature while maintaining a positive temperature gradient in the melt. Because the crystal growth is usually performed in multi-zone furnaces there are many degrees of freedom, and numerical modeling is required for the optimization of the furnace and the growth process. The main advantages of the VGF process include its scalability, low stress, high mechanical strength, and a defect rate that is orders of magnitude lower than conventional compound semiconductor crystal

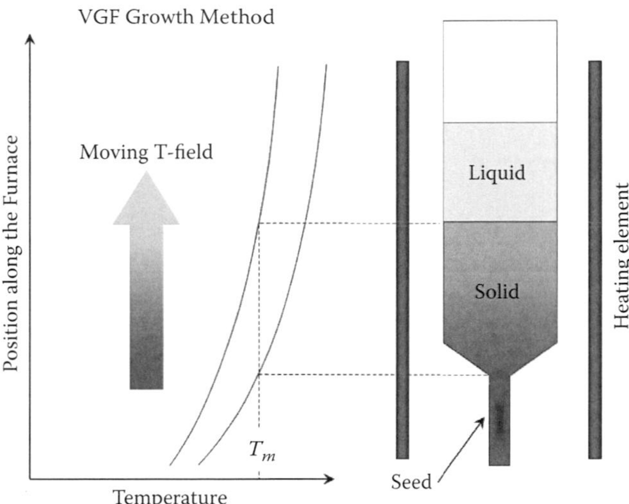

Figure 2.28 Essential elements of the vertical freeze growth (VGF) method. Left: the furnace temperature profile. Right: the material state along the charge. T_m is the melting temperature.

growth techniques. For example, for high-quality GaAs, EPD values of 30 cm^{-2} can be achieved, as opposed to values of over 500 cm^{-2} for LEC growth.

In addition to producing high-purity bulk crystals, the techniques discussed previously are also used to produce crystals with specified electrical properties, such as high-resistivity materials or n- or p-type materials. For example, high-resistivity or semi-insulating (SI) substrates are extremely useful in device isolation and for high-speed devices. However, it is difficult to produce high-resistivity Si substrates by bulk crystal growth, and resistivities are usually <10 Ω cm. Carrier-trapping impurities such as chromium and iron can be added to the melt to produce material with resistivities of ~10^3 Ω cm.

2.7 Discussion

In Table 2.4 we summarize the properties of growth techniques based on the two most common bulk crystal growth methods, Czochralski and Bridgman–Stockbarger. In terms

TABLE 2.4
Summary of the Properties of the Two Most Common Bulk Crystal Growth Techniques

Parameter	Czochralski	Bridgman–Stockbarger
Commonly used for	Si, Ge, III–V semiconductors	III–V and II–VI semiconductors
Precursor	Uses seed crystal	Uses seed crystal
Crystal growth	Vertical	Vertical or horizontal
Crucible	yes	yes
Growth speed	1–2 mm/min	1 mm/hr
Contamination	Oxygen contamination problems	Contamination from crucible
Comments	Axial resistivity is poor	Precise temperature gradient required
	Heat up/cool down times long	Crystal perfection better than seed

of specific growth systems, a number of comparative studies have been carried out with a view to maximizing ingot size while simultaneously minimizing dislocation defect densities. For example, in a comparison of the LEC, VCz, and VGF techniques for growing InP, Kawarabayashi et al., [48] concluded that, overall, the VCz technique was the most promising method for growing large-diameter crystals (>3 inches) with low dislocation densities. In fact, the EPDs were over an order of magnitude lower than those measured with conventional LEC. Their conclusions are summarized in Table 2.5.

TABLE 2.5
Comparison of Properties of Large-Diameter InP Crystal Growth by Conventional LEC, VCz, and the VGF Methods

Parameter	Conventional LEC	VCz	VGF
Size	O	O	Δ
EPD	×	O	O
Technology maturity	O	Δ	×

Key: O ≡ superior, Δ ≡ medium, × ≡ inferior
Source: From [48], © 1994 IEEE.

2.8 Epitaxy

The word epitaxy is derived from the Greek words επι (*epi* meaning "on") and ταξιζ (*taxis* meaning "order"). Thus, epitaxy refers to the ordered growth of one crystal upon another crystal. Epitaxial films may be grown from gaseous or liquid precursors. There are three main modes of epitaxial growth: (a) monolayer, (b) nucleated, and (c) nucleation followed by monolayer. Monolayer growth occurs when the deposited atoms are more strongly bound to the substrate than they are to each other. The atoms aggregate to form monolayer islands of deposit which enlarge, and eventually a complete monolayer coverage has taken place. The process is repeated for subsequent layer growth. In case of nucleated growth, the initial atoms deposited aggregate as small three-dimensional (3D) islands that increase in size as further deposition continues until they touch and intergrow to form a continuous film. This mode is favored where the forces of attraction between the deposited atoms are greater than the forces between them and the substrate. In the final mode, growth starts with the formation of a single or few monolayers on the substrate followed by subsequent nucleation of 3D islands on top of these monolayers.

2.8.1 Substrates

In order to structure the crystal correctly, the films must be grown on a substrate of material which is itself a structurally perfect crystal. Because the substrate acts as a seed crystal, the deposited film will attempt to replicate the lattice structure and orientation of the substrate, provided the shape and size of the lattices are not too dissimilar. Obviously the ideal substrate is a highly crystallographic form of crystal of the material to be grown—the better the quality of the substrate, the better the quality of the resultant epitaxial film.

If a film is deposited on a substrate of the same composition, the process is called homoepitaxy; otherwise it is called heteroepitaxy. For the elemental semiconductors Si and Ge, bulk crystal growth techniques are highly mature and high-quality substrates are readily available. For most compound semiconductors (with the possible exception of InP), they are not, and thus it is necessary to grow layers on substrates of another crystalline material, hence heteroepitaxy. This is not ideal because, even if the substrate has the same crystal structure, the lattice spacing can be quite different, introducing stress into the film. Unless the mismatch between crystals is elastically strained, the net result will be a bowing of the wafer, which can present complications during wafer processing.

In Figure 2.29 we show the effects of strain for two cases: when the lattice parameter for the substrate is less that that of the film, then the film will be in compression and the wafer will bow downward (Figure 2.29a). If the converse is true, then the film is under tension and the film will bow upward (Figure 2.29b). The degree of strain and introduction

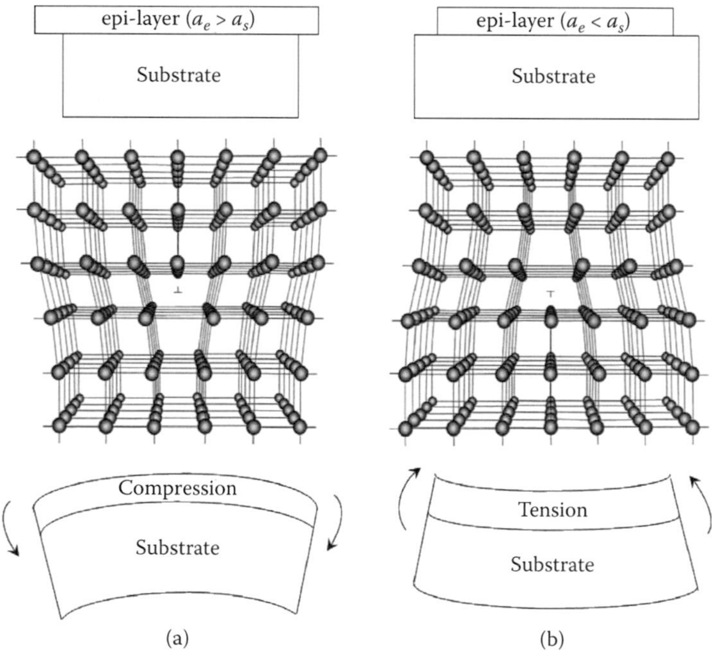

Figure 2.29 Examples of lattice mismatch between the substrate and the grown film showing the effects of bowing, which will occur if one or both crystals are inelastically strained. If the lattice parameter for the substrate a_s is less that that of the epi film case, a_e, (a) then the film will be in compression and the wafer will bow downward. If the converse is true (b), then the film is under tension and the film will bow upward.

of dislocations not only depends upon the lattice mismatch but also on the epi-layer thickness—if the strain energy is less than required to create a dislocation, the layer remains strained; if not, it partially or totally relaxes by the formation of dislocations. In the worst cases, the epi-layer will delaminate from the substrate. In practice, stable alloys can only be formed if the change in lattice constant is kept below 15%. If this is the case, it is possible to grow highly crystalline films. For example the nitrides GaN, AlN, and InN are grown on SiC or sapphire (Al_2O_3) substrates, and AlGaInP is grown on a GaAs substrate. Thus it is clear that semiconductor technology is critically dependent on the availability of high-quality substrates with the largest diameters possible. In summary, the most cost-effective way to produce large-area crystalline substrates is to use bulk crystal growth techniques, because deposition rates are much higher than for epitaxial techniques, and many substrate wafers can be cut from a single melt-grown ingot.

2.8.2 Strain and Electronic Properties

Strain in a crystalline semiconductor creates a proportional distortion in its key material properties. Specifically, the elastic strain energy stored in the material alters the band structure affecting its electronic properties. This includes the width and shape of the energy bands and the effective masses of the carriers. In practice, a lattice mismatch of less than 2% can modify the band structure by 100 meV or more. Strain can also have practical benefits. For example, the effective mass of an electron in a strained region is reduced, increasing its mobility. Inducing strain into the charge control region of, say, a transistor, can result in faster switching times. In this way, lattice-mismatched heterostructures containing strained films to enhance performance are rapidly growing in importance in semiconductor device technology.*

* In the semiconductor industry, the straining of a silicon crystal is a well-known technique to increase charge carrier mobility, thereby enhancing device performance. To be most effective, the strain in a silicon channel of a CMOS (complementary metal oxide semiconductor) device should be compressive to improve the hole conduction of PMOS (p-type metal oxide semiconductor) transistors and tensile to improve the electron conduction of NMOS transistors (n-type metal oxide semiconductor).

2.8.3 Lattice Matching

Unfortunately, it is not possible to produce high-quality substrates for most compound semiconductors, and although an iterative approach may work, it is usually not cost effective, because the original substrate is lost after each growth. In addition, the requirement that the lattice parameters differ by no more than a few percent imposes a significant constraint when forming higher-order alloys, since changing the composition of an alloy changes its average lattice constant because of the different atomic bonding radii of the constituent elements. This problem is best appreciated by reference to a graph of the energy gap versus lattice constant. This is also known as a phase diagram.

In Figure 2.30 we plot bandgap energy versus the cubic lattice constant for the most common III–V binary compounds. For a possible range of ternary alloy systems, a solid line is generated between the starting binary materials. In the case of a quaternary compound, the boundary is laid out by four intersecting lines. Thus, we see for the GaAs-AlAs system, there is

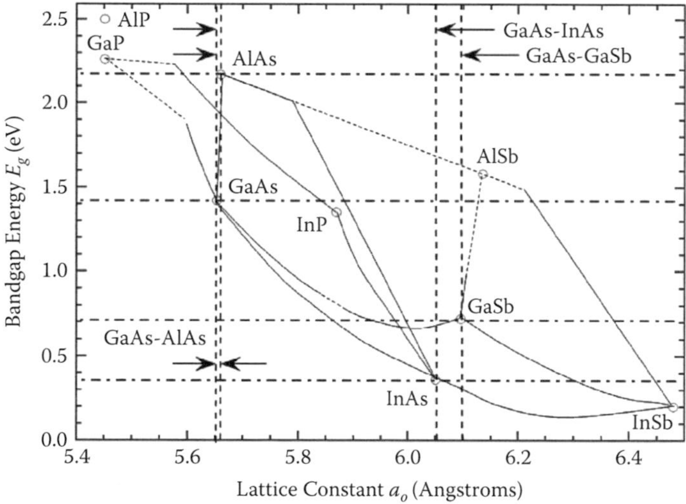

Figure 2.30 Bandgap energy, E_g, versus lattice constant, a_o, for the most common III–V ternary alloys at room temperature and their relationship to the participating binary (from [49]). The solid interconnect lines represent direct bandgap compounds, while the dotted interconnect lines represent indirect bandgap compounds.

little variation in lattice constant,* which allows the continuous formation of solid solutions of Al in GaAs over the full range of Al substitution. However, for the GaAs–InAs and GaAs–GaSb systems, the incorporation of In to GaAs (to form the $Ga_{(1-x)}In_{(x)}$ As alloy) or Sb to GaAs (to form the $GaAs_{(1-y)}Sb_y$ ternary alloy) results in major shifts of the average lattice constant, and stable alloys cannot be produced across the entire range of In or Sb substitution. Fortunately, lattice-mismatched layers can be grown pseudo-morphically (that is, defect-free) on a substrate with a different lattice constant if the epi-layer thickness is kept below a certain critical value, h_c, given by

$$h_c = a_e/14f \tag{2.4}$$

where f is the lattice misfit† given by $(a_e - a_s)/a_e$, and a_e and a_s are the lattice constants of the epi-layer and substrate, respectively. Below the critical value the lattice is elastically strained and its structure preserved. Above this thickness, plastic deformation occurs, leading to displacement of epi-layer atoms.

2.8.4 Bandgap Engineering

Unlike Si and Ge, whose electronic and chemical properties are "fixed," those of compound semiconductors can be modified by bandgap or, alternatively, wavelength engineering [50]. In this case, the ability to tailor the bandgap to a particular wavelength has profound optoelectronic applications. A specific example of how this is useful for radiation detector applications is CdZnTe. In Figure 2.31 we show the phase diagram for a range of II–VI materials, including CdTe and ZnTe. By altering the zinc fraction, x, in $Cd_{(1-x)}Z_xTe$, the range of possible alloys moves along the line between CdTe and ZnTe, making it possible to optimize the noise performance of a CdZnTe radiation detector for a given operating temperature. Specifically, increasing the zinc fraction, x, increases the bandgap energy E_g, which can be described empirically [51] by

$$E_g(x) = 1.510 + 0.606x + 0.139x^2 \text{ eV} \tag{2.5}$$

* The actual mismatch between GaAs and AlAs is 0.127%.

† Also known as the strain.

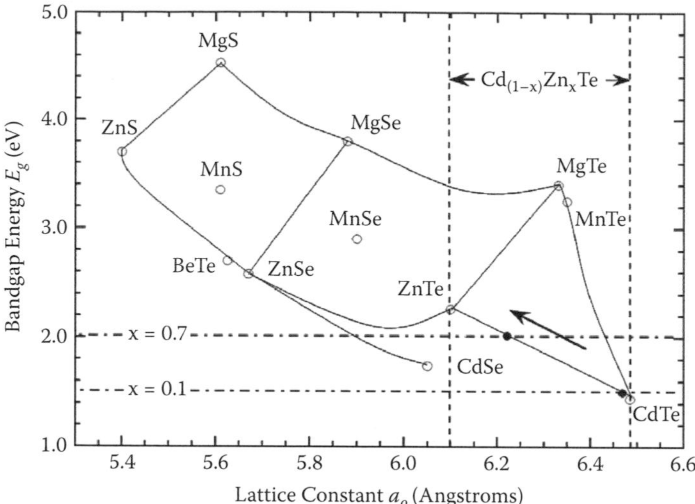

Figure 2.31 Bandgap energy, E_g, of II–VI compounds as a function of lattice constant a_o (from [49]). The dotted lines illustrate how altering the zinc fraction, x, in $Cd_{(1-x)}Zn_xTe$ alters the bandgap energy. Two cases are shown; $x = 0.1$, which provides optimum energy resolution at T = 243 K, and $x = 0.7$, which provides optimum energy resolution at room temperature (see text).

While an increase in E_g increases Fano noise due to carrier generation statistics, it simultaneously reduces shot noise due to thermal leakage currents. A trade-off can lead to a noise minimum at a given operating temperature and therefore an optimization of the spectral performance [52]. For example, a zinc fraction of 10% is optimum for operation at −30°C, whereas 70% is optimum at room temperature. Alloying of CdTe with Zn or ZnTe has additional benefits in that it increases the energy of defect formation [53] and mechanically strengthens the lattice, resulting in a lowering of defect densities [54].

2.9 Growth Techniques: VPE, LPE, MBE, and MOCVD

The four most commonly used epitaxial growth techniques are discussed below. These are vapor-phase epitaxy (VPE), liquid-phase epitaxy (LPE), molecular beam epitaxy (MBE), and metal organic chemical vapor epitaxy (MOCVD). MOCVD is also sometimes referred to as MOVPE (metal organic vapor-phase epitaxy). Although epitaxial silicon is usually grown by VPE (a modification of chemical vapor deposition, CVD), LPE and MBE are used extensively for growing group III–V compound semiconductors.

2.9.1 Liquid-Phase Epitaxy (LPE)

LPE is a relatively simple epitaxial growth technique which was widely used until the 1970s, when it gradually gave way to more sophisticated techniques, such as MBE and MOCVD. It is a much less expensive technique than MBE or MOCVD, but offers less control in interface abruptness when growing heterostructures. The method is commonly used for the growth of compound semiconductors and particularly ternary and quaternary III–V compounds on GaAs and InP substrates. Very thin, uniform, and high-quality layers can be produced. LPE refers to the growth of semiconductor crystals from a liquid solution at temperatures well below their melting point. This is made possible by the fact that a mixture of a semiconductor and a second element has a lower melting point than the pure semiconductor alone. Thus, for example, the melting point of a mixture of GaAs and Ga is considerably lower than 1238°C, the melting point of pure GaAs. The actual melting point of the mixture is determined by the proportion of the constituents Ga and GaAs. For example, in the growth of GaAs, LPE is commenced by placing a GaAs seed crystal in a solution of liquid Ga and GaAs which is molten at a temperature below the melting point of the seed. As the solution is cooled, a single crystal GaAs begins to grow on the seed, leaving a Ga-rich liquid mixture with an even lower melting point; further cooling causes more GaAs to crystallize on the seed.

The practical implementation of the technique is illustrated in Figure 2.32, in which we show an LPE system for producing complex heterostructures. Its principal components are:

- a boat to contain the melts and to separate them both from each other and the substrate
- a reducing atmosphere to prevent the formation of oxides during growth

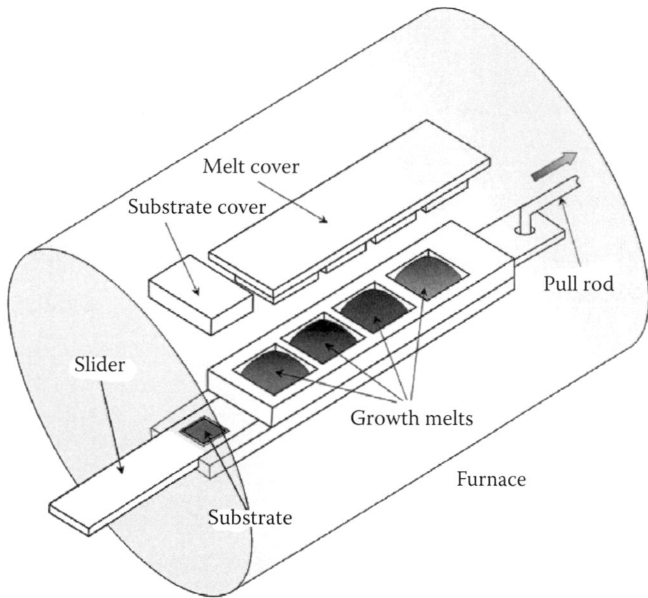

Figure 2.32 A schematic of LPE growth system suitable for growing heterostructures. The slider can be moved so that it is aligned with the different melts.

- a furnace with high thermal mass and lateral temperature uniformity
- a system to control the temperature of the furnace and hence temperatures of the melts and substrate
- a means of both bringing the melt solution and the substrate into contact at the start of the growth process and of separating the two at its conclusion

The boat consists of an upper fixed portion with a number of reservoirs to contain the melts, and a movable slider containing a recess for the substrate. The slider is pulled in sequence under several different melts to grow multiple layer structures. The liquid may also contain dopants that are to be introduced into the crystal. Because LPE is very close to being an equilibrium growth technique, it produces material with a very low native defect density. However, it is difficult to grow alloy systems that are not miscible or even to grow heterostructures with atomically abrupt interfaces. Nevertheless, heterostructures with interfaces graded over 10–20 Å can be grown. The strengths of LPE are that it is a simple, low-cost technique with a high throughput. Films can be deposited at a rate of 0.1 to 1 μm/minute. Additionally, it uses easily handled solids instead of toxic gases.

2.9.2 Chemical Vapor Deposition (CVD)/ Vapor-Phase Epitaxy (VPE)

A large class of epitaxial techniques rely on delivering the components that form the crystal from a gaseous environment. If the atoms or molecules emerging from the vapor can be deposited on the substrate in an ordered manner, epitaxial crystal growth can occur. Perhaps the most important is vapor-phase epitaxy (VPE), which is a derivative of the chemical vapor deposition (CVD) technique. It is used mainly for homoepitaxy and does not require the additional apparatus present in techniques such as MOCVD for precise heteroepitaxy. In this process, a number of gases are introduced in an induction-heated reactor where only the substrate is heated. The temperature of the substrate typically must be at least 50% of the melting point of the material to be deposited. Three basic forms of the technique are used, which are illustrated in Figure 2.33. These are the horizontal reactor, the vertical (or pancake reactor) and the multi-barrel reactor. The horizontal reactor

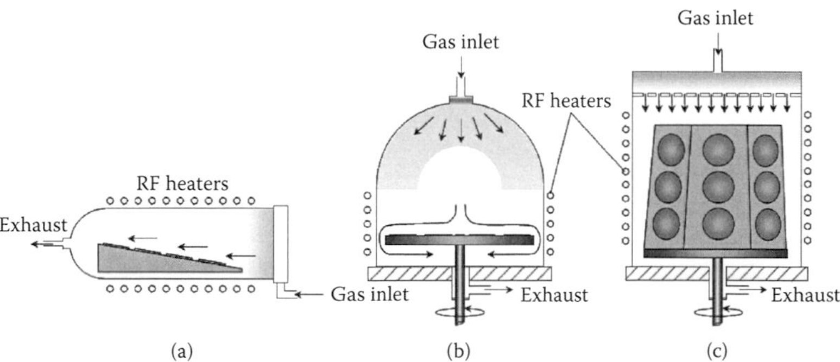

Figure 2.33 Reactors for VPE growth (adapted from [55]). (a) Horizontal reactor, (b) vertical reactor, and (c) multi-barrel reactor. In all cases, the substrate temperature must be maintained uniformly over its area.

is the simplest implementation and is shown in Figure 2.33 (a). Because the flow of gases is parallel to the surface of the wafers, growth is uniform laterally across a wafer but nonuniform in the upstream and downstream directions. The vertical reactor (Figure 2.33b) improves uniformity because the gas flows at right angles to the surface of all of the substrates, which are also rotated during growth to further improve uniformity. However, one disadvantage is that turbulence can occur in the chamber due to the presence of stagnation points. The multi-barrel reactor (Figure 2.33c) utilizes parallel gas flow and is particularly suitable for high-volume production.

2.9.2.1 *Doping in Vapor Deposition Systems*

For many optoelectronic applications, devices require layers with different carrier concentrations. This can be achieved by introducing the appropriate amounts of dopants (solid, liquid, or gas) into the gas phase. Gas dopants are preferable because they provide a finer control on the concentration of the layer. The concentration is altered by controlling the partial pressure of the dopant gas as it enters the reactor chamber. As in CVD, impurities change the deposition rate. Additionally, the high temperatures at which deposition is performed may allow dopants to diffuse into the growing layer from other layers in the wafer ("autodoping"). Conversely, dopants in the source gas may diffuse into the substrate.

2.9.3 The Multi-Tube Physical Vapor Transport (MTVPT) Technique

In view of the difficulty of growing high-vapor-pressure compounds, such as CdTe, CdZnTe, and HgTe, from the melt, Mullins, Brinkman, and Carles [56,57] have proposed the multi-tube physical vapor transport (MTVPT) technique, for growing II–VI materials directly from the vapor phase. This has several advantages over traditional liquid-phase techniques, namely improved structural perfection, purity, and an inherent resistance to "self poisoning." In this technique, control of vapor transport is critical to the growth, and this is achieved by separating the source and growth regions. A schematic cross

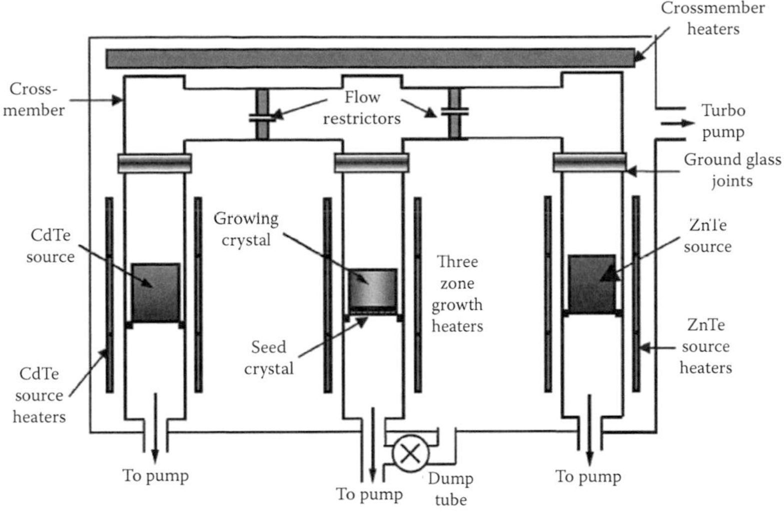

Figure 2.34 Schematic of the essential components of the Multi-Tube Physical Vapor Transport (MTPVT) growth system used to grow CdZnTe. (Image courtesy of Kromek®.)

section through a MTPVT system is shown in Figure 2.34. It is essentially a seeded vapor-phase growth system, in which the source materials are sublimed and allowed to condense on a cooler seed crystal. Cadmium telluride and zinc telluride source tubes, fabricated from quartz and heated by independent tubular furnaces, are connected to a third, similarly heated, quartz growth tube by a demountable quartz cross-member. The cross-member is heated to prevent condensation of sublimed source material before it reaches the growth region. In the growth tube, the seed wafer is located on a quartz pedestal separated from the inner wall of the tube by an annular gap in a Markov–Davydov configuration [59], ensuring wall-free growth. Accurate control of the temperature profile permits the vapor to condense and grow on a seed crystal in the center of the growth tube, while an annulus around the seeds allows the removal of excess components and volatile impurities.

In contrast to liquid-phase methods, vapor-phase techniques offer lower growth temperatures and the potential to grow ternary and multinary alloys of uniform composition. Indeed, the flexibility of vapor-phase growth has been widely exploited in the epitaxy of a wide range of optoelectronic materials and low-dimensional structures. In the context of the growth of bulk single crystals of the relatively weak II–VI compounds, particularly CdTe, lower growth temperatures have shown improved structural quality of material, as well as a reduction in defects and in contamination by impurities from the growth system [58].

2.9.4 Molecular-Beam Epitaxy (MBE)

At its simplest, MBE is a refined form of vacuum evaporation. The technique was pioneered in the late 1960s at Bell Telephone Laboratories by Arthur [60] and Cho [61]. In MBE, a source material is heated to produce an evaporated beam of particles. These particles travel through a very high vacuum (10^{-8} Pa) to the substrate, where they condense. MBE has a lower throughput than other forms of epitaxy (~1 to 300 nm/min), but is a much more precise delivery system for the growth of monoatomic layers. However, the slow deposition rates require proportionally better vacuum in order to achieve the same impurity levels as other deposition techniques.

A schematic diagram of an MBE system is shown in Figure 2.35. A critical feature is the extensive cryopaneling surrounding both the substrate station and the evaporation sources. The low temperature (usually 77 K) reduces the arrival rate of unwanted species and provides heat dissipation for both the evaporation sources and the substrate heater. The molecular beams are produced by evaporation or sublimation from heated liquids or solids contained in the crucibles, called effusion cells or Knudsen cells. At the pressures used in MBE equipment, collision-free beams emanate from the sources and interact chemically only with the substrate to form an epitaxial film. To improve growth uniformity, the substrate is also rotated. The composition of the grown epilayer and its doping level depend on the relative arrival rates of the constituent elements and dopants, which in turn depend on the evaporation rates of the corresponding sources. Simple mechanical shutters in front of the beam sources are used to interrupt the beam fluxes, to start and to stop the deposition or doping. Typically, shutter speeds are of the order of ~0.1 to 0.3 seconds, which is less than the time taken to grow a monolayer (~1

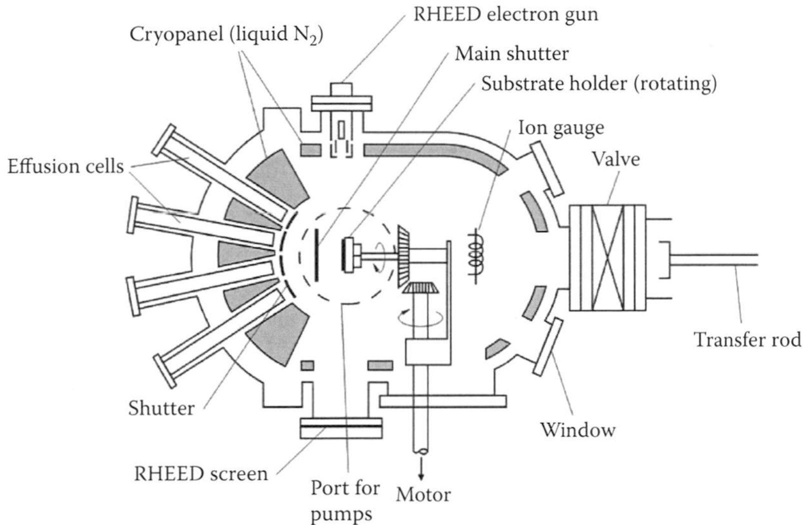

Figure 2.35 Schematic diagram of MBE system (from [62]).

second). This means that changes in composition and doping can be abrupt on an atomic scale. Because MBE is carried out in ultra-high-vacuum (UHV) conditions, it has a unique advantage over other epitaxial growth techniques, in that the physical and chemical properties of the films can be monitored and controlled *in situ* by surface-sensitive diagnostic techniques, such as reflection high energy electron diffraction (RHEED), Auger electron spectroscopy (AES), spectroscopic ellipsometry (SE), and reflection anisotropy spectroscopy (RAS). In RHEED [63], which is the most commonly used technique, a collimated electron beam at, or near, 10 keV impacts the crystal surface at a grazing angle. The electrons diffracted by the surface lattice interfere and form beams, which result in streaks or lines on a phosphor screen. The spacing between these lines is proportional to the surface reciprocal lattice spacing.

To summarize, the principal advantage of MBE is its precise control of film thickness, composition, and doping, with atomically abrupt interfaces readily obtainable. Compared to MOCVD, the process uses easily handled solids and no toxic gases, and its chemistry is relatively straightforward. The main disadvantage of MBE is its high expense, because films are grown one layer at a time.

2.9.5 Metal Organic Chemical Vapor Deposition (MOCVD)

Metal organic chemical vapor deposition (MOCVD) is a CVD process using organic sources and is widely used for heteroepitaxy. The use of metal organic compounds was first described by Manasevit in 1968 [64] and takes advantage of the fact that metal organic compounds are halide free and are often much more volatile and less thermally stable than metal halides, making it easier to get them to react at the deposition site. Like MBE, MOCVD is also capable of producing monolayer abrupt interfaces between semiconductors. However, unlike MBE, the gases used are not made of single elements, but are complex molecules which contain the elements needed to form the crystal. Because no UHV is needed, throughput is higher than in MBE and lends itself to industrial scale production. For a review, see Dapkus [65].

The growth process involves the forced convection of the metal organic vapor species over a heated substrate. The molecules striking the surface release the desired species, which then chemically react at the surface, producing growth. For group III–V semiconductors, the chemical processes involved are quite simple, in that an alkyl compound for the group III element and a hydride for the group V element decompose in the 500°C to 800°C temperature range to form the III–V compound semiconductor. For example, gallium arsenide could be grown in a reactor on a substrate by introducing trimethylgallium ($Ga(CH_3)_3$, often abbreviated to TMG) and arsine (AsH_3), via the reaction,

$$Ga(CH_3)_3 + AsH_3 \rightarrow GaAs + CH_4 \tag{2.6}$$

This growth process is illustrated in Figure 2.36. Similarly indium phosphide is often synthesized using trimethylindium ($In(CH_3)_3$, often abbreviated to TMI) and phosphine (PH_3), via the reaction,

$$In(CH_3)_3 + PH \rightarrow InP + CH_4\uparrow \tag{2.7}$$

A typical MOCVD system is shown in Figure 2.37 [66]. The reactor is a chamber made of a material that does not react with the chemicals being used, usually stainless steel

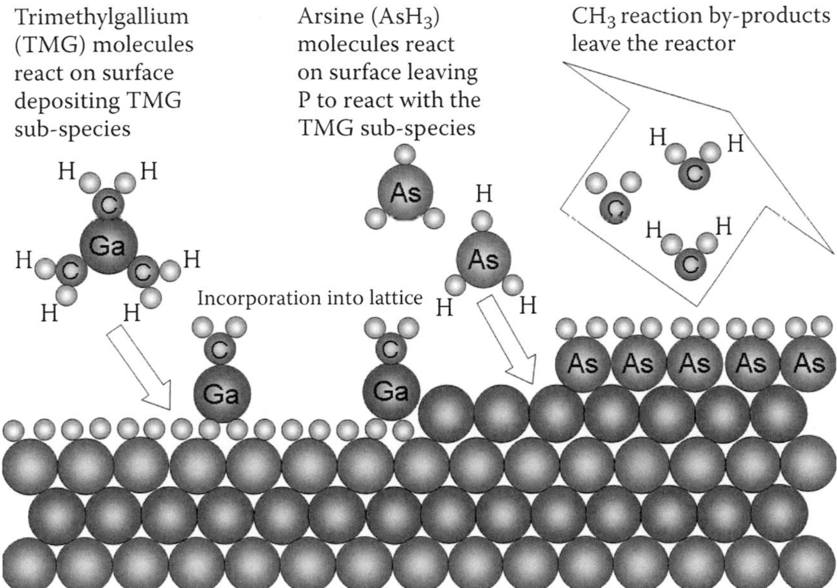

Figure 2.36 Figure illustrating the principle behind the MOCVD growth of GaAs.

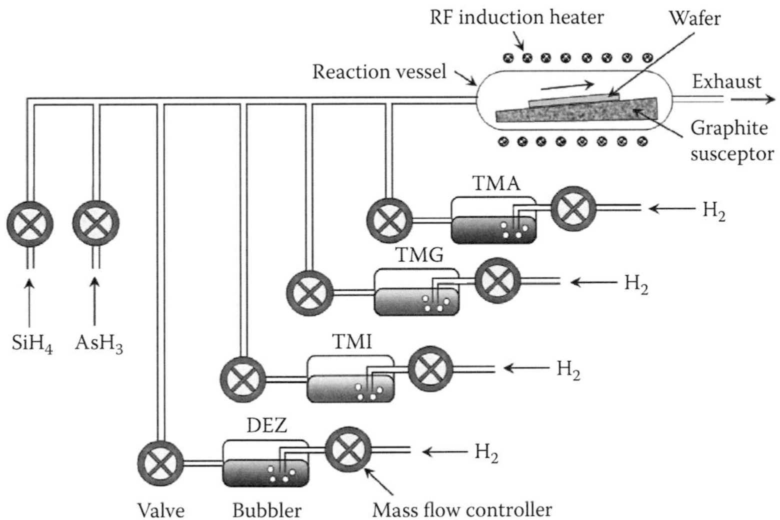

Figure 2.37 Schematic of the essential components of a typical MOCVD reactor system for the growth of a range of III–V compounds (modified from [66]). Here, the organometallic precursors are: TMA = trimethylaluminum, TMG = trimethylgallium, TMI = trimethylindium, DEZ = diethylzinc, AsH_4 = arsine, and SiH_4 = silane.

or quartz. It must also withstand high temperatures. This chamber is composed of reactor walls, liner, a susceptor (which holds the substrate), gas injection units, and temperature control units. To prevent overheating, cooling water must be flowing through the channels within the reactor walls. Special glasses, such as quartz or ceramic, are often used as the liner in the reactor chamber between the reactor wall and the susceptor. The susceptor is also made from a material resistant to the metal organic compounds and is used to transfer heat from the RF heater to the substrate. Graphite is commonly used. For growing nitrides and related materials, a special coating on the graphite susceptor is necessary to prevent corrosion by ammonia (NH_3) gas. Gases are introduced via a complicated switching system using devices known as "bubblers." In a bubbler a carrier gas (usually nitrogen or hydrogen) is bubbled through the metal organic liquid, and picks up some metal organic vapor, and transports it to the reactor. The amount of metal organic vapor transported depends on the rate of carrier gas flow and the bubbler temperature.

There are several varieties of MOCVD reactors—the most common being the atmospheric MOCVD and the low-pressure MOCVD (LPMOCVD) reactors. In an atmospheric MOCVD the growth chamber is essentially at atmospheric pressure, which alleviates the problems associated with vacuum generation, but at the expense of a larger gas flow rate. In LPMOCVD the growth chamber pressure is kept low. It uses less gas than in the atmospheric case, but growth rate is then lower—in fact even slower than in MBE. The principal advantages of MOCVD are that excellent uniformity in layer thickness, composition, and carrier concentration can be achieved over a large-area wafer. This technique also easily lends itself to the growth of abrupt heterointerfaces, allowing complex heterostructures to be grown layer by layer. The main disadvantage is that it uses large quantities of extremely toxic gases.

In Table 2.6, we summarize the relative advantages and disadvantages of the main epitaxial growth techniques.

TABLE 2.6
Advantages and Disadvantages of the Main Epitaxial Growth Techniques

Growth Method	Features	Advantages	Disadvantages
Liquid-Phase Epitaxy (LPE)	Growth from supersaturated solution onto substrate	Simple low-cost equipment and high throughput. Very high-quality material. No toxic gases and easily handled solids. Good control of impurities. High growth rates (0.1–1 micron/min). Suitable for selective growth.	Difficult to grow abrupt heterostructures. Limited substrate areas. Poor control over the growth of very thin layers. Re-dissolution of the grown material. Poor uniformity over large areas. High growth temperatures for some compounds (e.g., 900°C for InP).
Vapor-Phase Epitaxy (VPE)	Uses metal halides as transport agents to grow	Relatively simple reactors. Extremely high-purity material.	No Al compounds, thick layers. Toxic precursor gases.
Multi-Tube Physical Vapor Transport (MTPVT)	Controlled growth of high volatility II–VI compounds directly from the vapor phase	Simple, low-temperature process. *In situ* compounding. High-quality material. Reusable quartz ware. Low cost.	Lattice-matched substrates difficult to find for some compounds.
Metal Organic Chemical Vapor Deposition (MOCVD)	Uses metal organic compounds as the sources	High-quality material. Atomically abrupt interfaces. Low-temp growth. High vapor pressure material growth possible.	Expensive and slow (~3 monolayer/sec = 3 μm/hr). Some sources very toxic. Tendency for C contamination.
Molecular Beam Epitaxy (MBE)	Deposit epi-layer in ultrahigh vacuum, *in situ* characterization	High-quality material. Most precise epi-growth technique. Atomically abrupt interfaces. No toxic gases. Easily handled solids. Relatively simple chemistry.	Expensive and slower (~1 monolayer / sec = 1 μm/hr). UHV required. Hard to grow materials with high vapor pressure. Run-to-run reproducibility of layer thickness and composition. Surface “oval defects”.

References

1. B.D. Cullity, *Elements of X-Ray Diffraction*, Addison-Wesley Pub. Co., Reading, MA (1956). Revised edition: B.D. Cullity, S.R. Stock, *Elements of X-Ray Diffraction*, 2nd Ed., Prentice Hall, New Jersey (2001).
2. A. Bravais, "Sur les polyédres de forme symétrique," *Compt. Rendus Séances Acad. Sci.*, Vol. 14 (1849) pp. 1–40.
3. N.W. Ashcroft, A.N.D. Mermin, *Solid State Physics*, 1st ed., Brooks/Cole Publishing Co. (1976), pp. 112–129.
4. W.B. Pearson, *A Handbook of Lattice Spacings and Structures of Metals and Alloys*, Vol. 2, Pergamon Press, Oxford (1967).
5. C. Hermann, ed., *Internationale Tabellen zur Bestimmung von Kristallstrukturen*, Gebruder Borntraeger, Berlin, Vols. I and II (1935).
6. T. Hahn, *International Tables for Crystallography, Volume A: Space Group Symmetry,* 5th ed., Springer-Verlag, Berlin, New York (2002).
7. W.H. Miller, *A Treatise on Crystallography*, Cambridge, England, printed for J. & J.J. Deighton; London: J.W. Parker (1839).
8. C. Dang Ngoc Chan, "Example of Crystallographic Planes and Miller Indices for a Cubic Structure," http://commons.wikimedia.org/wiki/File:Indices_miller_plan_exemple_cube.png.
9. W. Hume-Rothery, R.W. Smallman, W. Haworth, *The Structure of Metals and Alloys,* The Institute of Metals, London (1969).
10. E. Mooser, W.B. Pearson, "The Chemical Bond in Semiconductors," *J. Electron.*, Vol. 1 (1956) pp. 629–645.
11. E. Mooser, W.B. Pearson, "Recognition and Classification of Semiconducting Compounds with Tetrahedral sp^3 Bonds," *J. Chem. Phys.*, Vol. 26 (1957) pp. 893–899.
12. L. Vegard, "Die Konstitution der Mischkristalle und die Raumfüllung der Atome," *Z. Phys.*, Vol. 5 (1921) pp. 17–26.
13. J.E. Bernard, A. Zunger, "Optical Bowing in Zinc Chalcogenide Semiconductor Alloys," *Phys. Rev.*, Vol. B 34 (1986) pp. 5992–5995.
14. Landolt-Börnstein, *Semiconductors–New Data and Updates for II-VI Compounds*, Vol. 44A, Springer-Verlag, Berlin (2009).
15. Landolt-Börnstein, *Semiconductors—New Data and Updates for I-VII, III-V, III-VI and IV-VI Compounds*, Vol. 44B, Springer-Verlag, Berlin (2009).

16. L.V. Azaroff, *Introduction to Solids*, McGraw-Hill Book Co., New York (1977).
17. E. Orowan, "Zür Kristallplastizität I-III," *Z. Physik*, Vol. 89 (1934) Teil I pp. 605–613; Teil II pp. 614–633, Teil II pp. 634–659.
18. M. Polanyi, "Über eine Art Gitterstörung, die einem Kristall plastisch machen könnte," *Z. Physik*, Vol. 89 (1934) pp. 660–664.
19. G.I. Taylor, "The Mechanism of Plastic Deformation of Crystals I-II," *Proc. Royal Soc.*, Vol. A145 (1934) Part I pp. 362–387; Part II pp. 388–404.
20. P. Lacombe, *Report of a Conference on Strength of Solids*, University of Bristol, England, The Physical Society, London (1948) pp. 91–94.
21. D. Blavette, E. Cadel, A. Fraczkiewicz, A. Menand, "Three-Dimensional Atomic-Scale Imaging of Impurity Segregation to Line Defects," *Science*, Vol. 286, no. 5448 (1999) pp. 2317–2319.
22. W.D. Callister, Jr., *Material Science and Engineering, An Introduction*, John Wiley & Sons, Inc., New York (1987).
23. F.C. Frank, J.H. van der Merwe, "One-Dimensional Dislocations. I. Static Theory," *Proceedings of the Royal Society of London. Series A, Mathematical and Physical Sciences*, Vol. A198 (1949) pp. 205–216.
24. M. Syväjärvi, "Epitaxial Growth and Characterization of SiC," PhD thesis (1999), Linköping University, LITH-Tek-Lic-694 (1998).
25. E.O. Hall, "The Deformation and Ageing of Mild Steel: III Discussion and Results," *Proc. Phys. Soc. London*, Vol. B64 (1950) pp.747–753.
26. N.S. Petch, "The Cleavage Strength of Polycrystals," *J. Iron Steel Inst.*, Vol. 174 (1953) pp. 25–28.
27. C. Szeles, M.C. Driver, "Growth and Properties of Semi-Insulating CdZnTe for Radiation Detector Applications," *Proc. of the SPIE*, Vol. 3446 (1998) pp. 2–9.
28. G.A. Carini, A.E. Bolotnikov, G.S. Camarda, G.W. Wright, L. Li, R.B. James, "Effect of Te Precipitates on the Performance of CdZnTe (CZT) Detectors," *Appl. Phys. Lett.*, Vol. 88, issue 14 (2006) pp. 143515-1–143515-3.
29. W.G. Pfann, "Principles of Zone Melting," *Transactions of the American Institute of Mining and Metallurgical Engineers*, Vol. 194 (1952) pp. 747–753.
30. W.G. Pfann, "Zone melting," *Science*, Vol. 135, no. 3509 (1962) pp. 1101–1109.
31. J. Czochralski, "Ein neues Verfahren zur Messung des Kristallisation-sgeschwindigkeit der Metalle," *Z. Phys. Chem.*, Vol. 92 (1918) pp. 219–221.

32. A. Voigt, K.-H. Hoffmann, R. Backofen, N. Botkin, A. Goepfert, O. Pykhteev, V. Turova, S. Vey, A. Zollorsch, "Viola," *Caesar Annual Report, Nanotechnology / Materials Science* (2005) 2_1_35.
33. W. Lin, K.E. Benson, "The Science and Engineering of Large-Diameter Czochralski Silicon Crystal Growth," *Annual Review of Material Science*, Vol. 17 (1987) pp. 273–298.
34. E.P.A. Metz, R.C. Miller, R. Mazelsky, "A Technique for Pulling Single Crystals of Volatile Materials," *J. Appl. Phys.*, Vol. 33 (1962) pp. 2016–2017.
35. J. Singh, *Semiconductor Devices: Basic Principles*, John-Wiley & Sons, New York (2001).
36. J.B. Mullin, B.W. Straugham, W.S. Brickell, "Liquid Encapsulation Techniques," *J. Phys. Chem. Solids*, Vol. 26 (1965) pp. 782–784.
37. J. Winkler, "*Beiträge zur Regelung des Czochralski-Kristallzüchtungsprozesses zur Herstellung von Verbindungshalbleitern*," PhD Thesis Technische Universität Dresden, Germany, Shaker Verlag Aachen (2007).
38. P. Rudolph, F.-M. Kiessling, "Growth and Characterization of GaAs Crystals Produced by the VCz Method without Boric Oxide Encapsulation," *Journal of Crystal Growth*, Vol. 292 (2006) pp. 532–537.
39. P.H. Keck, M.J.E. Golay, "Crystallization of Silicon from a Floating Liquid Zone," *Physical Review*, Vol. 89, Issue 6 (1953) pp. 1297–1297.
40. W. Zulehner, "Historical Overview of Silicon Crystal Pulling Development," *Materials Science and Engineering*, Vol. 73 (2000) pp. 7–15.
41. P.W. Bridgman, "Certain Physical Properties of Single Crystals of Tungsten, Antimony, Bismuth, Tellurium, Cadmium, Zinc, and Tin," *Proc. Amer. Acad. Sci.*, Vol. 60, no. 9 (1925) pp. 303–383.
42. D.C. Stockbarger, "The Production of Large Single Crystals of Lithium Fluoride," *Review of Scientific Instruments*, Vol. 7 (1936) pp. 133–136.
43. U. Lachish, "CdTe and CdZnTe Crystal Growth and Production of Gamma Radiation Detectors," (2000) http://urila.tripod.com/crystal.htm.
44. R. Sudharsanan, K.B. Parham, N.H. Karam, "Cadmium Zinc Telluride Detects Gamma-Rays," *Laser Focus World*, June (1999) pp. 199.
45. N. Audet, M. Cossette, "Synthesis of Ultra High Purity CdTe Ingots by the Travelled Heater Method," *J. Electron. Mater.*, Vol. 34, no. 6 (2002) pp. 683–686.

46. N. Audet, B. Levicharsky, A. Zappettini, M. Zha, "Composition Study of CdTe Charges Synthesized by the Travelled Heater Method," *IEEE Trans Nucl. Sci.*, Vol. 54, no. 4 (2006) pp. 782–3693.
47. H.C. Ramsperger, E.H. Melvin, "The Preparation of Large Single Crystals," *J. Opt. Soc. Am.*, Vol. 15, Issue 6 (1927) pp. 359–363.
48. S. Kawarabayashi, M. Yokogawa, A. Kawasaki, R. Nakai, "Comparisons between Conventional LEC, VCZ and VGF for the Growth of InP Crystals," *IEEE Proceedings of the Sixth International Conference on Indium Phosphide and Related Materials*, Santa Barbara, CA (1994) pp. 227–230.
49. A. Owens, A. Peacock, "Compound semiconductor radiation detectors," *Nucl. Instr. and Meth.*, Vol. A531 (2004) pp. 18–37.
50. W. Faschinger, in *Wide Bandgap Semiconductors*, ed. S. Pearton, William Andrew Publishing, New York (1999) pp. 1–37.
51. D. Olego, J. Faurie, S. Sivananthan, P. Raccah, "Optoelectronic Properties of $Cd_{1-x}Zn_xTe$ Films Grown by Molecular Beam Epitaxy on GaAs Substrates," *App. Phys. Lett.*, Vol. 47 (1985) pp. 1172–1174.
52. J.E. Toney, T.E. Schlesinger, R.B. James, "Optimal bandgap variants of $Cd_{1-x}Zn_xTe$ for high-resolution X-ray and gamma-ray spectroscopy," *Nucl. Inst. and Meth.*, Vol. A428 (1999) pp.14–24.
53. A.W. Webb, S.B. Quadri, E.R. Carpenter, E.F. Skelton, "Effects of Pressure on $Cd_{1-x}Zn_xTe$ Alloys ($0 < x < 0.5$)," *J. Appl. Phys.*, Vol. 61 (1987) pp. 2492–2494.
54. J.F. Butler, C.L. Lingren, F.P. Doty, "$Cd_{1-x}Zn_xTe$ Gamma Ray Detectors," *IEEE Trans. Nucl. Sci.*, Vol. 39 (1992) pp. 605–609.
55. M. Quick, J. Serda, *Semiconductor Manufacturing Technology*, Prentice Hall, Columbus, NJ (2001).
56. A.W. Brinkman, J. Carles, "The Growth of Crystals from the Vapour," *Progress in Crystal Growth and Characterization of Materials*, Vol. 37, Issue 4 (1998) pp. 169–209.
57. J.T. Mullins, J. Carles, N.M. Aitken, A.W. Brinkman. "A Novel 'Multi-Tube' Vapour Growth System and Its Application to the Growth of Bulk Crystals of Cadmium Telluride," *J. Crystal Growth*, Vol. 208 (2000), pp. 211–218.
58. J.T. Mullins, B.J. Cantwell, A. Basu, Q. Jiang, A. Choubey, A.W. Brinkman, B.K. Tanner, "Vapour Phase Growth of Bulk Crystals of Cadmium Telluride and Cadmium Zinc Telluride on Gallium Arsenide," *J. Electron Materials*, doi: 10, 1007/511664-008-0442-3.
59. E.V. Markov, A.A. Davydov, "Growing Orientated Single Crystals of Cadmium Sulfide from the Vapour Phase," *Inorg. Mater.*, 11 (1975) pp. 1504–1506.

60. J.R. Arthur, J.J. LePore, "GaAs, GaP, and $GaAs_xP_{1-x}$ Epitaxial Films Grown by Molecular Beam Deposition," *J. Vac. Sci. Technol.*, Vol. 6 (1969) pp. 545–548.
61. A.Y. Cho, "Film Deposition by Molecular-Beam Techniques," *J. Vac. Sci. Technol.*, Vol. 8, no. 5 (1971) pp. 31–38.
62. H. Ibach, H. Lüth, *Solid-State Physics: An Introduction to Principles of Materials Science*, Springer Verlag, Berlin (2003).
63. A. Ichimiya, P.I. Cohen, *Reflection High Energy Electron Diffraction*, Cambridge University Press, Cambridge, UK (2004).
64. H.M. Manasevit, "Single-Crystal Gallium Arsenide on Insulting Substrates," *Appl. Phys. Lett.*, Vol. 12, No. 4 (1968) pp. 156–159.
65. P.D. Dapkus, "Metal organic Chemical Vapor Deposition," *Annual Review of Materials Science*, Vol. 12 (1982) pp. 243–269.
66. http://britneyspears.ac/physics/fabrication/schmocvd.gif

3
Detector Fabrication

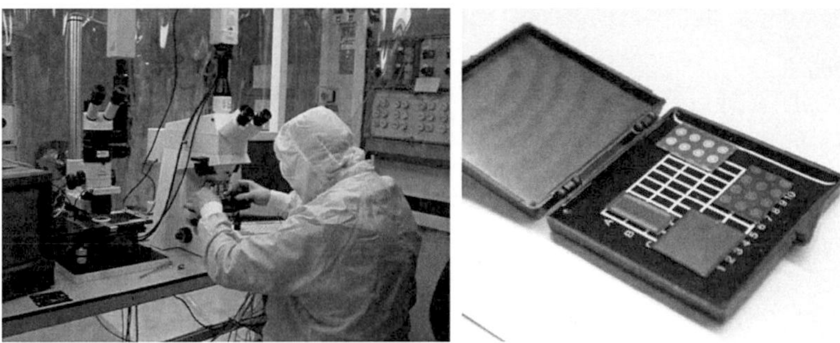

Frontispiece Left: detector assembly. Right: a selection of epitaxial GaAs diodes and arrays prior to packaging. (Image courtesy of Oxford Instruments Analytical Oy).

Once an ingot of semiconducting material has been produced, it must pass through several processing steps before it can be fabricated into a detector. These are, in order:

1. Thermal annealing
2. Cutting
3. Lapping and polishing
4. Etching
5. Cleaning

Following these preparatory steps, the crystal is contacted and characterized. Figure 3.1 shows the entire sequence of producing a detector, from the initial material synthesis to detector

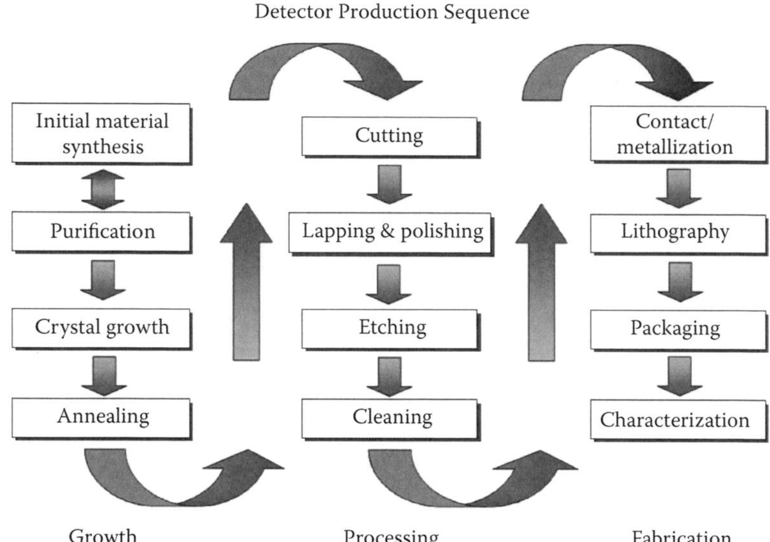

Figure 3.1 The entire sequence for producing a detector can be separated into three phases or stages: the growth phase, the material processing stage, and the detector fabrication stage.

characterization. The type and degree of processing depend on the mechanical properties of the material being processed. For example, group III–V materials are generally mechanically robust with hardness values of more than 500* kgf mm^{-2} on the Knoop microhardness scale (see Appendix E). These materials can therefore withstand a lot of fairly robust mechanical treatment without introducing a disproportionately large number of defects into the crystal, whereas groups II–VI and I–VII materials tend to be mechanically soft (Knoop hardness values of <~100† kgf mm^{-2}). For these latter materials, chemical treatments are preferred because mechanical processing, such as lapping and polishing, can introduce more damage than is removed.

* Approximately harder than steel.
† Approximately softer than tin.

3.1 Mechanical Processing Overview

To prepare the wafers, ingots are first thermally annealed and then sliced into individual detector blanks or wafers with multiple-blade inner-diameter saws. The sliced wafers are mechanically lapped under pressure using a counter-rotating machine to achieve flatness and parallelism on both sides of the wafer. Most lapping operations use slurries* of either aluminum oxide or silicon carbide. The edges of the individual wafers are also rounded by the use of wet automatic grinders, and in some cases a flat edge is added to denote crystal orientation. Between processes, wafers are stored in water or methanol in plastic containers. After lapping, wafers are etched with a solution containing nitric, acetic, and hydrofluoric acids. Etching may be done in manual etch tanks or by automated etching machines. The etching process removes external surface damage and reduces the thickness of the wafer. Next, the wafers are polished using an aqueous mixture of colloidal silica and sodium hydroxide. The wafers are mounted onto a metal carrier plate that is attached by vacuum to the polishing machine. The polishing process usually involves two or three polishing steps with progressively finer slurries, which again decreases wafer thickness and results in a mirror-like finish. Each of these processes is described below.

3.1.1 Thermal Annealing

Thermal annealing is an important postgrowth treatment and is necessary before further processing can be carried out. For some crystals, strains that develop during growth can be relieved or significantly reduced by thermal annealing. During the annealing process, crystals are heated to a specific temperature at a given rate. They remain at this temperature for a prescribed time and are then slowly cooled to room temperature. A general rule of thumb is to anneal at a

* In lapping operations, a slurry is a colloid—a thick suspension made up of an abrasive power and a liquid which may act as both a chemical corrosive and as a lubricant.

temperature $2T_m/3K$, where T_m is the melting temperature of the material. At this temperature, the defects (dislocations) become sufficiently mobile for the crystal to reconstitute itself.

Annealing can be carried out in either a high vacuum or in the presence of an inert atmosphere, such as argon or nitrogen. In the case of compound semiconductors, where the vapor pressure of one of the components can be significantly different from that of others, a deficit of the most volatile component can occur. To prevent deviations from stoichiometry, crystals are then annealed in an atmosphere containing an excess of this component. For example, CdS can be annealed in a sulfur atmosphere, GaAs in an As atmosphere, and CdTe in a Cd atmosphere.

3.1.2 Cutting

After thermal annealing, the crystal must be cut into the appropriate size for its end use. For X-ray detector applications, thin wafers of thickness 200 μm to 1 mm are required while gamma-ray applications use blocks of size a few millimeters to a few centimeters. Two types of sawing machine are commonly used. One is a disc saw (Figure 3.2. left), and the other is a wire saw (Figure 3.2 right). In both cases, the cutting surface is impregnated with diamond powder. This operation is carried out using wet lubricants. A good crystal cutter has features such as variable speed and tension and is equipped with a goniometer to enable the crystal to be cut at any angle. Soft materials are commonly cut by cleaving the crystal with a razor blade. However, cleaving introduces mechanical defects into the bulk and roughens the surface. For extremely soft materials, such as PbI_2, these plastic deformations can easily extend several hundred microns into the crystal. For such materials, a string saw is the preferred cutting method. In this case, the "string" is typically a cotton thread soaked in a diluted etchant, such as KI.

3.1.3 Lapping and Polishing

Crystal surfaces "as-cut" are very rough. For almost all detector applications, the crystal surface must be optically flat to

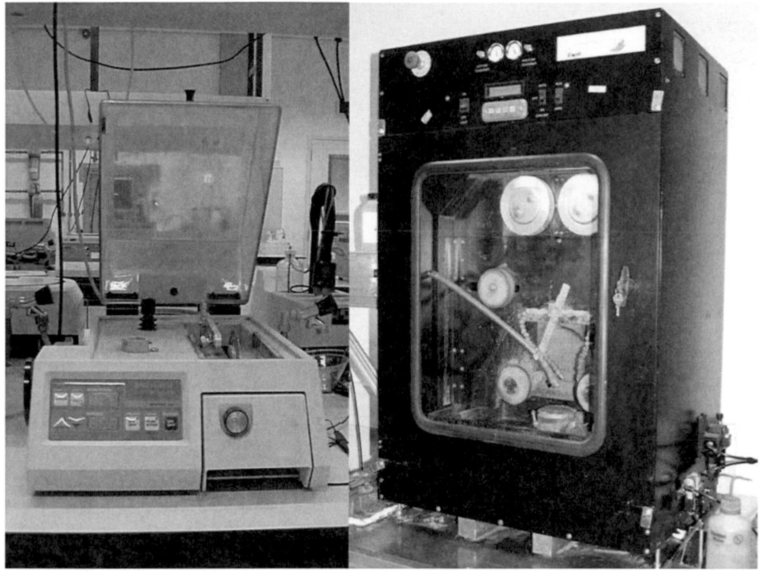

***Figure 3.2* (See color insert.)** Left: a diamond disk saw. Right: a wire saw used for cutting ingots into slices prior to detector preparation. (Images courtesy of the European Space Agency and Kromek®.)

ensure uniformity and reproducibility in its mechanical and electrical characteristics. To achieve such a finish, crystals are usually lapped and polished using a purpose-built machine as shown in Figure 3.3. Lapping is a machining operation, in which two surfaces are rubbed together with an abrasive between them. The crystal to be processed is one surface and is known as the workpiece. In the lapping machine, it is held in a metal chuck between two rings while pressed onto the second surface (known as the lap), which is usually a rotating wheel. Abrasive slurry flowing onto the wheel serves two functions—abrasion and lubrication. A weight on top of the workpiece provides the load, and a micrometer is used to gauge how much material is removed. Depending on the machine, the rings may be kept stationary while the disk rotates, or they may be moved in a reciprocating sliding action. Crystals are lapped first in coarse carborundum (SiC) powder, followed by fine carborundum powder. Polishing then takes place in several steps, first, in coarse alumina (Al_2O_3) emulsion (3 μm), then in fine (0.05 μm) alumina emulsion.

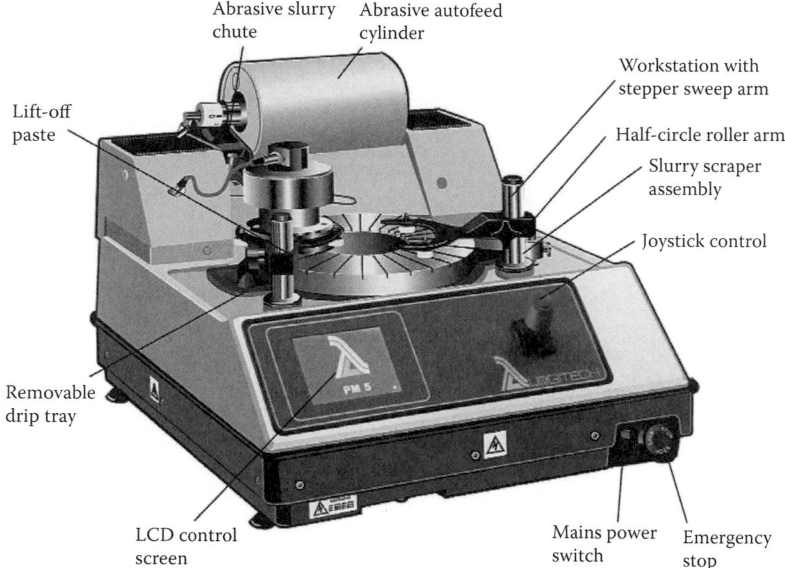

Figure 3.3 Essential elements of a lapping and polishing machine (from [1], image courtesy Logitech Ltd.). The crystal being lapped and polished is held in the metal chuck between two rings. In operation, the rings stay in one location as the lapping plate rotates beneath them. The abrasive is applied onto the plate in the form of slurry, which also aids in lubrication.

This procedure results in an optically flat surface with a mirror finish. For soft materials, where mechanical damage can extend several hundred microns into the crystal, it is usually recommended that the crystal be cut into thick slices and polished to the required thickness to remove the damaged layer.

3.1.4 Etching

Even though finer polishing agents may reduce the visibility of defects, considerable damage can still exist in the form of filled-in scratches, microcracks, pits, and dislocations, particularly in softer materials. These defects will affect the electronic properties of the material and so, prior to further processing, the polished surface is sometimes etched in order to remove the damaged layer, which for soft materials may be several hundred microns thick. Chemical etchants are commonly used, different for different crystals, although a bromine–methanol solution is commonly used for most materials. A comprehensive list of etchants for over 50 different metals, semiconductors, and clean room materials is given in reference [2], along with some etch rates. For group III–V materials, a discussion of chemical etchants and their properties may be found in Faust [3].

Besides removing damaged layers, etchants are also useful for revealing bulk defects present in the crystals, because they preferentially attack defect sites, revealing dislocations and other atomic arrangement faults. Dislocation densities may be assessed in this manner by measuring the etch pit density (EPD). This is determined by applying an etchant, usually a molten alkali, such as KOH, to the crystal surface. When a dislocation line intersects the surface of a metallic material, the associated strain field locally increases the relative susceptibility of the material to acidic etching, and an etch pit of regular geometrical format results. If the material is strained (deformed) and repeatedly re-etched, a series of etch pits can be produced which effectively trace defects. The EPD is determined directly by measuring the number of "pits" in the surface using an optical contrast microscopy. The advantage of this technique is that it is simple, inexpensive, and sensitive, and requires no special knowledge, unlike transmission electron microscopy (TEM). The disadvantage is that the there has been no systematic development of the technique, which makes etch patterns difficult to interpret.

Wet chemical etching (Figure 3.4) is one of the oldest methods used for the characterization of single crystals. It was originally employed in the nineteenth century for defining the crystallographic orientation of natural crystals, although at that time it was not understood that etch pits were formed due to the presence of crystallographic defects. Not only can etching proceed preferentially along certain crystallographic axes, but also the shapes of the etch pits can be used for determining the orientation of the crystal surface. For example, in Si etch pits are elliptical in shape along the [000] crystallographic axis, whereas they are triangular or pyramidal in shape along the [111] axis (see Figure 3.5).

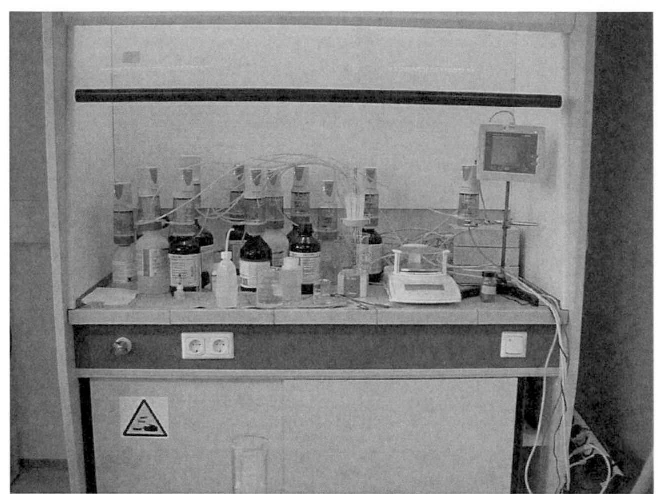

Figure 3.4 **(See color insert.)** Etching cabinet with automatic titration system for preparing different etch solutions. The cabinet also contains services such as a deionized water supply and compressed air system to clean and dry the crystals. (Image courtesy of the European Space Agency.)

3.1.5 Cleaning

Finally, the wafers are cleaned to remove any particles or residue remaining on the exterior surfaces of the polished product. Various cleaning steps and solutions containing ammonia, methanol, hydrogen peroxide, hydrofluoric acid, hydrochloric

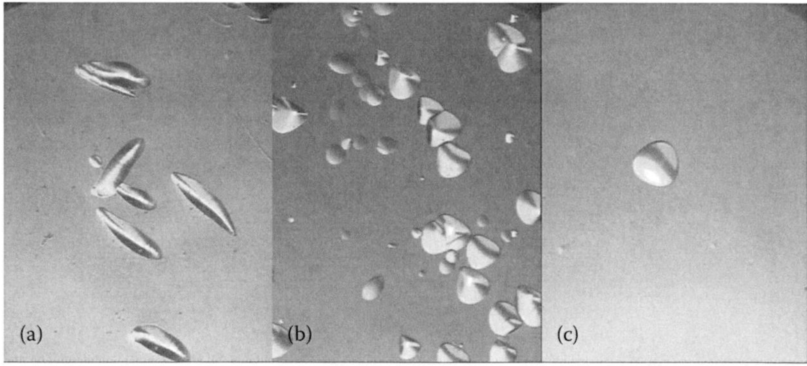

Figure 3.5 **(See color insert.)** Dislocation etch pits in Si, showing the effects of preferential etching along different crystallographic directions. In the [100] orientation, the etch pits appear (a) elliptical in shape, whereas along the [111] direction they can assume (b) triangular or (c) pyramidal shapes.

acid, and deionized water are used, depending on the semiconductor being processed.

3.1.6 Contact Deposition

After, cleaning, the contacts can be applied. The fabrication of stable and laterally uniform contacts can be a major problem, the practical aspects of which are discussed in Chapter 4. Physically, contacting can be realized by several different methods, for example applying metal-bearing paints and pastes such as Aquadag, melting metals directly on the semiconductor surface, evaporation, sputtering, molecular beam epitaxy, ion implantation, and others. For simple geometries, contacts can be applied by evaporation directly through a shadow mask, usually fabricated out of a thin metal foil such as brass. For more complex geometries, photolithographic techniques are used.

3.1.7 Lithography

For precise control of contact topology such as that required for pixilated detectors, contact features are generally defined using a shadow mask created by photolithography*. The technique is generally used to process complete wafers when multiple detectors are being produced. Photolithography is a process to remove selectively parts of a thin film, or even the bulk of a substrate. It uses light to transfer a geometric pattern from a photomask to a light-sensitive chemical, known as photoresist, on the wafer. The photoresist consists of three components: a resin base material, a photoconductive compound, and a solvent to control the mechanical properties of the photoresist. After exposure to light, a series of chemical treatments then etches the exposure pattern from the mask into the material beneath the photoresist. The entire process is illustrated in Figure 3.6.

The process begins by cleaning the metallized detector blank or wafer and baking it to remove any H_2O which may oxidize

* The word lithography derives from the Greek λίθος, meaning stone, and γράφειν, meaning to write.

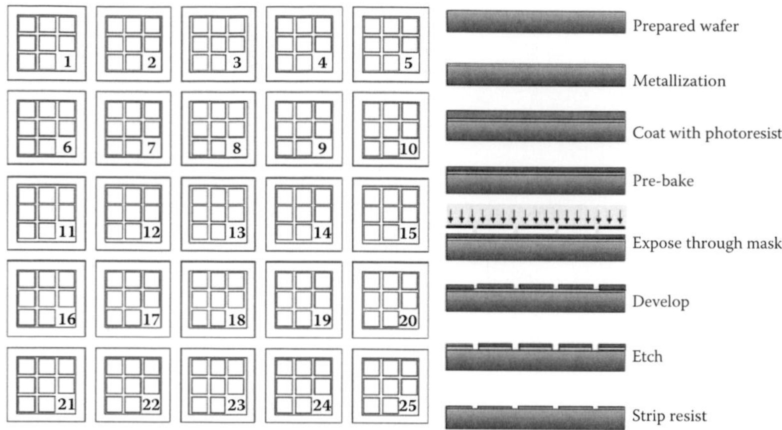

Figure 3.6 Left: a photolithographic mask used to produce a number of 3 × 3 InAs pixel arrays with guard rings (Image courtesy of Oxford Instruments Analytical Oy). Right: example of a typical sequence of lithographic processing steps for forming the contacts to an array, illustrated for a positive resist.

the sample. It is then covered with photoresist by dipping or spin coating to produce a uniformly thick layer. The photoresist is a viscous, liquid solution that is dispensed onto the wafer. For spin coating, the wafer is spun at 1200 to 4800 rpm for 30 to 60 seconds, which produces a layer between 0.5 and 2.5 μm thick. The photoresist-coated wafer is then prebaked to sensitize the photoresist to UV light by driving off excess solvent. This is typically carried out at 100°C for 5 to 30 minutes. After prebaking, the photoresist is exposed to a pattern of intense light, typically ultraviolet. Positive photoresist, the most common type, becomes chemically less stable when exposed; negative photoresist becomes more stable. This chemical change allows the unstable photoresist to be removed by a special solution, called “developer” by analogy with photographic developer. Exposure systems typically produce an image on the wafer using a photomask. The light shines through the photomask, which blocks it in some areas and lets it pass in others. Generally, three forms of lithography are used depending on the application. In order of complexity these are: contact, proximity, and projection. A contact printer is the simplest exposure system. It sets the photomask in direct contact with the wafer and exposes it to a uniform light source. Although it offers simplicity and high resolution (approximately the wavelength of the radiation), the mask and wafer are particularly susceptible to damage precisely because they are in contact. However, it is particularly attractive for research and small-scale production processes, because it uses inexpensive hardware and can still achieve high optical resolution.

Proximity printing reduces mask damage by setting the mask a fixed distance above the wafer (of the order of 20 mm). Unfortunately, the resolution limit is significantly increased, making proximity printing inappropriate for complex lithography. In addition, diffraction at the pattern edges causes light divergence.

For the needs of the semiconductor industry, projection printing is the most common method of exposure. Projection lithography derives its name from the fact that an image of the mask is projected onto the wafer. Projection lithography has become a viable alternative to contact/proximity printing since the advent of computer-aided lens design and improved optical materials allowed the production of lens elements of sufficient quality to meet the exacting requirements of the semiconductor industry. Unlike contact or proximity masks, which cover an entire wafer, projection masks (also called "reticles") show only one die. Projection exposure systems (steppers) project the mask onto the wafer many times to create the complete pattern. The smaller imaging field simplifies the design and manufacture of the lens, but at the expense of a more complicated reticle and wafer stage. Also, the depth of focus (a few micrometers) restricts the thickness of the photoresist and places strict demands on wafer flatness. In view of its complexity and significant capital costs, the technique is only suitable for large-scale production.

For both contact and proximity lithography the mask covers the entire wafer, and this requires that the light intensity be uniform across an entire wafer and that the mask is precisely aligned to features already on the wafer. As modern processes use increasingly large wafers and multiple layering, these conditions become progressively more difficult to fulfill.

After exposure, the wafer is then "hard baked," typically at ~150°C for ~25 minutes. The hard bake solidifies the remaining photoresist to make a more durable protecting layer for future ion implantation, wet chemical etching, or plasma etching. In the etching step, a liquid ("wet") or plasma ("dry") chemical agent removes the layers in the areas that are not protected by photoresist. These will become the contacts which are then either sputtered or evaporated onto the surface and adhere through the windows opened up during the etching procedure. In semiconductor fabrication, dry etching techniques are generally preferred, because they can be made anisotropic, in order to avoid significant undercutting of the photoresist pattern. This is essential when the width of the features to be defined is similar to or less than the thickness of the material being etched, for example, when the aspect ratio approaches unity.

Dopants may be introduced by either thermal diffusion in a gaseous ambient or by ion implantation. In this process ions are accelerated to a potential of 20 to 100 keV, depending on the desired penetration depth, and are initially distributed interstitially; however, in order to activate them electrically as donors or acceptors, they must be moved into substitutional lattice sites. This is achieved by heating (thermal annealing).

Once etching is completed the photoresist is no longer needed and must be removed from the substrate. This usually requires a liquid "resist stripper," which alters the resist chemically so that it no longer adheres to the substrate.

3.2 Detector Characterization

Following contacting, the detector is now ready for characterization. This usually consists of several steps, and the results will determine the potential of the detector. Any characterization of a semiconductor should in principle include:

1. Chemical composition
2. Crystallographic structure
3. Electrical properties
4. Electronic properties
5. Characterization of defects
6. Performance testing

To carry out this analysis, a wide variety of techniques have been developed, some of which are described below.

3.2.1 Chemical Analysis

When growing new semiconductors, it is sometimes found that the material has good crystallographic qualities but poor transport properties. Quite often this can be traced to an impurity in the precursors used to grow the crystal acting as a majority carrier mu-tau "killer." The impurity levels needed to do this may only be part per million, so they are not always listed in the manufacturer's assay of the stock material. If an impurity is found, then the options are either to go to a purer material or change vendors. There are several techniques to detect impurities to this level, the most sensitive of which are: glow-discharge mass spectrometry (GDMS) and inductively coupled plasma, mass and optical emission spectroscopy (ICP-MS and ICP-OES). A comparison of the two techniques is given in Table 3.1.

TABLE 3.1
Comparison of the Strengths and Weaknesses of the GDMS and ICP Techniques

Technique	Pros	Cons
Inductively coupled plasma (ICP) spectroscopy	Can determine elemental trace compositions to very low detection limits (typically sub-ppb) with high accuracy and precision	The sample to be analyzed must be digested and introduced to the plasma in the form of a liquid spray
	Many elements (up to 70 in theory) can be determined simultaneously in a single sample analysis	Emission spectra are complex and interelement interferences are possible
	Instrumentation suitable for automation.	
Glow discharge mass spectroscopy (GDMS)	Full periodic table coverage (except H) Sub-ppb to ppt detection Linear and simple calibration Capability to analyze insulators Depth profiling possible	Sample unhomogeneity Samples must be vacuum compatible Not suited for organic materials/polymers Smallest sample ~1 cm in diameter

3.2.1.1 Inductively Coupled Plasma Spectroscopy (ICP-MS and ICP-OES)

Inductively coupled plasma spectroscopy techniques are so-called "wet" sampling methods whereby samples are introduced in liquid form for analysis. Plasma mass spectroscopy (MS) is highly sensitive and capable of the determination of an elemental range from 0 to 240 amu, to a sensitivity of below one part in 10^{12}. It utilizes an inductively coupled plasma to ionize the sample material and a mass spectrometer to separate and detect the ions. The technique is also capable of monitoring isotopic speciation for the ions of choice. The sample to be analyzed is normally dissolved and then mixed with water before being sprayed into the plasma.

Plasma optical emission spectroscopy (OES) is a fast multi-element technique with a dynamic linear range and moderately low detection limits (~0.2–100 ppb). The instrument also uses an inductively coupled plasma source to dissociate the sample into its constituent atoms or ions, exciting them to a level at which they emit light of a characteristic wavelength. A quantitative determination takes place on the basis of the proportionality of the intensity of the radiation and the elemental concentrations in both calibration and analysis samples. Up to 60 elements can be screened per single sample run of less than one minute, and the samples can be analyzed in a variety of aqueous or organic matrices.

3.2.1.2 Glow-Discharge Mass Spectrometry (GDMS)

Glow-discharge mass spectrometry (GDMS) is a technique that is capable of providing trace-level elemental quantification for a wide range of solid and powder materials. The sample to be analyzed forms the cathode in a low-pressure (~100 Pa) gas discharge or plasma. Argon is typically used as the discharge gas. Positive argon ions are accelerated toward the cathode (sample) surface with energies from hundreds to thousands of eV, resulting in erosion and atomization of the upper atom layers of the sample. Only the sputtered neutral species are capable of escaping the cathode surface and diffusing into the plasma where they are subsequently ionized. The atomization and ionization processes are thus separated in space and time, which simplifies calibration and quantification, and ensures the near matrix independence of this technique as well. Although the analysis can be more time consuming than in solution-based analytical methods, the sensitivity, the ease of calibration, of flexibility, and the capability to analyze a wide variety of sample forms and matrices are impressive. Additionally, besides bulk element compositions, it is also possible to collect depth profiling information with very high sensitivity. The technique has a larger dynamic range and better detection limits than the ICP techniques and does not require sample dissolution.

3.2.2 Crystallographic Characterization

The physical and electronic properties of a particular semiconductor medium are intimately dependent on its crystal structure—the study of which in known as crystallography deriving from the Greek words κρύσταλλος (*crystallon* meaning "frozen drop") and γραφω (*grapho* meaning "write"). Crystallographic investigations not only give valuable information on the quality of the growth method, but are also key to understanding transport properties, particularly for asymmetric materials. Based on such information it is possible to prescreen new materials or gauge the performance of existing materials for detector applications. Before the development of X-ray diffraction (XRD) techniques, crystallographic studies were based on geometry, which involved measuring the angles of crystal faces relative to theoretical reference axes (crystallographic axes) and establishing the symmetry of the crystal in question. Present-day crystallography is largely based on the interpretation and analysis of X-ray diffraction data, in which the crystal structure of the material acts as a diffraction grating, so-called "Bragg reflection."

3.2.2.1 Single-Crystal X-Ray Diffraction

Single-crystal X-ray diffraction, or Laue technique, is a nondestructive analytical technique commonly used for the study of crystal structures. It provides detailed information on the lattices of crystalline materials, including unit cell dimensions, bond lengths, bond angles, as well as details on site-ordering. XRD is based on the observation that, when a crystal is

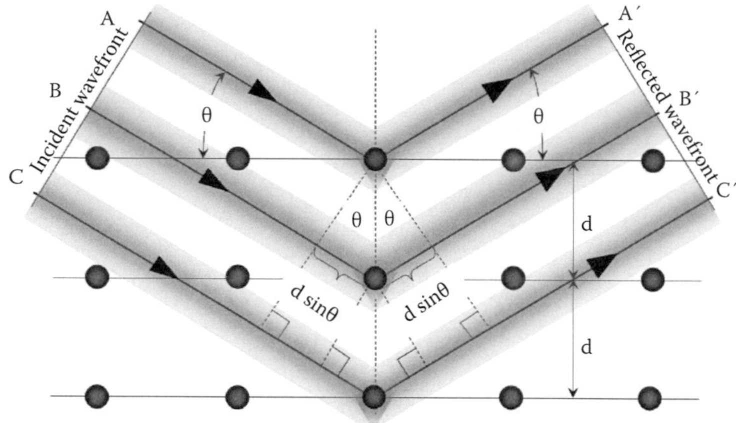

Figure 3.7 Schematic illustrating Bragg reflection. The diffracted X-rays exhibit constructive interference when the distances between paths AA′ and BB′ and CC′ differ by integer numbers of wavelengths (λ).

illuminated with X-rays of a wavelength similar to the spacing of the atomic lattice planes, intense diffracted X-rays are produced for certain incident angles. The strongest diffracted beam will leave the crystal at the same angle as the incident beam when the scattered X-rays interfere constructively, and for this to happen the differences in the travel paths must be equal to integer multiples, n, of the wavelength, where n is known as the order of the reflection. The process is illustrated in Figure 3.7. A general expression relating the wavelength of the incident X-rays, λ, angle of incidence, θ, and the spacing between the crystal lattice planes of atoms, d, is given by the Bragg relationship [4],

$$n\lambda = 2d\sin\theta \tag{3.1}$$

In single-crystal XRD measurements, the X-rays are generated by an X-ray tube, conditioned to produce monochromatic radiation, and collimated to a plane parallel beam at the sample. The sample is a single crystal of size 1 mm or larger which has been polished to a high degree. It is mounted in a Eulerian cradle which can be orientated in three directions of space. The primary beam defines the frame of reference. All accessible reflex positions can be adjusted and measured with either a two-dimensional or a single monolithic detector. By changing the geometry

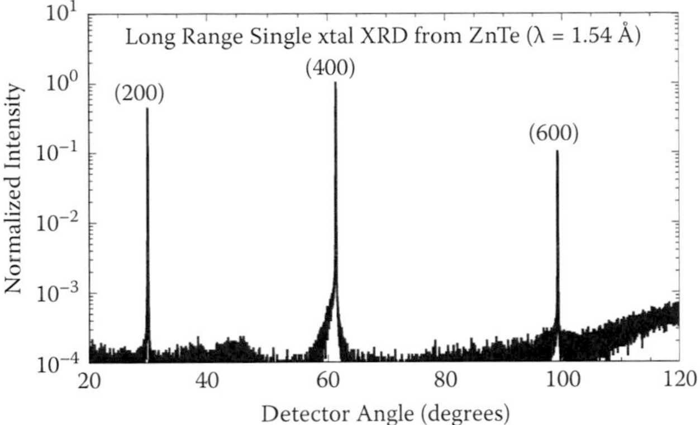

Figure 3.8 A single-crystal X-ray diffraction pattern from a ZnTe crystal. The peaks in the pattern are due to diffraction of X-rays off the various crystallographic planes of atoms within the crystal. In this case, the (200), (400), and (600) reflections are clearly evident. The angular position of each peak (2θ) can be used to calculate the spacing between planes within the structure using Equation (3.1). The derived lattice spacing was 6.18 ± 0.02 Å. The wavelength of the incident X-ray beam was 1.54 Å.

of the incident rays, the orientation of the target crystal relative to the detector, all possible diffraction directions of the lattice can be attained. Generally, for most materials the scattering angles are known, so that it is straightforward to position the detector approximately and then finely adjust the orientation of the crystal until a diffraction peak is observed. For crystalline material, the diffracted waves consist of sharp interference maxima (peaks) with the same symmetry as in the distribution of atoms. The widths of the peaks give information on the distribution of lattice plane spacing (i.e., microstrains) and on the grain size. In Figure 3.8, we show a diffraction pattern measured from a ZnTe crystal which has a cubic structure. The angular position of each peak (2θ) can be used to calculate the spacing between planes within the structure using Equation (3.1) ($d = 89.37/2\theta$).

X-ray diffraction methods are divided into single-crystal diffraction and powder diffraction—the relative advantages and disadvantages of each are summarized in Table 3.2. Single-crystal XRD is generally used to determine large-scale crystalline structures ranging from simple inorganic solids to complex molecules, such as proteins. It is an extremely precise technique; however, it is difficult to set up and is time consuming.

TABLE 3.2
Summary of the Pros and Cons of Single-Crystal and Powder XRD

Single-Crystal XRD		Powder XRD	
Pros	**Cons**	**Pros**	**Cons**
Nondestructive	Must have single polished crystal	Can be rapid (<10 min.)	Requires tenths of a gram of material
No separate standards required	Sample setup time consuming	Minimal sample preparation on spectrometer	Sample must be ground into a powder
Provides very detailed structural information	Data collection takes ~4 hours	In most cases unambiguous material identification	Must have access to a standard reference file of inorganic compounds (d-spacings, *hkl*s)
		Data interpretation relatively simple	For unit cell determinations, indexing of patterns for nonisometric crystal systems is complicated
		XRD spectrometers widely available	

Powder XRD is used to characterize crystallographic structure, crystallite size, and orientation in polycrystalline or powdered solid samples. It is widely used to identify unknown substances, by comparing diffraction data against a database maintained by the International Centre for Diffraction Data [5]. The main advantages of powder XRD over single-crystal XRD are that it easy to use, simple to set up, and gives rapid and accurate results. However, the data are more difficult to interpret.

3.2.2.2 *Powder Diffraction*

Powder XRD is perhaps the most widely used XRD technique for characterizing materials because it is rapid, easy to use, and gives accurate results. As the name suggests, the sample is a powder, consisting of fine grains of single-crystalline material. The grain size used is typically around 8 μm, and the sample volume is about 1 mm^3 or more. For a random assortment of grains, enough grains will be orientated so that an adequate number of crystallites are always in a reflection position for every single Bragg reflex. A basic powder diffractometer consists of a source of monochromatic radiation, a sample holder, and an X-ray detector situated on the circumference of a graduated circle centered on the specimen. Divergent slits located between the X-ray source and the specimen, and further divergent slits located between the specimen and the detector, limit scattered (nondiffracted) radiation, reduce background noise and collimate the radiation to produce a plane parallel beam at the sample. The detector and specimen holder are mechanically coupled with a goniometer so that a rotation of the detector through 2θ degrees occurs in conjunction with the rotation of the specimen through θ degrees, in a fixed 2:1 ratio. For typical powder patterns, data are collected at 2θ from ~5° to 70° angles. The statistics can be improved by a suitable azimuthal rotation of the sample. Powder diffraction data can be collected using either transmission or reflection geometries.

Because the crystalline domains are randomly oriented in the sample, a 2D diffraction pattern shows concentric rings of scattering peaks corresponding to the varying *d* spacing in the crystal lattice. The positions and the intensities of the peaks are used for identifying the underlying structure or phase of the material. Phase identification is particularly important. For example, the diffraction lines of graphite will be different from those of diamond, even though they both are made of carbon atoms. It is these differences which give rise to their totally different material properties. An example of a powder scan of $Cd_{1-x}Zn_xTe$ is given in Figure 3.9.

3.2.2.3 *Rocking Curve (RC) Measurements*

Another type of scan that is closely related to a θ–2θ scan is a rocking curve (RC) scan. By passing over a maxima the intensity distribution displays a narrow peak whose angular width is a measure of the crystalline quality of the material.

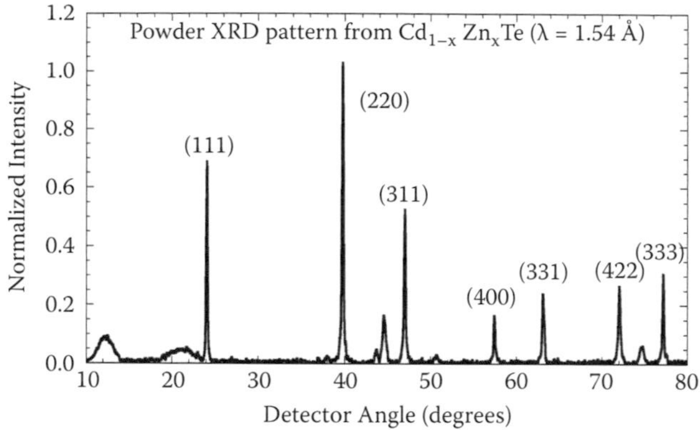

Figure 3.9 A powder XRF scan of a $Cd_{1-x}Zn_xTe$ crystal. The wavelength of the incident X-ray beam was 1.54 Å. Each of the crystal planes has been identified. The derived lattice spacing was 6.46 Å. By comparing this value with that expected from pure CdTe and using Vegard's law [6], the Zn fraction x, was determined to be 4.4%.

This peak is known as the rocking curve and is obtained in practice by keeping the scattering angle fixed and "rocking" the crystal from side to side. Rocking curves measure the amount of mosaicity in the crystal, which is an angular measure of the degree of long-range order of the unit cells. Lower mosaicity indicates better ordered crystals and hence better X-ray diffraction. From a rocking curve measurement it is possible to determine the mean spread in orientation of the different crystalline domains of a nonperfect crystal. If the crystalline particles are very small, it is possible to determine their size from a RC scan. RC scans also measure the underlying curvature in the lattice planes as well as giving indirect information on strains which manifest themselves in the broadening of the peak. In order to obtain the rocking curve, one need not perform a θ–2θ scan first, because the position for diffraction is usually well known. In this case the detector is moved to the diffraction peak. Then, data are acquired by varying the orientation of the sample by an angle Δθ around its equilibrium position, while keeping the detector position fixed. For RC scans, the detector does not need to have a small angular acceptance, because we are not measuring a scattering angle or lattice parameter. RC scans can also be very useful for determining the efficacy of crystal processing

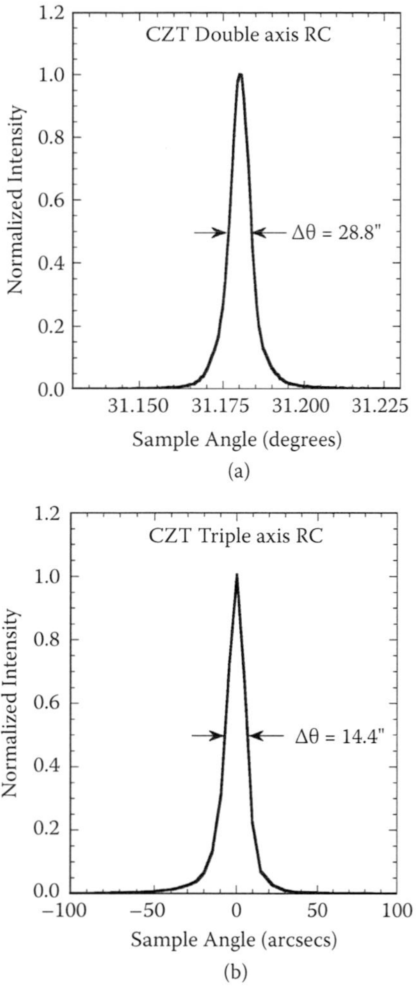

Figure 3.10 Examples of (a) double- and (b) triple-axis X-ray rocking curve (RC) scans of a $Cd_{1-x}Zn_xTe$ crystal, taken before and after various post-processing treatments have been applied to remove surface damage. The wavelength of the incident X-ray beam was 1.54 Å. Triple-axis RC is generally only used on the highest quality crystals and can give quantitative information on mosaicity and strain in the crystal.

techniques. For example, in Figure 3.10 we show double- and triple-axis XRD rocking curves for a CdZnTe crystal both before and after various postprocessing treatments have been applied to remove surface damage. Triple-axis RC is generally only used on the highest quality crystals for which intrinsic width is very small (<14 arc seconds) and can give quantitative information on mosaicity and strain in the crystal.

3.2.2.4 XRD and Detector Performance

Goorsky et al. [7] used triple-axis XRD to study the relationship between crystalline perfection and the spectral performance of radiation detectors for a number of compound semiconductors (GaAs, $Cd_{1-x}Zn_xTe$, and HgI_2). In all cases, mosaicity was found to be inversely related to detector performance—the lower the mosaicity, the better the spectral performance. In GaAs, low-angle grain boundaries were attributed to impaired detector performance, while in large HgI, detectors, detector performance was more intimately related to deviations from stoichiometry. Interestingly, in HgI_2 detectors, no differences in crystallinity were found between detectors which displayed polarization effects and those which did not.

3.2.3 Electrical Characterization

3.2.3.1 Current-Voltage (I-V) Measurements

Current-voltage measurements are used to determine the electrical characteristics of semiconductor devices and test structures by measuring the current flowing across the device as a function of applied voltage. These measurements yield information on the barrier formation, bulk resistivities, and contact performance in terms of specific resistivities. They are usually carried out on a probe station which utilizes manipulators to allow the precise positioning of thin needles onto the surface of a semiconductor device. The essential functional elements are illustrated in Figure 3.11. Depending on the accessory package, probe stations can measure currents as low as 1 FA over temperatures ranging from –70°C to +200°C or above. If the device is being electrically stimulated, the signal is acquired by the probe and displayed on an oscilloscope. Probe stations are extensively used in research and prototyping, because it is often faster and more flexible to test a new electronic device or sample with a probe station than to wire bond and package the device before testing.

Figure 3.11 A Micromanipulator 6000 series probe station. The sample is held on the central 4-inch-diameter chuck by vacuum suction. It can be raised to contact the micromanipulator needles (two shown) with a positional accuracy of a few microns using the microscope. The chuck can also rotate through 360° and can move ±2 inches in X and Y and ±0.7 inches in Z (the vertical direction).

The detector response is ultimately dependent on the physics of the metal–semiconductor interfaces and the size of the barrier height, which will determine whether it is either ohmic or Schottky. Ohmic contacts are formed when the work function of the metal contact matches the work function of the semiconductor, leaving little or no potential barrier at the interface. In this case the metal–semiconductor contact has a very low resistance which is independent of the applied voltage and may be expressed as $R = V/I =$ constant. Its current-voltage characteristic is therefore linear and symmetric about the origin. The slope of the I-V characteristic can be used to derive the resistivity of the bulk material. This quantity is important because it determines how much bias can be applied to the detector before breakdown is induced by excessive leakage current. Increased bias enhances charge collection efficiency and decreases response time. For radiation detectors, resistivities tend to be in the region of 10^6 to 10^{12} Ω cm.

Schottky contacts, on the other hand, display asymmetric current-voltage characteristics which allow high current to flow across when forward biased but block current flow under reverse bias. This behavior is controlled by bias-dependent changes of the potential barrier height in the contact region, as described in Chapter 4. The current voltage characteristic is generally described by the empirical diode equation,

$$I = I_o \left(\exp\left(\frac{qV}{nkT} \right) - 1 \right) \tag{3.2}$$

assuming that any series resistance in the device is negligible; otherwise V in the expression should be replaced by $V - IR$, where R is the series resistance. Here, I_0 is the reverse bias saturation or leakage current, T is the absolute temperature, k is Boltzmann's constant, q is the electronic charge, V is the bias across the detector, and n is the ideality factor, a dimensionless quantity introduced to include contributions from other current-transport mechanisms (see Chapter 4). The saturation current is given by

$$I_o = aA^{**}T^2 \exp\left(\frac{q\varphi_{\beta o}}{nkT} \right) \tag{3.3}$$

where, a is the anode area, A^{**} is the effective Richardson constant, and $\phi_{\beta o}$ is the barrier height. The analysis of I-V data usually proceeds with a least-squares fit to Equation (3.2), treating n and $\phi_{\beta o}$ as adjustable parameters. Most radiation detectors are operated under reverse bias to allow the application of a larger bias and, therefore, larger electric field. The slope of the reverse part of the characteristic gives the resistivity of the device. For successful operation as a radiation detector this should be $>10^6$ Ω cm.

Expression (3.2) is actually an approximation. Rhoderick [8] pointed out that such a relationship is physically implausible, because whatever mechanism is responsible for making n exceed unity must affect the reverse current, as well as the forward current. The means that the second term on the right-hand side should also contain n. Crowell and Rideout [9] have shown that the current/voltage relationship which results from tunneling through the barrier can be written in the form

$$I = I_o \exp\left(\frac{qV}{nkT}\right)\left\{1 - \exp\left(-\frac{qV}{kT}\right)\right\} \tag{3.4}$$

which has the desired property that the reverse current depends on n. Rhoderick [8] noted that Equation (3.4) is similar to current-voltage relationships resulting from other transport mechanisms, such as the voltage dependence of the barrier height and electron-hole recombination in the depletion region. Based on this, Missous and Rhoderick [10] argued that Equation (3.4) is, in fact, a generic form of current transport, and that a plot of

$$\log\left[I/\left\{1 - \exp\left(-\frac{qV}{kT}\right)\right\}\right] \tag{3.5}$$

against V should be linear for all values of V, including reverse voltages. The effect is dramatically demonstrated in Figure 3.12, in which we show current-voltage data plotted for an expitaxial Al/GaAs Schottky diode [10]. Figure 3.12a graph shows a conventional plot using Equation (3.2), while Figure 3.12b shows a plot using Equation (3.5). The success of this modified plot in linearizing the characteristic is immediately apparent, the graph being linear over the whole range from –1 V to +0.5 V. This allows a more accurate determination of the ideality factor, n, than Equation (3.2), which in turn, facilitates a more accurate calculation of the saturation current I_0 and barrier height. However, it should be pointed out that the ideality factor for this particular diode is very nearly unity. For this reason we have also plotted data from a non-epitaxial Al/GaAs diode in Figure 3.12b. For this diode we estimate n to be ~1.18. The calculated curve using Equation (3.5) overlies the curve for the epitaxial diode. The departure from ideal behavior is obvious and is predominantly due to generation/recombination in the depletion region for reverse bias. As pointed out by Missous and Rhoderick [10], it

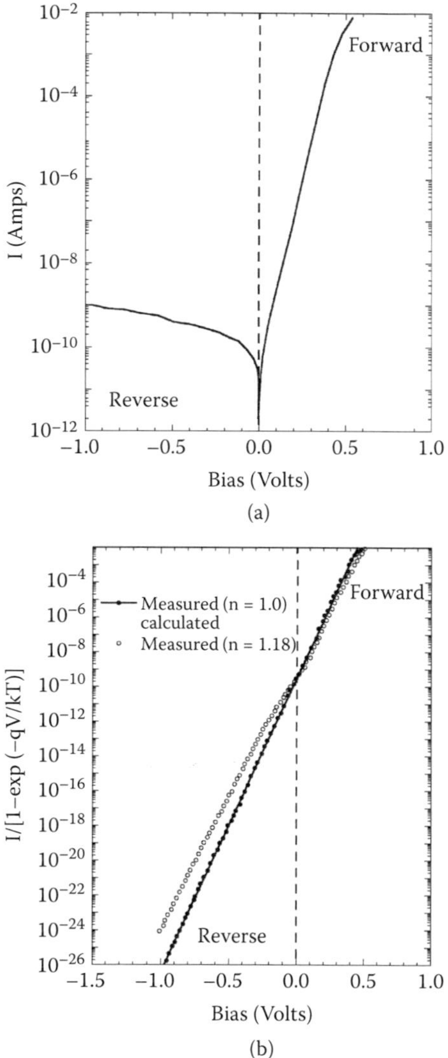

Figure 3.12 Current-voltage data for an epitaxial Al/GaAs Schottky diode, modified from Missous and Rhoderick [10]. (a) Conventional logarithmic plot of I versus V. (b) Logarithmic plot of $I/\{1 - \exp(-qV/kT)\}$ versus V. For comparison, we overplot data for a nonideal diode with an ideality factor of ~1.18. Note the calculated curve for this diode overlies the epitaxial diode calculated curve.

is probably easier to interpret the reverse portion of the characteristic with the aid of a conventional plot.

3.2.3.2 *Contact Characterization*

The formation of laterally stable and uniform contacts is of critical importance for the operation and reliability of compound semiconductor devices. In fact, the contacts are often the limiting factor affecting performance, as discussed at length in Chapter 4. In Figure 3.13 we show a typical detector construction which consists of a sandwich of a semiconductor between two metal contacts. R_c and R_{bulk} represent the resistances of the contact and the bulk semiconductor, so that the total impedance across the device is $R_{c1} + R_{bulk} + R_{c2}$. The contact resistance regions in which the contacts are formed may have a significant extent if a high degree of doping is used

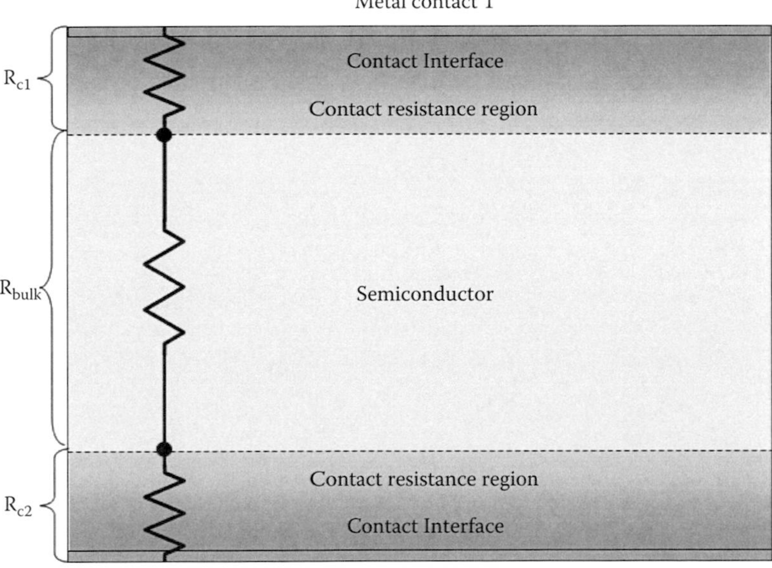

Figure 3.13 **(See color insert.)** Typical detector construction, which consists of a sandwich of a metal contact, semiconductor, and further metal contact. R_c and R_{bulk} represent the resistances of the contact and the bulk semiconductor. The contact resistance region is that region over which the contact is formed and may have a finite extent depending on how the contact was formed.

to reduce the Schottky barrier width (see Chapter 4) or if they are formed by alloying the metal with the semiconductor.

Contacts are usually characterized by their *contact resistance*, R_c, measured in ohms. This is the total resistance of the metal–semiconductor interface and is dependent on the area and the geometry of the contact. However, in reality the entire contact interface may not be available for conduction because of the physical properties of the interface, such as the barrier height of the metal, surface roughness, or preparation technique. All these factors are in principle independent of the contact geometry, but in practice may strongly influence the value of the contact resistance. Hence, a more useful figure of merit for a contact is its *specific contact resistance*, ρ_c, which is the contact resistance of a unit area contact and has units of Ω cm^2. It is particularly useful when comparing contacts of different sizes and is usually defined as the slope of the I-V curve at $V = 0$,

$$\rho_c = \left(\left. \frac{\partial J}{\partial V} \right|_{V=0} \right)^{-1} = \lim_{\Delta A \to 0} R_c \Delta A \tag{3.6}$$

where J is the current density in units of amps cm^{-2}, and A is the area of the contact. Most compound semiconductor detectors have a sandwich (contact–semiconductor–contact) structure; the contact resistance, R_c, of a contact is then given by,

$$R_c = \rho_c / A \tag{3.7}$$

Depending on the semiconductor material and on the contact quality, ρ_c can vary anywhere between 10^{-3} Ω cm^2 and 10^{-8} Ω cm^2. A typical current density in a sandwich-type device can be as high as 10^4 amps cm^{-2}. Hence, a specific contact resistance of, say, 10^{-5} Ω cm^2 (a value appropriate for LEDs and laser diodes [11]) would lead to a voltage drop of the order of 0.1 V. For the contact be ohmic in character, the voltage drop across the semiconductor must be much greater than this value. For III–V compounds, where biases can be of the order of volts, this can be a problem. However for II–VI compounds, where biases tend to be of the order of kV, it would not.

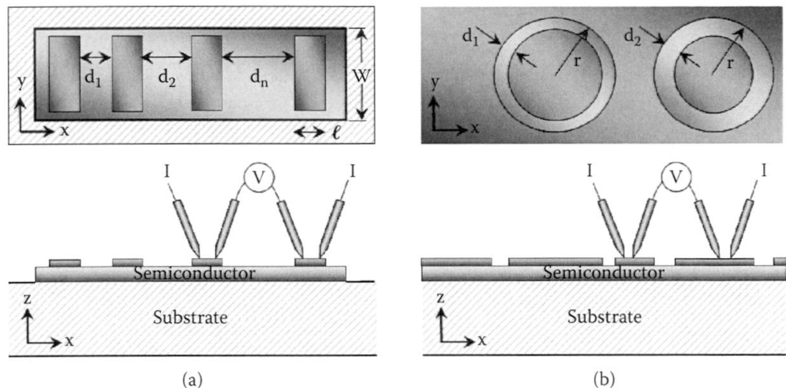

***Figure 3.14* (See color insert.)** Contact resistance test patterns (from [12]). (a) Measurement configurations for the transfer length method (TLM) and (b) circular transfer length method (CTLM). For the TLM measurement, the semiconductor has been etched away around the contacts to form a mesa in order to restrict current flow to adjacent contacts. For both TLM and CTLM, measurements are usually carried out using a four-probe technique as illustrated in the cross-sectional views.

3.2.3.3 Measuring Contact Resistance

Measuring contact resistances directly is extremely difficult because of the resistive contributions from leads, probe contact resistance, and bond wires. The transfer length method (TLM) is the classical approach for determining contact and sheet* resistances in a single-layer material. The technique involves fabricating a series of identical contact pads of width w and length l, separated by varying distances: $d_1, d_2, d_3, \ldots d_n$ (see Figure 3.14a) and applying probes to successive pairs of contacts. The contact and specific contact resistances are determined through the linear relationship between the resistance and the gap spacing between the contacts. For very small resistances, the TLM structure may require mesa etching to restrict current flow to adjacent contact pads and thus negate so-called "current crowding." One way to eliminate

* The sheet resistance (R_s) is a measure of the resistance of thin films that have a uniform thickness and is commonly used to evaluate the outcome of semiconductor processes, such as doping or metallization. Its utility is that, unlike resistivity measurements, it can be directly measured using the four-probe method and for thin films is simply related to the resistivity, $\rho = R_s/t$, where t is the film thickness.

this problem and the need for additional mesa etching is to use the circular transfer length method (CTLM). In this case, the use of concentric circular contacts ensures current confinement in the direction perpendicular to the contact; this is implicit in the design (see Figure 3.14b). In both cases, a four-point measurement* may be used to avoid the influence of external series resistances. One pair of probes carries the current and another pair sense the voltage. The measured resistance between two contacts separated by the distance d_i can be determined from the relationship:

$$R_{\text{meas}} = R_c\, x + R_s\, y \tag{3.8}$$

where R_c (Ω mm) is the contact resistance, R_s (Ω/□) is the sheet resistance† of the semiconductor, and x and y depend on the geometry of the contact. For a TLM structure,

$$x = 2/w \tag{3.9}$$

and

$$y = d_i/w \tag{3.10}$$

where w is the width of the pads. For a CTLM structure,

$$x = \frac{1}{2\pi} \ln\left(\frac{1}{r} + \frac{r}{r - d_i}\right) \tag{3.11}$$

and

$$y = \frac{1}{2\pi} \ln\left(\frac{r}{r - d_i}\right) \tag{3.12}$$

Figure 3.15 illustrates graphically the measured resistance as a function of distance d_i. Fitting Equation (3.8), we can derive the values of R_s and R_c from the slope and y-axis intercept of the curve, respectively. The specific contact resistance, ρ_c (Ω cm^2) is given by,

$$\rho_c = -\frac{R_c^2}{R_s} \tag{3.13}$$

* Originally developed by Wenner [13] in 1915 to measure the Earth's resistivity.

† Here given in units of ohms per square (which is used exclusively for sheet resistance precisely because it cannot be misinterpreted for bulk resistance).

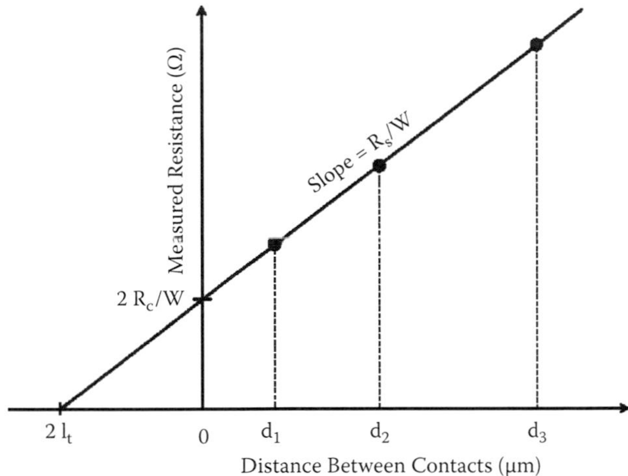

Figure 3.15 Evaluation of the contact and sheet resistances using TLM and CTLM measurements.

The intercept of the fitted line* with the x-axis is equal to twice the transfer length l_t (also known as effective contact length), which is defined by,

$$l_t = R_c / R_s \tag{3.14}$$

and is a measure for the "ohmic quality" of the metal contact. A shorter l_t means a better ohmic contact. Note, the validity of the above analysis depends on the assumption that the transfer length l_t is much smaller than the contact length l [14].

3.2.3.4 Capacitance-Voltage (C-V) Measurements

This technique is used with detectors that utilize a metal–semiconductor junction (Schottky barrier) or a p-n junction in which a depletion region is created. In normal operation, this region is empty of conducting electrons and holes, but may contain ionized donors and electrically active defects or traps. The depletion region with its enclosed ionized charges behaves

* Note: for CTLM structures, Marlow and Das [14] have pointed out that effects of finite contact radii and nonzero metal resistance can introduce large errors in the determination of the contact resistance if a linear fit is used to represent the voltage drop versus gap length data.

like a capacitor. By varying the voltage across the junction it is possible to vary the depletion width, w_d. The dependence of the depletion width upon the applied voltage provides information on the semiconductor doping profile and electrically active defect densities. Measurements are usually carried out on a probe station and may be done either using DC alone or DC together with a small amplitude AC signal.

Consider a Si p-n diode with $N_a = 10^{16}$ cm^{-3}, $N_d = 10^{17}$ cm^{-3}, and area 10^{-4} mm^2. The space charge Q_{ac} per unit area of the depletion region under the influence of an applied bias V is given by,

$$Q_{ac} = qN_d w_d = \sqrt{2q\varepsilon N_d (V_{bi} - V - kT/q)} \tag{3.15}$$

where w_d is the depletion layer width, ε is the permittivity of semiconductor, and V_{bi} is the built-in potential, defined in Chapter 4. The depletion layer capacitance per unit area is given by

$$C = \frac{|\partial Q|}{\partial V} Q_{ac} = \sqrt{\frac{q\varepsilon_a N_d}{(V_{bi} - V - kT/q)}} = \frac{\varepsilon}{w} \tag{3.16}$$

Thus, C is related simply to the number of carriers and depletion layer width.

In Figure 3.16 we show a plot of capacitance measured with a capacitance meter versus applied bias, from which we see that the capacitance does indeed drop as the square root of $|V - V_{bi}|$ as the depletion region width increases.

Equation (3.16) can be written in the form

$$\frac{1}{C^2} = \frac{2(V_{bi} - V - kT/q)}{q\varepsilon N_d} \tag{3.17}$$

In the case of uniform doping concentration, N_d is constant within the depletion region, and a straight line results from plotting $1/C^2$ against V (see Figure 3.16, right ordinate). The slope of this line is obtained by differentiating Equation (3.17) with respect to V,

$$\frac{d(1/C^2)}{dV} = -\frac{2}{q\varepsilon N_d} \tag{3.18}$$

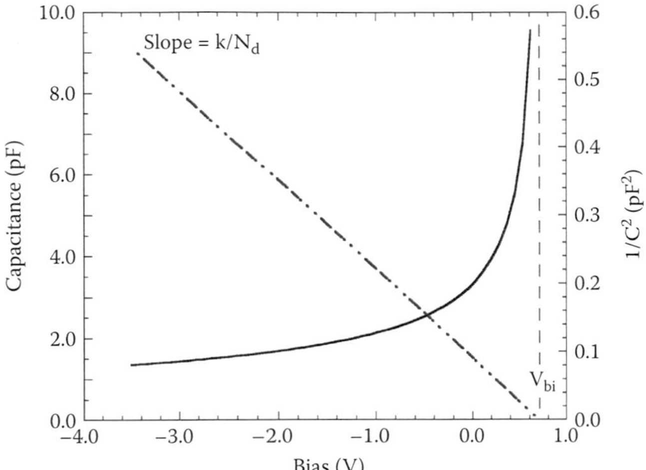

Figure 3.16 Capacitance (left ordinate—solid line) and $1/C^2$ (right ordinate—dashed line) versus bias voltage of a p-n diode with $N_a = 10^{16}$ cm^{-3}, $N_d = 10^{17}$ cm^{-3}. The area of the diode is 10^{-4} cm^2.

from which the doping concentration can be calculated; thus,

$$N_d = -\frac{2}{q\varepsilon \frac{d(1/C^2)}{dV}} \tag{3.19}$$

The intercept of this line on the horizontal axis occurs at V_{bi}, the built-in potential (see Chapter 4). In this case, $V_{bi} = +0.7$ eV. If N_d is not constant; the doping profile can be deduced from iterating Equation (3.19) with bias. The depletion layer width at a particular bias is simply given by

$$w = \varepsilon / C_d \tag{3.20}$$

Here C_d is the detector capacitance at the operational bias V.

3.2.4 Electronic Characterization

The carrier properties, particularly the transport properties, are probably the most important intrinsic parameters of a semiconducting material and dictate the type of detector and

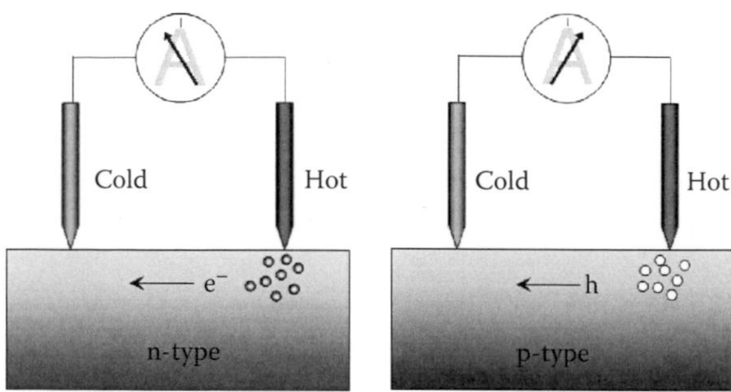

***Figure 3.17* (See color insert.)** Illustration of the hot probe technique for determining the majority carrier type in semiconductors. Carriers diffuse more rapidly near the hot probe. This leads a flow of majority carriers away from the hot probe and a resultant electrical current toward (p-type) or away from (n-type) the hot probe.

application that it can reasonably fulfill. For example, materials with poor transport properties would not make planar detectors with good stopping power, because if the active depth is too thick the generated charge due to interacting particles would not be transported to the electrodes and therefore collected.

3.2.4.1 Determining the Majority Carrier

In some materials the majority carrier may not be known. This information is crucial when selecting a contact material. The hot or thermoelectric probe technique is a simple method for determining whether a semiconductor is n-type or p-type. Two probes make contact with the semiconductor surface as illustrated in Figure 3.17. One is heated to 25°C to 100°C greater than the other. At the hot probe, the thermal energy of the majority carrier is higher than at the cold probe, so carriers will tend to diffuse away from the probe, driven by the temperature gradient. As they diffuse away from the hot probe, they leave behind the oppositely charged, immobile donor atoms, which results in a current flow toward the hot probe for p-type material and away from the hot probe for n-type material. Thus, a measurement of either the polarity of the short-circuit current or the open circuit voltage reveals the material type. The hot probe technique is most effective over the 10^{-3} to 10^{3} Ω cm resistivity range.

3.2.4.2 *Determining Effective Mass*

The effective mass of carriers can be determined by cyclotron resonance measurements [15]. In this experiment the frequency is measured at which there is strong attenuation of electromagnetic signals passing through a sample due to absorption between Landau levels created by a magnetic field. If the resonance frequency is ω, the effective mass is simply given by

$$m^* = qB/\omega \tag{3.21}$$

where B is the applied magnetic field.

3.2.4.3 *The Hall Effect*

The Hall effect measurement is a widely used technique for determining basic carrier properties, such as majority carrier type, concentration, and mobility. It is named after Edwin Hall, who in 1879 placed a thin layer of gold in a strong magnetic field. He connected a battery to the opposite sides of this film and measured the current flowing through it. He discovered that a small voltage appeared across this film which was proportional to the strength of magnetic field multiplied by the current [16]. Hall effect measurements commonly use two sample geometries: (1) a long, narrow Hall bar and (2) the nearly square or circular van der Pauw configuration [17]. Each has advantages and disadvantages.

To understand the Hall effect, consider the rectangular sample geometry shown in Figure 3.18. A voltage V_x is applied across the sample, and a magnetic field B_x is applied normal to the sample. Electrons and holes (shown by filled and open circles, respectively) flowing in the semiconductor will experience a force, bending their trajectories so that they build up on one side of the sample, creating a potential, V_H, across it, the so-called *Hall voltage*, as illustrated in Figure 3.18. On the assumption that all the conduction electrons have the same

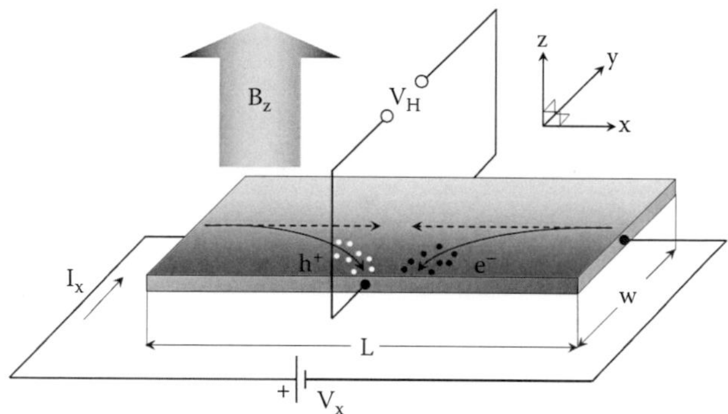

***Figure 3.18* (See color insert.)** Schematic illustrating the sign convention and terminology for the Hall effect.

drift velocity v_n and the same relaxation time τ_e, the resulting Lorentz force [18] acting on any electron is given by:

$$\vec{F} = -q(\vec{\varepsilon} + \vec{\upsilon}_n \times \vec{B}) \tag{3.22}$$

$$\vec{F} = m_e \frac{d\vec{\upsilon}_n}{dt} + m_e \frac{\vec{\upsilon}_n}{\tau_c} \tag{3.23}$$

where $m_e\upsilon_n/\tau_e$ is a damping term. In the steady state, the derivative of υ_n in Equation (3.23) is zero. Combining Equations (3.22) and (3.23) and using the relation,

$$\vec{J} = -qn_o\vec{\upsilon}_n \tag{3.24}$$

for electrons, we get:

$$\vec{\varepsilon} = \frac{\vec{J}}{q\mu_n n_o} + \frac{\vec{J}}{qn_o} \times \vec{B} \tag{3.25}$$

where $\mu_n = q\tau_e m_e$ is the mobility. Considering that $B_x = B_y = 0$ and $J_y = J_z = 0$, Equation (3.25) can be written as two scalar equations:

$$\varepsilon_x = J_x/\sigma \tag{3.26}$$

and

$$\varepsilon_y = -\frac{J_x B_z}{q n_o} = -\mu_n B_z \varepsilon_x \tag{3.27}$$

where $\sigma = q\mu_n n_o$ is the conductivity and Equation (3.26) is Ohm's law. Equation (3.27) expresses the fact that along the y direction the force on one electron due to the magnetic field ($q\mu_n B_z \varepsilon_y$) is balanced by a force ($-q\varepsilon_y$) due to the Hall field. Equation (3.27) is generally witten as

$$\varepsilon_y = R_H J_x B_z \tag{3.28}$$

where $R_H = -1/qn_{\mathrm{eo}}$ is the Hall constant. For a p-type semiconductor,

$$\varepsilon_y = R_H J_x B_z = \mu_p B_z \varepsilon_x \tag{3.29}$$

with, $R_H = 1/qn_{\mathrm{ho}}$ where n_{ho} is the equilibrium concentration of holes in the sample. It is thus seen that R_H has a negative sign for electrons and a positive sign for holes. For the rectangular bar geometry shown in Figure 3.18, the Hall constant and the Hall mobility can be expressed by:

$$R_H = \frac{V_H d}{I_x B} \tag{3.30}$$

and

$$\mu_H = \frac{V_H L}{V_x B w} \tag{3.31}$$

where V_H is the measured Hall voltage, V_x is the applied voltage along the length, L, and I_x is the current. For a semiconductor with comparable concentrations of electrons and holes, R_H is given by

$$R_H = -\frac{\left(\mu_h^2 n_h - \mu_{ne}^2 n_o\right)}{q(\mu_p n_{ho} - \mu_n n_{eo})^2} \tag{3.32}$$

and the interpretation of R_H and μ_H is ambiguous. However, if the material is strongly extrinsic, Equation (3.31) reduces to

$$R_H = 1/qn_{eo} \quad \text{and} \quad \mu_H = \mu_n \tag{3.33}$$

for a highly n-doped material, and

$$R_H = 1/qn_{ho} \quad \text{and} \quad \mu_H = \mu_p \tag{3.34}$$

for a highly p-doped semiconductor. Thus, in summary:

1. The results are only simply interpreted for strongly extrinsic material.
2. The sign of R_H gives the majority carrier type.
3. The measured values of R_H and μ_H determine the majority carrier concentration and the mobility, respectively.

3.2.4.3.1 Hall Effect Measurements

Practical layouts and contact configurations for Hall measurements are shown in Figure 3.19. The simplest arrangement for measuring the Hall voltage is the rectangular geometry shown in Figure 3.19a. However, a number of spurious voltages are included in this measurement due to the finite contact size. These spurious voltages are eliminated if four readings are taken by reversing the direction of the bias current I_x and the magnetic flux B. The true Hall voltage is then obtained by taking the average of the four readings. The contacts used for measuring the Hall voltage should be infinitesimally small, so that they do not perturb the current flow. In practice, the

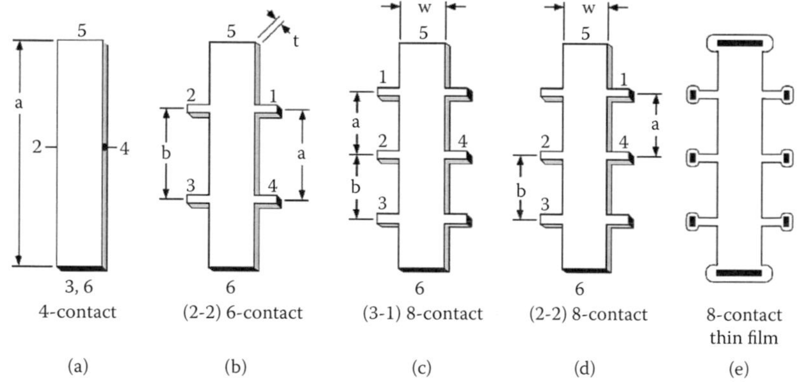

Figure 3.19 Hall measurement layouts and contact configurations (from [19], courtesy of Lake Shore Cryotronics Inc.). For measurements, the magnetic field is applied alternately into and out of the pads.

bridge-shaped geometry shown in Figures 3.19b,c,d,e is often used to reduce the distortion of the current. The projecting pads on the pattern allow a large area to be used for contacting the sample without a severe distortion of current flowing through it. The Hall voltage is measured between contacts 1 and 2 and then 3 and 4. The average is then taken as the final value.

In some cases, it may not be convenient to cut the specimen in the form of a rectangular bar, in which case the van der Pauw geometry is often used.

3.2.4.3.2 Van der Pauw Method

Van der Pauw [17] showed how to determine the resistivity, carrier concentration, and mobility of an arbitrary, flat sample as shown in Figure 3.20, if the following conditions are met:

- The sample thickness must be much less than the width and length of the sample.
- The contacts are at the circumference of the sample (or as close to it as possible).
- The contacts are sufficiently small. Any errors given by their nonzero size will be of the order D/L, where D is the average diameter of the contact and L is the distance between the contacts.
- The sample is of uniform thickness.
- The surface of the sample does not have isolated holes or islands.

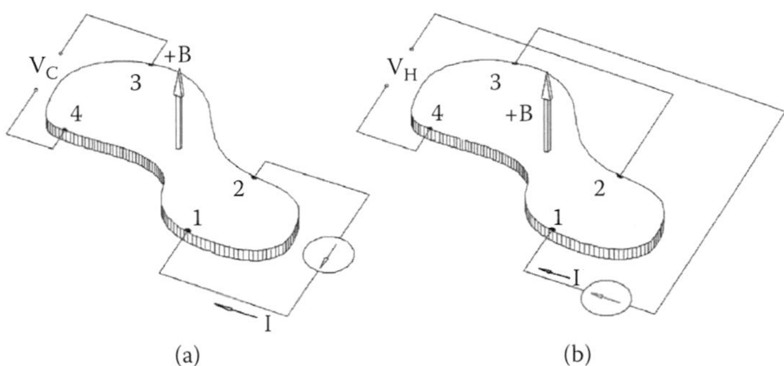

Figure 3.20 Van der Pauw arrangement [17] for (a) measuring resistivity and (b) the Hall coefficient in an arbitrarily shaped sample (from [19], courtesy of Lake Shore Cryotronics Inc.).

The method involves passing current through two adjacent points on the perimeter of the shape and measuring the voltage across two other points on the perimeter of the sample. With reference to Figure 3.20a, the resistivity of the sample is given by

$$\rho = -\frac{\pi t}{\ln(2)} \frac{V_{43}}{I_{12}} + \frac{V_{14}}{I_{23}} \tag{3.35}$$

where t is the sample thickness, V_{23} is $V_2 - V_3$, and I_{12} refers to the current that enters the sample through contact 1 and leaves through contact 2. Two voltage readings are required with the van der Pauw sample, whereas the resistivity measurement on a Hall bar requires only one. This same requirement also applies to Hall coefficient measurements, so that equivalent measurements take twice as long with van der Pauw samples.

In the presence of a magnetic field normal to the sample surface, a current I_{ac} is established between two opposite contacts (*1* and *3*), and the voltage is measured between the contacts *2* and *4*, as shown in Figure 3.20b. The resistance defined as

$$R_{13,24} = V_{24}/I_{13} \tag{3.36}$$

is first measured with the magnetic field, $-B$, $R_{13,24}$ $(+B)$ and then with the opposite direction of the field, $-B$, $-R_{13,24}$ $(-B)$. If all the variations in the current and magnetic field polarity are considered, then the Hall constant is given by:

$$R_H = (d/8B)\,(R_{13,24}(+B) - R_{13,24}(-B) + R_{24,13}(+B) - R_{24,13}(-B))$$
$$+ R_{31,42}(+B) - R_{31,42}(-B) + R_{31,42}(+B) - R_{31,42}(-B)) \tag{3.37}$$

and the Hall mobility

$$\mu_H = -\frac{R_H}{\rho_o} \tag{3.38}$$

where ρ_o is the zero-field resistivity given by Equation (3.35).

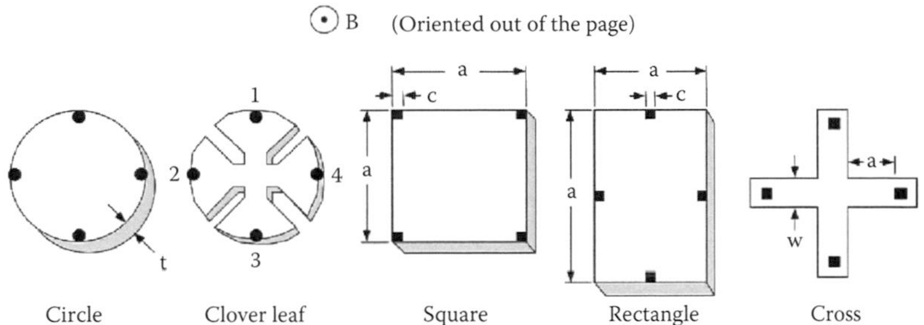

Figure 3.21 Common van der Pauw geometries (from [19], courtesy of Lake Shore Cryotronics Inc). The cross appears as a thin film pattern, and the others are bulk samples. Note: the contacts are shown in black, and the magnetic field is out of the page.

In practice, arbitrary shapes are not used because there are a number of "standard" geometries as shown in Figure 3.21. Because they are all symmetrical about two axes (see Figure 3.21), the van der Pauw relation simplifies to

$$R_s = \pi\,(R_{34,12} + R_{13,24})/2\ln 2 \tag{3.39}$$

where,

$$R_{34,12} = (V_3 - V_4)/I_{12} \tag{3.40}$$

and

$$R_{13,24} = (V_1 - V_3)/\mathrm{I}_{24} \tag{3.41}$$

Here the numbering of contacts goes from left to right and top to bottom. Compared to the conventional Hall measurements using bar geometries, the van der Pauw method has the advantage that only four contacts are required, simple geometries can be used, and there is no need to measure sample widths or distances between contacts. However, its disadvantages are that measurements take twice as long, and errors due to contact size and placement can be significant when using simple geometries.

3.2.5 Evaluating the Charge Transport Properties

Of the basic transport properties, carrier mobilities and mobility-lifetime or mu-tau ($\mu\tau$) products are the most useful and form the starting point for choosing a material for a particular energy range or application. They can be estimated as follows. Consider a simple planar detector of thickness, d. A bias, V, is applied across the detector giving a constant and uniform electric field, $E = V/L$, throughout the device. For the electric field strengths usually encountered in such detectors, the drift velocity v_d of the carriers is proportional to the electric field strength, E, with the constant of proportionality being the mobility μ; thus $v_d = \mu E$. The instantaneous current is $I = qnv_d$, where q is the charge on an electron and n is the number of carriers. The duration of the transient current pulse is determined by the distance the carriers must travel. Note that a consequence of the Shockley–Ramo theorem [20,21] is that charge induction on the contacts begins as soon as the carriers are created (or move) and not when the charges are deposited on the electrodes. There will clearly be two distinct current pulses, one from holes, I_h, and one from electrons, I_e. For most materials, the electrons will produce a much higher current for a shorter time, because they generally have much higher mobilities ($t \propto \mu^{-1}$ and $I \propto \mu$).

3.2.5.1 Estimating the Mobilities

The use of the Hall method for determining mobilities is not suitable for highly insulating materials because of the difficulty in measuring nA or smaller currents. Various other experimental techniques can be used. Most are based on time-of-flight (TOF), such as the canonical Haynes–Shockley experiment [22] which measures the transit times of injected carriers across a bar of semiconducting material. However, while this technique has a high pedagogical value it requires complicated sample preparation and a dedicated measurement system. Consequently it is little used outside undergraduate teaching laboratories.

For an existing detection system, a simpler procedure can be employed in which a detector is stimulated to produce transient ionization events near one contact and the response to the event measured on the other contact. Analysis of the transient pulse shape then provides information on the carrier transit time and also on trapping times. The stimulation can

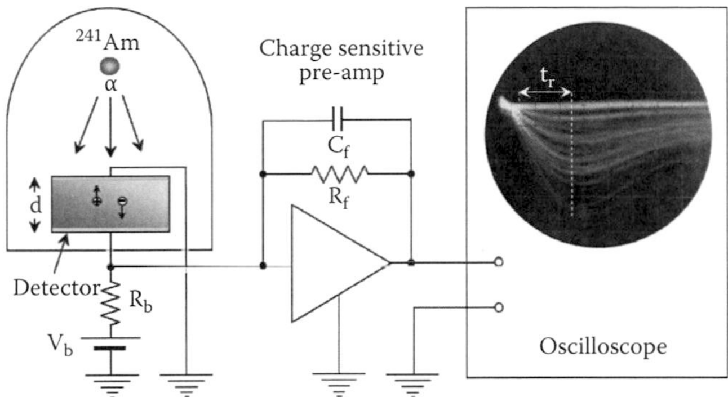

Figure 3.22 Experimental setup for determining carrier mobilities in a detector by the time-of-flight method. The apparatus is set up for electron measurements. For hole measurements, the source should be placed under the detector or the charge signal on the cathode monitored and the bias reversed. For materials suitable for detector applications, rise times will typically be in the range 0.1 to 10 μs.

be in the form of laser pulses, alpha particles, or gamma rays. An attractive feature of this technique is that the motion of both electrons and holes can be observed separately by simply reversing the polarity of the applied bias or directing the ionization to the other contact. For example, with reference to Figure 3.22, electron mobilities can be determined by measuring the anode signal rise time, t_{re}, on a digital oscilloscope for near cathode events. For trapping lengths greater than the detector thickness, the rise time of signal reflects the carrier drift time, thus $t_r \cong t_d$, across the detector. Thus, since $v_d = \mu E$, $v_d = d/t_d$ and $E = V/d$, then,

$$\mu_e = \frac{d^2}{V_b t_{de}} = \frac{d^2}{V_b t_{re}} \tag{3.42}$$

In a sense, this method is a derivative of the Haynes–Shockley technique [22], but less complicated. While inherently simple, it has several drawbacks—the most notable being the difficulty in determining the start and end times of the events for long transit times, particularly if thermal detrapping becomes significant. The latter can introduce a slow exponential component onto the rise time signal. Generally, most experimenters use the 10% and 90% or 10% and 95%

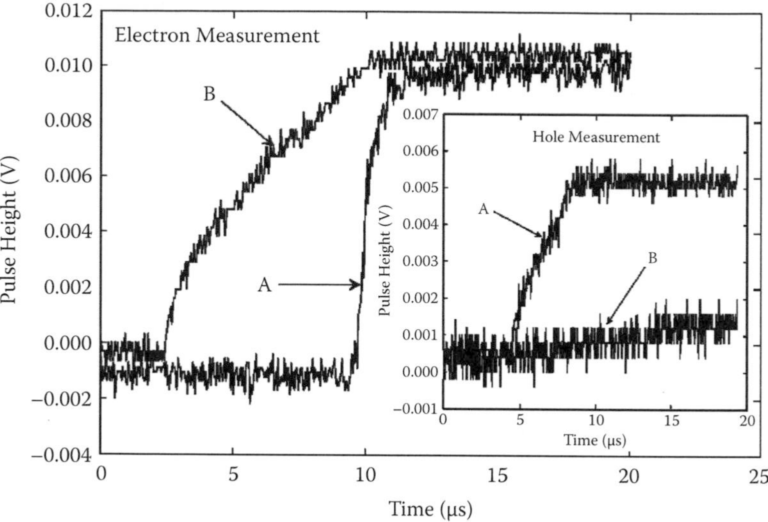

Figure 3.23 Measured drift time of electrons for a near cathode event in a 1-cm-thick HgI_2 pixelated detector (adapted from [23], © 2003 IEEE). Waveform *A* is the anode (pixel) signal, while waveform *B* is the cathode signal. Note that the pixel signal is generated only when the electrons are close to the pixel. From the figure, the electron drift time is estimated to be 7 μs. The inset shows pulse waveforms from an event near the anode. In this case, holes generate the signal. However, because they cannot travel the full thickness of the detector, the signal from the cathode is indistinguishable from noise (curve B), while a significant charge can be induced between the anode plane and pixel (curve A).

amplitude points when determining rise times. Even so, 5% accuracies in the electron mobility determination can be achieved.

By way of practical example, Figure 3.23 shows the anode and cathode signals collected from a 1 cm², 1-cm-thick HgI_2 pixel detector [23] following stimulation by gamma radiation. The detector consists of four 1 mm² pixels patterned into a large anode plane—each pixel separated from the plane by 1 mm. The bottom of the detector consists of a single 1 cm² planar cathode. From Figure 3.23, we derive an electron drift time of ~7 μs. The external electric field is 2500 Vcm^{-1} from which we estimate the electron mobility, μ_e, to be 60 $cm^2V^{-1}s^{-1}$. In principle, hole mobilities can be estimated in a similar manner by measuring the rise time of the cathode signal for a near anode event. However, because hole mobilities are generally much poorer, this can only be achieved realistically using thin detectors or between test structures on thick detectors. For the thick detector considered above, the hole signal from the cathode is indistinguishable from the noise, as illustrated by curve B in the inset of Figure 3.23. Fortunately, for pixel detectors hole mobility measurements can usually be carried out between pixels and/or guard rings. This is illustrated by the waveform shown by curve A of the inset. This was obtained by applying a positive bias to the anode plane relative to a pixel and recording the pixel's response. From the figure, we deduce a rise time of 5 μs. In this case the external electric field was set to 2500 Vcm^{-1}, which for a 1 mm gap, yields a hole mobility of $\mu_h = 8$ cm^{2}V^{-1}s^{-1}.

It is important to note that the mobilities derived from drift or conductivity measurements, μ_d, can be quite different from those derived from Hall effect measurements, μ_H. The ratio μ_H/μ_d is usually close to unity for direct bandgap semiconductors, but can be greater than unity for indirect bandgap materials.

3.2.5.2 *Estimating the Mu-Tau (μτ) Products*

The mu-tau products of a material are usually derived using the Hecht equation [24], which relates the electron and hole mean drift lengths to the amount of charge collected from the electrodes—usually expressed in terms of the charge collection efficiency (CCE). The CCE is defined as the ratio of the induced charge at the contact, Q, divided by the total charge created as electron–hole pairs in the material, Q_o. Assuming a uniform field across the detector and negligible trapping, the CCE can be expressed as

$$CCE = \frac{Q}{Q_o} = \frac{\lambda_e}{d}\left[1-\exp\left(-\frac{(d-x_o)}{\lambda_e}\right)\right]+\frac{\lambda_h}{L}\left[1-\exp\left(-\frac{x_o}{\lambda_h}\right)\right] \tag{3.43}$$

where d is the detector thickness, x_o is the distance from the irradiated electrode to the point of charge creation, and λ_e and λ_h are the carrier drift lengths in the applied electric field, E, given by $\lambda_e = \mu_e\tau_e E$ and $\lambda_h = \mu_h\tau_h E$. Here, μ_e and μ_h are the electron and hole mobilities, and τ_e and τ_h are the corresponding lifetimes. It follows from Equation (3.43) that the CCE depends not only on λ_e and λ_h, but also on the location where the charge was created. In the case when the interaction depth of the incident event is very close to one of the contacts, the induced signal will be due almost exclusively to the drift of one of the carriers. For example, in a material where electrons are the majority carrier, the signal will be only sensitive to electrons for events close to the cathode ($x_o \sim 0$). Equation (3.43) then reduces to

$$CCE = \frac{Q}{Q_o} = \frac{\mu_e \tau_e V}{d^2}\left[1 - \exp\left(-\frac{d^2}{\mu_e \tau_e V}\right)\right] \quad (3.44)$$

Practically, the mu-tau product is then derived by recording the collected charge (generated by laser light, low-energy X-ray or alpha particles) as a function of applied bias and best-fitting the results to Equation (3.44) with $\mu_e\tau_e$ the fitting parameter.

In Figure 3.24 we show a typical Hecht plot obtained from a 20 mm^2, 0.5-mm-thick, TlBr planar detector measured using ^{241}Am alpha particles [25]. Pulse height spectra were obtained from −10 V to −350 V and the electron mobility-lifetime

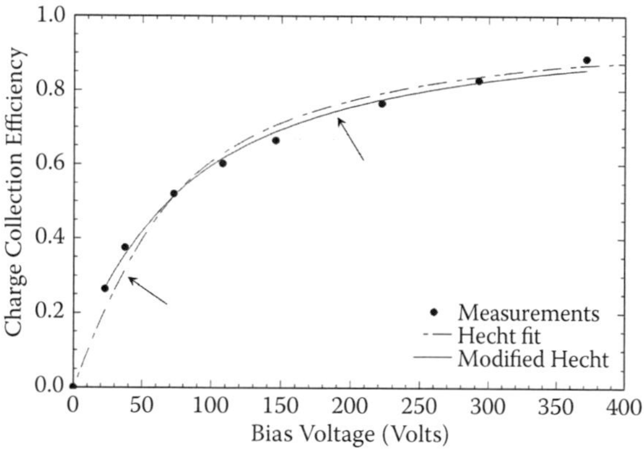

Figure 3.24 Charge collection efficiency (CCE) versus applied bias relationship for a 20-mm^2, 0.5-mm-thick TlBr planar detector. From a best fit of Equation (3.44), the mobility-lifetime product of electrons was estimated to be 2.8×10^{-3} cm^2/V (adapted from [25], © 2009 IEEE). The deviations from the fitted curve indicated by the arrows are most likely due to the assumptions implicit in the Hecht equation—no de-trapping or surface recombination. The dashed line shows the improvement in fit using a modified Hecht equation in which the fitted bias is offset by a delta amount.

product derived by fitting Equation (3.44). A value for $\mu_e\tau_e$ of $(2.8 \pm 0.2) \times 10^{-3}$ cm^{2}V^{-1} was obtained. Based on the measured electron mobility of (1040 ± 20) cm^2 V^{-1}s^{-1}, the effective electron lifetime* in this material is calculated to be τ_e = 6 μs. From Figure 3.24, we see that the curvature of the best-fitted Hecht curve is different from that defined by the data points as indicated by the arrows. Veale et al. [26] have pointed out that, for data showing enhanced CCE values at low biases, a better fit to the data can be achieved by modifying the Hecht equation to include a voltage offset term, such that $V = V_b - V_o$, where V_b is the applied bias and V_o is a constant offset voltage. This is equivalent to the intercept of the x-axis at CCE = 0. Physically, it can be thought of as an internal voltage of opposite polarity, produced by a polarization field within the device. The dashed line in Figure 3.24 shows a best-fit Hecht function with a voltage offset which is clearly a better fit the data. In this case, the derived $\mu_e\tau_e$ is $(2.34 \pm 0.03) \times 10^{-3}$ cm^{2}V^{-1}.

In principle, the mu-tau product for holes can be determined by carrying out the same analysis as for the electrons. However, it is not always possible to isolate the hole signal because of its much poorer mu-tau product. This is illustrated in Figure 3.25 in which we show measured energy-loss spectra in a 4-mm-thick HgI_2 detector [23] from both the anode and cathode. It can be seen that the anode hole signal peak appears at a much lower channel number than the cathode electron signal due to significant hole trapping. This means that, in practice, there will be an insufficient range of peak-channel positions with bias to fit a Hecht function. In this situation, the hole mu-tau can be determined using the relative electron and hole peak pulse height distributions at the nominal bias. The hole mu-tau product can then be calculated from the ratio

$$\frac{Q_a}{Q_c} = \mu_e\tau_e \frac{\left[1 - \exp\left(-d^2/\mu_e\tau_e V\right)\right]}{\left[1 - \exp\left(-d^2/\mu_h\tau_h V\right)\right]} \tag{3.45}$$

where the ratio of Q_a and Q_c is derived from the ratio of the relevant pulse height peak channels shown in Figure 3.25.

* In effect, the trapping time.

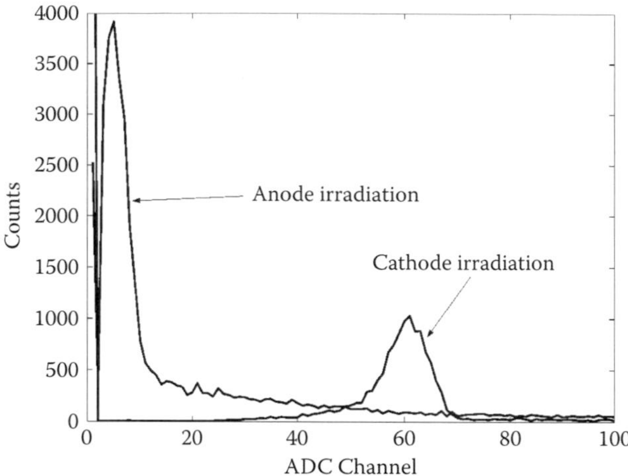

Figure 3.25 Anode and cathode pulse height distributions from a 4-mm-thick HgI_2 pixel detector (from [23], © 2003 IEEE). The effect of significant hole trapping is observed in the much reduced hole pulse height distribution.

Rearranging Equation (3.45) we get

$$\mu_h \tau_h = \frac{-d^2}{V\, In\left[1 - \left(\dfrac{Q_a}{Q_c}\right)\mu_e \tau_e\right]} \tag{3.46}$$

Alternately, when no hole peak can be discerned, it is still possible to evaluate the hole mobility-lifetime product using the average charge collection efficiency model of Ruzin and Nemirovski [27]. Consider the uniform illumination of the cathode by photons. For a high enough energy X-ray source, the electron and holes will be generated uniformly throughout the detector volume, and both will contribute to the signal. The average charge collection efficiency is given by:

$$\langle Q \rangle = \frac{N_0 q}{d(1-\exp(-Md))}\left\{(\lambda_e + \lambda_h)(1-\exp(-Md)) + \frac{\lambda_e M \exp(-d/\lambda_e)}{M - 1/\lambda_e}\right.$$
$$\left. \times (\exp(-d(M - 1/\lambda_e)) - 1) + \frac{\lambda_h M}{M + 1/\lambda_h}(\exp(-d(M + 1/\lambda_h)) - 1)\right\} \tag{3.47}$$

where M is the absorption coefficient of the photons of energy E, d the detector thickness, λ_e the mean free path for electron ($=\mu_e\tau_e$E), and λ_h the mean free path for hole ($=\mu_h\tau_h$E). The average charge deposited on the electrode can be determined from the mean pulse heights of events as a function of bias. By fitting Equation (3.47) to the data and fixing the mu-tau products for the electrons derived from a Hecht analysis, it is possible to extract the hole mobility-lifetime product, $\mu_h\tau_h$.

3.2.5.3 Limitations of the Hecht Equation

While the Hecht procedure is relatively straightforward, it suffers from several limitations in that it assumes (a) a uniform internal electric field, (b) carrier trapping is permanent (that is, no de-trapping), and (c) surface recombination effects are negligible. In polarizable materials these may not be valid, and the net effect of a Hecht analysis is to underestimate mu-tau products by factors of 3 or more. For II–VI materials, it is now widely recognized that highly localized defects on the surface are extremely efficient recombination centers and can be a dominant mechanism for controlling carrier lifetime. This can lead to a significant decrease in spectroscopic performance and affect a Hecht analysis by reducing the effective carrier lifetimes. The magnitude of the effect is also dependent on how the crystal was processed during detector fabrication (i.e., contacting, mechanical polishing, chemical etching, and passivation) and as a result, resistivities (of which recombination is a manifestation) can vary by many orders of magnitude. As a result, conditioning of the surfaces has remained little more than a “black art” and is not generally considered in the data analysis. Thus, many reported mobility-lifetime values must be regarded with reservation, since both mu-tau products and surface recombination velocities are strongly dependent on surface treatment.

As an example of the importance of considering surface effects, Figure 3.26 shows the electron charge collection efficiency as a function of bias voltage for a $10 \times 10 \times 2$ mm^3 CdZnTe detector [28]. The sample was polished and rinsed in methanol, treated in a standard 5% bromine-in-methanol

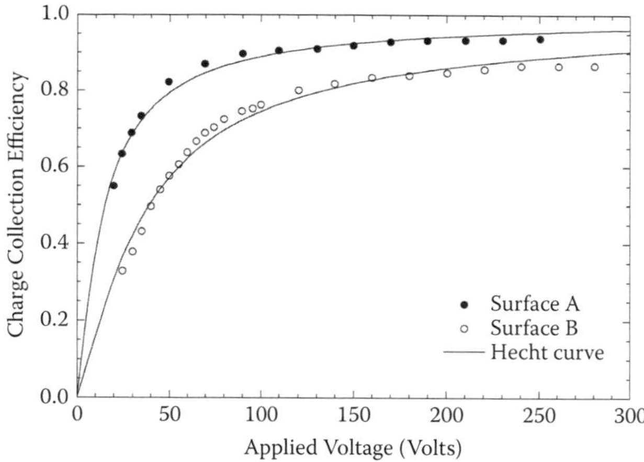

Figure 3.26 Charge collection efficiencies measured for two sides of a CdZnTe sample as a function of bias for the 59.54 keV photopeak of ^{241}Am (from [28]). The fitted parameters of a Hecht curve are given in Table 3.3.

etching solution for two minutes. Au contacts were then deposited by sputtering, and Pd leads were attached to contacts using a colloidal graphite suspension in water. Finally, the devices were covered with a protective coating. The detector was then irradiated on both surfaces A and B. The solid lines are theoretical fits with the Hecht equation. The fitted electron mobility-lifetime products for the two surfaces are listed in Table 3.3, from which we see that the electron μτ product

TABLE 3.3
Mobility-Lifetime (μτ) Products Derived from the Hecht Equation (Equation 3.44) and Taking Surface Recombination (S/μ) into Account for a 10 × 10 × 2 mm³ CdZnTe Crystal (Equation 3.49)

	Charge Collection Efficiency		
	Equation (3.44)	Equation (3.49)	
Surface	μτ (cm²/V)	μτ (cm²/V)	S/μ (V/cm)
A	1.7×10^{-3}	2.5×10^{-3}	17
B	6.4×10^{-4}	2.1×10^{-3}	110

Source: From [28].

for surface A is twice as large as that of B. Cui et al. [28] attribute the difference to surface recombination effects—in short, the two surfaces are different. While this cannot affect mobility, it will modify the carrier lifetime, resulting in a measured mu-tau product which is different from the bulk value. The bulk and surface contributions to the measured lifetime can be described by the equation

$$\frac{1}{\tau} = \frac{1}{\tau_b} + \frac{1}{\tau_s} \tag{3.48}$$

Here τ_b and τ_s are electron lifetimes in the bulk and surface. Following the approach of Many [29] who considered the surface recombination and bulk trapping times in photoconducting CdS, Equation (3.44) can be written as

$$CCE = \frac{\mu\tau_b V}{d^2}\left[1 - \exp\left(-\frac{d^2}{\mu\tau_b V}\right)\right]\left(\frac{1}{1 + d\,S/V\mu}\right) \tag{3.49}$$

where S is the electron surface recombination velocity (equal to d/τ_s), which is a measure of the rate of recombination between electrons and holes at the surface. For $S = 0$, Equation (3.49) reduces to the Hecht equation. Fitting Equation (3.49) to the data shown in Figure 3.26 results in much closer values of $\mu\tau_b$ for irradiation on the two surfaces, but different values for S/μ. The results are listed in Table 3.3.

3.2.5.4 *Measuring the Charge Collection Efficiency*

Ionizing radiation absorbed in the sensitive volume excites electron–hole pairs in direct proportion to the energy deposited. The amount of charge created is given by

$$Q_o = q(E_o/\varepsilon) \tag{3.50}$$

where q is the electronic charge, E_o is the energy deposited, and ε is the energy consumed to create an electron–hole pair. This charge induces a mirror charge on an electrode as it moves through the device. Some charge is lost because of trapping and the collected charge *is* now dependent on the path length, x, through which the carriers travel and, thus, the interaction location. The collected charge $Q(x)$ is given by

$$Q(x) = Q_o CCE(x) \tag{3.51}$$

where CCE(x) is defined as the charge collection efficiency. The signal from the electrode is fed to a charge sensitive preamplifier, which converts it to a voltage signal proportional to its charge gain G_c (equal to the inverse of its feedback capacitance). The preamplifier output voltage is then amplified and shaped by a voltage amplifier voltage of gain G before being fed to a multi-channel analyzer (MCA). The input voltage at the MCA is given by

$$V(x) = Q_x G_c G = (Q_x G)/C_f \tag{3.52}$$

The MCA then converts this signal to a channel number, $Ch(x)$ given by

$$Ch(x) = mV(x) + c \tag{3.53}$$

where m is the conversion gain of the MCA, in number of channels per volt, and c is an intercept term which corrects for any voltage offsets in the amplifier. Combining Equations (3.48)–(3.51) and solving for CCE(x),

$$CCE(x) = \left(\frac{\varepsilon}{qE_{inc}}\right)\frac{(Ch(x) - c)}{mG} \tag{3.54}$$

Since all terms on the right-hand side of Equation (3.52) are known or measurable, the CCE can be determined for a collected spectrum.

3.2.6 Defect Characterization

Crystal defects are invariably introduced throughout the growth and detector fabrication process. These defects may arise from impurities, grain boundaries, or for example, interfaces and result in the creation of traps and recombination centers which capture free electrons and holes. Even at very low concentrations, these trapping centers can dramatically alter device performance. For example, defects in the electronically active part of an integrated circuit (roughly within the first 5 μm to 10 μm) can easily destroy its performance. Therefore, an understanding of which trapping centers are present in a semiconductor is necessary in order to devise mitigation techniques, improving detector performance.

3.2.6.1 Thermally Stimulated Current (TSC) Spectroscopy

Before the introduction of deep level transient spectroscopy (DLTS), thermally stimulated current (TSC) spectroscopy [30] was a popular technique to study active defects or traps in semiconductors and insulators. The energy levels associated with traps are first filled by optical or electrical injection, usually at a low temperature. The levels are then emptied by heating the sample to a higher temperature, resulting in the emission of electrons or holes. The sample is then scanned in temperature at a given rate and the emitted current recorded. The resultant curve consists of a series of peaks which gives information on the trap energy levels. For each peak, the trap depth, E_T (also known as the activation energy), can be determined from the approximate relationship [31],

$$E_T = kT_m In \frac{T_m^4}{\beta} \tag{3.55}$$

Here, k is Boltzmann's constant, T_m is the TSC peak temperature, and β is the heating rate for the thermal scan.

Although DLTS has replaced TSC for investigating traps in Schottky or p-n junctions, conventional capacitance-mode DLTS cannot be used for wide-bandgap materials or insulators due to the difficulty of filling the levels by a change in electrical bias. For these materials TSC is still used. Recently, however, the sensitivity of the technique has been markedly improved with the introduction of a new class of TSC spectrometers in which the emitted light is measured as a function of both temperature and wavelength [32].

3.2.6.2 *Deep Level Transient Spectroscopy*

Deep level transient spectroscopy (DLTS) is a pulsed bias capacitance transient technique which is used to investigate energetically "deep" trapping levels in semiconductor space charge structures [33–35]. These may be either p-n junctions or Schottky barriers. By monitoring capacitance transients produced by pulsing the junction at different temperatures, a spectrum is generated which exhibits a peak for each deep level on a flat baseline. The height of the peak is proportional to trap density, and its sign allows one to distinguish between minority and majority traps. The position of the peak on the temperature axis leads to the determination of fundamental parameters governing thermal emission and capture (activation energy and cross section).

DLTS relies on the fact that the RF capacitance of a sample (usually measured at 1 MHz under reverse bias) depends on the charge state of deep levels in the space charge region. The RF capacitance of a sample having a homogeneous doping concentration is given by

$$C_o = A\sqrt{\frac{q\varepsilon_s(N_d - N_a)}{2(V_r - V_d)}} \tag{3.56}$$

assuming it is fully depleted. Here A is the sample area, $N_d - N_a$ is the total net charge density in the space charge layer, V_r is the reverse bias, ε_s $(= \varepsilon\varepsilon_o)$ is the permittivity of the semiconductor material, and q is the electronic charge. If the sample is a p-n junction, $N_d - N_a$ refers to the lower doped side of the junction. V_d is the built-in diffusion voltage of the space charge structure, which is the crossing point of the extrapolated $1/C^2$ plot versus V_r with the V_r-axis (see Figure 3.16). If A is measured in units of mm^2, $N_d - N_a$ in cm^{-3}, C_o in pF, and V_r and V_d in volts, Equation (3.56) can be rewritten as

$$N_d - N_a = 1.41 \times 10^{12}\,\frac{C_o(V_r - V_d)}{A^2\varepsilon_s} \tag{3.57}$$

If charged trapping levels exist in the space charge layer, their space charge has to be added to $N_d - N_a$. On the assumption of a donor-like trap level of concentration N_t in an n-type sample biased under a reverse bias V_r, the capacitance change by recharging these levels is

$$\Delta C = A\sqrt{\frac{\varepsilon_s q(N_d - N_a)}{2(V_r - V_d)}} - A\sqrt{\frac{\varepsilon_s q(N_d - N_a + N_t)}{2(V_r - V_d)}} \cong C_o \frac{N_t}{2(N_d - N_a)} \tag{3.58}$$

The last identity holds approximately if $N_t \ll N_d - N_a$. In this case, the trap concentration can be calculated from the change in capacitance ΔC,

$$N_t = \frac{2\,\Delta C}{C_o}(N_d - N_a) \tag{3.59}$$

The practical implementation of DLTS may be found in Lang [33]. The main advantage of DLTS is that it is a very sensitive technique, measuring defect concentrations down to a level of 10^{10} cm^{-3}. It is also nondestructive. The main disadvantages are:

1. It cannot be used for insulating materials.
2. The defects must be electrically active.
3. Their concentration must be <10% of the doping concentration.
4. The sample must have a depletion region (Schottky or p-n junction).
5. The identification of levels usually requires comparison with other techniques.

3.2.6.3 Photo-Induced Current Transient Spectroscopy (PICTS)

For insulating or very high-resistivity materials, photo-induced current transient spectroscopy (PICTS) [36] is widely used. It was first proposed by Hurtes et al. [37] as a "modification" to DLTS to accommodate high-resistivity materials, and relies on measuring the current released when filled traps empty following optical excitation. The sample is irradiated from the cathode side, and the photo-generated charges drift in the bulk semiconductor and fill the deep levels. After equilibrium between generation, collection, trapping/de-trapping is reached, the optical excitation is abruptly stopped. The photocurrent transient at the end of the light pulse consists of a rapid drop followed by a slow decay. The initial rapid drop is due to electron–hole pair recombination, and the slow decay is due to carriers thermally ejected from the traps [36,38].

Consider a saturated semiconductor, with no re-trapping of the emitted carriers. The current at time t after excitation is given by,

$$I(t) \propto \xi \mu \tau e_n \exp(-e_n t) \tag{3.60}$$

where ξ is the electric field, μ and τ are the mobility and the lifetime of the carriers, and e_n the emission rate of the trapped carriers. In the case of a discrete trap, e_n can be written as

$$e_n(T) \propto \sigma_n T^2 \exp\left(\frac{-E_t}{kT}\right) \tag{3.61}$$

where T is the sample temperature, σ_n and E_t are the cross section and the energy of the trap, respectively. To carry out a PICT analysis, semiconductor samples are excited with a pulsed light source (usually a LED) of wavelength above the bandgap energy, and the thermal relaxation time is calculated from the time dependence of the photo-induced current as a function of sample temperature [38,39]. Around certain temperatures, peaks in the PICTS signal can be seen. The temperatures corresponding to the peak of the PICTS current signal are then plotted as a function of the relaxation times on an Arrhenius plot and the energy and emission cross section of the trap is determined from the intercept of the curve, using Equation (3.61). The measurements are repeated at different sample temperatures to probe different trapping levels. The main advantage of PICTS is that measurements are relatively simple to carry out. However, it is difficult to determine trap densities, and measurements can be unreliable if the trapped carriers recombine or are re-trapped. In addition, where DLTS measurements can be carried out, it is found to be much more sensitive than PICTS. For example, in a study of deep trapping levels in undoped, semi-insulating and p-type CdTe single crystals, Kremer and Leigh [40] found that DLTS measurements revealed several trapping levels in all samples, whereas the PICTS data was generally only sensitive to one level.

3.2.7 Photon Metrology

Synchrotron light sources are widely used in materials science, protein crystallography, and biomicroscopy applications. They provide a unique, stable source of high-intensity photons, extending over a broad energy range, from the far infrared to the gamma-ray region. However, they have also proven invaluable for carrying out detailed metrology of radiation detectors by making available highly collimated, monochromatized beams of synchrotron radiation [41]. Light sources are only accessible at synchrotron research facilities and a number of specialized laboratories (for example the Physikalisch-Technische Bundesanstalt (PTB) radiometry laboratories in Berlin, Germany [42]) have been established specifically to carry out photon metrology from the UV to the X-ray range using primary source standards in conjunction with primary detector standards.

3.2.7.1 Synchrotron Radiation

Classically an electron moving in a magnetic field will execute a spiral trajectory and radiate as a dipole [43]. The emission is isotropic at the Larmor frequency

$$\nu_L = \frac{eB}{2\pi m_o c} = 2.8 \text{ MHz per Gauss} \tag{3.62}$$

where B represents the magnetic field component perpendicular to the particle velocity vector. If the electron is nonrelativistic, the radiation is isotropic and is emitted only at the Larmor frequency. This is known as cyclotron radiation. In the relativistic case, synchrotron radiation is emitted in a relativistically narrow cone of angle, $\theta \sim \gamma^{-1}$, where γ is the particle energy in units of its rest energy (typically 10^3–10^4). The frequency distribution is no longer discrete as in the nonrelativistic case, but is an asymmetric distribution with a maximum of the envelope at

$$\nu_m = 2/3\gamma^2\nu_L \tag{3.63}$$

or in terms of energy

$$E_m = 5 \times 10^{-9} \gamma^2 B \text{ (keV)} \tag{3.64}$$

For the magnetic fields used to steer particles in accelerators, E_m will be in the UV to X-ray range, and indeed synchrotron radiation was first observed emanating from early electron accelerators over 60 years ago [44]. At first, this phenomenon was considered an inconvenient waste product of particle acceleration. It was only in the early 1970s that it was exploited as a resource when it was realized that highly collimated, monochromatic photon beams make an excellent tool for probing the electronic structure of matter from the sub-nanometer to the millimeter level. This led to the construction of a number of dedicated synchrotron facilities based on storage rings.

The unique properties of synchrotron radiation can be summarized as follows:

- Production mechanism is unique in that it can be precisely described
- High brightness and high intensity that is many orders of magnitude greater than that of X-rays produced in conventional X-ray tubes
- Wide tunability in energy/wavelength (from sub-eV up to MeV) by monochromatization
- Small angular divergence of the beam (high collimation)
- Product of source cross section and solid angle of emission is small (low emittance)
- High level of polarization (linear or elliptical).
- Pulsed light emission with durations of 1 ns or less

Which attributes are exploited by the various scientific disciplines and how are described in [45].

3.2.7.2 *Light Sources*

At present, there are over 50 synchrotron light sources operating worldwide [46], which can be broadly grouped into three categories or generations. The fourth generation is currently under development. In chronological order, these are:

- First-generation synchrotron-radiation light sources, which were essentially parasitic on other programs, such as high energy physics. Examples include the Synchrotron Ultraviolet Radiation Facility (SURF) in Maryland, USA, and the 6-GeV Deutsches Elektronen-Synchrotron (DESY) in Hamburg, Germany.
- Second-generation synchrotron-radiation light sources, which are dedicated synchrotron radiation facilities but are not designed for low emittance or with straight sections for insertion devices. Examples include the Daresbury SRS in the United Kingdom and HASYLAB at DESY in Germany.
- Third generation synchrotron-radiation light sources, which are dedicated synchrotron radiation facilities designed for low emittance and with many straight sections for incorporating insertion devices. Examples include the ESRF in France, the ALS in the United States, BESSY II in Germany, and the Diamond Light Source in the United Kingdom.

Each generation differs from the previous generation by innovation and is improved by at least an order of magnitude in performance, usually quantified by the flux and the brilliance of the source. The flux is defined as

$$\Phi = \frac{N_p}{0.1\%\, mrad} \text{(photons/(second, 0.1\% energy spread, mrad horizontally))} \tag{3.65}$$

where N_p is the number of photons emitted per second for a given stored beam current. The brilliance, B, is the peak flux density in phase space,

$$B = \frac{N_p}{0.1\%, mm^2, mrad^2} \tag{3.66}$$

The flux is a function only of the electron current and energy, while the brilliance takes into account the phase space defined by diffraction effects and the electron beam emittance.

Considerable effort is now underway developing fourth generation light sources, which will most likely combine a hard X-ray (wavelength less than 1 Å) free-electron laser (FEL) with a very long undulator in a high-energy electron linear accelerator. Such a device would have a peak brightness many orders of magnitude beyond that of the third generation sources, as well as pulse lengths of 100 fs or shorter, and be fully coherent.

3.2.7.3 *Synchrotron Radiation Facilities*

In general, a synchrotron facility consists of several systems: an electron gun, a linear accelerator, a booster ring, a storage ring, and a set of beamlines which feed specialist experimental stations as illustrated in Figure 3.27. Synchrotron radiation

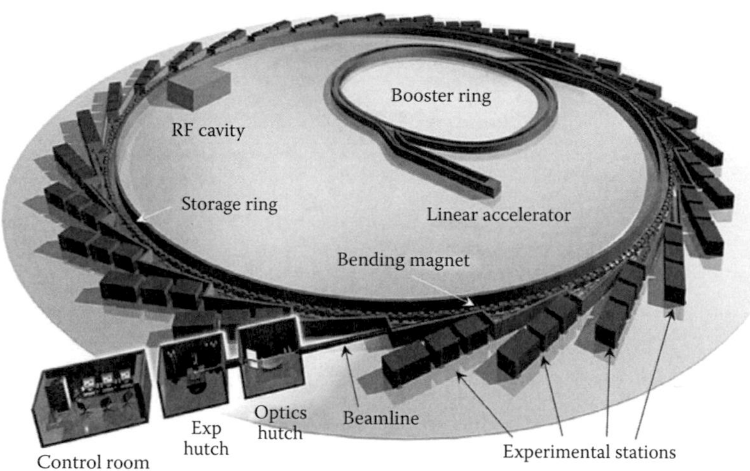

Figure 3.27 Schematic of the third generation 3-GeV Diamond Light Source [47] at Harwell, Oxford, UK, showing its major components. The ring is not truly circular, but is shaped as a twenty-four-sided polygon with a beamline at each vertex. Each beamline is optimized to support an experimental station which specializes in a specific area of science, including the life, physical, and environmental sciences. A station is generally comprised of an optics hutch, experimental hutch, and a control room. (Image courtesy of Diamond Light Source 2011.)

is produced whenever relativistic electrons are accelerated or decelerated. To achieve this, the electrons are first generated in an electron gun by a heated cathode before being accelerated through a hole at the end of the gun by a powerful electric field. They are then fed into a linear accelerator, where they are accelerated from an initial energy of ~50 keV to ~50 MeV and injected into a booster ring, in which the electrons are confined to travel in a circle. They are continuously energized by microwaves, before being injected at GeV energies into the main storage ring. The trajectory of the electrons is steered in the storage ring by a series of bending magnets. These define the ring and maintain the electron trajectory within the ring. However, whenever high-energy, relativistic, electrons are forced to travel in a curved path by a magnetic field, synchrotron radiation is emitted in a fan beam in a plane parallel to the electrons' orbit in the ring. The synchrotron light produced at the bending magnets is extracted by a suitable beamport and directed to an experimental station by a beampipe. An RF cavity, located in a straight section of the ring, is used to replenish synchrotron energy losses at the bending magnets.

For third-generation machines, long straight sections are incorporated into the storage ring for the inclusion of insertion devices. These are periodic arrays of magnets designed to produce a series of deflections of the primary electron beam in the straight-line section of the orbit. They consist of one array of magnets above the electron beam path and one coaligned array below. The poles are alternating so that, instead of one magnet deflecting the electron beam and generating a single fan of light, an entire array of alternating magnets now deflect the beam such that the electrons follow a wiggling or undulating path. Each deflection produces a kink in the electron trajectory adding to the intensity of the light from that point. There are two basic types of insertion device: wigglers and undulators. A wiggler can be considered to be a concatenation of N bending magnets and its brilliance scales as N, emitted over a wide bandwidth; in an undulator, the magnets are arranged such that the emitted radiation adds in phase and its brilliance scales as N^2, emitted over a narrow bandwidth. A comparison of the on-axis brilliance for bending magnets, wigglers, and undulators is given in Figure 3.28.

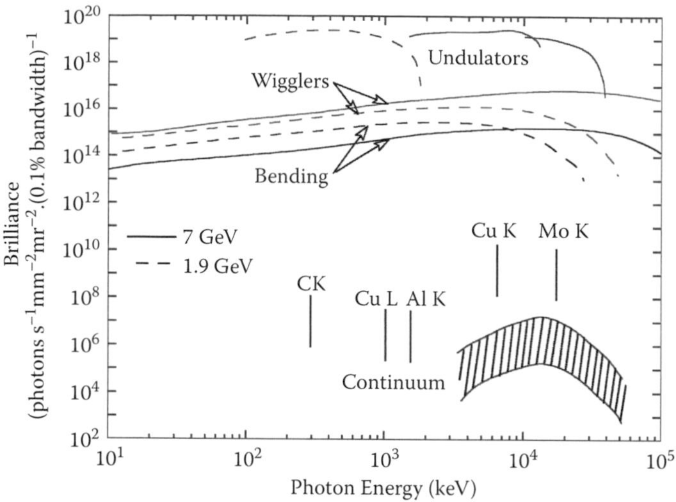

Figure 3.28 Spectral brilliance for several synchrotron radiation sources and conventional X-ray sources. The data for conventional X-ray tubes should be taken as rough estimates only, because brightness strongly depends on operating conditions (adapted from [48], courtesy LBNL).

3.2.7.4 Properties of the Beam

The beam emerging from the bending magnets is known as "white light" and has a well-defined energy spectrum extending from the microwave through to the UV, VUV, and soft and hard X-ray regions of the electromagnetic spectrum. The spectrum is usually characterized by its critical energy, E_c, which is defined as the energy at which half the radiant power is carried by photons above E_c. The critical energy is given by

$$E_c(\text{keV}) = 0.665E^2/\rho \tag{3.67}$$

where E is electron beam energy in GeV, and ρ is the bending radius. Critical energies generally range from ~10 keV for second-generation machines to ~25 keV for third-generation machines. A useful rule of thumb is that for general synchrotron work, the practical working energy range is $4E_c$, while for single photon counting detector work it is $8E_c$.

White light direct from an extraction point is extremely intense and has a brilliance (defined as the number of photons per second per mm^2 per mrad^2 in $\Delta\lambda/\lambda = 10^{-3}$) of ~$10^{15}$ photons s^{-1}mm^{-2}mrad^{-2} (second generation machines) up to 10^{20} photons s^{-1}mm^{-2}mrad^{-2} (third generation machines). To put the comparison into context, a rotating anode X-ray generator has a brilliance of 10^9 photons s^{-1}mm^{-2}mrad^{-2} in $\Delta\lambda/\lambda = 10^{-3}$. The white light directly from the synchrotron is, in fact, so intense that the raw beam will seriously damage a detector. Fortunately, it is possible to lose many orders of magnitude of intensity by, (1) monochromatizing the beam using a double crystal monochromator, (2) detuning the rocking curve of the first, or downstream, crystal, and (3) greatly reducing the area of the beam on the detector. For the type of photon metrology carried out on detectors, incident flux rates at the sample need to be in the range ~10^2 to 10^4 photons s^{-1}.

3.2.7.5 Beamline Design

Beamlines are usually tailored for particular experimental disciplines using subsystems to filter, intensify, or otherwise manipulate the light to generate a specific set of characteristics suitable for the needs of the experimental station, which are normally application specific.

A typical beamline layout suitable for detector metrology is shown in Figure 3.29—the X-1 beamline at the Hamburger

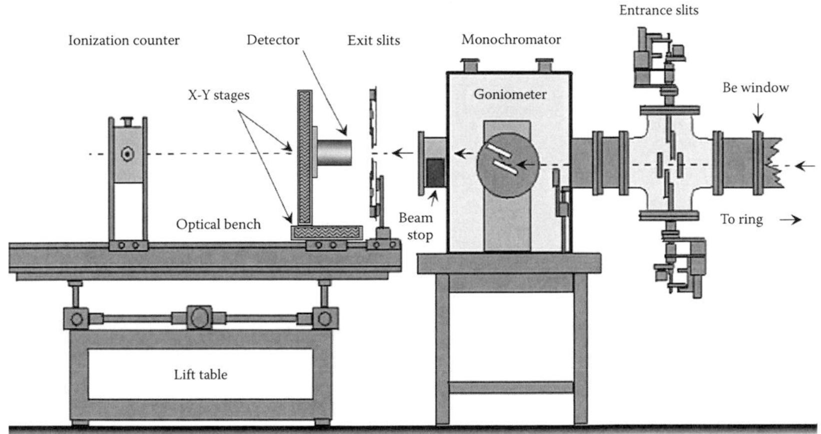

Figure 3.29 The X1 hard X-ray beamline at the HASYLAB synchrotron radiation source at DESY [49].

Synchrotronstrahlungslabor (HASYLAB) radiation facility [49] located at DESY in Hamburg, Germany. This beamline utilizes a double Si crystal monochromator to produce highly monochromatic X-ray beams across the energy range 10 keV to 100 keV. Depending on the energy range of interest (and the amount of acceptable harmonic pollution) a choice of Si(111), Si(311), and Si(511) crystal pairs can be selected. Because mechanically, the range of adjustable monochromator angles and crystal separation is finite, the order of the reflection usually sets the useable energy range in a particular system. The range of Si(511) is generally larger than Si(311), which in turn is larger than Si(111). Also the energy resolution tends to be higher for high-order crystals, ranging from 10^{-4} for Si(111) to 10^{-5} for Si(511). To cover the entire energy range 10–100 keV, a [511] reflection is usually used, yielding an intrinsic energy resolution of ~1 eV at 10 keV, rising to 20 eV at 100 keV.

The white-beam extracted from the synchrotron passes through a set of entrance slits which serve to define the beam profile incident on the first, or upstream, monochromator crystal. In normal operation, the slit width is set to ~10 mm in the horizontal plane and ~(0.1–1) mm in the vertical plane. The vertical width is much smaller, because it defines the energy resolution of the system. The upstream crystal can rotate in the vertical plane and translate in the horizontal plane along the beam axis, and is used to direct the Bragg reflected beam onto the second or downstream crystal. The second monochromator crystal is free to rotate about a single axis and thus, in combination with the translation and rotation on the first crystal, allows the diffracted beam to pass through a fixed exit point, which is parallel with the incident beam but displaced in the vertical plane. The monochromator serves two purposes. The first is to select the photon energy that will be incident on the sample. The second is to reduce significantly the flux of incident photons (by some six orders of magnitude), because the pass-band is only a few eV. The flux can be further reduced by detuning the parallelism of the two crystals. This is achieved in practice by using a piezoelectric piston to rotate finely the upstream crystal. Detuning this crystal has the added advantage of suppressing higher-order harmonics which may otherwise contaminate or compromise the measurements. Note that it is usually not possible to reduce the intensity by simply inserting absorbers. In the case of monochromatic radiation, they "amplify" the higher harmonics relative to the fundamental and generate non-negligible secondary radiation. In the case of white light, in order to reduce the highest-energy component of the white light continuum to an acceptable level for detector operation, the absorber becomes so thick that all spectral information is lost.

Following the monochromator, the diffracted beam passes through a pair of precision stepper-driven exit slits positioned immediately in front of the detector (see Figure 3.29), which defines the beam size at the detector and reduces stray light contributions. For the majority of detector measurements, this is usually set to a size of 50×50 μm^2. The monochromator, slits, and X-ray stages are all controlled from the station computer which is preprogrammed to carry out an extensive set of operations, such as energy scans for X-ray absorption fine structure (XAFS) measurements and 1D and 2D spatial scans for detector characterization. The station computer also logs housekeeping data, such as ring current and motor settings, as well as experimental data.

3.2.7.6 Installing the Detector

Detectors are usually mounted on an X-Y stage capable of positioning a detector to a precision of <1 μm in each axis over a range of ±10 cm. The mechanical interface between the X-Y stage and the detector is a CNC machined plate with a number of predefined locations for installing different detectors. The detectors are located by means of guide pins and quick release screws. A reference detector can be installed at one of a number of locations on the interface, all of which are precisely located with respect to principal detection axis. This is defined by the position of the detector under test. After the detector has been mounted (see Figure 3.30), a laser assembly is attached to the front of the detector. The laser beam axis is precisely aligned with the center of the detector aperture and points toward the center of the X-Y slits. The reference detector can now be used to precisely locate the beam with respect to the principal detection axis. This is achieved by carrying

Figure 3.30 Photograph showing the installation of a detector ready for characterization on beamline X1 at the HASYLAB synchrotron research facility at DESY. A laser attached to the front of the detector is used to align its aperture with the center of the slits. A coaligned reference detector is then used to establish precisely the position of the beam with respect to the detector principal axis prior to scanning.

out a series of X, Y scans across the beam, in which the reference detector count rate is recorded as a function of position. The center of the beam is located by centroiding. Next the X-Y stage is driven to position the beam at the center of the detector under test, and X-Y scans are again carried out. Once the center of the detector has been located the X-Y coordinates are now zeroed, and measurements can begin.

The reference detector also serves several other functions. Firstly, it is robust, so it is used to adjust the beam structure and flux between experiments before a sensitive test detector is exposed to the beam. The detector is also fully spectroscopic, so it can check on the spectral purity of the beam which may have been contaminated by scattering and fluorescence backgrounds caused by slight misalignments of the beam. For new experimental detectors, a lack of a signal does not necessarily mean the detector has malfunctioned—it could mean there is no beam. The reference detector is also used to check this.

3.2.7.7 *Harmonic Suppression*

The monochromator transmits not only the desired fundamental energy, but also higher harmonics of that energy. The suppression of these higher harmonics is an important consideration for X-ray experiments, because in the presence of any absorbing material, such as the path length in air or detector windows, the fundamental can be easily suppressed, effectively amplifying the higher harmonics. Bragg's law of X-ray reflection states that

$$n\lambda = 2d\sin\theta \tag{3.68}$$

where d is the spacing of the atomic planes of the crystal parallel to its surface, θ is the angle of the crystal with respect to the incident white light beam, λ is the wavelength of the diffracted X-ray, and n is an integer. The fundamental X-ray energy corresponds to $n = 1$, and X-rays of higher harmonic energies correspond to $n > 1$. Harmonic X-rays that are diffracted from the crystal depend on the crystal lattice and the cut of the crystal. It is useful to use a crystal that does not diffract the second harmonic ($n = 2$), because the intensity of the second harmonic is usually much greater than the intensities of the higher harmonics. Si crystals with a diamond structure (space group Fd $\bar{3}$ m($O_h{}^7$)) will not allow the harmonics that satisfy the equation $h + k + l = n$, where n is twice an odd number. Hence for example, Si(111) crystals do not diffract the second or sixth harmonic. When working at high X-ray energies it is possible that the energy of higher-order harmonics exceeds the maximum energy produced by the machine, in which case no harmonic rejection is needed. Common methods of reducing harmonic X-ray content include detuning the second crystal or using a harmonic rejection mirror. To detune the monochromator, the angle of the second crystal is slightly offset, or misaligned, with respect to the angle of the first crystal using a piezoelectric transducer. This has the effect of reducing the harmonic content much more than the fundamental, since the angular widths of the harmonics are narrower. For example, when two Si(111) crystals are detuned by 50% on the rocking curve, the intensity of the third harmonic is reduced by a factor of 10^3. For detector work, harmonics rarely present a problem, because to operate in single photon counting mode, the monochromator has to be detuned by as much as 95% at some energies. Another common method for removing harmonics X-rays is to use a harmonic rejection mirror. This mirror is usually made of Si for the lower XUV energies, Rh for X-ray energies, and Pt for the high X-ray energies. The mirror is placed at a grazing angle in the beam such that the X-rays with the fundamental energy are reflected toward the sample, while the harmonics are not. Slits placed downstream of the mirror are also used to block the direct beam containing the harmonics.

3.2.7.8 Extending the Energy Range

Most conventional synchrotron experiments use the increased brilliance produced by insertion devices purely to increase the flux incident within a very small area, for example, in protein crystallography to compensate for weak diffraction from very small samples. For detector metrology, wigglers are particularly useful because the increased flux, coupled with the increased spectral bandwidth, allow measurements to be carried out outside the useful energy range of bending magnets (subject, of course, to a suitably "sized" monochromator). Even though the photon flux falls off exponentially above the critical energy, for single photon counting applications there may still be significant flux at energies as high as an MeV. For example, Owens et al. [50] reported measurements over the energy range 6 keV to 800 keV,* carried out on the ID15 high-energy scattering beamline [51] at the ESRF. To achieve usable photon fluxes over such an extreme energy range, two insertion devices were used, an asymmetrical multipole wiggler (AMPW) followed by a superconducting wavelength shifter (SCWS). The spectral resolution was typically around 20 eV at 30 keV, rising to ~1 keV at 1 MeV. Higher-order harmonics were not an issue since the exponential fall in intensity above the critical energy ensured that the flux was vanishingly small.

---

* It is possible to generate higher energies. However, above 800 keV, photon grazing angles become so shallow that it is exceedingly difficult to separate the monochromatic beam from the white beam without recourse to a custom-built collimation system.

3.2.7.9 Detector Characterization

The use of highly collimated and tunable photon beams allows a wide range of detector measurements to be carried out, of which we list a few examples. We will primarily consider monolithic detectors and material characterization. For the specific characterization of area detectors (for example, measurements of QE, DQE, MTF, LPI, etc.), the reader is referred to the review of Ponchut [52].

For X-ray and gamma-ray applications, the energy resolution of a detector, its linearity, and uniformity of response can be directly measured at any energy within the energy range of the monochromator. For example, in Figure 3.31 we show the measured energy-loss spectra at 30 keV and 98 keV from a 250 × 250 μm², 40-μm-thick GaAs pixel detector which perfectly illustrates the high quality of data that can be obtained

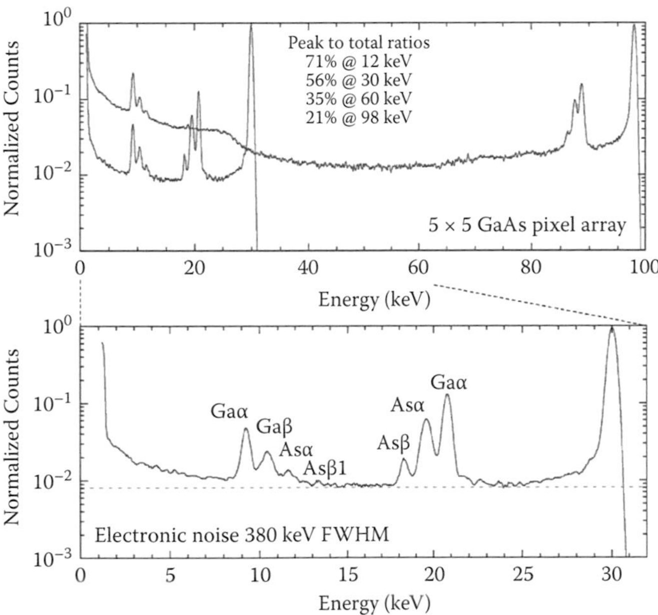

Figure 3.31 Top: measured energy-loss spectra at 30 and 98 keV of a 250 × 250 μm² GaAs pixel detector. For the latter distribution, a Compton edge is apparent near 30 keV, below which is a Compton continuum. The measurements were carried out at a temperature of −40°C, a bias of 100 V, and a shaping time of 2 μs. Bottom: an expansion of the 30 keV data.

with a synchrotron radiation source. From the figure, we can also see that, apart from the Compton feature below 30 keV in the 98 keV data, the level of continuum is extremely low, being only 1% of the photopeak amplitude. The lower curve in Figure 3.31 shows an expansion of the 30 keV data. The set of peaks centered around 20 keV are the so-called escape peaks of Ga and As—the characteristic Kα and Kβ X-rays emitted during photoelectric absorption of the incident X-rays which have escaped the active detector volume (see Section 5.4.1). The set of peaks near 10 keV are fluorescent Ga and As X-rays produced in the conductive GaAs substrate and detected in the active volume. The energy resolution function determined from the pulse height data in Figure 3.31 is shown in Figure 3.32. The resolution ranges from 400 eV FWHM (full width at half maximum) at 6 keV to 700 eV FWHM at 98 keV. The electronic noise of the system, as measured by a precision pulser, was 380 eV FWHM, and the other sources of noise were derived by best-fitting the expected resolution function described in Chapter 5 to the experimental data. Because the noise contribution due to the beam divergence at the monochromator is the order of tens of eV, which is many orders of magnitude less than other sources of noise, it is clear that the broadening measured is due

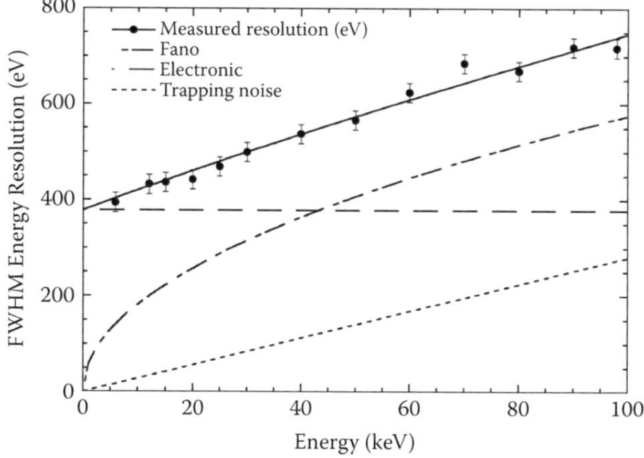

Figure 3.32 The energy-dependent FWHM energy resolution of a 250 × 250 μm² GaAs pixel detector under pencil beam illumination. The best-fit energy resolution function together with its key components is also shown.

solely to the detector. The resolution was also measured under full area illumination using an ^{241}Am radioactive source. The resolutions for the 59.5 keV nuclear line and 60 keV synchrotron pencil beam radiation were found to be the same within statistical uncertainty, thus indicating uniform crystallinity and stoichiometry. The above data show that electronic noise dominates the response of this detector over most of the energy range, although the resolution is nearly Fano limited above 40 keV with a best fit Fano factor of 0.14 ± 0.04. Based on these results, the conventional resistive feedback preamplifier was replaced with a resistorless feedback design. A substantial improvement in electron noise was achieved (from 30 electrons rms to 19 electron rms), resulting in a room temperature resolution of 266 eV FWHM at 5.9 keV and 219 eV FWHM with only modest cooling at –30°C.

In addition, crosstalk in pixel detectors can easily and quickly be assessed as well as gain maps across a detector or array. As an example, in Figure 3.33 we show the X-ray spatial response in the form of a gain map of a 4 × 4 GaAs pixel array [53]. The pixel sizes are 350 × 350 μm² with an inter-pixel gap of 50 μm. The array thickness is 325 μm. The map was produced using a 15 keV pencil beam of size 20 × 20 μm², normally incident on the pixels. The beam was raster scanned across this area with 10 μm spatial resolution in both dimensions.

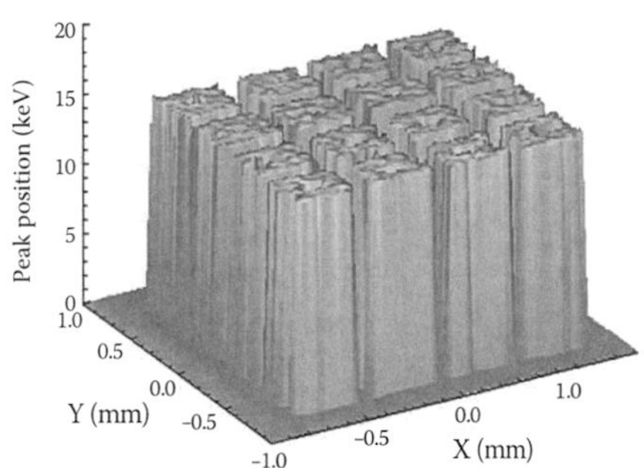

Figure 3.33 Left: photograph of a 4 × 4 GaAs pixel array and its associated front-end electronics. The inset shows a blow-up of the array before wire bonding. The pixel size is 350 μm × 350 μm, and the interpixel gap is 50 μm. Right: a surface plot of the spatial variation of the gain (the fitted centroid position) across the array measured at HASYLAB using a 15 keV, 20 × 20 μm² pencil beam (from [53]). The spatial sampling in X and Y was 10 μm.

Spectra were accumulated at each position for a fixed time interval, and the total count rate above a 3 keV threshold, the peak centroid position, and the FWHM energy resolution were determined by best fitting. Typically for this type of characterization, it is found that for 3% counting statistics the centroid (essentially the gain) can be located to a precision of ~0.1% and the fitted FWHM energy resolution to a precision of ~20%. The measurements show that, apart from a slight decrease in these parameters directly under the bond wires, their spatial distributions were very uniform over the surface of each pixel and the entire array. In fact, the average nonuniformity is typically no worse than a few percent and is consistent with a flat response. In other words, the variations seen in each distribution are consistent with the expected statistical variations. The fact that both the count rate and centroid responses are zero in the inter-pixel gaps, with no evidence of crosstalk, implies that it is possible to replicate isolated and identical pixels, and thus in principle megapixel arrays, also.

3.2.7.10 *Probing Depth Dependences*

In addition to measuring basic response functions such as energy resolution and linearity, it is also relatively easy to directly measure the energy-dependent efficiency function using a calibrated reference detector. The active depletion depth can be determined using the same method [54]. Alternatively, fine spatial scans at various energies can probe the depth-dependent structure in the detector. For example, at low energies the spacial morphology and lateral uniformity of the contacts can be mapped, while at high energies the spatial uniformity of the internal electric field can be probed. In fact, for detectors with good charge collection efficiency, these measurements can be carried out simultaneously by exciting the higher harmonics of the Bragg reflection. This can be archived by “tuning-up” the rocking curve and inserting thin pieces of absorber into the beam.

3.2.7.11 *Defect Metrology*

Fine spatial scanning can also reveal defect information such as dislocations and inclusions via their impact on local charge

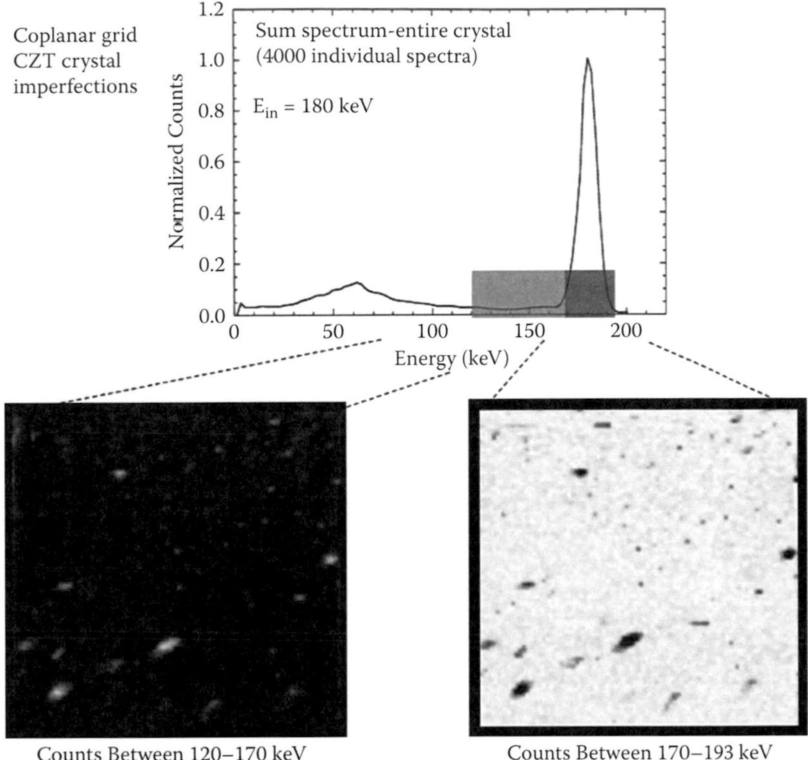

Figure 3.34 Defect diagnostics carried out on a 15 × 15 × 10 mm^3 CdZnTe coplanar grid detector [50]. The figure shows spectrally resolved count rate maps obtained by raster scanning a 20 × 20 μm^2, 180 keV normally incident beam in 40-μm steps across the detector. Crystal defects are plainly evident in the detector count rate response and account for ~2% of the active area of the detector (light color corresponds to photopeak events, black to a lack of events). The lower right-hand image shows that the counts in the defects do not originate from the photopeak and, in fact, mainly emanate from energies 120 to 170 keV (see lower left-hand image). In this energy region there are virtually no events from the rest of the detector area.

collection efficiency. This is illustrated in Figure 3.34, in which we show a spatial map in two spectral ranges of a large 15 × 15 × 10 mm^3 CdZnTe coplanar grid detector [50]. The map was obtained by raster scanning a 20 × 20 μm^2 region with a normally incident 180 keV beam in 40-μm steps across the detector. From the figure, crystal defects are plainly evident in the detector count rate response and account for ~1% of the active area of the detector. What is interesting is that spatially resolved spectroscopy shows that spectra acquired within the defects are essentially the same as those acquired outside the defects, but shifted to lower energies. In other words, the effect of the defects is not to distort the spectra but to reduce the CCE. Coupled with cross-edge, differential absorbtometry, the type of inclusion can even be determined. For example, den Hartog et al. [55] used such a technique to probe defects in a CdZnTe ring detector and showed that localized enhancements of Te could be recognized down to a level of < 0.1% by composition. Furthermore, by using spatial and depth information, they tentatively identify one extended defect to be a decorated Te grain boundary with a volume of extent 40-µm thick, located 45 µm below the surface.

3.2.7.12 Pump and Probe Techniques

Another powerful technique that can be exploited at synchrotron facilities is the so-called "pump and probe" technique in which an intense finely focused monochromatic X-ray beam is used to "pump" a region of interest on a detector, and a much weaker "probe" beam is used to map the resulting creation and subsequent dispersal of charge. As an example of the power of such a technique, Kozorezov et al. [56] used it to study the onset and subsequent evolution of polarization effects in TlBr radiation detectors.

Polarization in semiconductor X-ray detectors can cause significant degradation in detector performance, particularly at high radiation levels. The term polarization in this context means time-dependent variations in the detector properties, such as count rate, charge collection, and resolving power that seem to be correlated with incident radiation fluence and material properties, such as purity, stoichiometry, high resistivity, and dielectric constant. To date, the treatment of polarization has been largely anecdotal and qualitative. Recently, Bale and Szeles [57] have presented a quantitative dynamical model of polarization which is based on the supposition that polarization is a consequence of strong carrier trapping by deep impurity levels. As a result, a space charge region with a high charge density can be created, which causes a significant change in the profile of the internal electric drift field. At some point in the bulk of a detector (the "pinch point") the polarization field cancels the electric field created by the detector biasing, leading to a catastrophic deterioration of detector performance. In this regard, the measurements of Camarda et al. [58] are particularly interesting. They used a finely collimated, high-intensity X-ray beam to study the correlation between microscopic defects (inclusions) and variations in the collected charges in CdZnTe detectors. They were able to demonstrate that single Te inclusions trap a significant amount of charge in an electron cloud, and that polarization effects could be correlated with these defects and their surrounding areas.

The model of Bale and Szeles [57] is simplified in that it considers only uniform illumination, which means that the profile of the internal electric field is one-dimensional and depends only on the coordinate normal to the plane of the detector. As such, many details of the effect are still not understood, particularly those relating to the mechanism of the polarization process itself—the extent of the polarized region, the energy dependence of the process and details of the trapping processes responsible. As reported by Kozorezov et al. [56], this work was recently extended in a series of "pump and probe" scanning measurements with tightly focused monochromatic pencil beams at the HASYLAB synchrotron research facility. By combining these 2D scans for a range of X-ray energies and intensities, a 3D picture of polarization phenomena within the bulk material could be constructed which could then be used to extend the model to three dimensions. These results confirm that local polarization close to the exposed spot results in distortion of electric fields both inside and around the affected volume and that the charging of deep traps by one of the carriers perturbs the internal electric field sufficiently to affect the collection of the other carrier. The relaxation of the polarized region was also studied by regularly repeating the probe and pulse scans over a period of three hours after the initial "pump."

An example of the correlative information that can be gained is shown in Figure 3.35, in which we show data obtained from the pulse height spectra obtained for scans at

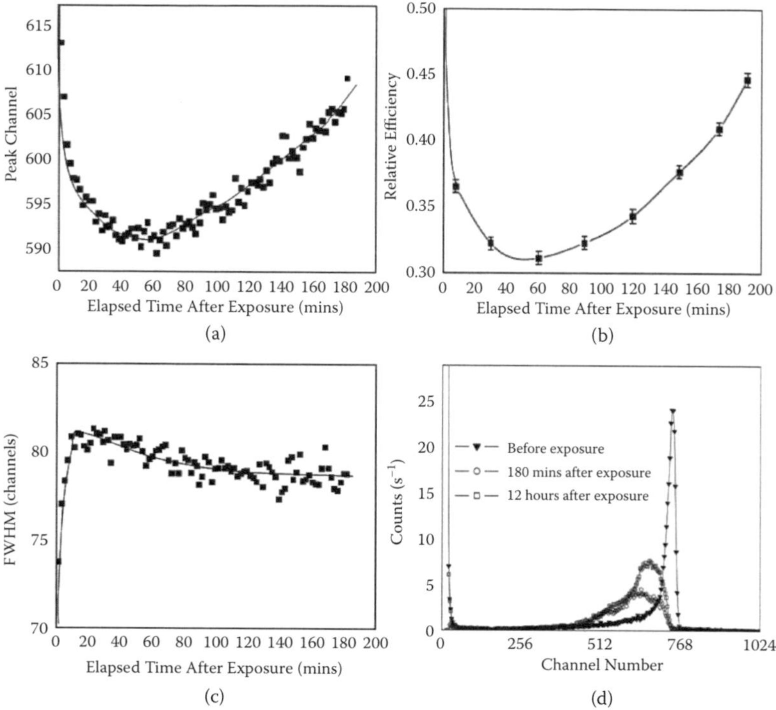

Figure 3.35 Pulse height data from repeated 60 keV "probe" scans of a polarized volume following a "pump" pulse of 2.4×10^6 60 keV photons. The figures show the evolution of the 60 keV energy-loss peak: (a) peak channel position (initial value 740), (b) relative efficiency (initial value unity), (c) FWHM energy resolution (initial value 27), and (d) pulse height spectrum as a function of elapsed time (from [56]).

60 keV incident energy. We see that, while the peak channel position and charge collection efficiency experience moderate signs of recovery, the spectral resolution and particularly the line shape remain degraded for much longer, leading to the conclusion that full relaxation of the polarized volume at –20°C is a very slow process. By modeling these results, it was found that depolarization proceeds through thermal recombination and cannot be achieved effectively through modulation of the bias voltage, as previously thought [59].

The model can be also be used as a diagnostic tool with which to understand and ultimately mitigate or eliminate, the effects of polarization in semiconductor detectors. For example, by simulating pump-and-pulse synchrotron data, trap densities, occupancy rates, and ionization energies can be derived together with a host of previously experimentally illusive transport properties, such as hole mu-tau products. Such data may aid in the identification of the particular traps involved. Even if material shortcomings cannot be avoided or compensated for during the growth process, it should now be possible to use the simulation to custom design susceptible detection systems to operate in specific radiation environments.

3.2.7.13 X-Ray Absorption Fine Structure (XAFS) Metrology

X-ray absorption fine structure measurements (XAFS) are routinely carried out at synchrotron facilities to probe both short- and long-range order in materials. XAFS is a generic term and can be broken down into structure originating far from an absorption edge and structure originating close to the edge, which arise from different processes. Extended X-ray absorption fine structure (EXAFS) is a diagnostic of short-range order by means of which details in the local geometry (atom types, bond lengths, and bond angles) around the photoabsorbing atom can be extracted from far-edge spectra. X-ray absorption near edge structure (XANES), on the other hand, is a diagnostic of long-range order through which details of atom types and how they are structured collectively (the coordination environment) can be extracted from near-edge spectra. XAFS measurements, both EXAFS and XANES, can also be applied to detector metrology. For example, several authors have noted the wide spread in the radiation detection properties of $Cd_{1-x}Zn_xTe$ crystals and have attempted to use the structural information embedded in XAFS to find a link between performance and structural perfection.

3.2.7.14 Structural Studies

Wu et al. [60] carried out XAFS studies around the Zn K-edge for a number of alloys with x ranging from 0.005 to 1.0 to assess the effects of zinc segregation and defect formation on the local atomic structure of CdZnTe. Ideally, the alloy $Cd_{1-x}Zn_xTe$ can be thought of as a CdTe crystal with Zn atoms randomly substituting a fraction x of Cd atoms. By Vegard's law [61], the substitution of Zn atoms should be accompanied by a change in the lattice constant linearly varying between CdTe ($x = 0$) and ZnTe ($x = 1$). However, a Fourier analysis of EXAFS data in wavevector phase space revealed a bimodal distribution of bond lengths, suggesting distortion of the Te sublattice. If true, this suggests that a linear interpolation of lattice constants with zinc fraction is, at best, an approximation.

Duff et al. [62] carried out a similar analysis on two CdZnTe crystals of predetermined spectral quality—one of high spectral performance and one of low spectral performance. Unlike previous analyses, the crystals used were intact, and the XAFS measurements were carried out in transmission. No significant differences were found in the local atomic structure of the three primary elements comprising the crystals. The derived Debye–Waller factors, which measure disorder in crystalline material, are essentially the same for each element. The authors conclude that spectral performance in CdZnTe is more intimately linked to other factors, such as the presence of secondary phases (precipitates and inclusions), which although known to limit local electron mobility, do not significantly degrade their bulk performance. In a follow-up study of secondary phases, Duff et al. [63] identified two dominant secondary phase morphologies. The first consists of numerous pyramidal-shaped empty voids of extent 20 μm. The other consists of 20-μm hexagonal-shaped bodies, which are composites of metallic Te layers that contain a teardrop-shaped core of polycrystalline CdZnTe.

3.2.7.15 Topographical and Surface Studies

An XAFS analysis of a semiconductor surface is a powerful tool with which to examine nanostructures such as interfacial layers and heterostructures. For example, Owens [64] described a method for determining the compositions, abundances, and thicknesses of dead layers above a silicon charge-coupled device (CCD) soft X-ray detector using the structural information embedded in XANES around the various K-edges. This was achieved fitting the expected efficiency function quantum efficiency, given by

$$QE = \frac{1}{P}[1 - \exp(\mu_d d)] \int_0^p \prod_{i=1,n} \exp[-a_i(x)(\mu_e)_i t_i] dx \quad (3.69)$$

to the measured quantum efficiency curve shown in Figure 3.36. Here, p is the pixel size, $(\mu_e)_i$ is the absorption coefficient of the *ith* component of the electrode structure, and t_i its thickness. The first term in brackets in Equation (3.69) accounts for interactions in the active depletion depth of the device, while the second product term accounts for absorption in the various overlying materials, such as gates, dielectrics, and passivation layers. Because of the complicated nature of the pixel structure (see Figure 3.37), the variation in material across a pixel is accounted for by a weighting factor, a_i.

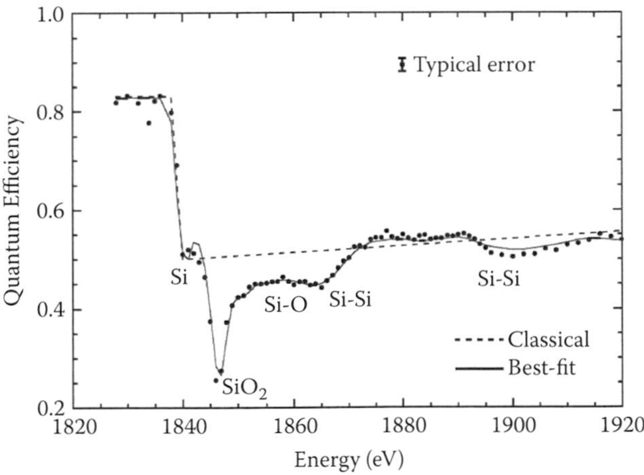

Figure 3.36 The measured quantum efficiency across the Si K-edge of an X-ray CCD [64]. Individual edges and bonds are identified. For comparison, we show our calculated values based on new linear absorption coefficients abstracted from photocurrent measurements along with the classical predictions of Cromer and Liberman [65].

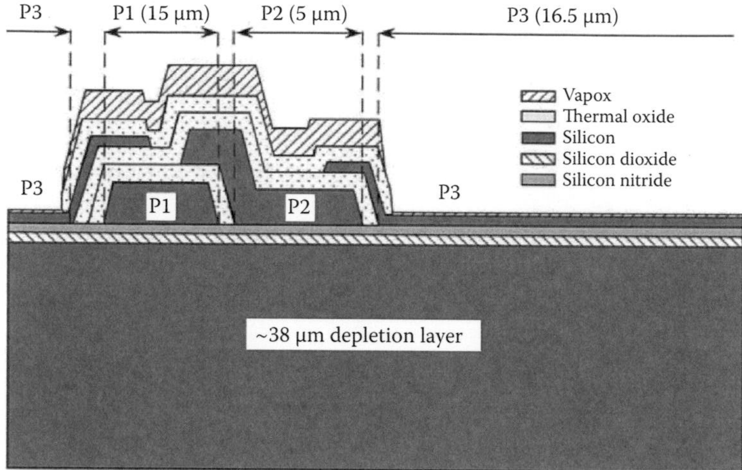

Figure 3.37 Detailed cross section through the overlying dead layers (electrodes, gate dielectrics, polysilicon gates, and passivation layers) above the active depletion region (from [64]).

The absorption cross sections (including XAFS) used in the calculation were derived from the photocurrent measurements of Owens et al. [66] using the following expression:

$$P(E) = \mu(E)[a_1 + a_2 E] \tag{3.70}$$

where P(E) is the total photo-yield at energy E, and μ(E) is the corresponding absorption coefficient. The constants a_1 and a_2 are determined by normalizing μ(E) to the classical values far enough above and below the edges to be free of the effects of XAFS. The applicability of Equation (3.70) has been verified by Owens et al. [67], who showed that the linear relationship holds over wide energy ranges to a precision of at least the few percent level. The derived linear attenuation coefficients for Si-c, Si-a, SiO_2, and Si_3N_4 are shown graphically in Figure 3.38.

Equation (3.69) was then fitted to the measured quantum efficiency using a nonlinear minimization routine allowing the thicknesses of overlying materials to be free parameters. The manufacturer's values were used as the initial inputs, which, although only known to an accuracy of ~±20%, can be

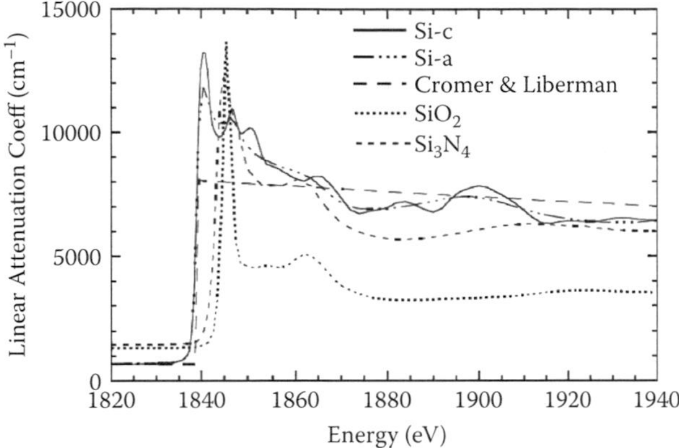

Figure 3.38 The derived linear attenuation coefficients across the Si K-edge. For Si, the letters *c* and *a* refer to crystalline and amorphous (from [66]). We also show the "classical" Si curve based on the calculation of Cromer and Liberman [65].

reproduced to precisions of a percent. The calculated quantum efficiency based on best-fit values is shown graphically in Figure 3.36, from which we can see there is excellent agreement with measurement. In fact, the residuals display no global systematic trends with energy yielding an average error of ~1.4%. The errors in the fitted layer thicknesses were typically <10%, which is about 50% of the value derived from process control metrology.

The above analysis illustrates the power of XANES metrology. In addition, in contrast to scanning electron microscopy (SEM), which only yields information on the linear extent of surface features, XANES can also provide elemental and bonding information—potentially to one precision of one atom in 10^{10}. By combining the structural information contained in the quantum efficiency measurement with X-ray photoelectron spectroscopy (XPS), it should be possible to isolate and image surface features using XANES tomography by focusing on a specific near edge structure. Such a technique is a powerful diagnostic tool with which to explore macroscopic surface structures, offering distinct advantages over traditional techniques, such as SEM, in that it is noninvasive and nondestructive.

References

1. PM5 Auto-Lap Precision Lapping & Polishing Machine Product sheet, Logitech Limited, Erskine Ferry Road, Glasgow G605EU, Scotland, UK (2005).
2. Department of Electrical and Computer Engineering, IMMERSE Web Team, "Wet Chemical Etching of Metals and Semiconductors," Brigham Young University (2009) http://cleanroom.byu.edu/wet_etch.phtml.
3. J.W. Faust, Jr., "Etching of the III–V Semiconductor Intermetallic Compounds" in *Compound Semiconductors, Vol. 1, Preparation of III–V Compounds*, eds. R.K. Willardson and H.L. Goering, Reinhold, New York (1962) pp. 445–468.
4. W.L. Bragg, "The Diffraction of Short Electromagnetic Waves by a Crystal," *Proc. of the Camb. Phil. Soc.,* Vol. 17 (1913) pp. 43–57.
5. J. Faber, T. Fawcett, "The Powder Diffraction File: Present and Future," *Acta Cryst.*, Vol. 58, part 3 no. 1 (2002) pp. 325–332.
6. L. Vegard, "Die Konstitution der Mischkristalle und die Raumfüllung der Atome," *Z. Phys.*, Vol. 5 (1921) pp. 17–26.
7. M.S. Goorsky, H. Yoon, M. Schieber, R.B. James, D.S. McGregor, M. Natarajan, "X-ray Diffuse Scattering for Evaluation of Wide Bandgap Semiconductor Nuclear Radiation Detectors," *Nucl. Instr. and Meth.*, Vol. A380 (1996) pp. 6–9.
8. E.H. Rhoderick, *Metal-Semiconductor Contacts,* Clarendon Press, Oxford, 2nd edition (1978) p. 87.
9. C.R. Crowell, V.L Rideout, "Normalised Thermionic-Field Emission in Schottky Barriers," *Solid State Electron.*, Vol. 12 (1969) pp. 89–105.
10. M. Missous, E.H. Rhoderick, "New Way of Plotting Current/Voltage Characteristics of Schottky Diodes," *Electronic Letts.*, Vol. 22, no. 9 (1986) pp. 477–478.
11. A.A. Bergh, P.J. Dean, "Light-Emitting Diodes," *Proc. of the IEEE*, Vol. 60 (1972) pp. 156–224.
12. S. Montanari, "Fabrication and Characterisation of Planar Gunn Diodes for Monolithic Microwave Integrated Circuits," PhD thesis, University of Aachen RWTH (2005).
13. F. Wenner, "A Method of Measuring Earth Resistivity," *Bur. Stand. (U.S.), Bull.*, Vol. 12 (1915) pp. 469–478.

14. G.S. Marlow, M.B. Das, "The Effects of Contact Size and Non-Zero Metal Resistance on the Determination of Specific Contact Resistance," *Solid-State Electron.*, Vol. 25 (1982) pp. 91–94.
15. G. Dresselhaus, A.F. Kip, C. Kittel, "Cyclotron Resonance of Electrons and Holes in Silicon and Germanium Crystals," *Phys. Rev.*, Vol. 98 (1955) pp. 368–384.
16. E.H. Hall, "On a New Action of the Magnet on Electric Currents," *American J. of Math.*, Vol. 2 (1879) pp. 287–292.
17. L.J. van der Pauw, "A Method of Measuring Specific Resistivity and Hall Effect of Discs of Arbitrary Shape," Technical report, *Philips Res. Reports*, Vol. 13 (1958) pp. 1–9.
18. H.A. Lorentz, "La Théorie Électromagnétique de Maxwell et son Application aux Corps Mouvants," *Arch. Ne´erl.*, Vol. 25 (1892) pp. 363–552.
19. "Appendix A: Hall Effect Measurements," Lake Shore 7500/9500 Series Hall System User's Manual, Lake Shore Cryotronics, Inc., Westerville, OH.
20. S. Ramo, "Currents Induced by Electron Motion," *Proc. IRE*, Vol. 27 (1939) pp. 584–585.
21. W. Shockley, "Currents to Conductors Induced by a Moving Point Charge," *J. Appl. Phys.*, Vol. 9 (1938) pp. 635–636.
22. J.R. Haynes, W. Shockley, "The Mobility and Life of Injected Holes and Electrons in Germanium," *Phys. Rev.*, Vol. 81, Issue 5 (1951) pp. 835–843.
23. J.E. Baciak, Z. He, "Spectroscopy on Thick HgI_2 Detectors: A Comparison Between Planar and Pixelated Electrodes," *IEEE Trans. Nucl. Sci.*, Vol. 50, no. 4 (2003) pp. 1220–1224.
24. K. Hecht, "Zum Mechanismus des lichtelektrischen Primärstromes in isolierenden Kristallen," *Z. Physik,* Vol. 77 (1932) pp. 235–245.
25. H. Kim, L. Cirignamo, A. Churliov, G. Ciampi, W. Higgins, F. Olschner, K. Shah, "Developing Larger TlBr Detectors—Detector Performance," *IEEE Trans. Nucl. Sci.*, Vol. 56, Issue 3 (2009) pp. 819–823.
26. M.C. Veale, P.J. Sellin, A. Lohstroh, A.W. Davies, J. Parkin, P. Seller, "X-Ray Spectroscopy and Charge Transport Properties of CdZnTe Grown by the Vertical Bridgman Method," *Nucl. Instr. and Meth.*, Vol. A576 (2007) pp. 90–94.
27. A. Ruzin, Y. Nemirovsky, "Statistical Models for Charge Collection and Variance in Semiconductor Spectrometers," *J. Appl. Phys.*, Vol. 82, no. 6 (1997) pp. 2754–2758.

28. Y. Cui, G.W. Wright, X. Ma, K. Chattopadhyay, R.B. James, A. Burger, "DC Photoconductivity Study of Semi-Insulating $Cd_{1-x}Zn_xTe$ Crystals," *Journal of Electronic Materials*, Vol. 30, issue 6 (2001) pp. 774–778.
29. A. Many, "High-Field Effects in Photoconducting Cadmium Sulphide," *J. Phys. Chem. Solids*, Vol. 26, Issue 3, (1965) pp. 575–578.
30. M.G. Buehler, "Impurity Centers in PN Junctions Determined from Shifts in the Thermally Stimulated Current and Capacitance Response with Heating Rate," *Solid-State Electron.*, Vol. 15 (1972) pp. 69–79.
31. K.H. Nicholas, J. Woods, "The Evaluation of Electron Trapping Parameters from Conductivity Glow Curves in Cadmium Sulphide," *British. J. Appl. Phys.*, Vol. 15 (1964) pp. 783–795.
32. P.D. Townsend, Y. Kirsh, "Spectral Measurement during Thermoluminescence—An Essential Requirement," *Contemporary Physics*, Vol. 30 (1989) pp. 337–354.
33. D.V. Lang, "Deep Level Transient Spectroscopy: A New Method to Characterize Traps in Semiconductors," *J. Appl. Phys.*, Vol. 45 (1974) pp. 3023–3032.
34. P.M. Mooney, "Defect Identification Using Capacitance Spectroscopy," in *Identification of Defects in Semiconductors*, Ed. M. Stavola, in *Semiconductors and Semimetals*, Academic Press, San Diego (1999) pp. 93–152.
35. P.M. Mooney, "Deep Donor Levels (DX Centers) in III–V Semiconductors," *J. Appl. Phys.*, Vol. 67 (1990) pp. R1–R26.
36. R.H. Bube, *Photoconductivity of Solids*, Wiley, New York (1960).
37. Ch. Hurtes, M. Boulou, A. Mitonneau, D. Bois, "Deep-Level Spectroscopy in High-Resistivity Materials," *Appl. Phys. Lett.*, Vol. 32, no. 12 (1978) pp. 821–823.
38. J.C. Balland, J.P. Zielinger, C. Noguet, M. Tapiero, "Investigation of Deep Levels in High-Resistivity Bulk Materials by Photo-Induced Current Transient Spectroscopy. I. Review and Analysis of Some Basic Problems," *J. of Phys. D: Appl. Phys.*, Vol. 19 (1986) pp. 57–70.
39. M. Tapiero, N. Benjelloun, J.P. Zielinger, S. El Hamd, C. Noguet, "Photoinduced Current Transient Spectroscopy in High Resistivity Bulk Materials: Instrumentation and Methodology," *J. Appl. Phys.*, Vol. 64 (1988) pp. 4006–4012.
40. R.E. Kremer, W.B. Leigh, "Deep Levels in CdTe," *J. of Crys. Growth.*, Vol. 86, Issues 1–4 (1990) pp. 490–496.

41. A. Owens, "Synchrotron Light Sources and Detector Metrology," *Nucl. Inst. and Meth.*, Section A(2012), doi: 10.1016/j.nima. 2011.11.077.
42. R. Klein, G. Ulm, M. Abo-Bakr, P. Budz, K. Bürkmann-Gehrlein, D. Krämer, J. Rahn, G. Wüstefeld, "The Metrology Light Source of the Physikalisch-Technische Bundesanstalt in Berlin-Adlershof," *Proc. of EPAC 2004*, Lucerne, Switzerland (2004) pp. 2290–2292.
43. J. Schwinger, "On the Classical Radiation of Accelerated Electrons," *Phys. Rev.*, Vol. 75 (1949) pp. 1912–1925.
44. F.R. Elder, A.M. Gurewitsch, R.V. Langmuir, H.C. Pollock, "Radiation from Electrons in a Synchrotron," *Phys. Rev. B.*, Vol. 71 (1947) pp. 829–830.
45. A. Hofmann, *The Physics of Synchrotron Radiation*, Cambridge University Press (2004).
46. H. Winick, D. Attwood, "Operating and Planned Facilities," in *X-Ray Data Booklet*, Section 2.3, LBNL/PUB-940 Rev. 3, Lawrence Berkeley National Laboratory (2009) p. 2–29.
47. Diamond Light Source, http://www.diamond.ac.uk/.
48. K. Kim, "Characteristics of Synchrotron Radiation," in *X-Ray Data Booklet*, Section 2.1, LBNL/PUB-940 Rev. 3, Lawrence Berkeley National Laboratory (2009) pp. 2–15.
49. DESY, "Beamline: X1 (RÖMO II) (2007) http://hasylab.desy.de/facilities/doris_iii/beamlines/x1_roemo_ii/index_eng.html.
50. A. Owens, T. Buslaps, V. Gostilo, H. Graafsma, R. Hijmering, A. Kozorezov, A. Loupilov, D. Lumb, E. Welter, "Hard X- and γ-Ray Measurements with a Large Volume Coplanar Grid CdZnTe Detector," *Nucl. Instr. and Meth.*, Vol. A563 (2006) pp. 242–248.
51. European Synchrotron Radiation Facility, "ID15—High Energy Diffraction and Scattering Beamlines," http://www.esrf.eu/UsersAndScience/Experiments/StructMaterials/ID15.
52. C. Ponchut, "Characterization of X-ray area detectors for synchrotron beamlines," *J. Synch. Rad.*, Vol. 13 (2006) pp. 195–203.
53. A. Owens, H. Andersson, M. Bavdaz, G. Brammertz, C. Erd, T. Gagliardi, V. Gostilo, N. Haack, I. Lisjutin, S. Nenonen, A. Peacock, H. Sipila, I. Taylor, S. Zataloka, "Development of compound semi-conductor arrays for X- and gamma-ray spectroscopy," *Proc. of the SPIE*, Vol. 4507 (2001) pp. 42–49.
54. C. Erd, A. Owens, G. Brammertz, D. Lumb, M. Bavdaz, A. Peacock, S. Nenonen, H. Andersson, "Measurements of the Quantum Efficiency and Depletion Depth in Gallium-Arsenide Detectors," *Proc. of the SPIE*, Vol. 4784 (2002) pp. 386–393.

55. R. den Hartog, A.G. Kozorezov, A. Owens, J.K. Wigmore, V. Gostilo, A. Loupilov, V. Kondratjev, M.A. Webb, E. Welter, "Synchrotron study of charge transport in a CZT ring-drift detector," *Nucl. Instr. and Meth.*, Vol. A648 (2011) pp. 155–162.
56. A. Kozorezov, V. Gostilo, A. Owens, F. Quarati, M. Shorohov, M.A. Webb, J.K. Wigmore, "Polarization Effects in Thallium Bromide X-Ray Detectors," *J. Appl. Phys.*, Vol. 108 (2010) pp. 1-1–1-10.
57. D.S. Bale, C. Szeles, "*Nature of Polarization in Wide-Bandgap Semiconductor Detectors under High-Flux Irradiation: Application to Semi-Insulating* $Cd_{1-x}Zn_xTe$," *Phys. Rev. B*, Vol. 77 (2008) pp. 035205–035221.
58. G.S. Camarda, A.E. Bolotnikov, Y. Cui, A. Hossain, R.B. James, "Polarization Studies of CdZnTe Detectors Using Synchrotron X-Ray Radiation," *IEEE Trans. Nucl. Sci.*, Vol. 55, no. 6 (2008) pp. 3725–3730.
59. K.G. Mckay, "Electron Bombardment Conductivity in Diamond," *Phys. Rev.*, Vol. 74 (1948) pp. 1606–1621.
60. Y.L. Wu, Y.-T. Chen, Z.C. Feng, J.-F. Lee, P. Becla, W. Lu, "Synchrotron Radiation X-Ray Absorption Fine-Structure and Raman Studies on CdZnTe Ternary Alloys," *Proc. of the SPIE*, Vol. 7449 (2009) pp. 74490Q-1–74490Q-11
61. L. Vegard, "Die Konstitution der Mischkristalle und die Raumfüllung der Atome," *Zeitschrift für Physik*, Vol. 5 (1921) pp. 17–26.
62. M.C. Duff, D.B. Hunter, P. Nuessle, D.R. Black, H. Burdette, J. Woicik, A. Burger, M. Groza, "Synchrotron X-Ray Based Characterization of CdZnTe Crystals," *J. of Elect. Mater.*, Vol. 36, no. 8 (1010) pp. 1092–1097.
63. M.C. Duff, D.B. Hunter, A. Burger, M. Groza, V. Buliga, J.P. Bradley, G. Graham, Z. Dai, N.E. Teslich, D.R. Black, H. Burdette, A. Lanzirotti, "Characterization of Spatial Heterogeneities in Detector Grade CdZnTe," paper presented at the 2011, MRS spring meeting, San Francisco.
64. A. Owens, "XANES Fingerprinting: A Technique for Investigating CCD Surface Features and Measuring Dead Layer Thicknesses," *Nucl. Instr. and Meth.*, Vol. A526 (2004) pp. 391–398.

65. D.T. Cromer, D. Liberman, "Relativistic Calculation of Anomalous Scattering Factors for X Rays," *J. Chem. Phys.*, Vol. 53 (1970) pp. 1891–1898.
66. A. Owens, G.W. Fraser, S.J. Gurman, "Near K-Edge Linear Attenuation Coefficients for Si, SiO_2 and Si_3N_4," *Rad. Phys. and Chem.*, Vol. 65 (2002) pp. 109–121.
67. A. Owens, S. Bayliss, G.W. Fraser, S.J. Gurman, "On the Relationship Between Total Electron Photoyield and X-Ray Absorption Coefficient," *Nucl. Instr. and Meth.*, Vol. A358 (1997) pp. 556–558.

4
Contacting Systems

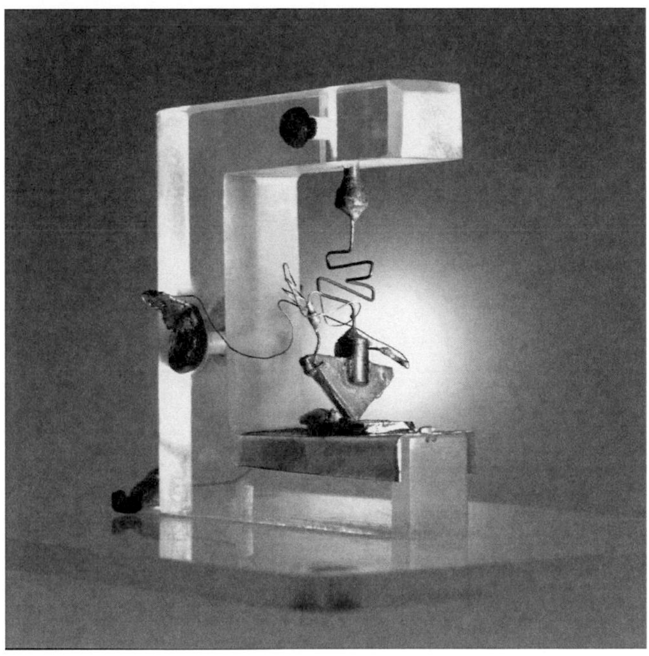

Frontispiece The first point contact transistor constructed at Bell Labs in 1947. The emitter and collector contacts are made of gold and are attached to the sides of a truncated wedge shaped piece of insulating material whose apex is in mechanical contact with a block of germanium. The other face of the semiconductor is in turn in mechanical contact with a gold plate that forms the base contact. Note that the "spring" providing the mechanical pressure is actually a bent paper clip. (Image courtesy of The Porticus Centre, Beatrice Consumer Products Inc., subsidiary of Beatrice Companies Inc.)

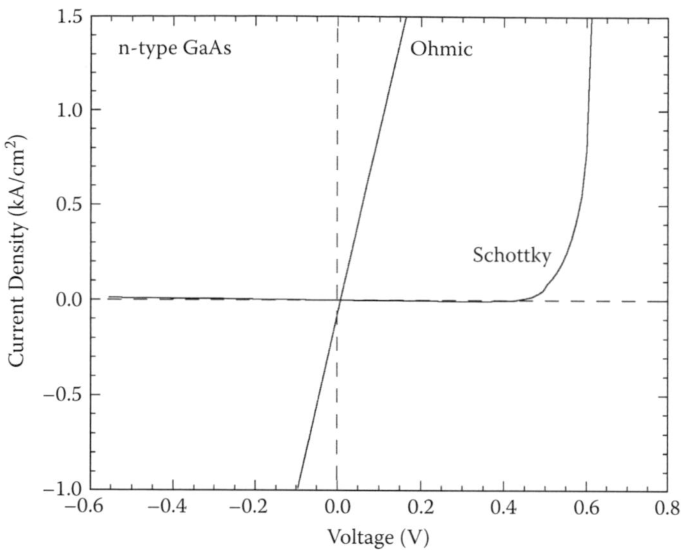

Figure 4.1 Examples of the current-voltage characteristics from ohmic and Schottky metal–semiconductor contacts to n-type GaAs doped to 10^{15} cm^{-3}.

Once charge is created in a semiconducting material, it must be extracted, and this is the function of the electrodes. However, the formation of stable and laterally uniform contacts can be a major problem—the practical aspects of which will be discussed in later sections. In general, metallization systems used are rectifying. However, depending on the actual metallization and semiconductor, they can display a wide range of behavior, ranging from very nearly ohmic to strongly rectifying. Based on the current-voltage characteristics they can be broadly classified into two groups (see Figure 4.1). These are:

1. Low-resistance or "ohmic" contacts.* Ideally, these have linear current-voltage characteristics; that is, there is an unimpeded transfer of either carrier from one material to another. The advantage of a true ohmic contact

* Also known as charge injecting contacts.

is that providing the contact resistance is small, it does not affect or modify the detector signal. From a practical point of view, Rhoderick [1] has proposed that this definition include all low-resistance contacts—meaning that the impedance of the contact is negligible compared to the impedance of the active region of the semiconductor, and the contact does not affect the I-V characteristics of the device.

2. Semipermeable contacts. These are rectifying or Schottky contacts that have current-voltage characteristics that are nonlinear and asymmetric. In this situation, the current is only allowed to freely pass in one direction, as in a diode. For low intrinsic doping the Schottky barrier acts as a diode with a lower turn-on voltage than a p-n junction diode and is therefore most frequently used for high-speed switching applications, mixers for heterodyne receivers, and as frequency multipliers for local oscillators. In practical detection systems they have the advantage that, by reverse biasing a detector, a higher bias can be applied without significantly increasing the leakage current.

   A subset of semipermeable contacting systems is the "blocking" or noninjecting contact, which is a rectifying contact that is universally used to reduce leakage currents in low-bandgap, low-noise systems by blocking the replacement of free carriers in the bulk after bias is applied.

Contacting can be achieved by several methods, such as applying metal-bearing paints and pastes, such as Aquadag,* melting metals directly on the semiconductor surface, evaporation, sputtering, molecular beam epitaxy, ion-implantation, and others. For a comprehensive review of contacting systems and compound semiconductor interfaces, see references [2–4].

* "Aqueous Deflocculated Acheson Graphite"—a colloidal dispersion of pure graphite in a water carrier, manufactured by Acheson Industries, a subsidiary of ICI.

4.1 Metal Semiconductor Interfaces

When the surfaces of a metal and semiconductor come into intimate contact, the valence and conduction bands of the semiconductor bend so that the Fermi levels in the metal and semiconductor equalize as shown in Figure 4.2. A so-called

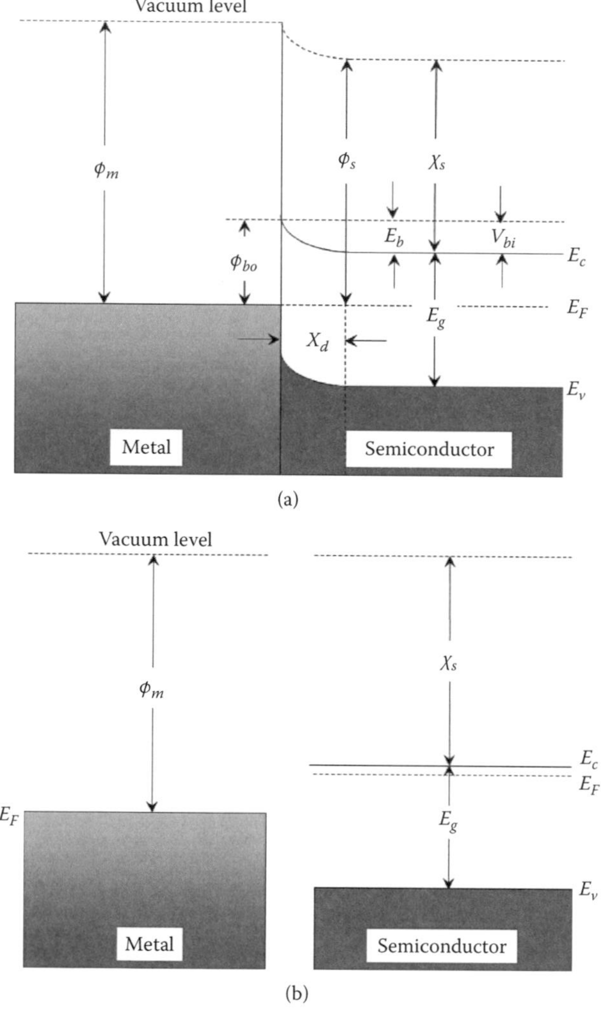

Figure 4.2 **(See color insert.)** Energy band diagram of a metal and n-type semiconductor (a) before and (b) after contact is made.

Schottky barrier is formed whose height is defined as the energy distance between the Fermi level, E_f, and the respective majority carrier band-edge at the interface, that is, the valence-band maximum, E_v, for p-type semiconductors or the conduction-band minimum, E_c, for n-type semiconductors. This barrier prevents the flow of electrons from the metal to the semiconductor, but promotes the flow of holes. The height of the barrier on the semiconductor side can, however, be altered by applying a potential across the junction.* It can raise the barrier height, reducing conduction, or lower it, increasing conduction. Thus, a rectifying contact has been formed where the junction conducts for one bias polarity, but not the other. Because thermally generated carriers are created with a distribution of energies, a small number will have enough energy to surmount the barrier and cross to the other material. This essentially is the leakage current and is a sensitive function of temperature. Metal semiconductor junctions are primarily majority carrier devices. Majority carriers flow from the semiconductor to the metal. Minority carrier injection into the semiconductor can usually be neglected.

4.2 Schottky Barriers

The rectifying properties of metal–semiconductor contacts were first investigated by Braun [5] in 1874, who observed that when metal sulfides were contacted by metal points, the electrical resistance varied with the magnitude and polarity of the applied voltage. While not explicitly mentioning a potential barrier, Braun attempted to explain these phenomena by a thin interface layer of extremely high resistance. In 1938, both Schottky [6] and independently Mott [7] pointed out that the observed direction of rectification could be explained by assuming that electrons have to pass over a potential barrier following the normal process of drift and diffusion through the semiconductor. They postulated that the height of the barrier equals the difference of the work

* Note the bias does not change the barrier height from the metal side.

function of the metal and the electron affinity of the semiconductor, thus

$$\phi_{bo} = \phi_m - \chi_s \tag{4.1}$$

where ϕ_{bo} is the contact barrier height encountered by electrons in the metal, ϕ_m = work function of the metal, and χ_s is the electron affinity of the semiconductor, given by

$$\chi_s = \phi_s - (E_c - E_f) \tag{4.2}$$

Here, ϕ_s is the work function of the semiconductor defined as the minimum energy needed to remove an electron from the Fermi energy level into vacuum. The electron affinity, in turn, is defined as the work required to remove an electron from an energy level at the bottom of the conduction band to an energy level corresponding to an electron at rest in vacuum outside the solid and beyond the range of the image force (see Section 4.2.1). The built-in potential V_{bi}, which is the potential across the junction in thermal equilibrium, encountered by electrons in the semiconductor, is given by

$$V_{bi} = \phi_m - \phi_s = \phi_{bo} - (E_c - E_f) \tag{4.3}$$

4.2.1 Image Force Reduction of the Schottky Barrier

Once the barrier is formed, a space charge region depleted of mobile carriers is created in the semiconductor adjacent to the metal layer. Obviously a thin layer of space charge with opposite polarity must also exist in the metal at the interface to complete the charge dipole and maintain charge neutrality. It can be shown that the electric field in the semiconductor is identical to that of the carrier itself and also another carrier with opposite charge at equal distance but on the opposite side of the interface. This charge is called the *image charge*. The difference between the actual surface charges and the image charge is that the fields in the metal are distinctly different. Image charges build up in the metal electrode of a metal–semiconductor junction as carriers approach the interface.

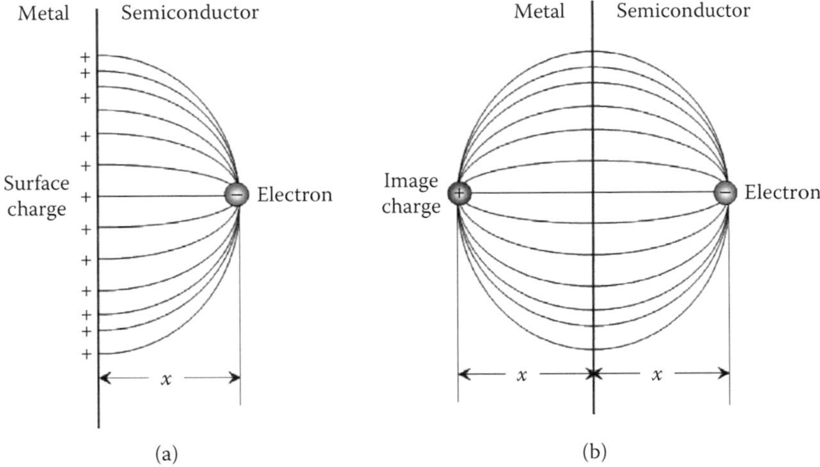

Figure 4.3 (a) Field lines and surface charges due to an electron in close proximity to a perfect conductor and (b) the field lines and image charge of an electron.

The potential associated with these charges reduces the effective barrier height and leads to a net increase in the current flow across the junction.

Consider an electron in vacuum, at a distance x from a metal surface (see Figure 4.3). From the Shockley–Ramo theorem [8,9], a positive charge will be induced on the metal at a distance $-x$ from its surface and will give rise to an attractive force between the two, known as the image force

$$F(x) = -qE_i(x) = -\frac{q^2}{4\pi\varepsilon_o\varepsilon_s(2x)^2} \tag{4.4}$$

where q is the unit electronic charge, ε_o is the permittivity of free space, ε_s is the dielectric constant of the semiconductor. This force has associated with it an image potential energy which corresponds to the potential energy of an electron at a distance x from the metal and is given by

$$\phi(x) = -\int_x^{\infty} E_i(x)\,dx - \frac{q}{16\pi\varepsilon_o\varepsilon_d x} \tag{4.5}$$

where ε_d is the image force dielectric constant which for most semiconductors is nearly equal to the static dielectric constant.* When Equation (4.3) is combined with the potential variation due to the electric field, we obtain the potential energy, V(x), at position x,

$$V(x) = -qE_{max}x - \frac{q^2}{16\pi\varepsilon_o\varepsilon_d x} \tag{4.6}$$

where the field due to the charge in the depletion region is assumed to be constant and equal to the maximum electric field at the interface, E_{max}, where

$$E_{max} = \sqrt{\frac{2qN_d(V_{bi} - V)}{\varepsilon_o\varepsilon_d}} \tag{4.7}$$

Here V_{bi} is the built-in potential, and V is the applied voltage. The potential energy due to the distributed charge of the ionized donors in the depletion region and a single electron reaches its maximum value at

$$x_{max} = \sqrt{\frac{q}{16\pi\varepsilon_o\varepsilon_s E_{max}}} \tag{4.8}$$

and the corresponding maximum value of the potential energy equals

$$V_{max} = q\phi - q\Delta\phi \tag{4.9}$$

where $\Delta\phi$ is the barrier height reduction given by [10]

$$\Delta\phi = \sqrt{\frac{qE_{max}}{4\pi\varepsilon_o\varepsilon_d}} \tag{4.10}$$

* The static dielectric constant primarily determines the depletion width through Poisson's equation. The image force dielectric constant is essentially the high frequency dielectric constant by virtue of the potential sensed by moving charges from the image force.

Thus, the actual barrier height is now $\phi_b - \Delta\phi$. For a uniformly doped semiconductor $\Delta\phi$ is given by

$$\Delta\phi = \left[\frac{q^2 E_b N_d}{8\pi^2 \varepsilon_o^3 \varepsilon_s \varepsilon_d}\right]^{\frac{1}{4}} = \left[\frac{q^3 N_d}{8\pi^2 \varepsilon_o^3 \varepsilon_d^2 \varepsilon_s}\left(V + V_d - \frac{kT}{q}\right)\right]^{\frac{1}{4}} \quad (4.11)$$

where E_b is the amount of energy by which the majority carrier band-edge is shifted at the interface to equalize the Fermi levels, V is the externally applied bias across the depletion region, and V_d is the diffusion potential associated with the barrier (that is ϕ_b minus the Fermi energy).

The barrier reduction tends to be small compared to the barrier height itself (typically, $\Delta\phi < 0.2\phi_b$). However, it is important because it depends on the applied voltage and leads to a voltage dependence of the reverse bias current, but note that the image force is not zero at zero bias because the electric field at zero bias is not zero.

4.2.2 Barrier Width

One consequence of the image force is that a depletion layer is created whose width is equal to that needed to support a potential change equal to the built-in potential. It we assume a uniform distribution of ionized impurities in the semiconductor, the shape of the potential energy barrier can be approximated by a one-dimensional solution of Poisson's equation:

$$\phi(x) = qV(x) = \frac{q^2 N_d x^2}{2\varepsilon_s \varepsilon_o} \quad (4.12)$$

for $0 \le x \le w$. Here, w is the depletion layer width, N_d is the ionized donor concentration in the semiconductor, ε_s the relative static dielectric constant, and ε_o the permittivity of free space. The width of the depletion layer is related to the amount of energy, E_b; the band has been shifted in the depletion region (see Figure 4.2) by

$$E_b = \phi_b - \phi_s - qV_{bias} = \frac{qN_d w^2}{2\varepsilon_s \varepsilon_o} \quad (4.13)$$

where, ϕ_b is the barrier height, ϕ_s is the position of the Fermi level relative to the conduction band edge, and V_{bias} is the applied forward bias. Thus the depletion width is given by

$$w = \sqrt{\frac{2\varepsilon_s\varepsilon_o(\phi_b - \phi_s - qV_{\mathrm{bias}})}{qN_d}} \tag{4.14}$$

and varies as $V_{\mathrm{bias}}^{1/2}$ and $N_d^{-1/2}$. Note, the dependences on impurity concentration and bias are stronger than for the corresponding reduction in barrier height by the image force, which varies as $V_{\mathrm{bias}}^{1/4}$ and $N_d^{1/4}$ (see Section 4.2.1). Typically, depletion widths are of the order of tens of nanometers for moderately doped semiconductors. For example, the unbiased depletion depth for a Cr–Si junction doped to 10^{17} cm^{-3} is ~60 nanometers, which increases to 260 nm under 5V reverse bias.

In reality, the shape of a metal–semiconductor potential barrier is not truly parabolic because of the interaction of the image force potential, which modifies the energy distribution function as follows [11]:

$$\phi(x) = \frac{q^2 N_d x^2}{2\varepsilon_s\varepsilon_o} - \frac{q^2}{16\pi\varepsilon_d\varepsilon_o(w - x)} \tag{4.15}$$

where ε_d is the relative dynamic (high frequency) dielectric constant of the semiconductor [12]. Consolidating the barrier and material properties into one term, we obtain [11],

$$\frac{\phi(x)}{E_b} = \frac{V(x)}{E_b} = \frac{x^2}{w^2} - \frac{1}{8\pi(E_b/E_{11})^{3/2}(1 - x/w)} \tag{4.16}$$

where $(E_b/E_{11})^{3/2} = Nw^3\varepsilon_d/\varepsilon_s$. The numerator, E_{11}, is a constant of the material, given by

$$E_{11} = \frac{q^2}{2\varepsilon_o}\left[\frac{N_d}{\varepsilon_s\varepsilon_d^2}\right]^{\frac{1}{3}} = 9.05\times10^{-7}\left[\frac{N_d}{\varepsilon_s\varepsilon_d^2}\right]^{\frac{1}{3}} \quad (eV) \tag{4.17}$$

and is a measure of the modification of the barrier shape by the image force. In fact, the term $(E_b/E_{11})^{3/2}$ in Equation (4.16) uniquely describes the deviation of the barrier shape from parabolic. The zero image force situation occurs when $(E_b/E_{11})^{3/2} \rightarrow \infty$.

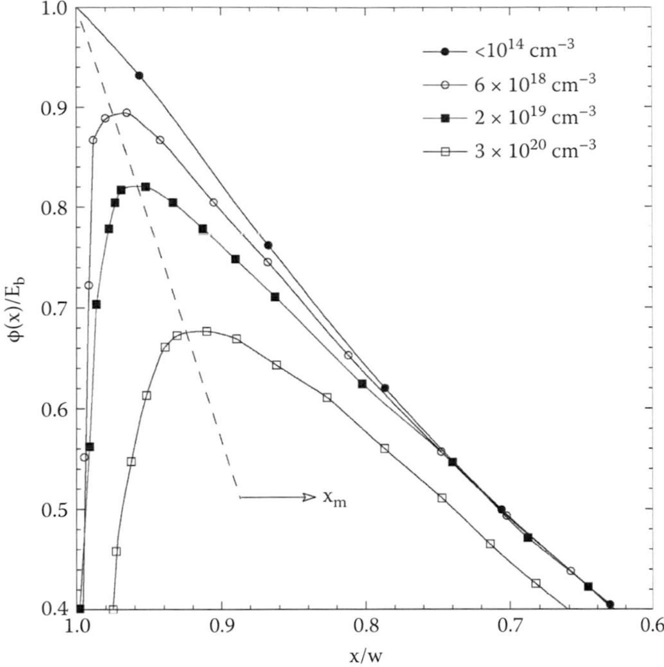

Figure 4.4 The effect of the image force on the shape of the potential barrier at a metal–semiconductor interface for n-type GaAs with a band-bending energy of 1 eV; x_m is the maximum height of the barrier.

In Figure 4.4 we plot the barrier potential energy distribution for n-type GaAs, which includes the modification due to the image force for a range of donor carrier densities. We note that, for a practical range of impurity donor concentrations,* the barrier height is only reduced by ~20%. We also see that, as the donor concentration increases, the barrier height not only decreases, but the peak of the potential energy distribution shifts into the semiconductor. The distance, x_m, of the potential energy maximum from the metal–semiconductor interface is

$$x_m = \frac{\Delta\phi}{2E_i} = \left(\frac{q}{16\pi\varepsilon_d\varepsilon_o E_i}\right)^{\frac{1}{2}} \tag{4.18}$$

* The practical doping limit for n-type GaAs is ~ 2×10^{19} cm^{-3}, although higher concentrations are easier to achieve for p-type material.

where E_i is the electric field at the metal–semiconductor interface. For Si, $\varepsilon_d = 12$, and x_m varies from 10 and 50 Å, depending on the applied field [12].

4.2.3 Measured Barrier Heights

It was soon found experimentally (see, for example, Schweikert [13]) that barrier heights in metal–semiconductor systems vary proportionally to the metal work function, but with a slope much smaller than unity, typically between 0.1 and 0.3. The departure of experiment from theory was first explained in terms of localized electronic surface states [14] or "dangling bonds" resulting from immobilized atoms with unfulfilled valence [15,16]. These arise naturally at the surface of a solid because the atoms have neighbors on one side only. As a consequence, at a metal–semiconductor interface the wavefunctions of the metal electrons decay exponentially into the semiconductor, forming a continuum of metal-induced gap states (MIGS) in the semiconductor forbidden energy band gap [17]. Physically these states may extend up to 1 nm into the semiconductor and determine the barrier height in an ideal, abrupt, defect-free, and laterally homogenous metal–semiconductor contact. However, in practice, there is limited agreement between experiment and theory, because surfaces are rarely ideal, because of defects introduced during interface formation [18], surface contamination, and oxide layers introduced during chemical processing.

4.2.3.1 Metal-Induced Gap States (MIGS)

The MIGS are continuously distributed in energy within the forbidden gap and are characterized by a "neutral level," ϕ_o, such that the surface states are occupied up to ϕ_o and empty above it, and hence the surface is electrically neutral. In general, the Fermi level and neutral level do not coincide. In this case, there is a net charge in the surface states. If, in addition, there is a thin oxide layer between the metal and the semiconductor,* the charge in the surface states together with its image charge on the surface of the metal will form a dipole layer. This dipole layer will alter the potential difference between the semicon-

* This is invariably introduced during chemical processing of the semiconductor. Such an oxide film is referred to as an interfacial layer.

ductor and the metal, which can be described by a simple modification [19] to the Schottky–Mott equation:

$$\phi_b = \gamma(\phi_m - \chi_s) + (1-\gamma)(E_g - \phi_o) - \Delta\phi_b \tag{4.19}$$

Here, E_g is the bandgap of the semiconductor in eV, ϕ_o is the position of neutral level (measured from the top of the valence band), $\Delta\phi_b$ is the reduction in barrier height due to image force lowering, and γ is a weighting factor,* which depends mainly on the surface state density and the thickness of the interfacial layer and is given by

$$\gamma = \frac{\varepsilon_o \varepsilon_i}{\varepsilon_o \varepsilon_i + q\delta D_s} \tag{4.20}$$

Here, D_s is the density of surface states per unit area per eV at the metal–semiconductor interface, ε_i is the relative permittivity of the interfacial layer, and δ is its thickness. The interface states can be classified broadly into two groups—intrinsic and extrinsic. The intrinsic states arise as a result of the discontinuity in the crystal structure of the solids at the interface, while the extrinsic defects are the results of chemical reactions or damage to the surface of the semiconductor during the metal deposition process. If there are no surface states, $D_s = 0$ and $\gamma = 1$, and Equation (4.19) reduces to the classical Schottky–Mott equation given by Equation (4.1), except for the image force-lowering term. However, if the density of states is large, ε becomes small, $\gamma \to 0$, and ϕ_b approaches a limiting value of $(E_g - \phi_o)$. When this occurs, the Schottky barrier height (SBH) is no longer dependent on the metal work function and the Fermi level and is said to be pinned relative to the band edges by the surface states.

4.2.3.2 Fermi Level Pinning

It was found experimentally that Schottky barrier heights for a wide range of metals in contact with a particular semiconductor fall within a narrow range. In fact, this range is so narrow that it is described empirically by the "one-third" rule, that the SBH is roughly $^1/_3$ of the bandgap in a p-type semiconductor and $^2/_3$ of the bandgap in an n-type semiconductor.

* Also known as the Schottky pinning factor.

Thus, the Fermi level appears to be pinned at $E_g/3$ from the valence band maximum. Experimentally, Fermi level pinning is also found to be dependent largely on the type of bonding. For example, pinning tends to be more prevalent in covalently bonded compounds than ionically bonded ones and therefore is related to the electronegativity of a compound, because covalent substances tend to be less electronegative than ionic. Thus, we would expect pinned surfaces to be more common in group IV and group III–V materials rather than II–VI or I–VII compounds. Experimentally, this is indeed found to be the case. In Figure 4.5a we plot the quantity $E_c–E_f$ for Au contacts

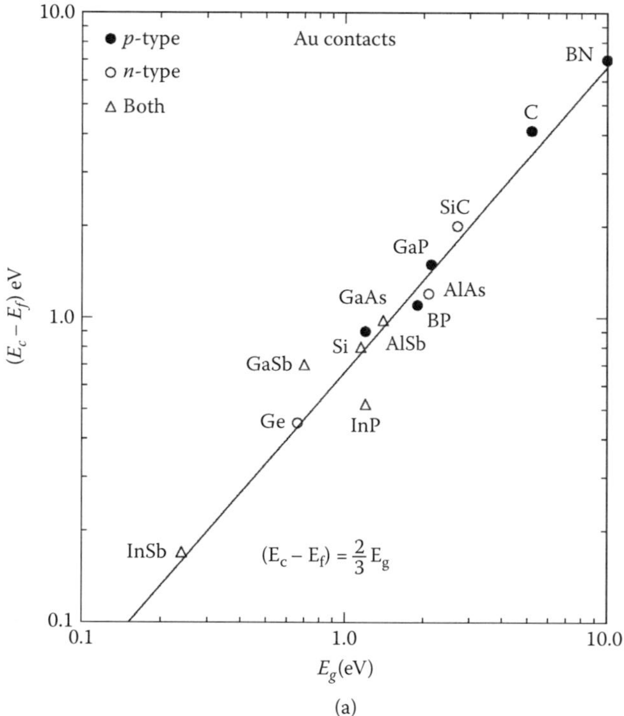

Figure 4.5 (a) The location of the Fermi level relative to the conduction band ($E_c–E_f$) barrier heights for Au contacts on various covalent semiconductors, plotted as a function of energy gap, illustrating the two-thirds rule for barrier height pinning for n-type semiconductors at the interface (adapted from [20]). (b) Experimentally determined barrier heights for various metals on an ionic semiconductor (ZnS) showing a clear dependence of the barrier height on work function. For contrast, we show data for a covalent semiconductor (GaAs), which has a much weaker dependence.

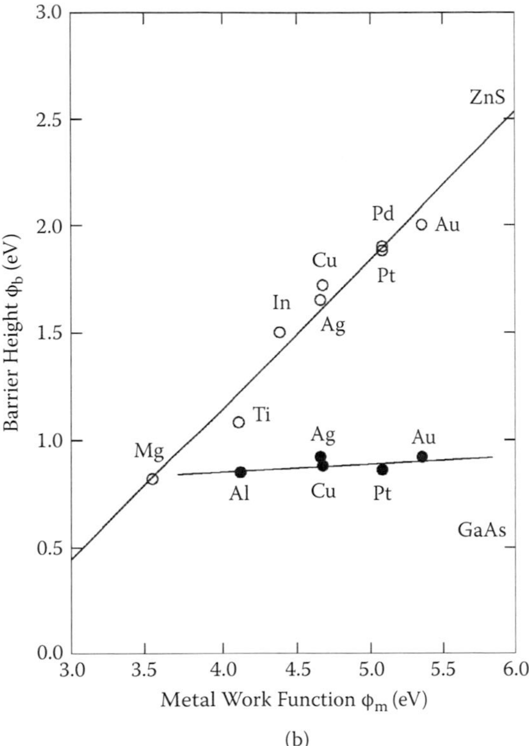

Figure 4.5 (Continued)

formed on various covalent semiconductors, as a function of bandgap energy. The data are well fitted by a straight line of slope $^2/_3$. In contrast, Figure 4.5b shows experimentally determined barrier heights for various metals on ZnS (a group II–VI ionic semiconductor) which display a marked dependence on metal work function, varying as $\sim\phi_m^{0.7}$. For comparison, we also show curves for a covalent compound (GaAs) which essentially show no variation; thus $\phi_b \sim \phi_m^{0.0}$. The difference has recently been attributed [21] to the observation that the penetration of MIGS in covalent semiconductors is deeper than in ionic semiconductors, and hence changes in metal work function are more effectively screened out. Calculation has shown [22] that the penetration depth of MIGS is typically 0.1 nm in ionic semiconductors and ~0.3 nm in covalent semiconductors.

4.3 Current Transport across a Schottky Barrier

The current across a metal–semiconductor junction is mainly due to majority carriers and, as we will see, is strongly influenced by the doping concentration in the semiconductor, N_d, and the junction temperature, T. There are three principal components to the overall current through the device [23,24]. These are thermionic emission (TE), thermionic field emission (TFE), and field emission (FE). The first component gives rise to the current rectification in metal–semiconductor diodes and is primarily dependent on the temperature and the barrier height. The other two components involve quantum mechanical tunneling through the barrier and are critically dependent on the impurity concentration level. The effect of impurity concentration on the structure of the metal–semiconductor interface is illustrated in Figure 4.6 for an n-type semiconductor.

4.3.1 Thermionic Emission (TE)

In a Schottky diode, TE is the dominant current transport mechanism. The current through the device is a two-step process. First, the electrons have to be transported through the depletion region by the processes of drift and diffusion, and second, they must undergo emission over the barrier into the metal. Figure 4.6a shows the interface band structure for an ideal Schottky barrier, which would be appropriate for a lightly doped semiconductor with a residual impurity doping concentration, $N_d < 10^{17}$ cm^{-3}. In this case, the depletion width is too wide to allow quantum mechanical tunneling, and therefore the only way for carriers to cross the barrier is by thermal excitation over it. The current through the barrier is dependent on the number of electrons that impinge on a unit area of the metal per second and is given by

$$I_{TE} = AA^*T^2 \exp\left(\frac{-q\phi_{bo}}{kT}\right)\left[\exp\left(\frac{-qV_{eff}}{nkT}\right) - 1\right] \tag{4.21}$$

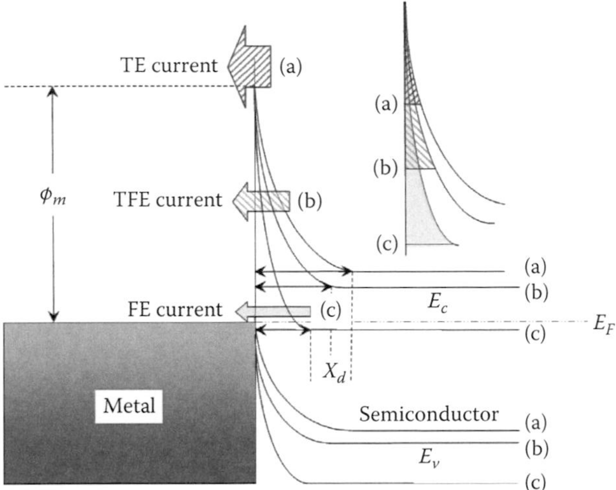

Figure 4.6 **(See color insert.)** Conduction mechanisms through a metal–semiconductor interface with different semiconductor donor levels for an n-type semiconductor, corresponding to (a) a low residual doping level of $<10^{17}$ cm^{-3}, (b) an intermediate doping level of 10^{17}cm^{-3} to 10^{18}cm^{-3}, and (c) a high doping level of $>10^{18}$cm^{-3}. It can be seen that, as the doping concentration increases, the barrier width decreases as the conduction band falls below the Fermi level. To the right of the barrier we show an additional schematic of the barrier in which the shaded regions give a visual indication of the fraction of carriers that can cross the barrier.

where A is the cross-sectional area of the metal–semiconductor interface, A^* is the modified Richardson constant, T is the temperature in K, k is Boltzmann's constant, q is the electronic charge, V_{eff} is the effective bias across the interface, and n is the ideality factor, which is a dimensionless quantity introduced to include contributions from other current-transport mechanisms. This mode of current transport is commonly referred to as thermionic emission current [24] and varies strongly with kT and barrier height. The ideality factor is a measure of the quality of the junction and is highly process dependent. For an ideal Schottky junction, $n = 1$; hence the emission is purely thermionic. In practice, however, n lies in the range $1 < n < 2$ due to the presence of non-ideal effects, mainly image force reduction of the barrier [24] and edge leakage currents. In fact, in the absence of other effects, image force lowering of the barrier will result in an ideality factor given by

$$n = \frac{1}{\left(1 - \frac{\Delta\phi_b}{V}\right)} \tag{4.22}$$

where $\Delta\phi$ is the image force lowering of the barrier given by Equation (4.10), and V is the applied bias. If tunneling effects become significant, the ideality factor can reach values >2. In this case, the junction performs better under reverse bias. In fact, as pointed out by Rideout [24], the deviation of the diode n-value from unity may be used as a measure of the relative contribution of thermionic emission and thermionic field tunneling to the conduction process.

4.3.2 Thermionic Assisted Field Emission (TFE)

For a moderately heavy doped semiconductor ($N_d \approx 10^{18}$ cm^{-3}), the depletion layer can become sufficiently thin that thermally assisted field emission (TFE) or alternately, thermally assisted tunneling, can take place in which hot carriers begin to tunnel through the top of the barrier. This is illustrated in Figure 4.6b. The derivation of an analytical form describing the current through the barrier rapidly leads to intractable expressions, because it is generally derived by expanding the tunneling probability in a Taylor series about the Fermi energy of the source material. Padovani and Stratton [25] approached the problem by assuming a simple parabolic barrier shape by ignoring image force barrier-lowering effects. It was found that the current due to TFE varies as

$$\exp\left(\frac{\phi_b}{E_{oo}\coth\left[\frac{E_o}{kT}\right]}\right) \tag{4.23}$$

where E_o is a constant given by,

$$E_o = E_{oo}\coth\left(\frac{E_{oo}}{kT}\right) \tag{4.24}$$

and E_{oo} is a tunneling parameter* related to the material properties of the semiconductor, given by the expression

$$E_{oo} = \left(\frac{qh}{4\pi}\right)\sqrt{\frac{N_d}{m^*\varepsilon_o\varepsilon_r}} = 1.85\times 10^{-11}\sqrt{\frac{N_d}{m_r\varepsilon_r}} \text{ (meV)} \quad (4.25)$$

Here, h is Planck's constant, N_d is the impurity doping concentration, and m^* is the effective mass of electron = $m_r m_o$.

Although TFE conduction has a complicated dependence on T and N_d, we can make some general statements. Since E_{oo} varies as $\sqrt{N_d}$, therefore E_{oo}/kT is proportional to $\sqrt{N_d}/T$, and as the temperature increases, the fraction of current transported due to thermionic emission also increases. However, if the doping is increased, the barrier width is reduced and the fraction due to quantum mechanical tunneling is enhanced.

For all except very low bias, the tunneling current, I_t, is given by

$$I_t = I_{to}\left[\exp\left(\frac{qV_{\text{bias}}}{E_o}\right) - 1\right] \quad (4.26)$$

where I_{to} is the tunneling saturation current which is a complicated function of temperature, barrier height, and semiconductor parameters. If we treat the material properties empirically, then we can incorporate them into the diode equation and arrive at the following approximate form for the combined TE and TFE currents [24]

$$I_t = I_w\left[\exp\left(\frac{qV_{\text{bias}}}{nkT}\right) - \exp\left(\left(\frac{1}{n} - 1\right)\frac{qV_{\text{bias}}}{kT}\right)\right] \quad (4.27)$$

which is applicable for both forward and reverse bias. When n equals unity, Equation (4.27) reduces to the classical diode equation given by Equation (4.21).

* The quantity E_{oo} is directly related to the expression derived for the barrier transmission derived from a Wentzel–Kramers–Brillouin (WKB) quantum mechanical treatment.

4.3.3 Field Emission (FE)

As the impurity concentration increases further, more and more thermal carriers begin to tunnel through increasingly wide regions of the barrier, until at some point ($N_d \geq 10^{19}$ cm^{-3}) significant numbers of carriers can even tunnel through the base as illustrated in Figure 4.6c. Conduction now takes place through the entire barrier, and the carrier flow is referred to as field emission (FE), or carrier tunneling emission and is the preferred mode of current transport in metal–semiconductor ohmic contacts. The current flow in this case is found to depend mainly on the barrier height and the tunneling parameter E_{oo}; thus

$$I_{FE} \propto \exp\left(\frac{\phi_b}{E_{oo}}\right) \tag{4.28}$$

with little dependence on temperature. Since E_{oo} is proportional to $\sqrt{N_d}$, the forward bias characteristic for field emission-dominated conduction is strongly dependent on the doping concentration. The transition from FTE to FE is illustrated in Figure 4.7. As with FTE, an analytic formulation of FE current transport is complicated by the necessity to use Taylor expansions. An additional complication is that the theories of FE and TFE are distinct and do not converge to one another as the fields and energies approach. However, Padovani and Stratton [25] showed that the transition between FTE and FE can be defined in terms of a constant, c_1, given by

$$c_1 = \frac{1}{2E_{oo}} \log\left[\frac{4(E_b - E)}{\xi}\right] \tag{4.29}$$

where E_b is the potential energy of the top of the barrier with respect to the Fermi level, E is the potential energy associated with an applied bias V, and ξ is the energy of the Fermi level with respect to the bottom of the conduction band. FTE dominates when $1/c_1 < kT$ and FE when $1/c_1 > kT$.

4.3.4 Relative Contributions of TE, TFE, and FE

A useful parameter indicative of the electron tunneling probability is the quantity kT/E_{oo}. When E_{oo} is high compared to

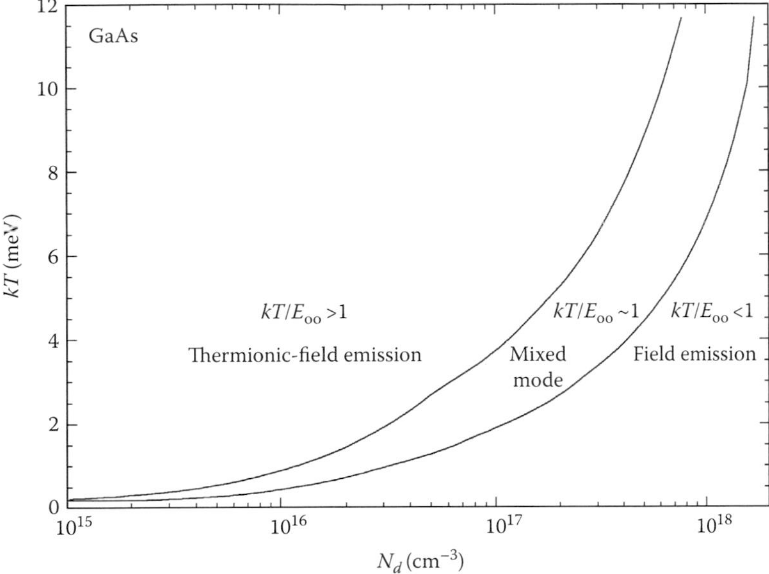

Figure 4.7 Ranges of temperature and donor concentration over which n-type GaAs Schottky diodes exhibit field and thermionic-field emission (adapted from [26], © 1982 IEEE.).

the thermal energy kT, the probability of electron transport by tunneling increases. Therefore the ratio kT/E_{oo} is a useful measure of the relative importance of the thermionic process to the tunneling process. For lightly doped semiconductors, E_{oo} will be small (Equation (4.25)), and $kT/E_{oo} \gg 1$. In this case, TE will dominate, and the contact is rectifying. For high levels of doping, $kT/E_{oo} \ll 1$, and FE now dominates, and the contact is ohmic. For intermediate levels of doping ($kT/E_{oo} \approx 1$) neither field nor thermionic emission accurately describes the conduction process. Note, that both TE and TFE are temperature dependent, while FE is not.

4.3.5 Estimated Contact Resistances for TE, FTE, and FE Current Modes

The current-voltage relations and therefore the expected contact resistances for the three main current transport mechanisms can be calculated by applying the semiclassical, quantum mechanical approximation of Wentzel–Kramers–Brillouin (WKB) to the simple energy band model shown in Figure 4.6. For thermionic emission, ρ_c is given by [23,27]

$$\rho_c = C_1 \exp\left(\frac{q\phi_b}{kT}\right) \tag{4.30}$$

where $C_1 = (k/qA)T$. For contacts with heavy doping in which the tunneling process (FE) is the dominant current transport mechanism, ρ_c is given by

$$\rho_c = C_2 \exp\left(\frac{q\phi_b}{E_{oo}}\right) = C_2 \exp\left[\frac{4\pi\sqrt{\varepsilon m^*}}{h}\left(\frac{\phi_b}{\sqrt{N_d}}\right)\right] \tag{4.31}$$

where C_2 has a weak temperature dependence. For the contacts in which TFE is the dominant transport mechanism, ρ_c is given by

$$\rho_c = C_3\left(\frac{\phi_b}{\sqrt{N_d}\coth(E_{oo}/kT)}\right) \tag{4.32}$$

where C_3 is a function of ϕ_b and T. These equations show that a reduction of ρ_c can be achieved by reducing the ϕ_b value and/or increasing the N_d value in the vicinity of the metal–semiconductor interface. In the case of TE and TFE, ρ_c can also be reduced by increasing the temperature.

4.3.6 Other Current Components

In addition to primary current mechanisms, image force lowering of the potential barrier and surface leakage currents are additional consequences of recombination in the depletion region,

4.3.6.1 Current Due to Image Force Lowering of the Potential Barrier

The current due to image force lowering of the potential barrier is difficult to calculate, because the image force reduces not only the barrier height but also the width, leading to thermionic, field assisted thermionic, and field components. Rideout and Crowell [11] have modeled current transport across a Schottky barrier for three cases of image force lowering and show that the inclusion of the image force can lead to a significant increase in the magnitude of the current density.

4.3.6.2 Generation–Recombination Effects

The generation and recombination of carriers within the depletion region give rise to a parallel component of the thermionic emission current transport mechanism which is equivalent to minority carrier injection [28]. Recombination is presumed to take place via Shockley–Read centers located near the middle of the bandgap. The current contribution, I_{gr}, due to this mechanism is given by [29]

$$I_{gr} = I_{gro}\left[\exp\left(\frac{qV_{\text{bias}}}{2kT}\right) - 1\right] \tag{4.33}$$

where I_{gro} is the generation–recombination saturation current, which in turn is given by

$$I_{gro} = \frac{qn_i wA}{2\tau_o} \tag{4.34}$$

Here, n_i is the intrinsic carrier concentration of the semiconductor (which is proportional to $\exp(-qE_g/2kT)$), w is the width of the depletion layer, and τ_o is the effective carrier lifetime within the depletion region. The relative importance of thermionic emission and of recombination in the depletion region depends on ϕ_b, E_g, T, and τ. Both generation and recombination components tend to be important in metal–semiconductor interfaces with high barriers, materials with low lifetimes (such as GaAs), at low temperatures, and/or junctions operated at low biases.

4.3.6.3 Surface Leakage Current

Surface leakage current, I_{leak}, is another parallel component of the total current. It is caused by leakage which flows near the edge of the contact instead of uniformly throughout the metal and across the metal–semiconductor interface. Its magnitude tends to be unpredictable, but can usually be reduced significantly by careful design and fabrication techniques. Because it is primarily a surface phenomenon, it bypasses the metal–semiconductor interface altogether and is often thought of as a large leakage resistor, R_{leak}, in parallel with it. For a given detector bias, V_{bias}, the leakage current can be expressed as

$$I_{leak} = V_{bias}/R_{leak} \tag{4.35}$$

4.4 Ohmic Contacts

All semiconductor devices need at least one ohmic contact, and it is often the quality of this contact that significantly affects the performance of semiconductor devices. As discussed previously, the term ohmic refers in principle to a metal–semiconductor contact that (a) is noninjecting, (b) has a linear I-V characteristic in both directions, and (c) has negligible contact resistance relative to the bulk or spreading resistance* of the semiconductor. In practice, the contact is usually acceptable if it does not perturb the performance of the device substantially and can supply the required current density with a voltage drop that is small compared to the drop across the active region, implying that the contact resistance should be small. A small contact resistance is important for other reasons. For example, the resistor capacitor (RC) time constant associated with the contact resistance may limit the frequency response of devices.

Consider the metal/n-type semiconductor junction shown in Figure 4.6. The conduction properties of the contact are determined by the current transport mechanisms discussed in the Schottky barrier section. However, a true Schottky contact is not the best system for extracting signals from a device, because it is difficult to obtain (a) a low specific contact resistance and (b) a linear and symmetric current-voltage

* Due to the non-linearity of the electric field around the contact.

relationship within the limits of its intended use. Ohmic contacts are thus preferred. Consequently, considerable effort has been expended in making metal–semiconductor contacts look electrically like ohmic contacts. There are two main methods to achieve this. The first is to lower the barrier height by choosing a suitable contacting material, and the second is to make the barrier so thin that quantum mechanical tunneling becomes the dominant carrier transport mechanism.

As the mode of conduction changes from TE to TFE to FE the exponential bias dependence of the current changes from [23]

$$\exp\left(\frac{\phi_b}{kT}\right) \rightarrow \exp\left(\frac{\phi_b}{E_{oo}\coth\left[\dfrac{E_{oo}}{kT}\right]}\right) \rightarrow \exp\left(\frac{\phi_b}{E_{oo}}\right) \tag{4.36}$$

and is reflected in the current-voltage behavior of the contact, which in turn changes from rectifying to ohmic in character. This is illustrated in Figure 4.8a in which we show the I-V characteristics of a Au/n-type GaAs Schottky barrier contact for progressively larger carrier concentrations. From Figure 4.8b, we also note that, as the impurity concentration increases, the magnitude of the current in the vicinity of zero bias is greatly increased, which strongly enhances ohmic behavior.

4.4.1 Practical Ohmic Contacts

As pointed out by Rideout [24], "It is probably a fair assessment, however, that ohmic contact technology has developed thus far more as a technical art than as a science." To achieve ohmic conditions, one needs to significantly reduce or modify the Schottky barrier, and toward this end several approaches can be tried.

4.4.2 Barrier Height Reduction

We have seen in previous discussions that there are several ways to reduce the barrier height, which we will expand upon below.

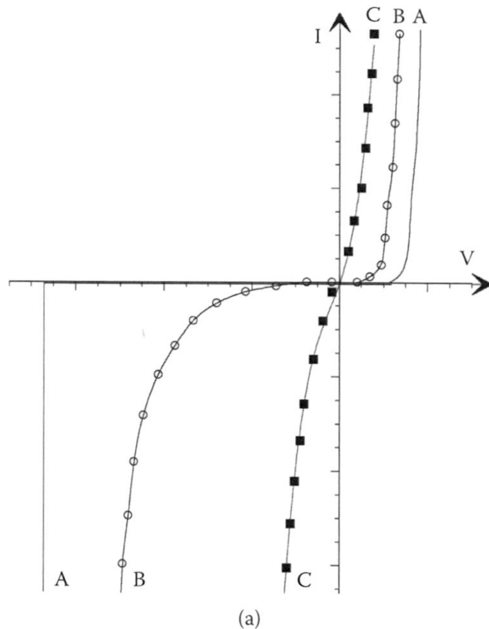

Figure 4.8 (a) Illustration of the current-voltage characteristics for a Schottky barrier contact (Au on n-type GaAs) for progressively higher carrier concentrations (from [30]). The curves are: (a) $N_d \leq 10^{17}$ cm^{-3}, for which thermionic emission dominates, (b) $N_d \approx 10^{17}$ cm^{-3}- 10^{18} cm^{-3}, for which thermionic field tunneling dominates, and (c) $N_d \geq 10^{19}$ cm^{-3}, for which field emission tunneling dominates. (b) Current-voltage characteristics as a function of carrier concentration (from [30]]. As can be seen the current can vary by over six orders of magnitude depending on the carrier concentration.

4.4.2.1 Choice of Metal

The simplest approach to reducing the Schottky barrier is to choose a metal that results in a negligible potential barrier height ϕ_b—the closer the barrier height is to zero, the more ohmic the contact. The simple theory of M-S interfaces predicts that the barrier height $\phi_b = \phi_m - \chi_s$, and so, naively, it might be expected that metals whose work functions are close to the semiconductor's electron affinity should most easily form ohmic contacts. This is found not to be the case, and in fact metals with high work functions form the best contacts to p-type semiconductors, while those with low work functions form the best contacts to n-type semiconductors. Unfortunately for wide bandgap

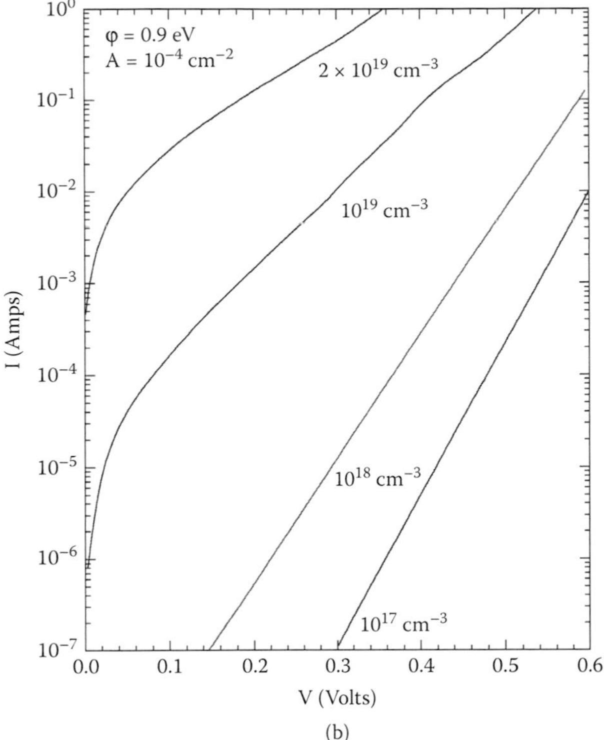

Figure 4.8 (Continued)

materials,* it is difficult to find such metals. The problem is illustrated in Figure 4.9a, in which we plot the distribution of work functions for the metallic elements commonly used for contacting metallizations. These are tabulated in Table 4.1 and are sourced from rows three through six of the periodic table [31]. From the figure it is clear that the work functions form a narrow distribution of mean 4.3 eV with a standard deviation of 0.9 eV, illustrating that there is very little spread in work-functions, with over 60% in the range 4 to 5 eV. In Figure 4.9b we show the measured barrier heights for Au contacts on both n-type and p-type semiconductors. The data are tabulated in Table 4.2. We note that both distributions are very similar, with

* For a general discussion and review of ohmic contacts to II–VI and III–V materials, the reader is referred to Kim and Holloway [32].

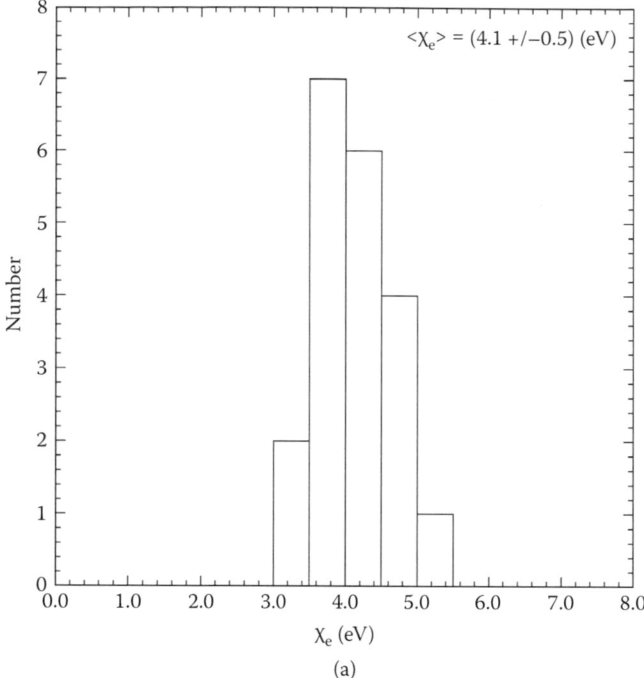

Figure 4.9 (a) The distribution of work function values for the metallic elements, from which it can be seen that they all cluster around a narrow range of values. (b) Measured barrier heights for Au/*n*-type and Au/*p*-type metal/semiconductor interfaces.

a mean of ~1 eV and a standard deviation of 1 eV. For ohmic contacts, resistivities should be of the order of $10^{-3}\ \Omega\ cm^2$, which requires barrier heights to be less than 0.4 eV. However, from the data, these are greater than 0.4 eV in over 90% of cases, indicating that the junctions are Schottky.

The previous discussion highlights the problem of finding metals with suitable work functions simply because they are tightly grouped around a single value. For III–V covalent compounds, the problem is further exacerbated by the fact that ϕ_b is essentially fixed by interface states; that is, the Fermi level is pinned. General procedures for forming ohmic contacts to pinned compounds may be found in Kim and Holloway [32]. For III–V compounds, the survey of Baca et al. [33] discusses contacting materials and properties as well as critical material issues pertaining to representative III–V compounds.

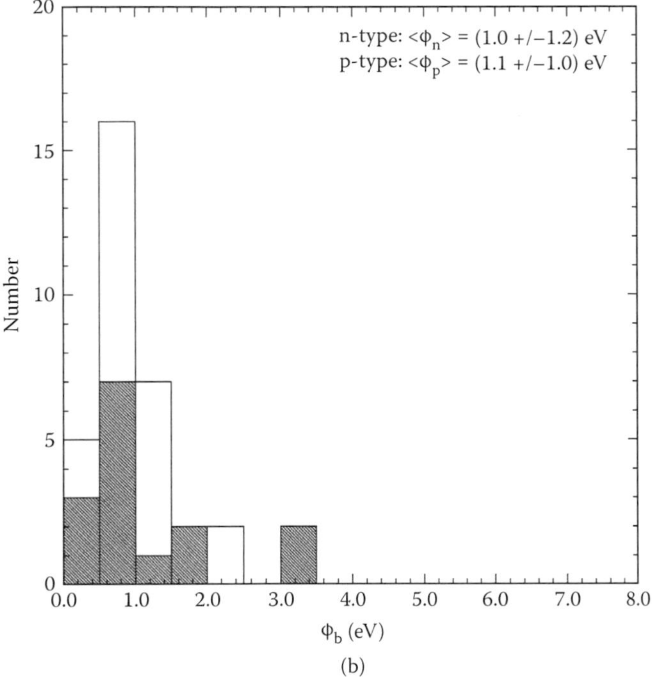

Figure 4.9 (Continued)

4.4.2.2 Doping Concentration

In Section 4.1, we noted that the Schottky barrier height is weakly dependent on the doping concentration ($\Delta\phi \sim N_d^{1/4}$) through the image force induced by the charge carriers. The amount of barrier lowering is given by [24]

$$\Delta\phi = \left[\frac{q^2 E_b N_d}{8\pi^2 \varepsilon_s \varepsilon_d^2 \varepsilon_o^3}\right]^{1/4} \tag{4.37}$$

where E_b is the band-bending potential (see Figure 4.2). Rearranging Equation (4.39) and solving for the doping concentration, N_{zb}, that will give zero barrier height, that is when $\Delta\phi$ equals E_b, we find that

$$N_{zb} = 1.8 \times 10^{19} \varepsilon_s \varepsilon_d^2 E_b^2 \text{ cm}^{-3} \tag{4.38}$$

TABLE 4.1
The Electronic Work Functions, ϕ_m, of a Range of Metallic Elements Used to Form Metal–Semiconductor Contacts

Metal	ϕ_m(eV)	Metal	ϕ_m(eV)	Metal	ϕ_m(eV)
Ag	4.63	Ga	4.32	Pd	5.41
Al	4.17	Hf	3.9	Pt	5.65
Au	5.1	Hg	4.49	Re	4.96
Ba	2.52	In	4.09	Rh	4.98
Be	4.98	Ir	5.27	Ru	4.71
Bi	4.22	K	2.30	Sb	4.63
Ca	2.87	Mg	3.66	Sm	2.7
Cd	4.08	Mn	4.1	Sn	4.42
Co	4.92	Mo	4.57	Ta	4.25
Cr	4.50	Nb	4.33	Tb	3.0
Cs	1.95	Ni	5.20	Ti	4.33
Cu	4.76	Os	4.83	W	4.61
Fe	4.74	Pb	4.25	Zn	3.63

Note: The data are ordered alphabetically.
Source: Taken from the compilation given in [31].

where E_b is expressed in units of eV. At first sight, Equation (4.38) offers a simple solution for obtaining an ideal zero barrier and therefore ohmic contact. However, in reality the derived impurity concentrations are of the order of 10^{22} to 10^{23} atoms cm^{-3}, which are orders of magnitude in excess of the solubility limits for known semiconductors. In addition, the reduction in barrier width with doping concentration is also effective in increasing conduction across the barrier, because barrier width reduction varies as $N^{1/2}$, as opposed to $N^{1/4}$ for barrier height reduction (see Section 4.2).

4.4.2.3 Annealing

In existing semiconductor/metal systems, annealing can create an alloy at the junction, which may lower the barrier height.

TABLE 4.2
Schottky Barrier Heights, ϕ_n and ϕ_p, between Au/*n*-Type and Au/*p*-Type Semiconductors

Semiconductor	ϕ_n	ϕ_p
Diamond		1.57
Si	0.78	0.30
Ge	0.50	0.07
3C-SiC	0.99	
6HSiC	1.32	
c-BN		3.10
h-BN		3.10
BP	1.40	1.85
AlAs	1.45	
AlSb		0.56
a-GaN	1.02	0.57
GaP	1.38	0.96
GaAs	0.92	0.50
GaSb	0.68	
InP	0.41	0.86
InAs	0 (ohmic)	0.47
InSb	0.14	0 (ohmic)
ZnO	0.77	
a-ZnS	2.10	
b-ZnS	2.10	
ZnSe	1.46	1.2
ZnTe		0.57
w-CdS	0.76	
w-CdSe	0.60	
CdTe	0.789	0.6

Note: A positive barrier height for an *n*-type semiconductor means the current flows more easily into the semiconductor, while a positive barrier height for a *p*-type semiconductor means the current flows more easily into the metal.
Source: Adapted from [34].

4.4.2.4 *Interface Doping*

It is observed experimentally that the barrier between a metal and a semiconductor is usually smaller for semiconductors with smaller energy gaps. Hence, another way to decrease the contact resistance is to place a layer of a narrow-gap highly doped semiconductor material between the active region of the device and the contact metal. Some of the best ohmic contacts to date have been achieved this way.

4.4.3 Barrier Width Reduction

A commonly used approach to produce an ohmic response is to fabricate a tunnel contact. The idea in this case is to control the barrier width through E_{oo} and therefore N_d, rather than its height. For a uniformly doped semiconductor, the width of the barrier is given by

$$w = \sqrt{\left(\frac{2\varepsilon_s\varepsilon_o}{qN_d}\right)\left(\left(\phi_b - V_n - V - \frac{kT}{q}\right)\right)} \tag{4.39}$$

which includes the modification due to the image force. In practice, width reduction is achieved by very heavy doping (>10^{19} cm^3) of the semiconductor immediately adjacent to the metal, so that the semiconductor depletion region becomes so thin (of the order of 3 nm or less) that even for a high barrier FE dominates, and the contact behaves as ohmic. This highly doped layer is formed either before the metal is deposited (for example, by diffusion, ion implantation, or epitaxy) or during the contact preparation. The latter is the most widely used procedure and consists of depositing a multicomponent metal structure on the semiconductor surface, followed by heating the system. The metal layers should contain a suitable dopant species—donor or acceptor atoms. A heat treatment is used to drive the dopant into the semiconductor to form an n^{++} or p^{++} layer, thus creating a thin highly doped tunneling region required for enhanced ohmic behavior. Metal contacts layers are usually prepared by vacuum deposition, electron beam, thermal evaporation, or sputtering. Alternatively, several plating techniques, such as electroplating, pulse plating, or electroless plating, can be used.

4.4.4 Introducing Recombination Centers

Another method which may be used to form an ohmic contact is to deliberately introduce recombination centers near the metal–semiconductor interface, for example, by damaging or straining the semiconductor surface. If the density of these centers is high enough, recombination in the depletion region will become the dominant conduction mechanism and will cause a significant decrease of contact resistance. This approach has been used to obtain ohmic contacts to Ge surfaces, but since a defect or a strained region can also gather impurities and/or generate underlying point defects, this method is not considered promising. The present consensus is that any damage of the semiconductor subsurface must be avoided in order to assure the long-term stability of the devices.

4.4.5 Desirable Properties of Ohmic Contacts

The important features for ohmic contacts are reliability and reproducibility. In particular, the contact material should not undergo electromigration under high fields. The thermal impedance of the contact must be low enough that sufficient heat can be removed from the device through the contact, if required, without damaging the contact or junction. Other desirable properties of ohmic contacts should include good adhesion to the semiconductor, smooth surface morphology (particularly where near micron device geometry is concerned), the ability to bond gold wires to connect the device to external circuitry, and finally contact reliability. With these in mind, the fabrication of a practical ohmic contact system often consists of a wetting agent to promote adhesion, followed by the dopant species, and finally a thick layer of Au for bonding purposes. Where in-diffusion of the top Au poses potential reliability hazards, often a diffusion barrier is inserted between the gold and the dopant layer. In addition to the previous considerations, special metallization schemes may also be required if

TABLE 4.3
Common Compound Semiconductor Contacting Materials

	Contact System	
Detector Material	Ohmic	Schottky
CVD Diamond	Ti/Pt	Al
SiC	Cr/Ni	Ti/Ni, Au, Ni_2Si
AlAs *n*-type	Au, Au-Ge	
AlAs *p*-type	Au, Au-Ge	
GaAs *n*-type	SnAu, AuGe[a]/Ni/Au	
GaAs *p*-type	ZnAu[a]	
GaAs	Ni/Au/Ge/Au	Au/Pt/Ti, Au,Pt
GaSb *n*-type	In	
GaSb *p*-type	In	
InSb *n*-type	In	
InP	Au/Ge/Ni	Au
GaN *n*-type	Ti/Al/Ni/Au, Al/Ti	
GaN *p*-type	Ni/Au	
GaN	Au	
GaP	Al	Au
CdTe	Pt, Au	In
CdZnTe	Pt, Au	
ZnTe	Au	
HgI_2	Pd, Aquadag	
PbI_2	Pd, Au, Aquadag	

Note: The slash symbol between elements denotes a deposition sequence with the right-hand element in contact with the semiconductor.

[a] Eutectic.

the device is to operate in unusual or harsh environments. In Table 4.3, we list common compound semiconductor contacting materials for a range of systems. The symbol "/" between elements denotes a deposition sequence with the right-hand element in contact with the semiconductor.

4.4.6 Nonideal Effects in Metal–Semiconductor Junctions

Simple Schottky barrier theory predicts that the barrier height, ϕ_b, between a metal and an n-type semiconductor equals the difference between the work function of the metal, Φ_m, and the electron affinity, χ. However, experimental results can yield vastly different values for the barrier height, which in part can strongly depend on the preparation of the semiconductor before metal deposition. This discrepancy is believed to be due to the presence of an interfacial layer or surface states caused by incomplete covalent (or "dangling") bonds at the surface of the semiconductor. A chemical reaction between the metal and the semiconductor can further affect the metal–semiconductor contact. Interfacial layers readily occur when a metal is deposited on a piece of semiconductor which has been exposed to air, because most semiconductors oxidize in air, yielding a thin (~3 nm) native oxide. Such an interfacial layer increases the built-in potential as obtained from a capacitance-voltage measurement. It also increases the ideality factor of the junction, decreases the measured barrier height as extracted from the temperature dependence of the saturation current, and limits the maximum current density. The effect is most significant for a thick interfacial layer with low dielectric constant on a highly doped semiconductor.

4.5 Contactless (Proximity Effect) Readout

Luke et al. [35] have proposed a novel technique based on proximity electrodes which eliminates the need for physical contacts in those applications where signal and not current output is significant. This technology is particularly valuable for those materials for which it is technologically difficult to form stable lateral contacts. The concept had been proposed in the 1960s for the readout of flash tubes in magnetic spectrographs [36], later for readout of resistive cathodes in gas ionization counters [37], and most recently for edge compensation in CdZnTe

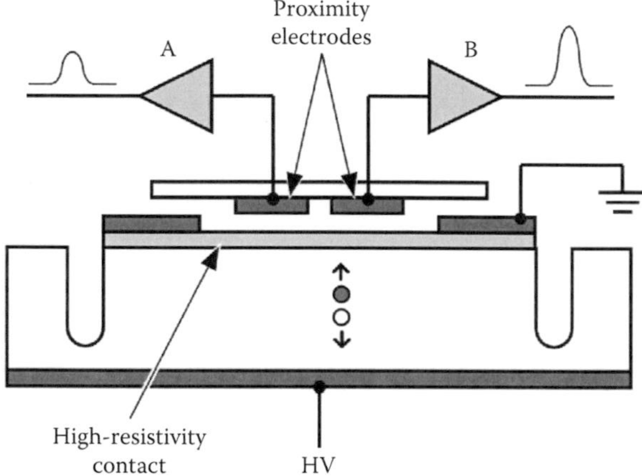

Figure 4.10 Noncontact readout of a detector using proximity electrodes (from [35], © 2009 IEEE).

coplanar grid detectors [38]. The implementation for semiconductor detectors is illustrated in Figure 4.10. In its basic form, the detector comprises a semiconductor crystal with two or more electrodes deposited on its surfaces. The electrodes serve two functions: first, as a means to apply a bias voltage across the detector to collect radiation-generated carriers (electrons and holes), and second, to serve as a readout electrode. Charge carriers drifting across the detector induce a charge signal at the electrode, which can then be measured by a charge-sensitive amplifier connected to the electrode. Although readout electrodes of a detector are generally formed on the detector itself, this is not a requirement for charge induction. Charge can be induced on any electrode, even if the electrode is not physically in contact with the semiconductor. By avoiding the need for hard-wired electrical connections between readout electronics and detectors, detector fabrication and assembly can be greatly simplified, especially in the case of pixelated detectors. In addition to eliminating physical or direct contacting, the use of a number of suitably placed proximity effect electrodes could make the readout position sensitive. Electrodes mounted near the sides of a detector could allow a simple "add-on" for sensing incomplete charge collection.

References

1. E.H. Rhoderick, *Metal-Semiconductor Contacts,* Oxford University Press, Oxford, 2nd edition (1978) p. 14.
2. E.H. Rhoderick, R.H. Williams, *Metal-Semiconductor Contacts,* Clarendon Press, Oxford, 2nd edition (1988).
3. L.J. Brillson, *Contacts to Semiconductors: Fundamentals and Technology*, Noyes Publ. Co., Park Ridge, NJ (1993).
4. C. Wilmsen, *Physics and Chemistry of III–V Compound Semiconductor Interfaces*, Plenum Press, New York (1985) p. 129.
5. F. Braun, "Uber die Stromleitung durch Schwefelmetalic," *Annalen der Physik and Chemie,* Vol. 153, No. 4 (1874) pp. 556–563.
6. W. Schottky, "Halbleitertheorie der Sperrschicht," *Naturwiss.*, Vol. 26 (1938) pp. 843–843.
7. N.F. Mott, "Note on the Contact Between a Metal and an Insulator or Semiconductor," *Proc. Camb. Phil. Soc.*, Vol. 34 (1938) pp. 568–572.
8. W. Shockley, "Currents to Conductors Induced by a Moving Point Charge," *J. Appl. Phys.*, Vol. 9 (1938) pp. 635–636.
9. S. Ramo, "Currents Induced by Electron Motion," *Proc. IRE*, Vol. 27 (1939) pp. 584–585.
10. S.M. Sze, "Metal–Semiconductor Contacts," in *Physics of Semiconductor Devices*, J. Wiley & Sons, Singapore, 2nd edition (1981) pp. 250–254.
11. V.L. Rideout, C.R. Crowell, "Effects of Image Force and Tunnelling in Metal Semiconductor Contacts," *Solid State Electron.*, Vol. 13 (1970) pp. 993–1009.
12. S.M. Sze, C.R. Crowell, D. Kahng, "Photoelectric Determination of the Image Force Dielectric Constant for Hot Electrons in Schottky Barriers," *J. Appl. Phys.*, Vol. 35 (1964) pp. 2534–2536.
13. H. Schweikert, *Verhandl. Phys. Ges.,* Vol. 3 (1939) p. 99.
14. J. Bardeen, "Surface States and Rectification at a Metal-Semiconductor Contact," *Phys. Rev.*, Vol. 71 (1947) pp. 717–727.
15. C. Kittel, *Introduction to Solid State Physics*, 8th edition, Wiley & Sons, New York (2005) pp. 488–489.
16. A. Stirling, A. Pasquarello, J.-C. Charlier, R. Car, "Dangling Bond Defects at Si–SiO_2 Interfaces: Atomic Structure of the Pb1 Center," *Phys. Rev. Lett.*, Vol. 85 (2000) pp. 2773–2776.
17. V. Heine, "Theory of Surface States," *Phys. Rev.*, Vol. 138, no. 6A (1965) pp. A1689–A1696.

18. W. Mönch, "Role of Virtual Gap States and Defects in Metal-Semiconductor Contacts," *Phys. Rev. Letts.*, Vol. 58 (1987) pp. 1260–1263.
19. A.W. Cowley, S.M. Sze, "Surface States and Barrier Height of Metal-Semiconductor System," *J. Appl. Phy.*, Vol. 36 (1965) pp. 3212–3220.
20. C.A. Mead, "Metal-Semiconductor Surface Barriers," *Solid State Electron.*, Vol. 9, (1966) pp. 1023–1033.
21. J. Tersoff, "Schottky Barrier Heights and the Continuum of Gap States," *Phys. Rev. Lett.*, Vol. 52 (1984) pp. 465–468.
22. R.T. Tung, "Charges and Dipoles at Semiconductor Interfaces," *Mat. Res. Soc. Symp. Proc.*, 719 (2002) F12.1.
23. A. Piotrowska, A. Guivarc'h, G. Pelous, "Ohmic Contacts to III-V Compound Semiconductors: A Review of Fabrication Techniques," *Solid State Electron.*, Vol. 26(3) (1983) pp. 179–197.
24. V.L. Rideout, "A Review of the Theory and Technology for Ohmic Contacts to Group III-V Compound Semiconductors," *Solid State Electron.*, Vol. 18 (1974) pp. 541–550.
25. F.A. Padovani, R. Stratton, "Field and Thermionic-Field Emission in Schottky Barriers," *Solid State Electron.*, Vol. 9 (1966) pp. 695–707.
26. E.H. Rhoderick, "Metal-Semiconductor Contacts," *Proc. Inst. Elec. Eng.*, Vol. 129, pt. 1 (1982) pp. 1–14.
27. M. Murakami, "Development of Refractory Ohmic Contact Materials for Gallium Arsenide Compound Semiconductors," *Sci. and Tech. of Adv. Mat.*, Vol. 3 (2002) pp. 1–27.
28. A.Y.C. Yu, E.H. Snow, "Minority Carrier Injection of Metal-Silicon Contacts," *Solid State Electron.*, Vol. 12 (1969) pp. 155–160.
29. D. Donoval, M. Barus, M. Zdimal, "Analysis of I-V Measurements on PtSi-Si Schottky Structures in a Wide Temperature Range," *Solid State Electron.*, Vol. 34(12) (1991) pp. 1365–1373.
30. C.A. Mead, "Physics of Interfaces," in *Ohmic Contacts to Semiconductors*, Ed. B. Schwartz, Electrochemical Society, New York (1969) pp. 3–16.
31. T.J. Drummond, "Work Functions of the Transition Metals and Metal Silicides," U.S. Government Report from Sandia National Laboratories (1999).
32. T.-J. Kim, P.H. Holloway, "Ohmic Contacts to II-VI and III-V Compound Semiconductors," in *Wide Bandgap Semiconductors,* Ed. S. Pearton, William Andrew Publishing, New York (1999) pp. 80–150, ISBN 0-8155-1439-5

33. A.G. Baca, F. Ren, J.C. Zolper, R.D. Briggs, S.J. Pearton, "A Survey of Ohmic Contacts to III-V Compound Semiconductors," *Thin Solid Films*, Vol. 308 (1997) pp. 599–606.
34. S. Adachi, *Properties of Group-IV, III-V and II-VI Semiconductors,* John Wiley & Sons, Wiley Series in Materials for Electronic & Optoelectronic Applications (2005) ISBN: 978-0-470-09032-9.
35. P.N. Luke, C.S. Tindall, M. Amman, "Proximity Charge Sensing with Semiconductor Detectors," *IEEE Trans. Nucl. Sci.*, Vol. 56, issue 3, no. 2 (2009) pp. 808–812.
36. C.A. Ayre, M.G. Thompson, "Digitisation of Neon Flash Tubes," *Nucl. Instr. and Meth.*, Vol. 69 (1969) pp. 106–108.
37. G. Battistoni, E. Iarocci, G. Nicoletti, L. Trasatti, "Detection of Induced Pulses in Proportional Wire Devices with Resistive Cathodes," *Nucl. Instr. Meth.*, Vol. 152 (1978) pp. 423–430.
38. P.N. Luke, M. Amman, T.H. Prettyman, P.A. Russo, D.A. Close, "Electrode Design for Coplanar-Grid Detectors," *IEEE Trans. Nucl. Sci.*, Vol. 44, no. 3 (1997) pp. 713–720.

5
Radiation Detection and Measurement

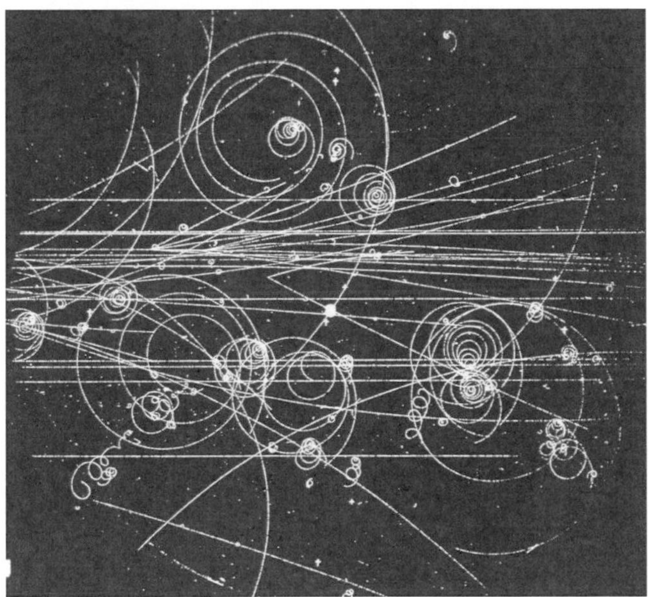

Frontispiece Particle tracks from the decay of 16 GeV pions captured in a liquid hydrogen bubble chamber. (Image courtesy of CERN Photolab.)

5.1 Interaction of Radiation with Matter

Detectors detect radiation by recording energy deposition into their active components. For most detectors this energy deposition is in the form of ionization produced in the detection medium (which may be solid, liquid, or gas) by charged particles. The choice of which detection medium to use depends to a large extent on what the detector will be used for. For example, in a tracking detector one wishes to detect the presence of a particle without affecting its trajectory, so the medium will be chosen to minimize energy loss and particle scattering (thus, low density). Conversely, if one wishes to measure the total energy deposition by calorimetry or spectroscopy, the absorber will be chosen to optimize energy loss, for example, by high density or high atomic number. Energy is then converted into an electrical signal, either directly or indirectly. In direct energy conversion, the incident radiation produces charge in the detector which is directly proportional to the energy absorbed and is collected by an electrode system. For example, in a gas counter the radiation ionizes the atoms/molecules of the gas, and the resulting charge is collected by electrodes. Similarly, in a semiconductor detector, the ionization produced by the radiation will create electron–hole pairs, which are swept toward the electrodes by an electric field. In indirect conversion, incident radiation excites atomic or molecular states that decay by the emission of light, as in the case of scintillation detectors. This light is then converted into an electrical signal using a photosensitive sensor, such as a photomultiplier tube.

The spectroscopic power of these systems or their ability to resolve different energies depends directly on how many "information carriers" are generated. For a scintillator it takes ~20 to 500 eV of energy deposition to create a single scintillation photon, whereas in a gas counter it takes 30 eV of energy to create an electron–ion pair, and in a semiconductor between 3 and 10 eV of energy to create an electron–hole pair.

5.2 Charged Particles

When charged particles pass through matter, they lose energy primarily by the ionization and excitation of the atoms and molecules of the medium and through nuclear interactions with the atoms and nuclei of the material. The former processes dominate at low and intermediate energies, and the latter dominates at the highest energies. It is the process of ionization we consider here. The basic theory was first developed by Bohr [1] using a classical approach and later using quantum mechanics by Bethe [2] and Bloch [3]. Physically, the electrostatic interactions between the traversing particles and the electrons of the atoms composing the medium result in the electrons being ionized. Eventually, the particles are slowed down and brought to rest by the small, but almost continuous, transfer of kinetic energy to the electrons. The energy loss per unit path length of a charged particle of charge z in any material can be described by the Bethe equation [2]

$$-\left(\frac{dE}{dx}\right)=\frac{4\pi e^4 z^2}{m_o v^2}NZ\left[\ln\frac{2m_o v^2}{I}-\ln(1-\beta^2)-\beta^2\right]\text{ergs cm}^{-1} \quad (5.1)$$

where $v = \beta c$ is the velocity of a particle whose mass, M, is much larger than m_o, the electron mass, in a medium containing N atoms cm^{-1} of atomic number Z, and I is a constant for the material derived from the Thomas–Fermi model [4,5], which is proportional to Z and close to the mean ionization potential. The quantity dE/dx is often referred to as the stopping power of the material. The various terms in the formula arise as follows. For nonrelativistic energies ($\beta \ll 1$) only the first term in the square brackets is significant, and the energy loss is proportional to $1/\beta^2$ (that is, it decreases with energy), whereas for relativistic particles the rate of energy loss increases with ln β. For small values of β, the two relativistic correction terms in the square brackets can be approximated by $\beta^4/2$. As β increases, the loss dE/dx decreases, mainly due to the $1/v^2$ term outside the brackets. As v approaches c, the ionization losses pass through a broad minimum and then increase logarithmically with energy. It is interesting to note that the minimum ionization rate occurs at a Lorentz factor of $\gamma \sim 3$, which corresponds to $E = Mc^2$. A good approximation for $(dE/dx)_{min}$ for many nuclear species is given by ~0.2 z^2 MeV (kg m^{-2})$^{-1}$. For singly charged particles at minimum ionization, the energy loss varies from approximately 2 MeV g^{-1}cm^{-1} for light elements to 1 MeV g^{-1}cm^{-1} for heavy elements. Thus we can see why light materials are more efficient per gram in stopping protons than high-mass materials.

The total energy loss in any material is calculated by solving Equation (5.1) iteratively for many differential thicknesses. In practice it is found that the solution converges to <5% for differential thicknesses less than a few millimeters. Note that the thicknesses of the active regions of a lot of semiconductor detectors rarely exceed a few millimeters. However, for X-ray detection, the active regions are usually less than 100 microns thick, and it has been known for some time that the measured distribution of energy losses caused by minimally ionizing particles in thin absorbers is far broader than that predicted by Landau [6] (see reference [7]). For protons, the effect is most pronounced for energies >1 GeV and absorbers <200 microns, which is particularly troublesome for X-ray-sensitive Si charge-coupled devices (CCDs) because the active regions are typically only ~30 μm to 100 μm thick [8]. The discrepancy between the predicted and measured energy loss distributions can be attributed to atomic electron binding effects and, to a lesser extent, delta-ray escape (escaping electrons). The question of energy loss by fast particles in thin layers was first solved by Landau [6], who derived the expected energy-loss distribution by solving the integral transport equation,

$$\frac{df}{dx}(x,\Delta) = \int_0^\infty w(E)\,[f(x,\Delta - E) - f(x,\Delta)]\,dE \qquad (5.2)$$

where $f(x,\Delta)$ represents the distribution function (the probability that the incident particle will lose an amount of energy Δ on traversing a layer of thickness x), and $w(E)\,dE$ denotes the probability per unit path length of a collision transferring energy E to an electron in the material. Hall [9] pointed out that Equation (5.2) may be conveniently solved by the Laplace transform method and the results written as

$$f(x,\Delta) = \frac{1}{2\pi i} \int_{c-i\infty}^{c+i\infty} e^{I}\, dp \tag{5.3}$$

where

$$I = d\Delta - x \int_{0}^{\infty} w(E)(1 - e^{-pE})\, dE \tag{5.4}$$

Unfortunately $w(E)$ is a complicated function of E and x and cannot be simply derived. Landau [6] was able to arrive at an approximate solution based on the free electron Rutherford cross section. However, this was later found to break down for certain regions of parameter space, notably highly relativistic particles and thin absorbers. Close agreement with experiment was eventually achieved by applying kinematic constraints on the energy transfer, corrections for the density effect, and taking into account the fact that the electrons in the material are not free. As expected, the revised formula converges to the Landau form for most of parameter space. By accounting for the above corrections semiempirically, Hall [9] was able to show that solution of Equation (5.3) reduces to a simple convolution of the Landau distribution, f_L, with a normal distribution:

$$f(x,\Delta) = \frac{1}{\sqrt{2\pi x \delta_2}} \int_{-\infty}^{+\infty} f_L(\Delta - y)\, e^{-y^2/2x\delta^2}\, dy \tag{5.5}$$

whose variance $x\delta_2$ may be derived from the "bound electron" corrections of Shulek et al. [10] or Blunck and Leisegang [11] or alternately from the photoabsorption ionization model of Hall [9].

5.2.1 Energy Loss of Secondary Electrons—Collisional and Bremsstrahlung

In practical detection systems we must also consider energy losses by electrons, because the secondary electrons produced by photoelectron, Compton, and pair processes transfer energy from the primary photons to the detection medium. For electrons and positrons, the Bethe–Bloch formula has to be modified to include radiative energy losses in the form of bremsstahlung as well as collisional losses, due to their small masses. Thus electrons lose energy through two processes. The energy loss per unit path length due to collisions is given by

$$-\left(\frac{dE}{dx}\right)_c = \frac{2\pi e^4 z^2}{m_o v^2} NZ\left[\ln\frac{2m_o v^2}{2I^2(1-\beta^2)} - (\ln 2)(2\sqrt{1-\beta^2} - 1 + \beta^2) + (1-\beta^2) + \frac{1}{8}(1-\sqrt{1-\beta^2})^2\right]^2 \quad (5.6)$$

Radiative losses occur because the electron is constantly scattered and therefore deflected along its trajectory. The deflections cause acceleration, and therefore the electron will radiate electromagnetically from any position along the electron trajectory. The energy loss per unit path length in this case is given by

$$-\left(\frac{dE}{dx}\right)_r = \frac{NEZ(Z+1)e^4}{137 m_o^2 c^4}\left[4\ln\frac{2E}{m_o c^2} - \frac{4}{3}\right] \quad (5.7)$$

These losses are most important for high energies and heavy absorbers. The total electron energy loss is therefore the sum of the collisional and radiative losses:

$$\left(\frac{dE}{dx}\right)_T = \left(\frac{dE}{dx}\right)_c + \left(\frac{dE}{dx}\right)_r \quad (5.8)$$

We can assess the relative importance of these two loss mechanisms in a detector by examining the ratio of the two,

$$\frac{(dE/dx)_r}{(dE/dx)_c} \approx \frac{EZ}{700} \quad (5.9)$$

where E is in units of MeV. For compound semiconductor detectors, the operational energy range of the detector will be

<1 MeV, and the average Z of the detection medium will be ~30 to 60. Therefore, radiative losses will be a factor of ~10 to 20 times lower than collisional losses and can generally be ignored.

5.3 Neutron Detection

Since neutrons are uncharged, they cannot be easily detected in conventional detectors and can only be detected indirectly by converting them into charged particles. In fact, they do not even interact directly with the electrons in matter, as gamma rays do. It is the ionization in the detecting medium caused by these secondary charged particles that is detected. For fast neutrons this can be achieved by inelastic scattering off detector nuclei, transferring some of their kinetic energy to the nuclei. If enough energy is transferred the recoiling nucleus ionizes the material surrounding the point of interaction. The maximum energy transferred to a nucleus of atomic weight A by a neutron of kinetic energy E is given by

$$E_{\text{max}} = \frac{4AE}{(A+1)^2} \tag{5.10}$$

Thus, it can be seen that this mechanism is only efficient for neutrons interacting with light nuclei. For a single scattering event the actual energy transferred to the recoiling nucleus lies between 0 and E_{max}, depending on the scattering angle, and has equal probability for any value in this range. Because the neutron is uncharged, the reactions are essentially billiard ball collisions with geometric cross sections. As such, scattering cross sections do not vary very much from nucleus to nucleus and typically have values of a few barns (10^{-24} cm^2).

Although neutrons do not interact with the charge distribution in atoms by means of long-range Coulomb interactions, they do interact with atomic nuclei through short-range nuclear force interactions. Slow or thermal neutrons ($E < 1$ eV) can be captured or absorbed by detector nuclei, exciting them to a more energetic state. The capture probability varies inversely as the neutron's velocity. The de-excitation products from these reactions, such as protons, alpha particles, gamma rays, and fission fragments, can then initiate the detection process. Because the energy of the captured neutron is small compared to the Q-value of the reaction, the reaction products carry away an energy corresponding to the Q-value, which means that any knowledge of the incident neutron energy is lost. Capture or absorption cross sections vary widely depending upon the distribution of nuclear energy levels, so that nuclear resonance effects may be encountered. For example, the absorption cross section for thermal neutrons on ^{12}C is about 0.0034 barns, whereas for ^{157}Gd it is 254000 barns. In Table 5.1, we list thermal neutron capture reactions commonly exploited in conventional neutron detectors. Most of these

TABLE 5.1
Nuclear Reactions Commonly Exploited for Thermal Neutron Detection

Reaction	Products	Reaction Q-Value (MeV)	Cross Section [12,13] (barns)
$^{3}He + {}^{1}n \Rightarrow$	^{3}H(0.191 MeV) + ^{1}p(0.573 MeV)	0.764	5333
	^{7}Li (1.015 MeV) + $^{4}\alpha$(1.777 MeV)	2.792 (to ground state)	
$^{10}B + {}^{1}n \Rightarrow$	{		3837
	^{7}Li*(0.840 MeV) + $^{4}\alpha$(1.470 MeV)	2.310 (first excited state)	
$^{6}Li + {}^{1}n \Rightarrow$	^{3}H(2.73 MeV) + $^{4}\alpha$(2.05 MeV)	4.780	940
$^{113}Cd + {}^{1}n \Rightarrow$	^{114}Cd + γ(0.56 MeV) + *conv. electrons*		20600
$^{155}Gd + {}^{1}n \Rightarrow$	^{156}Gd + γ(0.09 0.20,0.30 MeV) + *conv. electrons*		60900
$^{157}Gd + {}^{1}n \Rightarrow$	^{158}Gd + γ(0.08,0.18,0.28 MeV) + *conv. electrons*		254000
$^{235}U + {}^{1}n \Rightarrow$	fission fragments	201	583
$^{239}Pu + {}^{1}n \Rightarrow$	fission fragments	160	748

materials are either used as converter coatings on gas counters or as dopants in scintillation materials. The table lists the Q-value, or the total energy available from each reaction to create charged particles, as well as the total cross-section.

For most compound semiconductor detection media both scattering and capture mechanisms are fairly inefficient. However, there are a number of boron compounds (BN, BP, B_xC), cadmium compounds (CdZnTe, CdTe, CdMnTe), and mercury compounds (HgI_2, HgCdTe) which may potentially be used for neutron detection by virtue of the high thermal neutron capture cross sections of ^{10}B (3837 barns), ^{113}Cd (20,600 barns) and ^{199}Hg (2150 barns). Note however, these are all isotopes—the effective cross sections for "normal" isotopic compositions are for B 749 barns, for Cd 2450 barns, and for Hg 384 barns.

5.4 X- and Gamma Rays

Unlike charged particles, a well collimated beam of X-rays or gamma rays shows a truly exponential absorption in matter. This is because photons can only be absorbed or scattered in a single event. Fano [14,15] categorized the possible processes by which the electromagnetic field of a gamma ray may interact with matter. With reference to Table 5.2, there are 12 ways of combining columns 1 and 2; thus in theory there are 12 different processes by which gamma rays can be absorbed or scattered.

TABLE 5.2
Possible Photon Interaction Mechanisms in Matter

Interaction	Effects
1. Interaction with electrons	a. Complete absorption
2. Interaction with nuclei	b. Elastic scattering (coherent)
3. Interaction with the electric field surrounding nuclei or electrons	c. Inelastic scattering (incoherent)
4. Interaction with the meson field surrounding nucleons	

Many of these processes are quite infrequent, and indeed some have yet to be observed. It turns out that gamma rays interact primarily through only four of the twelve processes listed. These are the photoelectric effect 1(a), elastic scattering 1(b), the Compton effect 1(c), and pair production 3(a).

5.4.1 Photoelectric Effect

The photoelectric process involves the complete transfer of the incident photon energy to an atomic electron. Photoelectrons may be ejected from any of the K,L,M, … shells of an atom, but a free electron cannot absorb a photon and become a photoelectron, because a third body (the nucleus) is necessary to conserve momentum. It is found that about 80% of photoelectric absorptions take place in the K-shell, providing that the incident photon energy, $h\upsilon$, exceeds the K-shell binding energy. Because the entire atom participates, the photoelectric effect may be visualized as an interaction of the primary photon with the atomic electron cloud, so that the entire photon energy is absorbed and an electron is ejected from the atom with a kinetic energy,

$$E = h\upsilon - \phi \tag{5.11}$$

where ϕ is the binding energy of the electron. It is this electron which goes on to produce the ionization that generates the information carriers in a detector. The remainder of the energy appears as characteristic X-rays or Auger electrons resulting from the filling of the vacancy on the inner shell by an electron from a higher energy shell. The Auger electrons have extremely short ranges and are reabsorbed; however, the characteristic X-rays (and particularly K X-rays) may escape in detectors of small size, giving rise to an escape peak. This is especially true for compound semiconductors for two reasons First, they have limited volumes because of poor transport parameters, and second. they are generally composed from high-Z material, so the fluorescent escape X-rays have high energies that are not contained in the detection volume. If the energy of the photon is sufficiently small so that relativistic effects are insignificant, but yet large enough so that the binding energy of the electrons in the K-shell may be neglected (that is, $E = h\upsilon$), then the cross section per atom for photoelectric absorption is given by [16],

$$\alpha_K^{\tau} = \sigma_T Z^5 \alpha^4 2^{5/2} \left(\frac{m_o c^2}{h\upsilon} \right)^{7/2} \text{ cm}^2/\text{atom} \tag{5.12}$$

where σ_T is the Thomson cross section, α is the fine structure constant (α = 1/137) representing the strength of the interaction between electrons and photons, $m_o c^2$ is the rest mass energy of the electron, and Z is the atomic number of the absorbing material. The most important feature of the cross section is the strong dependence on the atomic number ($\sim Z^5$) and on the energy of the incident photon ($\sim E_\gamma^{-3.5}$), from which it follows that this process is especially efficient in the absorption of low-energy photons by heavy atoms.

5.4.2 Coherent Scattering—Thomson and Rayleigh Scattering

If the energy of the incident photon is insufficient to produce ionization or excitation of the atom, either Thomson or Rayleigh scattering can occur, depending on whether the photon interacts with free or bound electrons. For both mechanisms, the photon is elastically scattered with no change in internal energy of the scattering atom, nor of the X-ray photon. For Thomson scattering [17], the total cross section is given by

$$\sigma_T = \frac{8\pi}{3} \left(\frac{e^2}{4\pi\varepsilon_o m_o c^2} \right)^2 = \frac{8\pi}{3} r_o^2 = 6.652 \times 10^{-25} \text{ cm}^2 \tag{5.13}$$

where ε_o is the permittivity of free space. The quantity r_o is the classical radius of the electron (=2.818×10^{-15} m), independent of energy. Note that both the differential and the total Thomson scattering cross sections are completely independent of the frequency (or wavelength) of the incident radiation.

The angular distribution for unpolarized incident radiation is given by

$$\frac{d\sigma_T}{d\Omega} = r_o^2\left(\frac{1+\cos^2\theta}{2}\right) \quad \text{cm}^2\text{sr}^{-1}\text{per electron} \qquad (5.14)$$

where θ is the scattering angle. Thus photons are scattered mainly in the forward and reverse directions. For polarized incident radiation, the cross section vanishes at 90° in the plane of polarization. In the case of Rayleigh scattering, the cross section per atom is a perturbation of the classical Thomson cross section:

$$\sigma_R = \frac{3}{8}\sigma_T \int_{-1}^{+1} (1+\cos^2\theta)\,[F(x,Z)]^2 d\cos\theta \quad \text{cm}^2\,\text{per electron} \quad (5.15)$$

where $F(x,Z)$ is the atomic form factor, and $x = \sin(\theta)/\lambda$. The form factor tends toward Z as $\theta \rightarrow 0$ and 0 as $\theta \rightarrow \pi$. Thus Rayleigh-scattered photons have an angular distribution much more peaked in the forward direction than Thomson scattering, especially at high energies. Rayleigh scattering also tends to be much more efficient in the forward direction than Thomson scattering, because the interaction probability varies as $\sim Z^2$; however, the interaction cross section does decrease significantly with increasing energy (as $\sim E^{-2}$).

For practical detector design, elastic scattering is never more than a minor contribution to the absorption coefficient and has little effect in photon detection other than to scatter photons into and out of the detection volume. It can, however, cause acute problems in imaging systems, leading to a blurring of images.

5.4.3 Incoherent Scattering—Compton Scattering

At energies much greater than the binding energy of the electrons, photons can be scattered as if the electrons were free and at rest. The energy of the incident photon is shared between the scattered photon, which may escape, and the kinetic energy of the recoil electron, which is most likely

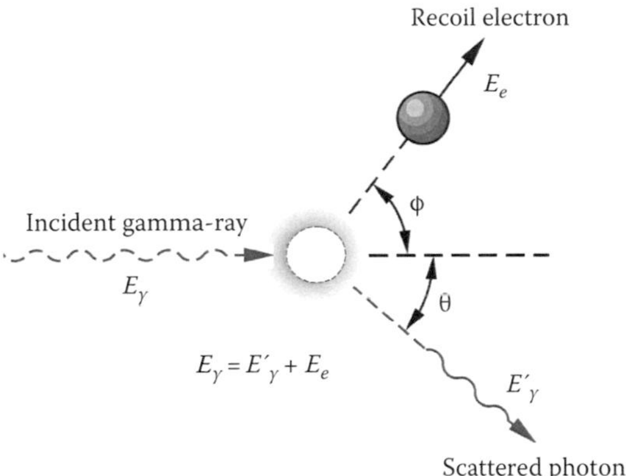

Figure 5.1 **(See color insert.)** Compton scattering geometry. The electron is assumed to be initially at rest and is ejected at angle ϕ by the incident photon with a kinetic energy E_e. The initial photon in turn is scattered through angle θ but with a reduced energy E_γ', which by conservation of energy is equal to $E_\gamma - E_e$.

absorbed. This is the Compton effect [18], which at 1 MeV is the dominant mode of interaction. Figure 5.1 shows the scattering geometry for a Compton interaction. The energy of the scattered photon is given by

$$E_\gamma' = h\upsilon' = \frac{E_\gamma}{1 + \left(E_\gamma / m_o c^2\right)(1 - \cos\theta)} \tag{5.16}$$

where E_γ is the energy of the incident photon, and θ is the angle through which it is scattered. For small scattering angles, θ, very little energy is transferred. The kinetic energy of the scattered electron is,

$$E_e = E_\gamma - E_\gamma' = \frac{E_\gamma(1 - \cos\theta)}{m_o c^2 \left(1 + \left(E_\gamma / m_o c^2\right)(1 - \cos\theta)\right.} \tag{5.17}$$

The kinetic energy of the electron has its maximum value when $\cos\theta = -1$ or $\theta = 180°$. This electron will produce the ionization which generates the information carriers in a detector. Note that the scattered photon may then continue to Compton scatter again in the detector, be completely absorbed via the photoelectric effect, or escape the detection volume totally. The latter is particularly true for small detectors.

The differential scattering cross section (that is, the probability of scattering a photon through angle θ within solid angle Ω) is given by the Klein–Nishina [19] formula. The energy scattered from unpolarized radiation by stationary, unbound electrons is given by

$$\frac{d\sigma_K}{d\Omega} = r_e^2\left(\frac{(1+\cos^2\theta)}{2}\right)\frac{1}{(1+\gamma(1-\cos\theta)^2)} \times\left[\frac{\gamma^2(1-\cos\theta)^2}{(1+\cos^2\theta)(1+\gamma(1-\cos\theta))}\right] \quad (5.18)$$

where γ is the ratio of the incident photon energy to the rest mass energy of the electron (E_γ/m_oc^2). Equation 5.18 can be written more simply in terms of the energies of the primary and scattered photons,

$$\frac{d\sigma_K}{d\Omega} = \frac{r_e^2}{2}\left(\frac{E'_\gamma}{E_\gamma}\right)^2\left(\frac{E'_\gamma}{E_\gamma}+\frac{E_\gamma}{E'_\gamma}-\sin^2\theta\right)\ \text{cm}^2\ \text{sr}^{-1}\ \text{per electron} \quad (5.19)$$

This is the most useful form for detector design unless polarization is important. The Klein–Nishina formula can be thought of as the classical Thomson cross section multiplied by a form factor. Whereas the Thomson cross section gives a symmetric angular distribution of scattered photons about 90°, the Klein–Nishina formula predicts a strongly forward-peaked cross section as γ increases. This is illustrated in Figure 5.2 in which we plot the differential angular distribution of Compton-scattered photons as a function of the angle of scattering, θ, for various incident photon energies.

For most detection systems where the energy of the incident particle is determined by detecting both the scattered photon and electron, a more useful quantity is the total Compton cross section obtained by summing the Klein–Nishina differential cross section over all angles and polarizations of the scattered

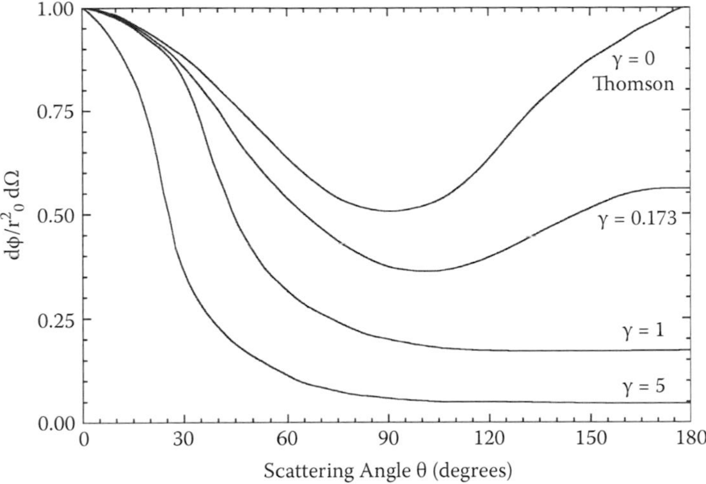

Figure 5.2 Differential angular distribution of Compton-scattered photons as a function of the angle of scattering, θ, for various incident photon energies, γ, expressed in units of the electron rest mass energy.

photon. The solution leads to a lengthy formula, which is given approximately by

$$\sigma_K^{\tau} = 8\pi r_e^2 \frac{(1+2\gamma+1.2\gamma^2)}{3(1+2\gamma)^2} = 3\sigma_T \frac{(1+2\gamma+1.2\gamma^2)}{3(1+2\gamma)^2} \text{ cm}^2 \text{ per atom} \tag{5.20}$$

For very low energies ($\gamma \rightarrow 0$) the formula reduces to the classical Thomson cross section; this represents the probability of removal of the photon from a collimated beam while passing through an absorber containing one electron cm^{-2}. As the photon energy increases and eventually becomes comparable with the rest mass energy of the particle, the Klein–Nishina formula predicts that forward scattering of photons becomes increasingly favored relative to backward scattering. The probability of interaction is independent of atomic number Z, but proportional to density, ρ.

5.4.4 Pair Production

In the Coulomb field of a nucleus, a photon of energy $E_\gamma > 2\, m_o c^2$ may be converted into an electron–positron pair such that

$$E_\gamma = (E_- + m_o c^2) + (E_+ + m_o c^2) \tag{5.21}$$

where E_- and E_+ denote the energies of the electron and positron, respectively. The excess energy above the threshold is randomly shared between the electron and positron. The information carriers in a detector are then generated by the ionization trail of the electron and positron. The electron is always completely absorbed. However when the positron slows sufficiently it will annihilate with an electron in the detector giving rise to two characteristic 511 keV gamma rays emitted at 180° to each other, which may then go on to interact in the detector medium via Compton or photoelectric absorption. The cross section is difficult to derive and is obtained from a quantum mechanical treatment. Marmier and Sheldon [20] summarize the useful forms of the total cross section as,

$$_a\kappa_p = \frac{r_o^2 Z^2}{137}\left[\frac{28\ln}{9}\left(\frac{2h\upsilon}{m_o c^2}\right) - \frac{218}{27}\right] \text{ cm}^2 \text{ per atom} \quad (5.22)$$

for $1 \ll E_\gamma \ll 1/\gamma\, Z^{1/3}$, and for complete screening,

$$_a\kappa_p = \frac{r_o^2 Z^2}{137}\left[\frac{28\ln}{9}\left(\frac{183}{Z^{0.3}}\right) - \frac{2}{27}\right] \text{ cm}^2 \text{ per atom} \quad (5.23)$$

for $E_\gamma / m_o c^2 \gg 1/\gamma\, Z^{1/3}$. Note that the latter expression is independent of energy. For both cases, the probability of interaction scales as Z^2, and unlike σ_p and σ_c, increases with energy. Because the useful operating energy range of a detectors is dictated by its active volume, or more specifically by its ability to contain the incident radiation, pair production effects can be ignored in compound semiconductor detectors because the small active volumes generally limit the operating energy ranges to < 500 keV.

5.5 Attenuation and Absorption of Electromagnetic Radiation

The probability of a photon traversing a given distance in an absorber without any interaction is the product of the probabilities of survival for each individual type of interaction. Thus a collimated photon beam of initial intensity, I_o, after

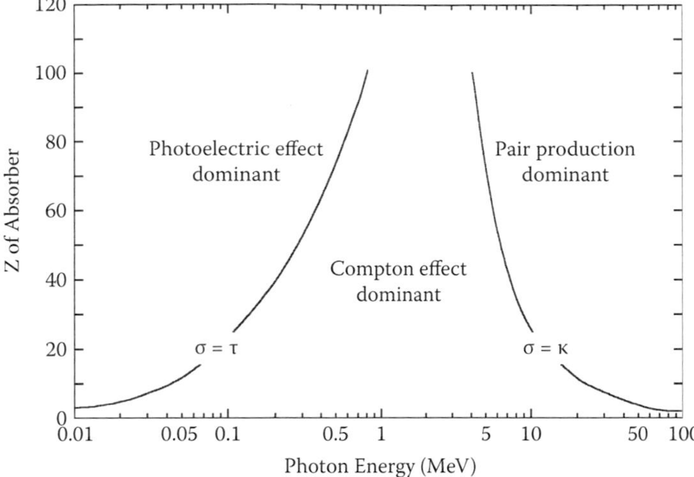

Figure 5.3 The lines delineate the regions where photoelectric effect, the Compton effect, and pair production dominate (from Evans [21]). Here the left line marks the region of space where the photoelectric and Compton cross sections are equal as a function of energy and Z. Thus, to the left of the line, the photoelectric effect dominates, while to the right, Compton dominates. The right-hand line marks the boundary in energy-Z space where the Compton and Pair effects dominate.

traversing a thickness x of absorber will have a residual intensity, I, of unaffected primary photons equal to,

$$I = I_o \exp(-\mu_o x) \tag{5.24}$$

where $\mu_o = \tau + \sigma + \kappa$ are the photoelectric, total Compton, and pair production cross sections. μ_o is known as the total linear attenuation coefficient, which is a measure of the number of primary photons which have interactions. In Figure 5.3, we show the regions in energy space where each of the three primary energy deposition processes dominate. The solid lines show the boundaries where the photoelectric (left) and pair production (right) become equal to the Compton cross section. We can see that Compton dominates the entire energy range for $Z < 10$, but becomes increasingly "squeezed" by the photoelectric and pair effects as Z increases. For the intermediate-mass materials used to fabricate detectors, photoelectric interactions dominate for energies < ~500 keV, Compton scattering dominates above ~500 keV and less than ~5 MeV, and pair production dominates above ~ 5 MeV.

Absorption coefficients are usually expressed as mass attenuation coefficients, which are the linear coefficients divided by the density, ρ g/cm^3. As Evans [21] points out, mass attenuation coefficients are of more fundamental value for detector work than the linear coefficients, because all mass attenuation coefficients are independent of the actual density or physical state of the absorber, whether gaseous, liquid, or solid form. This is because the fundamental interactions are expressible as cross sections per atom, ${}_a\tau$, ${}_a\sigma$, and ${}_a\kappa$, and when these are multiplied by the number of atoms per gram, the mass attenuation coefficient is obtained directly. Thus for mixtures or compounds the mass attenuation coefficient is simply the sum of μ_o/ρ of the different elements multiplied with the corresponding weight fractions w_i, thus,

$$\left(\frac{\mu}{\rho}\right)=\sum_i w_i\left(\frac{\mu_o}{\rho}\right) \quad \text{cm} \tag{5.25}$$

5.6 Radiation Detection Using Compound Semiconductors

In terms of radiation detection, semiconductor detectors are usually operated in one of two modes, either by exploiting the photoconduction properties of semiconductors or by operating the device as a solid state ionization chamber. The former method was widely used in early work but has now been largely superseded by the latter, which offers many advantages, not least the combined ability to resolve precisely the spectral and temporal signatures of a wide range of radiation types.

5.6.1 Photoconductors

The simplest type of semiconductor radiation detector is the photoconductor, in which incident radiation of energy $E > E_g$ creates additional charge carriers in a semiconductor

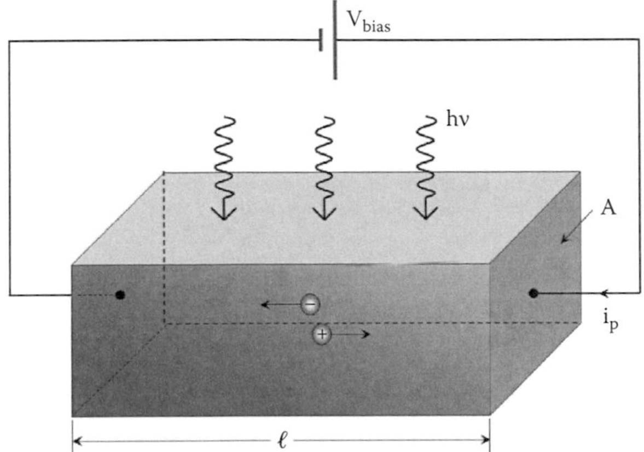

Figure 5.4 Schematic of a photoconductor detector consisting of a highly conductive semiconductor bar. Radiation incident on the bar creates additional electron–hole pairs, which lead to an increase in conductivity and consequently current though the device.

resulting in a change in conductivity (Figure 5.4). The semiconductor itself is usually doped to increase conductivity during this process to reduce Johnson noise, improve contacting, and extend response to longer wavelengths. These devices have traditionally been used for optical applications, mainly in the far infrared, because optical photon absorption depths are compatible with typical device thicknesses, and photon energies are compatible with the bandgap energies of common semiconductors. Photoconductors differ from crystal counters,* in that they usually have a limited bandwidth and are primarily integrating devices, because their large intrinsic conductivity makes them insensitive to individual particles or quanta. A schematic of such a device is shown in Figure 5.4. A comprehensive treatment of the photoconductive process is given in Rose [22]. Light incident on the crystal generates electron–hole pairs, which increases the carrier concentration and, in turn, its conductivity. If a small bias is now applied across the crystal via two contacts, a measurable current can be induced to flow

* An early name for a semiconductor radiation detector operated as a "solid state ionization chamber."

through it. The conductivity σ in the absence of illumination may be determined from the current density J, velocity v, charge density ρ, electric field ξ, and mobility μ. The current density is given by

$$J = \sigma\xi = \rho v \tag{5.26}$$

Because $v = \mu\xi$, the conductivity can be expressed in terms of the mobility,

$$\sigma = \mu\rho \tag{5.27}$$

Now, the current density, ρ, is equal to qN, where N is the number of charge carriers. If μ_e is the electron mobility, μ_h the hole mobility, N_e the density of electrons in conduction band, and N_h the density of holes in the valence band, the conductivity is given by the sum of electron and hole charge components is

$$\sigma = e(\mu_e N_e + \mu_e N_h) \tag{5.28}$$

Consider an incoming photon flux ϕ γ s^{-1}. The number of carriers in equilibrium is

$$\phi\eta\tau \tag{5.29}$$

where η is the quantum efficiency, and τ is the mean lifetime before recombination. (typically, $\tau \sim 1$/impurity concentration). The number of carriers per unit volume is

$$n = p = \frac{\varphi\eta\tau}{A\ell} \tag{5.30}$$

and the resistance is related to the conductivity by

$$R = \frac{\ell}{\sigma A} \tag{5.31}$$

Substituting Equations (5.28) and (5.30) for σ and A, we obtain

$$R = \frac{\ell^2}{q(\mu_e n + \mu_h p)\phi\eta\tau} \tag{5.32}$$

which is useful for optical applications because it relates the resistivity of the detector to the photon arrival rate. The time for an electron to drift from one electrode to the other, $\tau_{t,}$ is given by

$$\tau_t = \frac{\ell}{\langle v_l \rangle} = \frac{\ell}{\mu\xi} \tag{5.33}$$

where v_l is the average drift velocity between the electrodes. The photoconductive gain is given by

$$G = \frac{\tau\mu\xi}{\ell} = \frac{\tau}{\tau_t} \tag{5.34}$$

Thus the gain is the ratio of the carrier lifetime to carrier transit time, implying that it is possible to have a gain greater than unity so that the absorption of a single photon produces τ/τ_d carrier pairs. This is understood by realizing that τ_d represents the time for the charge carriers to traverse the conductor and enter the external circuit. Current continuity and charge conservation require that new carriers enter the semiconductor from the external circuit and eventually recombine after time τ. Thus, the charge carriers effectively traverse the semiconductor τ/τ_l times resulting in gain. The quantity G quantifies the probability that a generated charge carrier will traverse the extent of the detector and reach an electrode; if $G \ll 1$ the majority of charge carriers recombine before reaching an electrode, and if $G \gg 1$ most reach the electrode. To optimize G, the detector should be thin as possible to reduce the transit time, but thick enough to maximize η. Alternatively, the efficiency can be improved by increasing the bias voltage or improving material quality by reducing defects and impurities.

5.6.2 The Solid-State Ionization Chamber

For energies much greater than the bandgap energy, semiconductor radiation detectors are usually operated as solid-state ionization chambers. This is most commonly realized by making a p-n junction or a metal–semiconductor junction (surface barrier diode) on very low-doped material. The junction is reverse biased so that a thick depletion layer is formed where the field is high and at the same time the leakage current is low. In Figure 5.5 we illustrate a simple planar detector geometry. The device can be thought of as a parallel plate capacitor of separation L with ohmic contacts. We assume that space charge within the detector is negligible, implying that the electric field, E, is constant in the material. Ionizing radiation absorbed in the sensitive volume excites electron–hole pairs in direct proportion to the energy deposited (that is, $n = E_o/\varepsilon$, where n is the number of electron–hole pairs generated, E_o is the energy deposited, and ε is the average energy consumed to create an electron–hole pair). Note that ε will be greater than ε_g, the bandgap energy—the difference being due to the energy lost in producing phonons which is required to conserve both energy and momentum. Remarkably, the ratio, $\varepsilon/\varepsilon_g$, is constant for nearly all materials and is independent of the type of radiation [24]. Applying an electric field across the

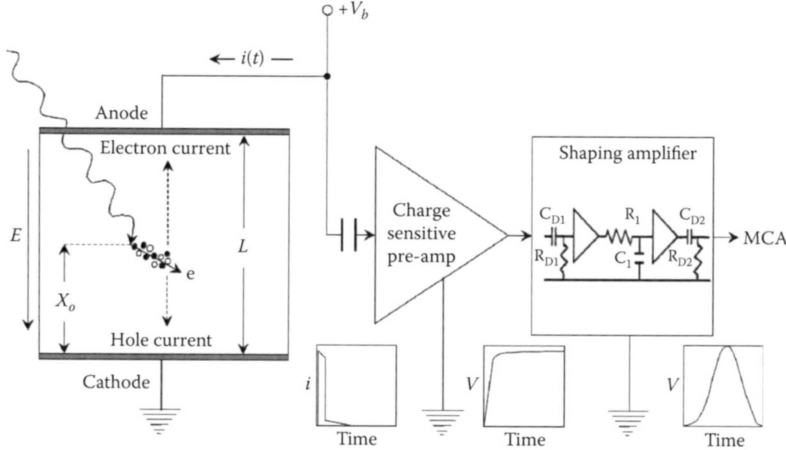

Figure 5.5 Schematic of a simple planar detection system (from [23]). For completeness, we also show the signal chain and the evolution of currents and voltages. Ionizing radiation absorbed in the sensitive volume generates electron–hole pairs in direct proportion to the energy deposited. These are subsequently swept toward the appropriate electrode by the electric field induced by the bias voltage V_b.

detector causes the liberated carriers to separate. The electrons drift toward the anode and the holes toward the cathode with drift velocities v_e and v_h. The motion of the carriers creates an induced charge on the electrodes and by current continuity into the electronics. The induced charge Q_{ind} is given by the Shockley–Ramo theorem [25,26], which makes use of the concept of a weighting potential. The weighting potential is defined as the potential that would exist in the detector with the collecting electrode held at unit potential, while holding all other electrodes at zero potential. The theorem states that, regardless of the presence of space charge, the change in the induced charge ΔQ_{ind} and the current i at an electrode caused by a charge q moving from x_i to x_f are given by

$$\Delta Q_{\text{ind}} = \int_{xi}^{xf} q\vec{E}_w \, d\vec{x} = -q[\Phi_w(x_f) - \Phi_w(x_i)] \tag{5.35}$$

and

$$i = \frac{dQ}{dt} = q\vec{v}\vec{E}_w \tag{5.36}$$

$\vec{v}$ is the carrier velocity, and Φ_w and $\vec{E}_w$ are the weighting potential and electric field at position x. Note that the charge considered here need not be physically collected by an electrode to produce a signal. The motion of the charges through the bulk produces the signal by induction, so the signal begins to form immediately when energy deposition starts. It is only when the last carrier arrives at the collecting electrode that charge induction ceases and the signal is now fully formed. The total induced charge will be the sum of the induced charges due to the electron and holes. The integration of these charges along their respective path lengths gives the total charge, Q. In the absence of trapping, $Q = Q_o$, the original charge created, which in turn is proportional to the energy of the incident photon. However, in any semiconductor some density of electron and hole traps is always present, and results in a loss of carriers and therefore of charge at the electrodes. For compound semiconductors, crystal growth techniques lead to a much higher density of traps than in the elemental semiconductors, and hence to shorter lifetimes. For example, in CdZnTe, typical lifetimes are $\tau_e = 3 \times 10^{-6}$ sec and $\tau_h = 5 \times 10^{-8}$ sec. Because the hole lifetime is much shorter than the hole transit time, the induced charge is significantly reduced and is now dependent upon the depth of interaction, x_o. For a uniform electric field and negligible de-trapping, the fraction of charge that is induced at the electrodes is best described by the charge collection efficiency (CCE), given by the Hecht equation [27]:

$$\mathrm{CCE} = \frac{Q}{Q_o} = \frac{\lambda_e}{L}\left[1 - \exp\left(-\frac{L - x_o}{\lambda_e}\right)\right] + \frac{\lambda_h}{L}\left[1 - \exp\left(-\frac{x_o}{\lambda_h}\right)\right] \quad (5.37)$$

where L is the detector thickness, x_o is the distance from the cathode to the point of charge creation, and λ_e and λ_h are the carrier drift lengths* in the applied electric field, E, given by $\lambda_e = \mu_e \tau_e E$ and $\lambda_h = \mu_h \tau_h E$. Here, μ_e and μ_h are the electron and hole mobilities, and τ_e and τ_h are the corresponding lifetimes. It follows from Equation (5.37) that the CCE depends not only on λ_e and λ_h, but also on the location where the charge was created. Since the interaction points of incident photons are essentially random at intermediate and high energies, being weighted by the classical exponential absorption law, the width of the peak in the energy spectrum broadens to an extent governed by the ratios λ_e/L and λ_h/L.

In characterizing materials, the most useful figure of merit is the mobility-lifetime product ($\mu\tau$), which is directly related to the drift length. Low $\mu\tau$ products result in shorter drift lengths and therefore smaller λ/L, which in turn limits the maximum size and energy range of detectors. Drift lengths range from ~1 m for Si and Ge to ~1 mm for GaAs, CdTe, and HgI_2 and ~1 μm for GaP, InP, and PbI_2. For the elemental semiconductors $\mu\tau$ is of the order of 1 cm^2V^{-1} for both electrons

* λ is also known as the trapping length, "schubweg," or charge collection distance and is the mean distance travelled by a charge carrier before trapping or collection occurs.

and holes, whereas for compound semiconductors it is typically a few times 10^{-4} cm^{2}V^{-1} for electrons and 10^{-5} cm^{2}V^{-1} for holes, becoming smaller with increasing Z (see Table 5.3). The degradation can usually be traced to trapping centers caused by impurities, vacancies, structural irregularities such as dislocations, or for the softer materials, plastic deformation caused by mechanical damage during fabrication.

5.6.2.1 Spectral Broadening in Radiation Detection Systems

In addition to broadening caused by poor charge collection, the width of the full energy peak, ΔE, is also broadened by the statistics of carrier generation (Fano noise) and by electronic noise. For most compound semiconductors, however, these components are generally far less important than the noise due to incomplete charge collection, except for thin detectors where $\lambda_e/L \gg 1$. The energy resolution, ΔE, of the system is defined as the full width at half maximum (FWHM) of the energy-loss distribution resulting from exposure to monoenergetic radiation. The width of the energy-loss spectrum, in turn, results from the convolution of the probability distributions of the various noise components. For most detection systems, three components tend to dominate.

$$\Delta E = f(\sigma_F^2, \sigma_e^2, \sigma_c^2) \quad \text{(keV)} \tag{5.38}$$

where, σ_F^2 is the variance of the noise due to carrier generation or Fano noise, σ_e^2 is the variance of the noise due to the leakage current and amplifier noise, and σ_c^2 is the variance of the noise due to incomplete charge collection due to carrier trapping.

5.6.2.1.1 Fano Noise

In semiconductors, only about one-tenth of the initial energy deposition results in the creation of electron–hole pairs, the remainder being lost to lattice vibrations, phonons, and plasmon production. Consequently, unlike scintillation or gas counter detection systems, the limiting spectroscopic resolution is not a simple function of total energy deposition, and until recently the only realistic descriptions were semiempirical, based on the so-called Fano factor [28]. This was originally introduced to quantify the departure of the observed

TABLE 5.3
Room-Temperature Properties of Wide Bandgap Compound Semiconductor Materials Suitable for Hard X-Ray and Gamma-Ray Detectors

Material	Crystal Structure	Growth Method	Atomic Number	Density (g/cm³)	Bandgap (eV)	E_{pair} (eV)	Resistivity (Ω-cm)	μτ(e) Product (cm²/V)	μτ (h) Product (cm²/V)
Si	cp	FZ	14	2.33	1.12	3.62	$<10^4$	>1	~1
Ge	cp	Cz	32	5.33	0.66	2.96	50	1	1
4H-SiC	hcp	CVD	14,6	3.29	3.26	7.8	$>10^3$	4×10^{-4}	8×10^{-5}
GaAs	cp	CVD	31,33	5.32	1.43	4.35	10^{10}	1×10^{-4}	4×10^{-6}
InP	cp	LEC	15,49	4.79	1.34	4.2	10^8	2×10^{-5}	1×10^{-5}
CdTe	cp	THM	48,52	5.85	1.48	4.43	10^9	3×10^{-3}	2×10^{-4}
$Cd_{0.9}Zn_{0.1}Te$	hcp	HPB	48,30,52	5.78	1.57	4.6	10^{11}	1×10^{-2}	2×10^{-4}
PbI_2	layered	B-S	82,53	6.2	2.32	4.9	10^{13}	1×10^{-5}	1×10^{-6}
HgI_2	layered	VAM	80,53	6.4	2.13	4.2	10^{13}	1×10^{-4}	4×10^{-5}
TlBr	cp	B-S	81,35	7.56	2.68	6.5	10^{12}	3×10^{-3}	6×10^{-5}

Note: The abbreviations are: FZ, float zone; Cz, Czochralski; CVD, chemical vapor deposition; LEC, liquid encapsulated Czochralski; THM, traveller heater method; B-S Bridgman–Stockbarger method; HPB, high pressure Bridgman; VAM, vertical ampoule method; cp, cubic close packed; hcp, hexagonal close packed.
Source: From [23].

statistical fluctuations in the number of charge carriers in a gas from that expected from pure Poisson statistics. It is generally expressed as

$$F = \frac{\sigma^2_{\text{exp}}}{\text{var}_{\text{poisson}}} \tag{5.39}$$

where F is the Fano factor, σ^2_{exp} is the experimentally observed variance, and var_{poisson} is the Poissonian variance (equal to the mean number of events, N). Obviously, for purely Poisson statistics σ^2_{exp} is equal to the variance and $F = 1$. In semiconductors, the Fano factor essentially describes the energy partition; that is, the fraction of the total energy that goes into the production of electron–hole pairs. The contribution due to Fano noise, ΔE_F, can be calculated from

$$\Delta EF = 2.355\sqrt{\sigma_F^2} = 2.355\sqrt{FE\varepsilon} \quad \text{keV} \tag{5.40}$$

where, ε is the energy to create an electron–hole pair, and E is the incident energy. The experimental determination of the Fano factor is extremely difficult for several reasons and has only been accurately determined for the elemental semiconductors Si and Ge (Si = 0.143 ± 0.001 [29], Ge = 0.62 ± 0.02 [30]). The first reason is that the determination is only accurate when trapping noise and electronic noise are so low that the resolution is already close to the Fano limit. This is because Fano noise is not normally distributed so a simple Gaussian decomposition of Equation (5.38) is incorrect (see discussion in [29]). The second reason is that Fano noise is both a function of the Fano factor and of the electron–hole pair energy, which is usually determined by assuming the Fano factor *a priori*. A further complication arises from the theoretical work of Fraser et al. [31], supported by the experimental work of Owens et al. [32] on Si, which shows that both quantities are energy and temperature dependent—and that the dependences are different. The same has been shown for Ge [33].

In Figure 5.6 we show the intrinsic energy resolution that is potentially achievable at 5.9 keV as a function of bandgap

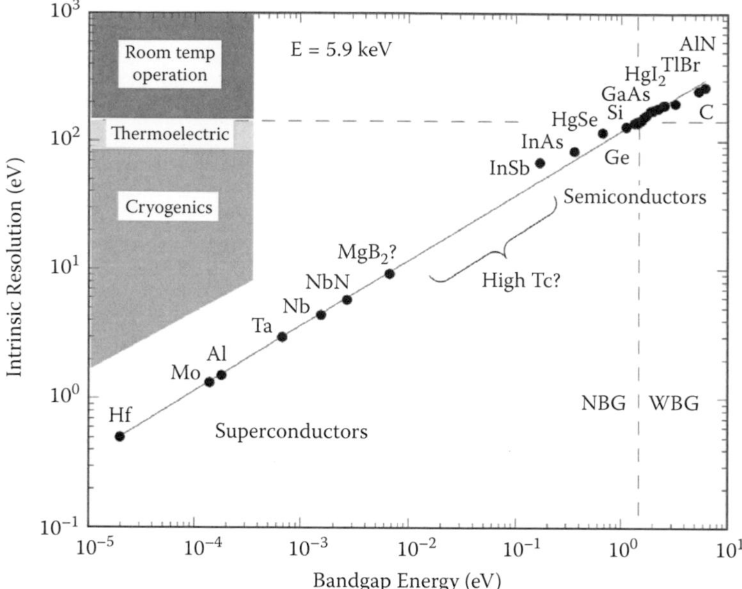

Figure 5.6 The limiting energy resolution achievable at 5.9 keV for a range of compound semiconductors as a function of bandgap energy at 5.9 keV (from [34]). For completeness we also include the superconductors. Curves are given for average values of the Fano factor (i.e., 0.22 for superconductors and 0.14 for semiconductors). NBG and WBG show the regions in which the narrow bandgap and wide bandgap semiconductors lie.

energy for a range of compound semiconductors, assuming an average value for the Fano factor of 0.14. For comparison, we also include some superconductors. The temperature ranges in which different compounds can be expected to operate with near Fano energy resolution are illustrated on the left-hand side of the figure. Cooling is required to suppress the thermal generation of carriers, which can swamp the limiting resolution, depending on the bandgap energy. We see that room-temperature operation can only be achieved for bandgaps above ~1.4 eV, while thermoelectric cooling can be used for Si down to Ge. At and below Ge, cryogenic cooling is required. We thus define wide bandgap (WBG) semiconductors as having a bandgap energy conducive to room temperature operation, >1.4 eV, and narrow bandgap (NBG) compounds having a bandgap energy below this value.

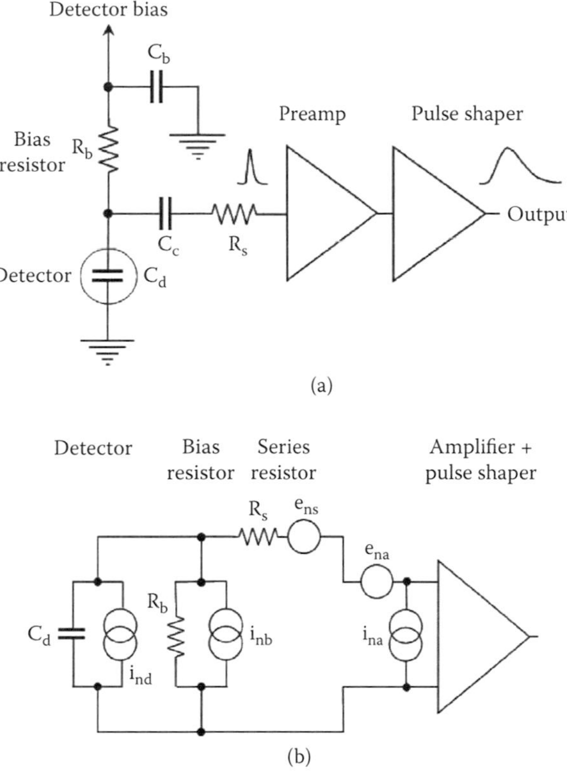

Figure 5.7 (a) A typical detector front-end detector-preamplifier-shaper circuit. (b) Equivalent circuit diagram (adapted from [35]).

5.6.2.1.2 Electronic Noise

The electronic noise term in Equation (5.38) arises largely from shot noise generated by the leakage current of the detector and noise generated in the amplification system. Both types of noise are random in amplitude and time over a wide frequency range, and their contribution to measured energy fluctuations depends on how the signal is processed. A typical detector front-end schematic is given in Figure 5.7a. The detector can be represented by a capacitor, because in essence it consists of two parallel conductors with a dielectric between them. The detector is coupled to the bias supply through a bias resistor R_b. The bias supply is decoupled by capacitor, C_b, whose primary function is to act as a filter for the supply. Under AC conditions this capacitor allows R_b essentially to shunt the detector. The signal from the detector is then fed to the preamplifier via the blocking capacitor, C_c, whose function is to pass the signal to the preamp unimpeded while isolating the preamp from the bias supply. The series resistance R_s represents the sum of all resistances present in the input signal path (such as contact and any parasitic resistances). The preamplifier buffers the signal and sends it to a pulse shaper which filters the overall frequency response to optimize signal-to-noise. At the same time, it limits the duration of the signal pulse to accommodate the expected signal rate. Generally, Gaussian or pseudo-Gaussian shaping is used, designed to avoid overshoot to a step function input while also minimizing the rise and fall time of the signal. The electronic noise expected from the front-end can be derived from its equivalent electrical circuit, shown in Figure 5.7b, in which various contributors to the noise are represented by current or voltage sources, depending on whether they are in parallel (current) or series (voltage) with the detector capacitance. By convention, those in parallel are known as parallel noise sources and those in series as series noise sources. The various noise sources add in quadrature. Because the energy deposited in the detector translates directly into charge, it is convenient to express the electronic noise as an equivalent noise charge (ENC), defined as the number of electron charges which, if applied to the input, would give rise to the same RMS output voltage. The ENC and FWHM energy resolution are simply related by

$$\Delta E_e = \frac{2.355\,\varepsilon\,ENC}{e} \tag{5.41}$$

The parallel component, Q_p, is comprised of the shot noise due to the detector leakage current, I, and Johnson noise from the bias resistor, R_b, which under signal conditions effectively shunts the detector. It is given by the expression,

$$Q_p^2 = \left(2eI + \frac{4kT}{R_b} + i_{na}^2 \right) \tau \tag{5.42}$$

Here e is the electronic charge, k the Boltzmann's constant, T the temperature, and $i^2{}_{na}$ represents the current noise due to the preamplifier. The leakage current is the sum of two main components, the bulk leakage current and the surface leakage current, which depends on detector design. Bulk leakage currents are generated internally within the detector volume and arise from a number of sources. First, because of the finite resistivity of the bulk material, a standing current will always be present. In junction-type detectors, bulk leakage also arises as a natural consequence of the junction because, while majority carriers are repelled from the boundaries of the depletion region under bias, minority carriers are attracted and free to diffuse across it. Minority carrier current is usually small and can generally be neglected compared to surface leakage currents. Bulk leakage current also arises due to the thermal generation of carriers within the depletion region, the rate of which depends on its volume. Clearly this component is largely dependent on the bandgap energy and can be significantly reduced by cooling. Surface leakage takes place at the surface of the detector and is particularly acute at the edges where potential gradients are the greatest. Its magnitude is very dependent on material processing, surface preparation, and passivation techniques as well as material type.

Elemental semiconductor detectors are usually junction devices operated in a reverse bias mode to reduce bulk leakage currents and maximize active depths. However, for most semiconducting materials the bulk resistivity is still too low to allow low-noise operation. Therefore, a Schottky barrier or blocking contact is usually employed. Assuming surface leakage components can be neglected, the leakage current in this case is

$$I = A^{*}T^{2}e^{-e\Phi/kT} \tag{5.43}$$

where A^{*} is a constant related to Richardson's constant, and Φ is the barrier height of the Schottky contact. We see that the leakage current is a strong function of temperature. For this reason, it is usually necessary to cool the detector to achieve low-noise performance. At the extremes, Fano-limited operation requires leakage currents in the pA range, whereas MeV alpha particles can barely be detected above the noise for leakage currents in the mA range.

The series noise contribution, Q_s, is largely due to the voltage noise due to the series resistance, R_s and noise due to the amplifier. It is given by the expression,

$$Q_s^2 = \left(4kTR_s + v_{na}^2\right) C_d^2 \frac{1}{\tau} + 4A_f C_d^2 \tag{5.44}$$

where $v^2{}_{na}$ is voltage noise due to the preamplifier, and A_f is a noise coefficient which depends on the particular amplifier design. Normally the input stage of the preamp is a field effect transistor (FET). In this case, the square of the shot noise of the first stage of the preamplifier is proportional to

$$\frac{T}{g_m \tau} C_d^2 \tag{5.45}$$

where g_m is the transconductance of the FET. The last term in Equation (5.44) represents 1/*f* or "pink" noise due to the amplifier. Note that the amplifier noise sources are not present at the amplifier input but originate within the amplifier and appear at the output. In order to take account of this component, the noise can be referred back to the input by dividing by the gain so that it looks like a voltage noise generator.

For simple RC-RC shaping, Spieler [35] gives the following formulation for the total ENC:

$$Q_n^2 = \left(\frac{\varepsilon^2}{8}\right)\left[\left(2eI + \frac{4kT}{R_b}\right)\tau + \left(4kTR_s + v_{na}^2\right) C_d^2 \frac{1}{\tau} + 4A_f C_d^2\right] \tag{5.46}$$

↑ current noise ↑ voltage noise ↑ 1/f noise

We see from Equation (5.46) that current or parallel noise increases with shaping time and is independent of detector capacitance. It can be reduced by (a) measuring at shorter shaping time, (b) reducing the current through the detector, or (c) choosing a feedback resistor with a high resistance or by avoiding it altogether and using a different reset mechanism. Voltage or series noise can be reduced by (a) measuring at a longer shaping time and (b) minimizing detector and stray

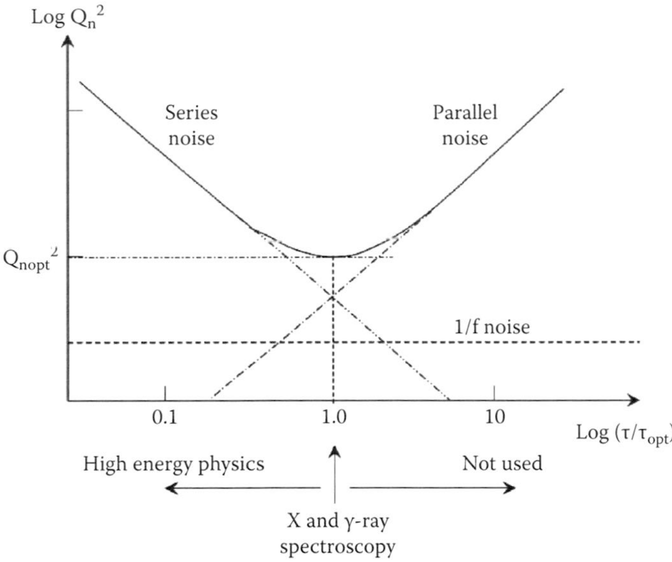

Figure 5.8 Equivalent noise charge versus shaping time. At low shaping times, series (voltage) noise dominates, whereas at high shaping times current (parallel) noise dominates.

capacitance. For a FET input stage voltage noise can be reduced by selecting a low-noise FET with a larger transconductance. The 1/*f* noise is independent of shaping time and can only be reduced by reducing the detector capacitance or the noise coefficient of the amplifier. As the shaping time τ is changed, the total noise goes through a minimum, where the current and voltage contributions are equal. This is illustrated in Figure 5.8, from which we see that at short shaping times voltage noise dominates, whereas at long shaping times current noise takes over. The noise minimum is flattened by the presence of 1/*f* noise. A spectrometer will normally be operated at $\tau = \tau_{opt}$, if event rate considerations do not require smaller shaping times, as in high-energy physics applications. A typical value of τ_{opt} is around 1 μs, but values of 10 μs or more are common in high-resolution spectroscopy (such as GeHP detectors).

Generally we know little of the preamplifier and shaper. For noise evaluation purposes, Equation (5.46) can be simplified if either the Johnson noise contribution is negligible for the bias resistor (therefore high resistance and low temperature), or the series resistance is negligible (which is usually true) or the remaining voltage noise terms can be treated as a single empirical constant. In this case, the FWHM energy resolution due to the electronic noise component, ΔE_e, reduces to

$$\Delta E_e = 2.355\sqrt{I\tau\varepsilon^2 A/e + \sigma_a^2} \text{ keV} \tag{5.47}$$

where A is a constant depending on the type of signal shaping, and σ_a^2 is the variance of the amplifier shot noise which depends on the actual design and specifically the capacitive loading of the front-end electronics. It is this term we can treat as an empirical constant. For the Gaussian shaping typically used in spectroscopy, $A = 0.875$.

From Equations (5.46) and (5.47), we note that for a given temperature and operating condition, electronic noise is independent of the energy of the incident radiation, which makes it relatively easy to measure directly. This is achieved by injecting a precise amount of charge into the input of the preamplifier and measuring the degree by which it is broadened. The empirical term in Equation (5.47) can then be determined by best-fitting.

5.6.2.1.3 Trapping Noise

The functional form of noise due to incomplete charge collection, ΔE_c, is much more difficult to predict, because it is intimately dependent on the trap density distribution as well as the charge diffusion and collection properties of the detector. For the case of $\lambda_e \neq \lambda_h$, which is invariably true for compound semiconductors, Iwanczyk et al. [36] have arrived at the following analytic form for the relative broadening on the assumption of a uniform electric field:

$$\left(\frac{\sigma_c}{E}\right)^2 = \frac{2\lambda_e^2\lambda_h^2}{L^3(\lambda_e - \lambda_h)}(e^{-L/\lambda_e} - e^{-L/\lambda_h}) - \frac{1}{L^4}\left[\lambda_e^2(e^{-L/\lambda_e}) + \lambda_h^2(e^{-L/\lambda_h} - 1)^2\right] - \frac{\lambda_e^3}{2L^3}(e^{-2L/\lambda_e} - 1) - \frac{\lambda_h^3}{2L^3}(e^{-2L/\lambda_h} - 1) \tag{5.48}$$

However, while Equation (5.48) can be used to calculate the width ΔE_c, the shape of the pulse height distribution can only be realistically evaluated by a more detailed approach such as that of Trammell and Walter [37], in which the individual pulse heights in infinitesimal slices through the detector are summed over the detector thickness. For low trapping, the summed pulse height distribution will be nearly symmetric at low and intermediate energies, and we can therefore assume ΔE_c is normally distributed. For this case, several semiempirical formulas have been proposed for the summed response. For example, Henck et al. [38] derived an expression for planar detectors in which $\Delta E_c \propto E^{-1/2}$, whereas Owens [30] predicted that $\Delta E_c \propto E$ for coaxial detectors. Because of the uncertainties in the functional form of ΔE_c and our imprecise knowledge of F, a semiempirical approach is generally used to describe the resolution function in the form,

$$\Delta E = 2.355\sqrt{F\varepsilon E + (\Delta E_e/2.355)^2 + a_1 E^{a2}} \text{ keV} \tag{5.49}$$

where the electronic noise component ΔE_e is measured directly using a precision pulser, and F, a_1, and a_2 are treated as semiempirical constants determined by best-fitting. Equation (5.49) is found to fit the resolution functions of compound semiconductors reasonably well, with the charge trapping exponent, a_2, varying between 2 and 3. Recently, the validity of the functional form of the trapping term has been rigorously tested by Kozorezov et al. [39], who derived a general analytical expression, which includes the effects of geometry. Interestingly, it can be expressed in the form $G(E)E^2$, in which the function $G(E)$ reduces to a constant when the $1/e$ absorption lengths become comparable with the detector thickness—which is invariably true for spectroscopic detectors. The relative magnitudes of the various noise components are illustrated in Figure 5.9. Here we show the measured ΔE for a 3.1 mm^2, 2.5-mm-thick CdZnTe detector as a function of energy deposition [40]. The total leakage current at the nominal bias was <1nA. In the low-energy region, electronic noise dominates the response, while at intermediate energies, Fano noise takes over. At high energies, trapping noise becomes the dominant component.

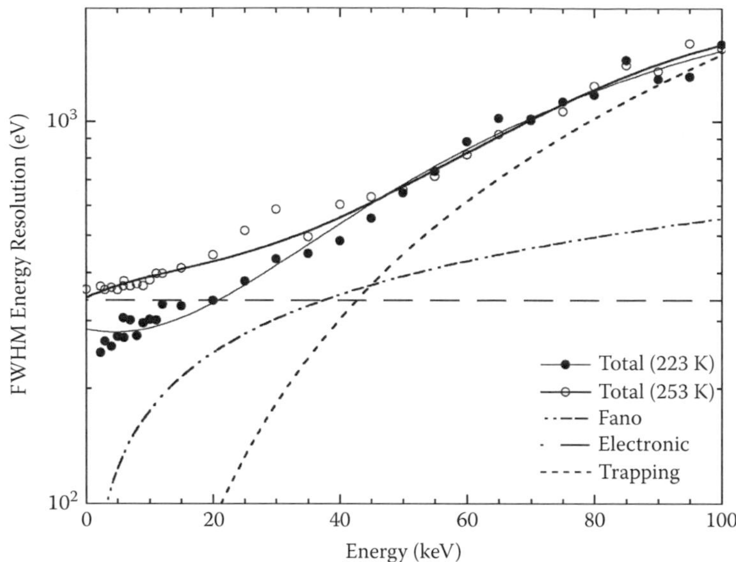

Figure 5.9 The energy resolution ΔE of a 3.1-mm², 2.5-mm-thick CdZnTe detector measured at two temperatures (from [40]). The solid line shows the best-fit resolution function to –20°C data. The individual components to the FWHM are also shown for this curve. These are: the noise due to carrier generation or Fano noise ΔF, electronic noise due to leakage current and amplifier shot noise, Δe, and incomplete charge collection or trapping noise Δc.

References

1. N. Bohr, "On the Theory of Decrease of Velocity of Moving Electrified Particles on Passing through Matter," *Phil. Mag.*, Vol. 25, no. 145 (1913) pp. 10–31.
2. H.A. Bethe, "Zur Theorie des Durchgangs schneller Korpuskularstrahlen durch Materie," *Ann. Physik.*, Vol. 397 (1930) pp. 325–400.
3. F. Bloch, "Bremsvermögen von Atomen mit mehreren Elektronen," *Z. Phys.*, Vol. 81 (1933) pp. 363–376.
4. L.H. Thomas, "The Calculation of Atomic Fields," *Proc. Cambridge Phil. Soc.*, Vol. 23, 5 (1927) pp. 542–548.
5. E. Fermi, "Un Metodo Statistico per la Determinazione di alcune Priorieta dell 'Atome." *Rend. Accad. Naz. Lincei*, Vol. 6 (1927) pp. 602–607.

6. L. Landau, "On the Energy Loss of Fast Particles by Ionisation," *J. Phys. (USSR)*, Vol. 8 (1944) pp. 201–205.
7. R. Bailey, C.J. Damerell, R.L. English, A.R. Gillman, A.L. Lintern, S.J. Watts, F.J. Wickens, "First Measurement of Efficiency and Precision of CCD Detectors for High-Energy Physics," *Nucl. Instr. and Meth.*, Vol. 213 (1983) pp. 201–215.
8. A. Owens, K.J. McCarthy, "Energy Deposition in X-Ray CCDs and Charged Particle Discrimination," *Nucl. Instr. and Meth.*, Vol. A366 (1995) pp. 148–154.
9. G. Hall, "Ionization Energy Losses of Highly Relativistic Charged Particles in Thin Silicon Layers," *Nucl. Instr. and Meth.*, Vol. 220 (1984) pp. 356–362.
10. P. Shulek, B.M. Golovin, L.A. Kulyukina, S.V. Medved, P. Pavlovich, "Fluctuations of Ionization Loss," *Sov. J. Phys.*, Vol. 4 (1967) pp. 400–401.
11. O. Blunck, S. Leisegang, "Zum Energieverlust schneller Elektronen in dünnen Schichten," *Z. Physik*, Vol. 128 (1950) pp. 500–505.
12. S.F. Mughabghab, M. Divadeenam, N.E. Holden, *Neutron Cross Sections, Volume 1, Neutron Resonance Parameters and Thermal Cross Sections, Part A, Z=1–60*, Academic Press, New York (1981).
13. S.F. Mughabghab, Neutron Cross Sections, Volume 1, *Neutron Resonance Parameters and Thermal Cross Sections, Part B, Z=61–100*, Academic Press, New York (1984).
14. U. Fano, "Gamma-Ray Attenuation, Part 1," *Nucleonics*, Vol. 11(8) (1953a) pp. 8–12.
15. U. Fano, "Gamma-Ray Attenuation, Part 2," *Nucleonics*, Vol. 11(9) (1953b) pp. 55–60.
16. W. Heitler, *The Quantum Theory of Radiation*, 3rd edition, Oxford University Press, Oxford (1954).
17. J.J. Thompson, "On the Scattering of Rapidly Moving Electrified Particles," *Proc. Camb. Phil. Soc.*, Vol. 75 (1910) pp. 465–471.
18. A.H. Compton, "A Quantum Theory of the Scattering of X-Rays by Light Elements." *The Physical Review*, Vol. 21, 5 (1923) pp. 483–502.
19. O. Klien, Y. Nishina, "Über die Streuung von Strahlung durch freie Elektronen nach der neuen relativistischen Quantendynamik von Dirac," *Z. Physik*, Vol. 52 (1929) pp. 853–868.
20. P. Marmier, E. Sheldon, *Physics of Nuclei and Particles, Volume I*, Academic Press, New York (1969).
21. R.D. Evans, *The Atomic Nucleus*, McGraw-Hill, New York (1955).

22. A. Rose, *Concepts in Photoconductivity and Allied Problems*, Interscience Publishers, New York (1963).
23. A. Owens, "Semiconductor Materials and Radiation Detection," *J. Synchrotron Radiation*, Vol. 13, part 2 (2006) 143–150.
24. C.A. Klein, "Bandgap Dependence and Related Features of Radiation Ionization Energies in Semiconductors," *J. Appl. Phys.*, Vol. 4 (1968) pp. 2029–2033.
25. W. Shockley, "Currents to Conductors Induced by a Moving Point Charge," *J. Appl. Phys.*, Vol. 9 (1938) pp. 635–636.
26. S. Ramo, "Currents Induced by Electron Motion," *Proc. IRE*, Vol. 27 (1939) pp. 584–585.
27. K. Hecht, "Zum Mechanismus des lichtelektrischen Primärstromes in isolierenden Kristallen," *Z. Physik*, Vol. 77 (1932) pp. 235–245.
28. U. Fano, "Ionization Yield of Radiations. II. The Fluctuations of the Number of Ions," *Phys. Rev.*, Vol. 72 (1947) pp. 26–29.
29. A. Owens, G.W. Fraser, K.J. McCarthy, "On the Experimental Determination of the Fano Factor in Si at Soft X-Ray Wavelengths," *Nucl. Instr. and Meth.*, Vol. A491 (2002) pp. 437–443.
30. A. Owens, "Spectral Degradation Effects in an 86cm^3 Ge(HP) Detector," *Nucl. Instr. and Meth.*, Vol. A235 (1986) pp. 473–478.
31. G.W. Fraser, A.F. Abbey, A. Holland, K. McCarthy, A. Owens, A. Wells, "The X-Ray Energy Response of Silicon Part A. Theory," *Nucl. Instr. and Meth.*, Vol. A350 (1994) pp. 368–378.
32. A. Owens, G.W. Fraser, A.F. Abbey, A. Holland, K. McCarthy, A. Keay, A. Wells, "The X-Ray Energy Response of Silicon (B): Measurements," *Nucl. Instr. and Meth.*, Vol. A382 (1996) pp. 503–510.
33. B.G. Lowe, "Measurements of Fano Factors in Silicon and Germanium in the Low-Energy X-Ray Region," *Nucl. Instr. and Meth.*, Vol. A399, (1997) pp. 354–364.
34 A. Owens, A. Peacock, "Compound Semiconductor Radiation Detectors," *Nucl. Instr. and Meth.*, Vol. A531 (2004) pp. 18–37.
35. H. Spieler, "Front-End Electronics and Signal Processing," *Instrumentation in Elementary Particle Physics AIP Conference Proceedings*, Vol. 674 (2003) pp. 76–100.
36. J.S. Iwanczyk, W.F. Schnepple, M.J. Masterson, "The Effect of Charge Trapping on the Spectrometric Performance of HgI_2, Gamma-Ray Detector," *Nucl. Instr. and Meth.*, Vol., A322 (1992) pp. 421–426.

37. R. Trammell, F.J. Walter, "The Effects of Carrier Trapping in Semiconductor Gamma-Ray Detectors," *Nucl. Instr. and Meth.*, Vol. 76 (1969) pp. 317–321.
38. R. Henck, D. Gutknecht, P. Siffert, L. De Laet, W. Shoenmaekers, "Trapping Effects in Ge(Li) Detectors and Search for Correlation with Characteristics Measured on the *p*-Type Crystals," *IEEE Trans. Nucl. Sci.*, Vol. NS-17 (1970) pp. 149–159.
39. A. Kozorezov, K. Wigmore, A. Owens, R. den Hartog, A. Peacock, H.A. Al-Jawhari, "Resolution Degradation of Semiconductor Detectors Due to Carrier Trapping," *Nucl. Instr. and Meth.*, Vol. A546 (2005) pp. 209–212.
40. A. Owens, M. Bavdaz, H. Andersson, T. Gagliardi, M. Krumrey, S. Nenonen, A. Peacock, I. Taylor, L. Tröger, "The X-Ray Response of CdZnTe," *Nucl. Instr. and Meth.*, Vol. A484 (2002) pp. 242–250.

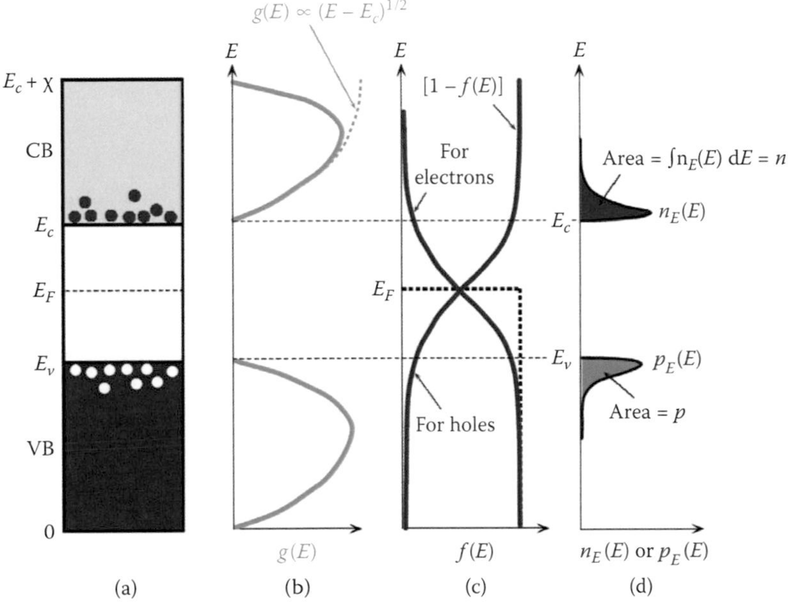

Color Figure 1.11 Schematic of the various distributions discussed in the text leading up to the energy density of electrons and holes in the conduction and valence bands, respectively (from reference [36]). The example given is for an intrinsic semiconductor. (a) The energy band diagram. (b) The density of states (number of states per unit energy per unit volume. The total number of states in the valence band and conduction bands is equal to the number of valence electrons. However in metals the total number of states in the valence band is much larger, which is the reason why electrons in metals need no activation to become mobile. (c) The Fermi–Dirac probability function (probability of occupancy of a state) and (d) the product of g(E) and f(E), which gives the energy density of electrons in the conduction band. The area under n_E(E) versus E is the electron concentration in the conduction band.

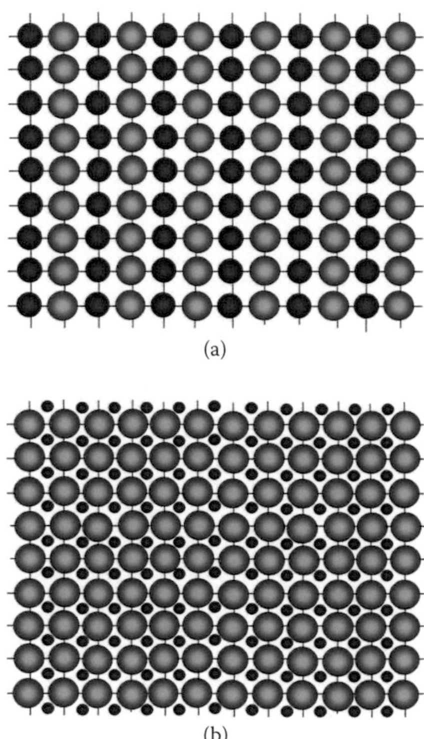

Color Figure 2.4 (a) Schematic of an ordered substitutional cubic lattice in which an atom of one element replaces an atom of the host element in an alternating sequence. The ability to form a stable lattice depends on whether the two species can satisfy the Hume-Rothery rules. (b) Example of an interstitial lattice in which the atoms of one element fit interstitially into the spaces in the lattice of the host element.

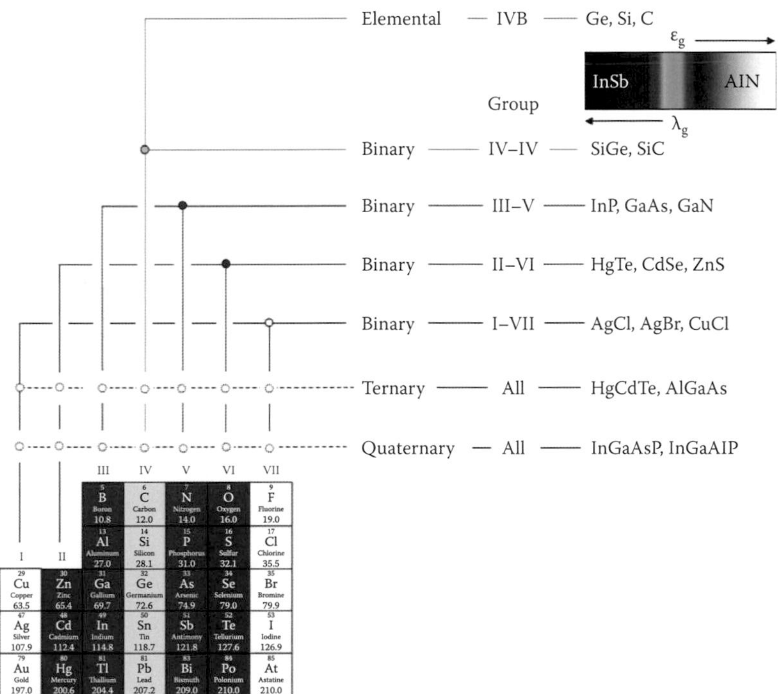

Color Figure 2.5 Diagram illustrating the relationship of the elemental and compound semiconductors. Examples of the compound type are given and are listed by increasing bandgap energy or alternatively, decreasing wavelength, from the infrared to the ultraviolet. InSb and AlN delineate the extremes of the range in which compound semiconductors lie (0.17 eV–6.2 eV).

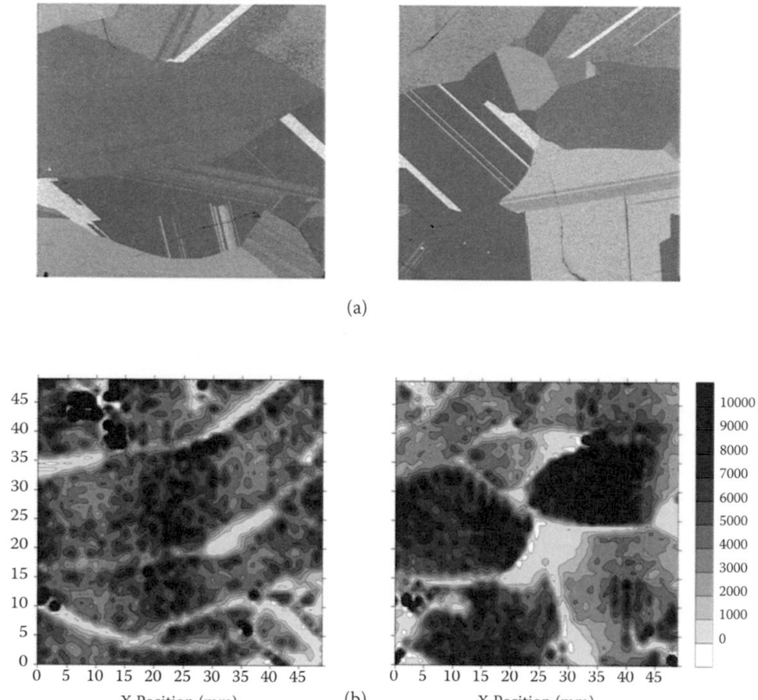

Color Figure 2.16 (a) Optical images of two 50 × 50 mm^2, 3-mm-thick slices of a CdZnTe crystal grown by the High Pressure Bridgman method (from [28]). Numerous grain boundaries and twins are apparent in the image. (b) The crystals count rate response, measured with a ^{57}Co radioactive source is shown in the lower images, illustrating poor charge collection at the grain boundaries. Interestingly, no correlation was found with the numerous twin boundaries observed inside the grains, indicating that twins have a negligible effect on the electric field and charge collection of semi-insulating CdZnTe devices.

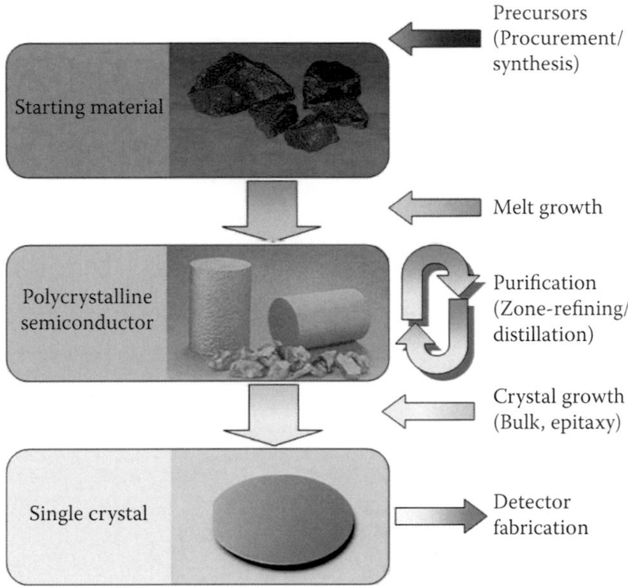

Color Figure 2.18 Steps required to produce single-crystal material for detector production.

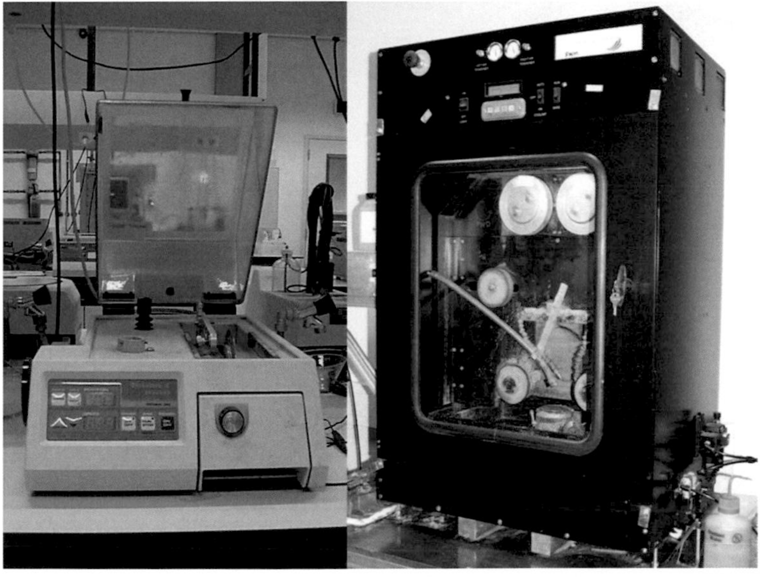

Color Figure 3.2 Left: a diamond disk saw. Right: a wire saw used for cutting ingots into slices prior to detector preparation. (Images courtesy of the European Space Agency and Kromek®.)

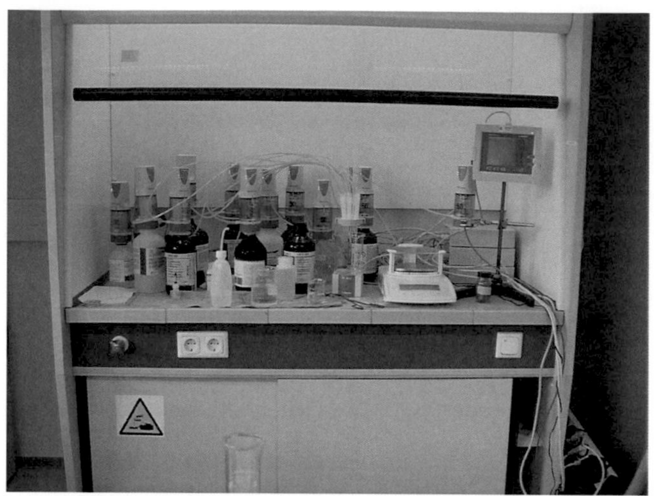

Color Figure 3.4 Etching cabinet with automatic titration system for preparing different etch solutions. The cabinet also contains services such as a deionized water supply and compressed air system to clean and dry the crystals (Image courtesy of the European Space Agency.)

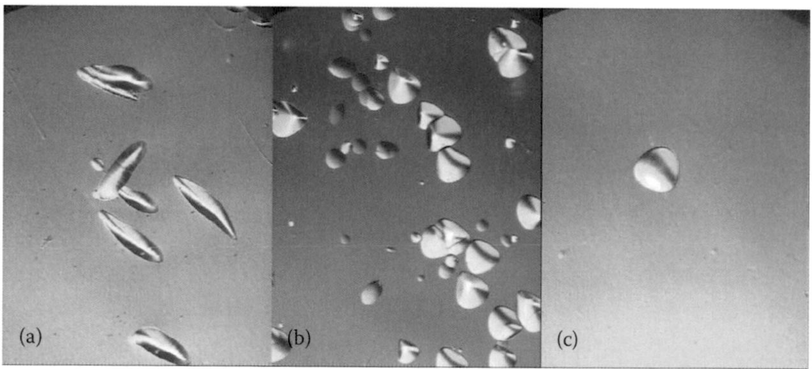

Color Figure 3.5 Dislocation etch pits in Si, showing the effects of preferential etching along different crystallographic directions. In the [100] orientation, the etch pits appear (a) elliptical in shape, whereas along the [111] direction they can assume (b) triangular or (c) pyramidal shapes.

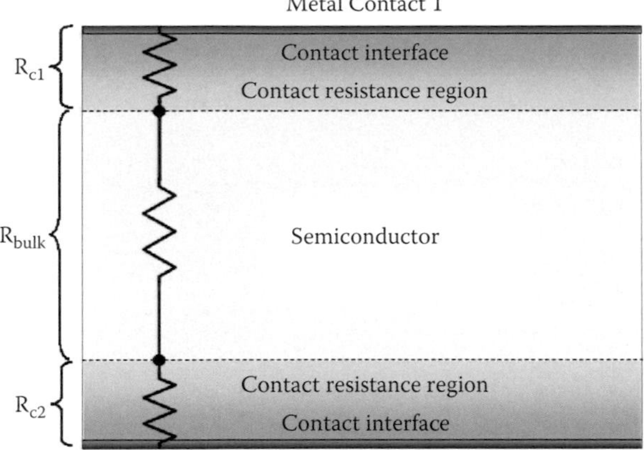

Color Figure 3.13 Typical detector construction, which consists of a sandwich of a metal contact, semiconductor, and further metal contact. R_c and R_{bulk} represent the resistances of the contact and the bulk semiconductor. The contact resistance region is that region over which the contact is formed and may have a finite extent depending on how the contact was formed.

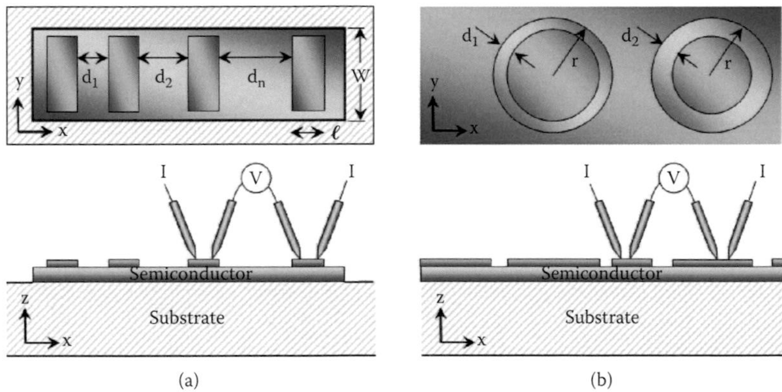

Color Figure 3.14 Contact resistance test patterns (from [12]), (a) measurement configurations for the transfer length method (TLM) and (b) circular transfer length method (CTLM). For the TLM measurement, the semiconductor has been etched away around the contacts to form a mesa in order to restrict current flow to adjacent contacts. For both TLM and CTLM, measurements are usually carried out using a four-probe technique as illustrated in the cross-sectional views.

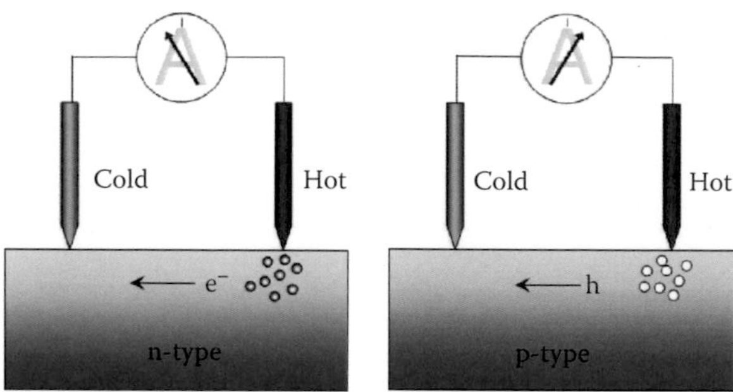

Color Figure 3.17 Illustration of the hot probe technique for determining the majority carrier type in semiconductors. Carriers diffuse more rapidly near the hot probe. This leads a flow of majority carriers away from the hot probe and a resultant electrical current toward (*p*-type) or away from (*n*-type) the hot probe.

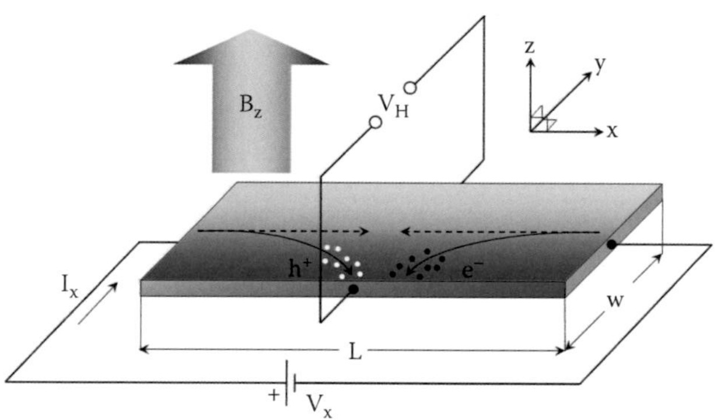

Color Figure 3.18 Schematic illustrating the sign convention and terminology for the Hall effect.

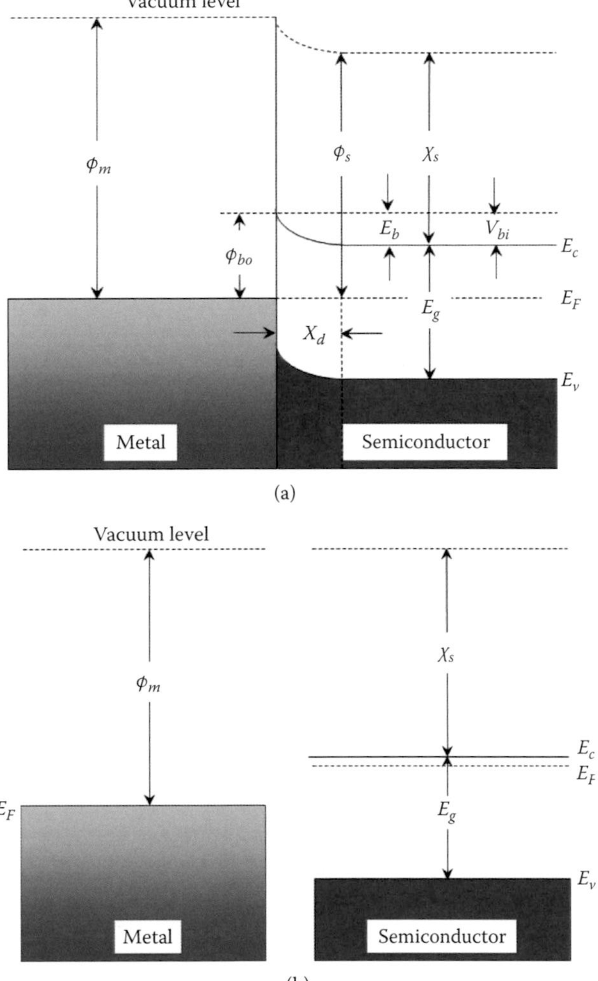

Color Figure 4.2 Energy band diagram of a metal and n-type semiconductor (a) before and (b) after contact is made.

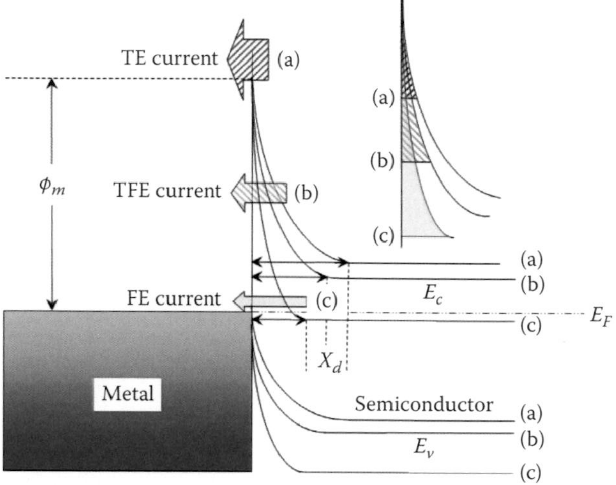

Color Figure 4.6 Conduction mechanisms through a metal–semiconductor interface with different semiconductor donor levels for an *n*-type semiconductor, corresponding to (a) a low residual doping level of $<10^{17}$ cm^{-3}, (b) an intermediate doping level of 10^{17}cm^{-3} to 10^{18}cm^{-3}, and (c) a high doping level of $>10^{18}$cm^{-3}. It can be seen that, as the doping concentration increases, the barrier width decreases as the conduction band falls below the Fermi level. To the right of the barrier we show an additional schematic of the barrier in which the shaded regions give a visual indication of the fraction of carriers that can cross the barrier.

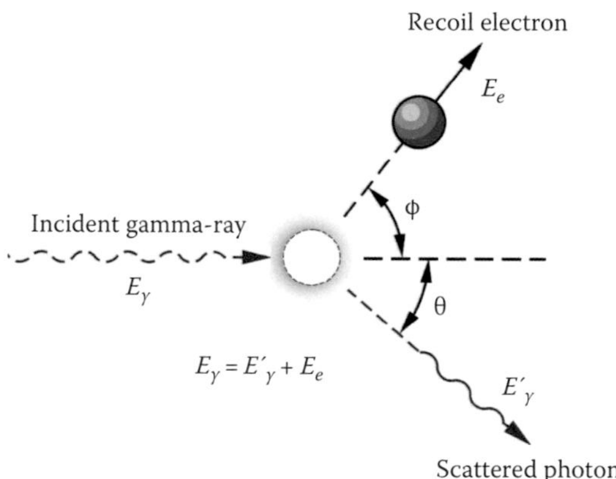

Color Figure 5.1 Compton scattering geometry. The electron is assumed to be initially at rest and is ejected at angle ϕ by the incident photon with a kinetic energy E_e. The initial photon in turn is scattered through angle θ but with a reduced energy E_γ', which by conservation of energy is equal to $E_\gamma - E_e$.

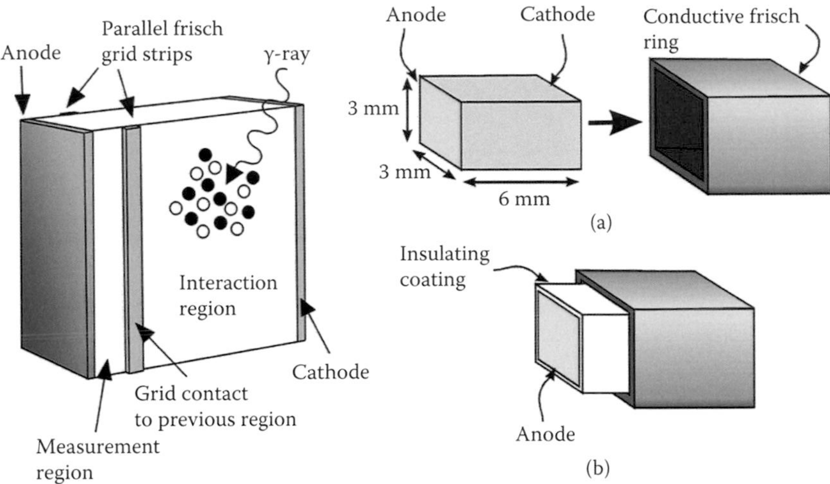

Color Figure 7.8 Two implementations of the Frisch grid concept. Left: a prototype semiconductor design (taken from [17]). Two parallel contact strips are fabricated on the sides of the detector between the anode and cathode planes and act as a pseudo-Frisch grid. Charge carriers are excited in the interaction region, and the electrons are drifted through the parallel grids by an applied electric field. Right: schematic of a capacitively coupled Frisch grid detector, consisting of (a) a bar-shaped detector placed inside an isolated but conductive ring (b) (from [23]). This implementation effectively eliminates leakage current between the grid and the anode, because the ring is not actually connected to the detector.

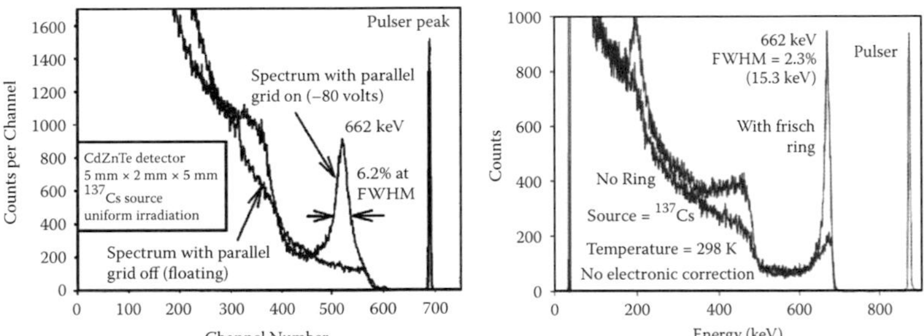

Color Figure 7.9 Left: response of a 5 × 2 × 5 mm^3 CdZnTe patterned Frisch grid detector to a ^{137}Cs radioactive source under full area illumination (taken from [17]). Shown are spectra taken with the grid turned off and on. No full energy peak is apparent when the parallel grid is off; however a full energy peak of 6.2% FWHM at 662 keV becomes obvious when the grid is activated. Right: ^{137}Cs spectra from a 3 × 3 × 6 mm^3 CdZnTe device with a 5 mm insulated capacitively coupled Frisch ring (from [26]). The thin-lined blue spectrum shows the response of the device with the ring connected, and the thick-lined red spectrum with it disconnected. With the ring connected, the FWHM energy resolution is 2.3% at 662 keV.

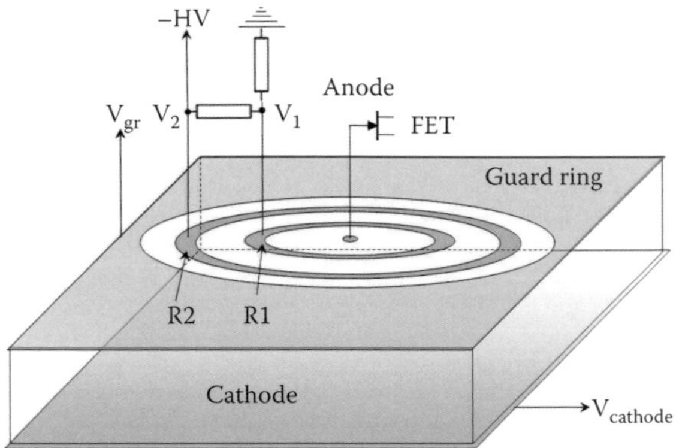

Color Figure 7.16 Schematic image of the prototype ring-drift detector. The crystal has dimensions 5 × 5 × 1 mm^3. The inner anode has a diameter of 80 μm, and the centers of the other two ring electrodes, R1 and R2, are 0.19 and 0.39 mm, away from the center of the anode. The guard ring extends beyond a radius of 0.59 mm.

6 Present Detection Systems

Frontispiece Evolution of GaAs X-ray detector technology at the European Space Agency, starting with test diodes, small-format arrays, and large arrays, toward an ultimate goal of spatially resolved spectroscopy for X-ray astrophysics and planetary space missions. (Images courtesy of Oxford Instruments Analytical Oy and Thales Alenia Space.)

6.1 Compound Semiconductors and Radiation Detection

Two and a half decades ago, Armantrout et al. [1] produced a rank-ordered listing of the most promising materials for further development as radiation detectors. However, out of a list of nine compounds, only CdSe, HgI_2, and CdTe were investigated, and of these only HgI_2 and CdTe (and its alloy CdZnTe) are still under active development. Unfortunately, it soon became clear that the development of semiconductors based on compounds would be far more difficult than those based on group IV elements, for three fundamental reasons, as follows:

1. Resistivities are not as high as expected from the bandgap, and indeed it is only possible to get resistivities above 1 MΩ for a few materials.
2. Carrier trapping greatly reduces charge collection, giving rise to poor spectroscopic capability or even none at all. The problem is further exacerbated by the large mismatch between the electron and hole transport properties. Generally, the mu-tau products for holes are at least an order of magnitude worse than for electrons.
3. Polarization effects degrade poor performance even further.

All these effects can be directly attributed to imperfect material or surface problems, which in turn arise from the difficulty in growing and processing monocrystalline material with exact stoichiometry and in its processing. In fact, of the fifty or so compounds available, less than half have been investigated as possible detection media. The situation is summarized in Table 6.1. The materials superscripted with "*" are those that have shown some response to radiation, usually MeV alpha particles. Those superscripted with "++" are those compounds which have shown a spectroscopic response to X-rays (thus $E/\Delta E > 1$), and of these, only CdTe, CdZnTe, and HgI_2 have matured sufficiently to produce commercially viable detection systems, and another four are tantalizingly close. Nevertheless, CdTe is now used in fewer applications because CdZnTe has largely superseded it in view of its higher resistivity, lower dislocation density, and lower susceptibility to polarization effects. We describe these materials below; their physical and electrical properties are listed in Appendix E. We normally quote results for simple monolithic detectors with planar electrodes, because these tend to be dominated by basic material properties and not by electrostatic effects

TABLE 6.1
Compound Semiconductor Materials Listed by Group and in Order of Increasing Bandgap Ranging from the Near IR to XUV Wavelengths

Bandgap Energy (eV)	Elemental Group IVB	Binary IV–IV Compounds	Binary III–V Compounds	Binary III–VI Compounds	Binary II–VI Compounds	Binary IV–VI Compounds	Binary n–VIIB Compounds	Pseudo-Ternary/ Ternary Compounds
0.00–0.25	Sn		*InSb		HgTe			HgCdTe
0.25–0.50			*InAs		HgSe	PbSe, PbS, PbTe		
0.50–0.75	++Ge		GaSb					InGaAs
0.75–1.00		SiGe						
1.00–1.25	++Si					SnS		
1.25–1.50			++GaAs, ++InP		++CdTe			AlInAs
1.50–1.75			AlSb	++GaTe	++CdSe			AlGaAs
1.75–2.00			*B_4C, BP, InN	++GaSe			*BiI_3	++CdZnTe, ++CdZnSe, InAlP, ++$TlGaSe_2$ ++Tl_6I_4Se
2.00–2.25		++SiC	AlAs		HgS		*InI, ++HgI_2	++CdMnTe

(continued)

TABLE 6.1 (CONTINUED)
Compound Semiconductor Materials Listed by Group and in Order of Increasing Bandgap Ranging from the Near IR to XUV Wavelengths

Bandgap Energy (eV)	Elemental Group IVB	Binary IV–IV Compounds	Binary III–V Compounds	Binary III–VI Compounds	Binary II–VI Compounds	Binary IV–VI Compounds	Binary n–VIIB Compounds	Pseudo-Ternary/ Ternary Compounds
2.25–2.50			++GaP, AlP		ZnTe, CdS		++PbI_2	++TlBrI, *$TlPbI_3$
2.50–2.75					++ZnSe		++TlBr	++HgBrI
2.75–3.00					MnSe			
3.00–3.25					MnTe			
3.25–3.50			*GaN		MgTe, MnS			
3.50–3.75					MgSe, ZnS			
3.75–4.00								
4.00–4.25								
4.25–4.50					MgS			
4.50–4.75								
4.75–5.00								
5.00–5.25								

5.25–5.50	*C	
5.50–5.75		
5.75–6.00		++BN
6.00–6.25		AlN
6.25–6.50		
6.50–6.75		
6.75–7.00		

Note: We include the elemental semiconductors for completeness. Materials subscripted with a "*"are those for which spectroscopic measurements (that is, $E/\Delta E > 1$) have been made at X- or gamma-ray wavelengths. Here, ΔE is the FWHM energy resolution at energy E. The compounds subscripted with a "++"are those which have shown some response to radiation.

such as the small pixel effect. Techniques to mitigate against poor transport will be presented in the next chapter.

6.2 Group IV and IV–IV Materials

This grouping contains the classical elemental semiconductors, Si, Ge, C (diamond), and gray tin (α-Sn) which crystallize in the diamond structure and are unique in the periodic table in that their outer shells are exactly half filled. Consequently, they only bond covalently. An examination of the properties of group IV elements shows that bandgap, hardness, and melting points all decrease with increasing Z, while charge carrier mobilities, densities, and lattice constants generally increase. These trends may be attributed to the progressive metallization of the elements with increasing Z within the group. One can also combine two different group IV semiconductors to obtain compounds such as SiC and SiGe whose physical and electronic properties are intermediate.

6.2.1 Silicon Carbide

Silicon carbide (SiC) is currently being explored as a high-temperature Si alternative that is also chemical and radiation tolerant. It has several distinct advantages over Si, in that it has twice the thermal conductivity and eight times the maximum breakdown electric field. The former property is important for producing thermally stable or high-power semiconductor devices, while the latter means that much higher biases can be applied, resulting in higher drift velocities and more efficient charge collection. It also has a high saturated electron drift velocity (almost twice that of Si), which ensures a low trapping probability as well as a high displacement threshold energy of 21.8 eV, which ensures a high radiation tolerance. The wide bandgap (three times that of Si) means that dark currents are extremely low, which in principle should allow high-temperature operation up to +700°C. Indeed, SiC n-MOS (n-type metal oxide semiconductor) devices have been successfully operated and thermally cycled up to 630°C [2].

SiC belongs to a family of materials which display a one-dimensional polymorphism called polytypism [3]. Polytypes differ by the stacking sequence of each tetrahedrally bonded Si–C bilayer, crystallizing into cubic, hexagonal, or rhombohedral structures. In this respect, SiC can be thought of as an ordered alloy. Although over 200 polytypes have been discovered, the main building block for each is a tetrahedron consisting of a carbon atom bonded to four silicon atoms and vice versa. This allows polytypes to be easily categorized using the notation proposed by Ramsdell [4]. For example, the most commonly encountered forms are 2H-SiC, 4H-SiC, 6H-SiC, 3C-SiC, 15R-SiC. Here, the number refers to the number of silicon carbide double layers in a unit cell, while the letters H, C, and R refer to the type of lattice (hexagonal, cubic, or rhombohedral). The 3C-SiC polytype has a zinc blende structure and is referred to as β-SiC. It is the only cubic polytype. All other polytypes have hexagonal structures (collectively referred to as α-SiC) of which the 2H-SiC form represents the basic wurtzite allotrope. SiC is a polar crystal, in that the outermost atoms in the [0001] direction are Si atoms, and in the $[000\bar{1}]$ direction, C atoms. As a result, the Si- and C-terminated faces have different chemical and growth behaviors. A summary of the properties of the three main polytypes used in the semiconductor industry is given in Table 6.2. The significantly higher electron mobility observed in 4H-SiC makes this the preferred polytype for radiation detector applications.

The growth of high-resistivity SiC for detector applications can be achieved using two different techniques, either as bulk material grown as a single crystal, or as an epitaxial layer. However, it should be noted that controlling the crystal structure of SiC presents a major growth problem, because many different polymorphs can grow under apparently identical conditions. Bulk SiC is currently the only route to produce thick wafers (100 to 500 μm thickness), although the quality is relatively poor. Conversely, epitaxial SiC grown onto a substrate wafer can produce high-purity material, but thicknesses are currently limited to ~150 μm. In any case, high residual doping concentrations ($\geq 5 \times 10^{13}$ cm^{-3}) currently limit depletion depths to ~100 μm. At the present time, only epitaxial

TABLE 6.2
Material Properties of Selected Silicon Carbide Polytypes

Parameter	SiC Polytype		
	3C	4H	6H
Lattice structure	face centered cubic	hexagonal	hexagonal
Lattice constants a, c (Å)	4.359	3.079,10.115	3.0817,15.117
Eg (eV) @ T< 5 K	2.20	3.29	3.02
Eg (eV) @ T=300 K	2.39	3.26	3.05
Ecrit (MV/cm)	2.1	2.2	2.5
Θ_K ($Wcm^{-1}K^{-1}$)@ 300 K	3.6	3.7	4.9
μ_e ($cm^2V^{-1}s^{-1}$)[a]	800	1000	400
μ_h ($cm^2V^{-1}s^{-1}$)	40	120	100
Sat. e-drift vel. (cm/s)	2.5×10^7	2.0×10^7	2.0×10^7
$\varepsilon_r(0)$	9.7	10.0	10.0

Note: The number in the heading refers to the number of silicon carbide double layers in a unit cell, while the letters H and C, refer to the type of lattice (hexagonal, cubic, or rhombohedral).

[a] Parallel to c-axis.

material is suited for detector use, and even then only for radiation having a mean penetration depth of less than 100 μm. Fortunately, both growth methods are undergoing rapid evolution, and the quality of SiC material continues to improve. However, it is likely that thick epitaxial SiC will provide the best route to detector-grade material, provided that a fast and cost-effective growth route can be commercially developed.

SiC radiation detectors were originally developed four decades ago for applications in the nuclear power industry, specifically for extreme environment instrumentation and control systems for direct monitoring of reactor cores and waste management. Babcock et al. [5,6] studied the response of small SiC p-n junction diodes to alpha particles up to 700°C [5]. Ferber and Hamilton [7] were later able to demonstrate neutron detection using ^{235}U converter layers. They found good agreement between flux profile measurements made with the diode and those made with conventional gold foil activation techniques in a low-power reactor. In addition, the diode response as a function of neutron flux was linear over four decades of reactor power output. X-ray detection with SiC diodes was first demonstrated by Bertuccio et al. [8], who fabricated Schottky junctions on epitaxial SiC. The devices had a junction area of 3 mm^2 and were grown on a 30-μm-thick *n*-type 4H-SiC layer with a dopant concentration of 1.8×10^{15} cm^{-3}. At 300 K, the reverse current density of the best device varied between 2 pA/cm and 18 pA/cm for mean electric fields of 40 kV cm^{-1} and 170 kV cm^{-1}. The devices showed a spectroscopic response to X-rays and gamma rays. At 60 keV, measured energy resolutions of 2.7 keV full width at half maximum (FWHM) were recorded at room temperature (RT), limited mainly by detector capacitance.

Since this time, X-ray performances have steadily improved [9]. For a 0.03 mm^2, 70-μm-thick device, Bertuccio et al. [10] recorded a FWHM resolution at 60 keV of 693 eV at 27°C, increasing to 1.1 keV at 100°C operating temperature. Phlips et al. [11] tested the radiation response of a 1-mm-diameter SiC PIN* diode originally developed for high-power applications. The device consisted of a 100-μm-thick 4H-SiC layer grown epitaxially on a SiC substrate. The measured room temperature resolution at 60 keV was ~550 eV FWHM, limited by an electronic noise of 28 electrons rms (root mean square). Later measurements by Bertuccio et al. [12], on gold-contacted 4H-SiC, 200-μm-diameter Schottky diodes show current densities of 17 pA/cm^2 at 340 K—more than two orders of magnitude lower than commercial silicon devices. Good spectroscopic performance was demonstrated at low energies, with energy resolutions of 196 eV FWHM recorded at 5.9 keV at +30°C and 233 eV FWHM at +100°C. The former value is reasonably close to the expected Fano-limited value of 160 eV, assuming a Fano factor of 0.1. In Figure 6.1, we show the measured spectral response to an ^{241}Am source, at room temperature and at 100°C. The electronic noise corresponds to 120 eV FWHM (6.5 electrons rms) at +27°C and 177 eV FWHM (9.6 electrons rms) at +100°C. At 26.3 keV, the resolution is ~400 eV and is essentially the same at operating temperatures of both 27°C and 100°C.

* A PIN (*p*-type–intrinsic–*n*-type) diode is essentially a refinement of the p-n junction in which a layer of intrinsic material is inserted between the *p* and *n* layers. As a result, PIN diodes have a high breakdown voltage, a low level of junction capacitance, and a larger depletion region than a simple junction—ideal for detector applications.

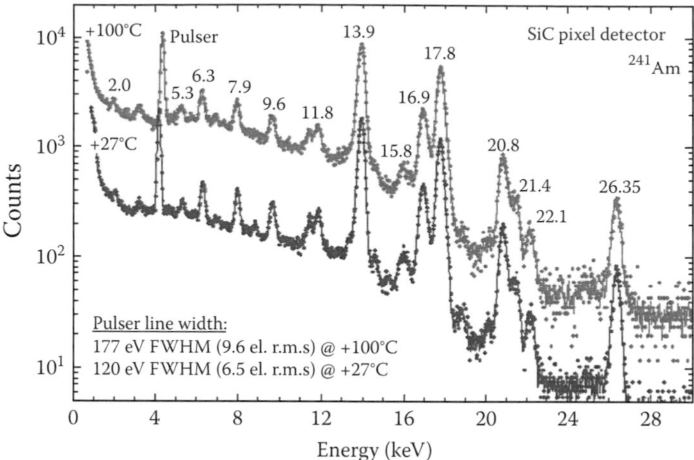

Figure 6.1 ^{241}Am X-ray and γ-ray spectra recorded by a 0.03-mm^2 70-μm-thick epitaxial 4H-SiC detector at +27°C and +100°C (from [12]). The spectrum at +100°C is displaced on the y-axis for clarity. At 26.3 keV, the FWHM energy resolution is 670 eV at 27°C and is essentially the same at 100°C operating temperature.

6.2.2 Diamond

Chemical vapor deposition (CVD) diamond has been proposed for use in hostile, hot, corrosive, and very high-radiation environments [13]. For example, in medicine, CVD diamond should be ideal for hadron therapy applications (such as dosimetry), in view of the high radiation gradients encountered and particularly its tissue equivalence, alleviating the need for dose corrections. However, for most applications the use of diamond has been limited by the inability to grow semiconductor-grade material and by the difficulty in doping this extremely stable material. For thin detectors, ion implantation offers a promising approach to doping, due to its ability to create nonequilibrium defects during the bombardment process.

Diamond is a carbon allotrope,* bonded tetrahedrally in the diamond cubic lattice structure (two interpenetrating face centered cubic lattices with a displacement of one quarter body diagonal). This bond structure, in conjunction with the low atomic number of carbon, gives diamond the highest atom den-

* Hexagonally bonded graphite is the other.

sity of any material, a feature which is responsible for many of its superlative properties. It is chemically inert, strong, extremely hard, and an excellent thermal conductor. Initial studies show that both natural and CVD material respond to radiation and are extremely radiation hard [14], but they are not spectroscopic to photons at this time [15], although Kaneko et al. [16] achieved an energy resolution of 0.4% FWHM for 5.486 MeV alpha particles with a single-crystal diamond dctector, having a size of $2.0 \times 2.0 \times 0.7$ mm^3. In fact, the resolution was so good that the peaks of 5.443 MeV and 5.389 MeV resulting from two of the excited states of ^{247}Np from ^{241}Am could be clearly resolved. The detector was fabricated from a single diamond crystal grown by plasma-assisted CVD onto a (100) surface of a type Ib diamond substrate. An Al Schottky contact and a Ti/Pt ohmic contact were applied to the crystal by evaporation and sintering.

6.3 Group III–V Materials

These are compounds which combine an anion from group V (from nitrogen or below) and a cation from group III (usually, Al, Ga, or In). Each group III atom is bound to four group V atoms and vice versa, so that each atom has a filled (8 electron) valence band. Although bonding would appear to be entirely covalent, the shift of valence charge from the V atoms to the III atoms induces a component of ionic bonding to the crystal. This ionicity causes significant changes in semiconducting properties. For example, it increases both the Coulomb attraction between the ions and the forbidden bandgap. When grown epitaxially (MBE, MOCVD, and variants). III–V materials usually take up a zinc blende (ZB) structure so that in their basic electronic and crystal structures they are completely analogous to the group IV elements. The stable bulk allotrope often has a wurtzite structure.

6.3.1 Gallium Arsenide

Gallium arsenide (GaAs) is a direct bandgap III–V material with a simple cubic lattice structure. It is widely used in the manufacture of red light-emitting diodes (LEDs), infrared windows, and laser diodes. Compared to many semiconductors it has a number of attractive attributes. For example, its density (5.32 g cm^{-3}) is more than twice that of Si, and thus it has better stopping power, especially in the hard X-ray energy range. Its bandgap (1.43 eV) is wide enough to permit room-temperature operation, but small enough that its Fano-limited spectroscopic resolution is close to that of Si. Its electron mobility (8500 $cm^2V^{-1}s^{-1}$) is about a factor of 6 higher than Si, allowing faster operation, while its a bulk resistivity (>10^3 Ω cm) is much larger than Si, allowing higher fields to be used for charge collection, which in turn allow thicker devices to be fabricated. In fact, high-frequency metal semiconductor field effect transistors (MESFETs) are now commercially available that have usable gain up to 40 GHz. From a manufacturing point of view, devices fabricated on semi-insulating material can be self-isolating, and therefore GaAs is ideally suited to integrated circuit fabrication and replication techniques. Unlike most of the II–VI compounds which are grown specifically for radiation detectors, bulk-grown semi-insulating GaAs is produced mainly for substrates for the electronics industry. The preferred growth method is the liquid encapsulated Czochralski (LEC) technique. For the best spectral performances epitaxial techniques are employed, both liquid-phase epitaxy (LPE) and vapor-phase epitaxy (VPE).

Harding et al. [17] reported particle detection with semi-insulating GaAs bulk photoconductivity devices. The first radiation detectors to show reasonable energy resolution at X-ray wavelengths were fabricated in the 1970s using liquid-phase epitaxy. Eberhardt et al. [18,19] reported measuring an energy resolution of 1 keV FWHM at 59.5 keV at 130 K with a 60 μm thick, 1.5 mm^2 surface barrier device. Since this time, performances have steadily improved, primarily due to improvements in both bulk material properties and detector fabrication techniques. Schottky barrier detectors fabricated using semi-insulating GaAs have consistently demonstrated reasonable room-temperature performances. For example, McGregor and Hermon [20] report measuring a RT energy resolution of 8 keV FWHM at 59.5 keV with a 130-μm-thick, 0.5 mm^2 device. Owens et al. [21] fabricated a series of detectors from ultra-high-purity GaAs grown by chemical vapor phase deposition (CVPD). From

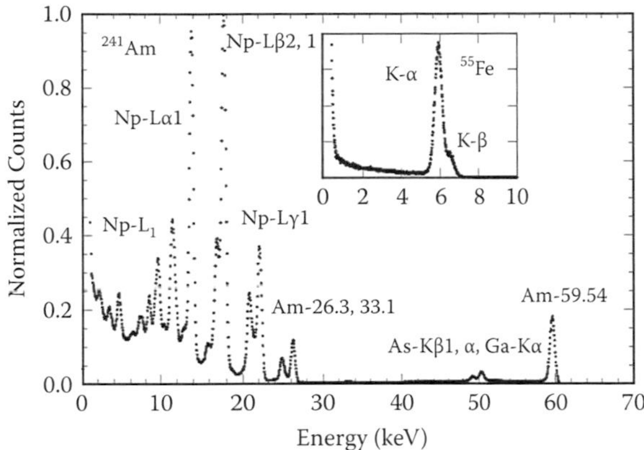

Figure 6.2 The response of a 1-mm-diameter, 40-micron-thick GaAs detector to ^{241}Am and ^{55}Fe radioactive sources (from [21]). The detector operating temperature was −40°C. The FWHM energy resolutions are 435 eV at 5.9 keV and 670 eV at 59.54 keV.

I/V characteristics, the typical current densities at −100 V were <0.04 nA/mm^2 at room temperature. This is a factor of 3 to 100 lower than normally obtained with CVPD detectors and a factor of 10^3 to 10^4 lower than bulk detectors. The measured energy resolutions of a 3.142 mm^2, 40-μm-thick device operated at −40°C were 435 eV FWHM at 5.9 keV and 670 eV FWHM at 59.54 keV (see Figure 6.2). At room temperature they were 572 eV and 780 eV, respectively. In Figure 6.3, we show the response of a pixel detector (250 × 250 × 40 μm^3), fabricated from the same base material to an ^{55}Fe radioactive source [22]. At 5.9 keV, a room-temperature (23°C) resolution of 266 eV FWHM was achieved, and 219 eV FWHM with only modest cooling (−31°C). The corresponding pulser resolutions were 163 eV (13 electrons rms equivalent) and 242 eV (19 electrons rms), respectively. The expected Fano noise at this energy is ~130 eV. The energy resolution of the 59.5 keV ^{241}Am nuclear line was measured to be 487 eV FWHM at room temperature.

Owens et al. [23] fabricated a 64 × 64 pixel array from the same base material. The pixel size was 170 μm and the pitch 200 μm. The array was back-thinned, contacted, and flip-chip bump-bonded onto a MEDIPIX 1 application-specific integrated circuit (ASIC) readout chip [24]. Its analogue front-end comprised a charge-sensitive preamplifier and a shaper.

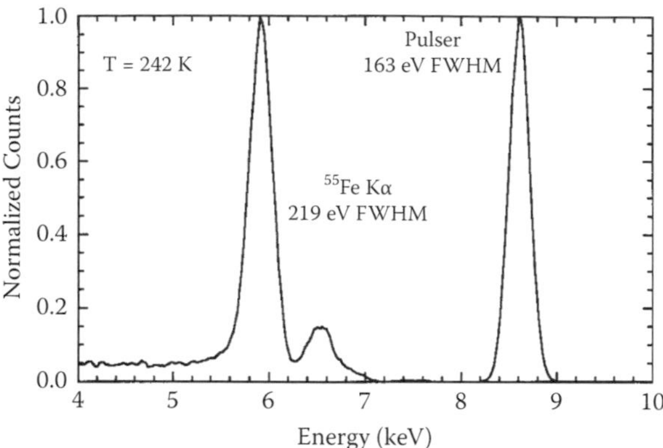

Figure 6.3 ^{55}Fe spectrum acquired with a 200 × 200 μm² GaAs pixel detector at −31°C. At room temperature the same pixel shows 242 eV FWHM on the pulser line and 266 eV FWHM at 5.9 keV (from [22]).

Incoming charge from a semiconductor sensor was amplified and compared with a threshold in a comparator. If the signal exceeded this threshold the event was counted. In X-ray tests, the bump yield was determined to be 99.9%. In Figure 6.4 we show an image of a "swatch," illustrating the imaging quality of the GaAs chip. The data were taken with a conventional X-ray tube using a tungsten target. The objects were mounted

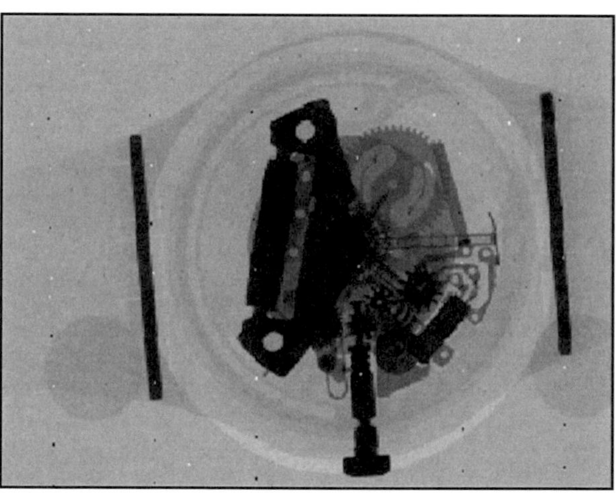

Figure 6.4 X-ray image of a "swatch," taken by a 64 × 64 GaAs pixel array using a conventional X-ray set operating at 55 kV with a W target and a 2.5 mm Al filter (from [23]).

on a sample holder, and the detector on a precision x–y stage. The detector was then scanned past the object in steps of 0.8 of the detector width using the so-called move and tile method. The resulting composite images have been flat-field corrected and a 3×3 median filter applied locally around defective pixels. After flat field corrections, the spatial uniformity of the array was commensurate with Poisson noise.

6.3.2 Gallium Phosphide

Gallium phosphide (GaP) is an indirect, wide-band gap (2.26 eV) semiconductor with a cubic (zinc blende) crystal structure. Its density is 4.1 g cm^{-3}, intermediate between Si and Ge. Very little information is available on its transport properties. It was originally investigated as an optical material and for high-temperature component applications. However, its main use since the 1960s has been in the manufacture of low and standard brightness red, orange, and green light-emitting diodes, the various colors being achieved by doping. Industrially, crystalline material is generally grown by the LEC method. At an early stage it was identified as a possible candidate for room-temperature gamma ray detection [25]. However, this application is unlikely to be realized, in view of its low electron and hole mobilities (<200 $cm^{-2}V^{-1}s^{-1}$) coupled with the fact that carrier lifetimes in III–V materials rarely exceed 100 ns. Most recently, Litovchenko et al. [26] noted that the IV characteristics of GaP LEDs are permanently altered in strong neutron and electron radiation fields, making them candidates for some dosimetry applications. Recent experiments using a commercial GaP Schottky diode [27] have shown a spectroscopic response to alpha particles. The device consists of a 10-nm-thick Au Schottky layer (anode) deposited on a 30-μm n-type GaP layer ($n_d < 10^{16}$ cm^{-3}) grown on an n-type GaP (100) substrate ($n_d \sim 5 \times 10^{17}$ cm^{-3}). A further metallization forms the rear contact (cathode). The measured energy resolution to 5.5 MeV alpha particles from an ^{214}Am source was 3.5% at room temperature. The device was also found to be responsive to X-rays in the range 11 to 100 keV. Although individual energies are not spectrally resolved, there is a proportionality of response to increasing X-ray energy.

6.3.3 Gallium Nitride

Although advances in growth technology have led to the realization of epitaxial growth of GaN layers, heterostructures, and nanocrystallites, its future development is currently limited by the lack of a suitably latticed matched substrate. Consequently, only thin layers can presently be grown. Until recently, much of the work focused on optoelectronic applications, because only GaN, by virtue of its direct bandgap, can emit efficiently in the blue region of the spectrum, making possible the solid state generation of white light. Besides optoelectronics, GaN is attracting considerable attention for high-temperature/high-power electronic device applications, because its material properties offer an order of magnitude improvement in power amplifier performance over, for example, GaAs and Si, particularly at microwave frequencies, as well as several system-level benefits. These include compact size, high power-per-mass ratio, lower combiner losses, high bandwidths, facilitated thermal management, and robustness.

The physical properties of GaN, such as its wide bandgap (3.39 eV), high density (6.15 g cm^{-3}), large displacement energy (~20 eV), and thermal stability, make it an ideal candidate as a high-temperature, radiation-hard particle and X-ray detection medium. Both PIN and Schottky devices have been fabricated by depositing epitaxial films on sapphire or silicon carbide substrates, despite the mismatch in lattice constants. For example, Vaitkus et al. [28] grew epitaxial GaN layers by metal organic chemical vapor epitaxy (MOCVD) on Al_2O_3 substrates. The structure consisted of a 2.5 μm layer of semi-insulating GaN on a 2-μm-thick n^*–GaN layer with Au Schottky contacts. The devices showed a spectroscopic response to 5.5 MeV alpha particles but no response to 60 keV photons. Owens et al. [29] produced a number of GaN PIN diodes by MOCVD. The devices consisted of a 2-μm GaN layer epitaxially grown an *n*–type $Al_xGa_{1-x}N$ nucleation layer, which in turn was deposited on a *p*–type 4H–SiC substrate. Au ohmic contacts were applied to both the top of the GaN layer and the bottom of the SiC substrate which act as the anode and cathode electrodes, respectively. A number of devices with contact radii varying from 0.4 mm to 0.7 mm were tested. All showed good diode behavior with reverse leakage currents in the tens-to-hundreds of μA range. C–V measurements showed that the GaN layer was fully depleted for nominal biases > 20V. The devices also showed spectroscopic responses to 5.5-MeV alpha particles with typical energy resolutions of ~25% FWHM at room temperature. No response to 60-keV photons was observed.

6.3.4 Indium Phosphide

Indium phosphide (InP) is widely used in optoelectronic [30] and high-speed microelectronic applications ([31] and references therein) and has even been proposed as a neutrino detector [32] because of the large neutrino capture cross-section on indium ($\nu + {}^{115}\mathrm{In} \rightarrow {}^{115}\mathrm{Sn}^{*} + e^{-}$). It is a group III–V direct bandgap material whose electronic properties are similar to Si and GaAs. It has a zinc blende crystal structure, with a single nondestructive phase transition below the melting point making it amenable to standard crystal growth techniques. It has one of the highest electron mobilities of any semiconductor material (~3 times that of silicon—i.e., 4600 $cm^2V^{-1}s^{-1}$), making it particularly suitable for applications where high count rate operation is desirable. Its bandgap of 1.35 eV implies that detectors should operate at room temperature with a Fano-limited spectroscopic resolution close to that of Si, and its relatively large density (approximately twice that of Si) ensures a high X-ray detection efficiency above 10 keV. From a materials point of view, InP is also of great interest because it is structurally suitable for the creation of integrated devices and micromachines. However, in spite of all these desirable attributes, very little work has been carried out on this material even though it was first proposed as a radiation detector over two and a half decades ago [1]. In fact, it has only recently been possible to resolve photon peaks clearly at hard X-ray and gamma ray wavelengths. The main difficulty in fabricating radiation detectors is related to the formation of an ohmic electrode system, due to a high concentration of surface states and the high chemical reactivity of the free InP surface. The net result is that the Fermi energy level lies close to the middle of the bandgap and is essentially pinned. It has not been possible to fabricate true blocking contacts.

Experimentally, InP was initially studied as a high-speed photoconductor for time-resolved measurements of synchrotron X-ray pulses [33,34]. The first single photon counting experiments were reported by Lund et al. [32] who fabricated a number of gamma ray detectors from semi-insulating, iron-doped InP. Doping with Fe increases resistivity by compensating for the background conductivity in bulk material. The detectors consisted of a ~2-mm-thick single crystal of InP with two identical gold contacts on opposite sides. The detectors were found to be sensitive but not spectroscopic to 662 keV photons. Suzuki et al. [35] fabricated a p-n junction diode by diffusing Zn into an n-type InP layer grown by liquid-phase epitaxy. The epitaxial layer was 25 μm thick and had a sensitive area of 3.4 mm^2 with Au contacts deposited on both sides. The detector was operated at room temperature and found to be sensitive to both α-particles and γ-rays. For 2.2 MeV α-particles, a measured resolution of 10% (FWHM) was obtained, and a resolution of 55% was obtained for 60 keV photons. Olschner et al. [36] fabricated a number of gamma ray detectors from single crystals of iron-doped, zinc-doped and copper-doped InP. All detectors were tested using ^{207}B, ^{57}Co, and ^{137}Cs radioactive sources. The detectors had active volumes ranging from 2.5×10^{-3} cm^3 to 1 cm^3 and thicknesses from 0.26 mm to 3 mm. They were all responsive to radiation, but in terms of performance the iron-doped detectors gave the best results, with peaks clearly discernable but not resolved for the ^{57}Co, ^{137}Cs, and ^{207}Bi sources. Jayavel et al. [37] grew Fe-doped InP single crystals using the LEC technique. The crystals were cut into 0.5-mm-thick wafers which were then lapped and polished to a thickness of 350 μm. An Au/Ge/Ni-alloyed ohmic contact was deposited on the back side of the sample and an Au Schottky contact was applied to the top, resulting in a response to the ^{57}Co and ^{137}Cs sources similar to that measured by Olschner et al. [36].

Dubecký et al. [38] fabricated a number of radiation detectors from semi-insulating InP wafers doped with Fe. After lapping and polishing, the detectors had thicknesses of ~200 μm. An Au electrode was evaporated onto the top of the detector samples and a nonalloyed AuGeNi eutectic onto the bottom. The Au forms a pseudo-blocking contact and the eutectic an

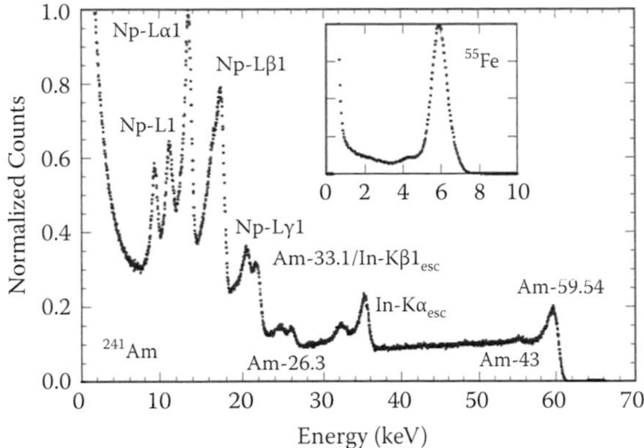

Figure 6.5 The measured response of a 3.1-mm^2, 200-μm-thick InP detector to ^{241}Am under full area illumination at –170°C. The inset shows its response to ^{55}Fe. The measured FWHM energy resolutions were 911 eV at 5.9 keV and 2.5 keV at 59.54 keV (from [21]).

ohmic contact. The devices were tested with ^{241}Am, ^{133}Ba, ^{57}Co, and ^{137}Cs radioactive sources. They were found to be responsive to X-rays at room temperature but not spectroscopic. At an operating temperature of 216 K, FWHM energy resolutions of 7 keV at 59.54 keV, 11 keV at 122 keV, and 93 keV at 662 keV were recorded. Owens et al. [39] reported FWHM spectral resolutions at –60°C of ~ 2.5 keV and 9.2 keV at 5.9 keV and 59.54 keV, respectively, with a small Fe-doped device of area 3.1 mm^2 and thickness 180 μm (Figure 6.5). At –170°C, these figures improved considerably to 911 eV at 5.9 keV and 2.5 keV at 59.54 keV [21].

Gorodynskyy et al. [40] and Yatskiv et al. [41] have argued that, while Fe doping increases material resistivity, it also introduces deep electron and hole traps which limit charge collection efficiency (CCE) and therefore energy resolution. Attempts to increase the CCE by lowering the doping concentration have the same result, because lowering the Fe doping concentration also lowers the resistivity, which in turn degrades the energy resolution. In fact, the best room-temperature CCEs obtained with bulk material are ~75% when measured with alpha particles emitted by an ^{241}Am source [42]. The authors proposed to increase the CCE by co-doping with Ti and Mn [40] or Ti and Zn [41] instead of Fe.

The presence of Ti produces high-resistivity material with much reduced hole trap capture cross sections compared to Fe doping [43]. Co-doping with Mn has the effect of suppressing the electron traps. Although a significant improvement in CCE was found using Mn-Ti co-doped SI material (91% for 5.5. MeV alpha particles at 230 K), an X-ray or gamma ray response could not be measured due to a high level of system noise. Subsequent work by Yatskiv et al. [41] using a Ti-Zn co-doped InP system has achieved a CCE of 99.9% and a spectral resolution of 0.9% FWHM for 5.5 MeV alpha particles at 230 K. No results were given for photons.

6.3.5 Indium Iodide

Indium iodide (InI) is a wide-bandgap, base centered orthorhombic crystal, with a layered structure. It has a relatively low melting point (351°C) and exhibits no solid phase transition between its melting point and room temperature. Therefore high-quality crystals may be obtained by using simple melt-based processes. Due to the high atomic numbers of its constituent elements (Z_{In} = 49 and Z_I = 53) and high density (5.31 g cm^{-3}), InI exhibits a photon stopping power similar to that of CdTe. Its bandgap is 2.0 eV, which offers the potential for low-noise operation at and above room temperature.

Squillante et al. [44] fabricated a radiation detector on a single-crystal wafer grown by the Bridgman process. Carbon electrodes were formed by painting a lacquer–graphite suspension onto InI slices. The contact area was 5 × 5 mm^2, and the thickness was 0.5 mm. The resistivity was found to be >10^{11} Ω cm. The device showed a clear, but not totally spectroscopic, response to 22 keV X-rays from a ^{109}Cd radioactive source at room temperature. At 120°C, the spectroscopic performance degraded, but the device was still able to function as a counter with a counting efficiency almost as high as at lower temperatures.

Onodera et al. [45] fabricated radiation detectors from InI crystals grown by the travelled molten zone (TMZ) method. Au electrodes of area 1 mm^2 were deposited onto the tops and bottoms of InI wafers of thickness ~0.4 mm. The responses of the devices to 22 keV X-rays from a ^{109}Cd radioactive source at room temperature were very similar to those observed by Squillante et al. [44], despite the resistivity of the device being two orders of magnitude lower (~3 × 10^9 Ω cm). All devices showed clear polarization effects which manifested themselves as a time-dependent deterioration of energy resolution and peak position at room temperature. The effects could be suppressed by cooling the detectors to −20°C.

Bhattacharya et al. [46] grew InI ingots using both a vertical Bridgman and a vertical gradient freeze technique using zone-refined commercially available material and material synthesized from vapor, respectively. The ingots were sliced into 2-mm-thick wafers and Pd metal electrodes were deposited by RF sputtering. The current-voltage characteristics were found to be linear in polarity of applied voltages, with resistivities of 2 × 10^9 Ω cm for the zone-refined and 1 × 10^8 Ω cm for the vapor-synthesized starting materials, respectively. Both detectors showed a spectroscopic response to alpha particles from an ^{241}Am source with a best recorded energy resolution of ~50% at 4°C operating temperature and a nonspectroscopic response to 662 keV gamma rays at room temperature.

6.3.6 Narrow-Gap Materials

Narrow gap describes a semiconductor with a forbidden energy gap of less than ~0.7 eV. Examples include InAs, InSb, PbS, PbSe, PbTe, Bi_2Te_3, and HgCdTe. They are used primarily in terahertz source, thermoelectric, and infrared applications. At X-ray and gamma ray wavelengths, several narrow-bandgap materials have attracted attention as possible replacements for Si and Ge. These materials, InAs and InSb in particular, offer the possibility of spectral resolution beyond that of the elemental semiconductors, closer to the high-temperature superconductors, despite the disadvantage of requiring substantially lower operating temperatures. Potentially, InAs can achieve twice the energy resolution of silicon and InSb three times. In addition, both have very high electron mobilities, permitting low bias operation, and both are already extensively used in the semiconductor industry. From a detector point of view, both InSb and HgCdTe have been successfully used for infrared focal planes [47].

6.3.6.1 Indium Arsenide

Indium arsenide, or indium monoarsenide (InAs) is a direct bandgap material with a cubic crystal structure, a bandgap of 0.35 eV, an electron mobility of ~33,000 $cm^2V^{-1}s^{-1}$ and a density of 4.68 g cm^{-3}. It has conventionally been used in infrared detection in the wavelength range of 1 to 3.8 μm and to produce infrared diode lasers for telecom applications. Theoretically, the energy resolution of InAs-based X-ray detectors can be expected to exceed that of Si by a factor of two by virtue of its low bandgap. Recently, Säynätjoki et al. [48] fabricated a 3 × 3 pixel array onto a commercially available substrate. The active pixel volumes were defined by diffusing Zn through a mask using metal organic vapor phase epitaxy (MOVPE). The pixels themselves were then defined by chemically etching mesas and contacts applied by metallization. The pixel sizes were 250 μm square, and the thickness of the active layer ~1 μm. Typically, reverse leakage currents were of the order of a few milliangstroms at liquid nitrogen temperature. The device was found to be responsive but not spectroscopic to 5.51 MeV alpha particles.

6.3.6.2 Indium Antimonide

Like InAs, indium antimonide (InSb) is commonly used to fabricate infrared detectors which are sensitive in the wavelength range 1–5 μm. It has a zinc blende crystal structure and a bandgap of 0.165 eV—almost 10 times smaller than silicon. Its density of 4.78 g cm^{-3} is twice that of Si. Its electron mobility is exceedingly high (78,000 $cm^2V^{-1}s^{-1}$ at 77 K), and although its hole mobilities are considerably lower (750 $cm^2V^{-1}s^{-1}$), they are still higher than those measured in HgI_2 and CdTe. All of the above are very attractive attributes for a potential high-resolution radiation detector. Although initially suggested as an X-ray detection medium by Harris [49] in 1986, very little work has been carried out until recently. Kanno et al. [50] fabricated both Schottky and p-n junction diodes on *p*-type InSb grown by the vertical Bridgman method. A 3-mm diameter, 10-micron-thick mesa was etched into the substrate and doped with Sn to form an *n*-type layer. An evaporation of Au–Pd formed a Schottky contact for the Schottky diode, and an Al–Sn evaporation formed the ohmic contact for the p-n device. The I/V characteristics show typical diode behavior with reverse leakage currents in the tens of microampere range. The estimated resistances of the Schottky and p-n junctions at 4.2 K were 50 kΩ and 250 kΩ, respectively. Both detectors were found to be responsive to 5.5 MeV alpha rays with FWHM energy resolutions of ~25% at 20 K. The Schottky diode detector operated up to 77 K, the p-n junction device up to 115 K. Although not spectroscopic, undoped Schottky devices were found to be responsive to 60-keV and 81-keV gamma rays [51].

Following on from this work, Sato et al. [52] fabricated a detector by depositing a ~115-μm-thick *p*-type InSb layer on a 0.4-mm-thick commercially available InSb substrate by liquid phase epitaxy (LPE). A 1-mm-diameter Al electrode was deposited on the epitaxial side of the device forming a rectifying contact. The substrate was indium soldered to a Cu plate which served as an ohmic contact. At 4.4 K, the resistance of the device was measured to be 680 kΩ, substantially higher than that achieved with Schotkky and p-n devices fabricated from material grown by the vertical Bridgman method. From the I/V characteristics, the typical reverse leakage currents were lower than in previous work, being of the order of ~4 μA. In addition, the breakdown voltage of approximately –2 V represents improvement on previous work by a factor of 2. The detector was spectroscopic to alpha particles with a measured FWHM energy resolution of 3.1% at 5.5 MeV. The measured energy loss spectrum is shown in Figure 6.6. The device also showed a response to gamma rays from a ^{133}Ba radioactive source, shown in the inset in Figure 6.6. While it is tempting to ascribe the apparent feature near channel 250 to 81-keV gamma rays from the source, transport simulations have indicated that it is more likely due to a "shoulder" of 356-keV decay gamma rays from the source [53].

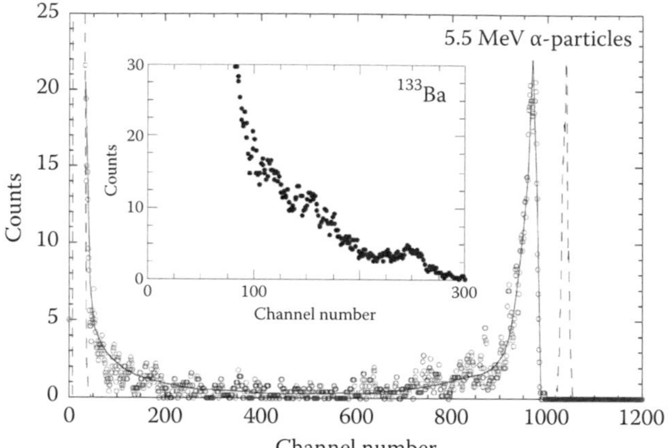

Figure 6.6 Energy spectrum of 5.5-MeV alpha particles recorded at 5.5 K by a 1-mm-diameter InSb detector grown by LPE (data courtesy of I. Kanno). The dotted line shows a pulser spectrum. The inset shows the response of the detector to gamma rays from a ^{133}Ba radioactive source. The amplifier gain is 1.75 times that of the gain used to acquire the alpha particle spectrum.

6.4 Group II–VI Materials

Group II–VI materials have attracted a lot of attention because of the wide range of compounds available and the possibility to engineer the bandgap over an almost continuous range. These are compounds which combine a group IIb metal (such as Zn, Cd, and Hg, in periods 3, 4, and 5, respectively) with a group VIa cation. The latter is usually S, Se, or Te. Structurally, it forms when atomic elements from one type bond to the four neighbors of the other type, as shown in Figure 2.4a. A major motivation for developing II–VI semiconductors is their broad range of bandgaps (from 0.15 eV for HgTe to 4.4 eV for MgS), high effective Z, and the capability for making MBE- and MOCVD-grown heterostructures, in the same way as for III–V systems. Additionally, all II–VI binaries have direct bandgaps, which make them particularly attractive for optoelectronic applications. Compounds generally crystallize naturally in a hexagonal or NaCl structure. Representative compounds are CdTe and HgTe. Pseudo-binary alloys with

Zn, Se, Mn, or Cd are also common—particularly for radiation detector and optoelectronic applications (for example, $Cd_{(1-x)}Zn_xTe$, $Cd_{(1-x)}Mn_xTe$, and $Hg_{(1-x)}Cd_xTe$). Group II–VI compounds typically exhibit a larger degree of ionic bonding than III–V materials, because their constituent elements differ more in electron affinity due to their location in the periodic table. A major limitation of II–VI compounds is the difficulty in forming *n*-type and *p*-type material of the same compound, preventing the formation of a p-n junction. In fact, it is only recently that *p*-type doping of ZnSe has been achieved [54]. In addition, it has been difficult to control the defect state density within the bandgap due to self-compensation. However, despite these limitations Group II–VI compounds are the most widely used for radiation detection. Group II–VI semiconductors can also be created in ternary and quaternary forms, although less common than III–V varieties. As with III–V materials, a major problem is controlling the multiplication of stoichiometric errors during the growth process, which leads to poor transport properties.

6.4.1 Cadmium Telluride

Cadmium telluride (CdTe) was one of the first compound semiconductors to be synthesized in the late 1800s. However, until the 1940s its only use was as a pigment, whereupon it was used in the production of photocells. It is a group II–VI material which crystallizes in a cubic zinc blende structure. In its bulk crystalline form it is a direct bandgap semiconductor with a bandgap of 1.56 eV at 300 K, which is high enough to allow room-temperature operation. It has a density of 5.85 g cm^{-3}. The large atomic numbers of its constituent atoms ($Z_{Cd} = 48$, $Z_{Te} = 52$) ensure a high stopping power and therefore high quantum efficiency, in comparison with Si and Ge. For example, in terms of efficiency, 2 mm of CdTe is equivalent to 10 mm of Ge for gamma ray detection.

CdTe is generally grown by the travelled heater method (THM). To achieve high resistivity (10^8 to 10^{10} Ω cm) THM-grown crystals are sometimes doped with Cl to compensate impurities. However, Cl doping can also introduce a number of other problems, namely polarization effects and long-term

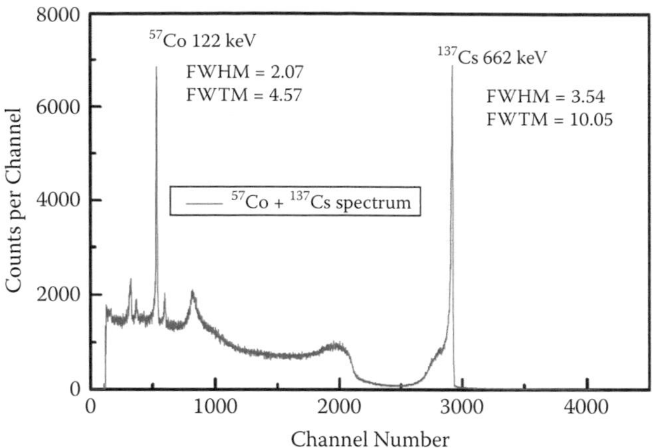

Figure 6.7 ^{57}Co and ^{137}Cs spectra taken with a 1-cm^2, 2.1-mm thick P–I–N CdTe detector cooled to –35°C (from [60]). The operating bias was –3000 V. The FWHM energy resolution at 662 keV was 3.5 keV.

reliability problems. As a rule, CdTe is easier to produce in large quantities than CnZnTe, because the latter is invariably grown by the high pressure Bridgman (HPB) method. Large wafers of CdTe can be produced up to 50 mm in diameter without grain boundaries with good reproducibility and homogeneity. CdTe is currently used primarily in the manufacture of photovoltaics, particularly solar cells, nanoinks, and nanorods. Other applications include electro-optical modulators, IR windows, and photorefractive materials.

Since the late 1960s, CdTe has also been regarded as a promising semiconductor material for hard X-ray and gamma-ray detection in view of its relatively high carrier mobilities (μ_n~10^3 cm^{2}V^{-1}s^{-1} and μ_p~100 cm^{2}V^{-1}s^{-1}) and high stopping power. Early experiments by Akutagawa et al. [55] showed that it was possible to resolve spectrally both alpha particles and photons using ~1-mm-thick detectors. In fact, a FWHM spectral resolution of ~40% for 160 keV photons was achieved at room temperature. Over the next three decades detector performances improved incrementally. For small ~25 mm^3 planar detectors, room-temperature energy resolutions are in the range 2 to 6 keV. Recently, however, high-resolution CdTe detectors with FWHM energy resolutions better than 1 keV at 60 keV have become available. This has been achieved by significantly lowering leakage currents by fabricating diode structures, either using a blocking electrode [56] or a P–I–N type structure [57,58]. Leakage currents of the order of several nA mm^{-2} have been obtained at room temperature, which compares well with commercially available CdZnTe detectors. Takahashi et al. [57] report a room-temperature FWHM energy resolution of 1.8 keV for the ^{241}Am nuclear line at 59.54 keV using a Schottky contact detector* of dimensions 2 × 2 × 1 mm^3. For comparison, the same sized detector with conventional ohmic contacts yielded a resolution of 3.3 keV. However, the main disadvantage of CdTe Schottky contact diodes is their susceptibility to polarization effects. While the conventional detector was stable, the Schottky diode was unstable on a time scale of minutes, and the spectra disappeared after 30 minutes. Polarization effects can be reduced or even eliminated by operating the detector at a much higher bias or reduced temperature. At a reduced temperature of –25°C, the performance of the Schottky diode detector improved considerably to 810 eV and was stable over 24 hours. A similar diode of thickness 0.5 mm, gave FWHM energy resolutions of 830 eV at 59.54 keV and 2.1 keV at 662 keV [56] when operated at a bias of 1400 V and a temperature of –40°C. This bias is roughly three times that of an equivalent CdTe detector with ohmic electrodes. These resolutions are close to those expected from liquid nitrogen-cooled high-purity Ge detectors.

Niemela et al. [58] and Khusainov et al. [59] produced CdTe diodes with a P–I–N structure. For a 16-mm^2, 1-mm-thick device, they achieved FWHM energy resolutions of 1.1 keV at 59.54 keV and 2.5 keV at 662 keV at an operating temperature of –30°C. Khusainov et al. [59] obtained similar results with a 7-mm^2, 1-mm-thick detector. At 662 keV and an operating temperature of –30°C, they measured a FWHM energy resolution of 2.5 keV. In Figure 6.7 we show ^{57}Co and ^{137}Cs spectra taken with a larger 11.3 × 9.1 × 2.13 cm^3 P–I–N detector cooled to –35°C [60].

In contrast to HgI_2, CdTe does not undergo an irreversible phase change below its melting point which could prevent high-temperature operation. The maximum temperature of CdTe is actually limited to ~120°C by the onset of migration of

* [2]Also known as M–π–n detectors.

the chlorine dopant used to compensate impurities. Mahdavi et al. [61] repeatedly operated eight sample CdTe detectors for extended time periods at temperatures up to 100°C. The detectors had simple planar geometries of size $5 \times 5 \times 2$ mm^3 with ohmic contacts. The most obvious change in performance was an exponential increase in leakage current with temperature. At 100°C the current was ~10 μA—about 1000 times the value at room temperature. This resulted in a progressive degradation in energy resolution with temperature. For example, for 122 keV photons, the FWHM energy resolution at room temperature was ~5 keV, rising to 10 keV at 60°C and 35 keV at 100°C.

6.4.2 Cadmium Zinc Telluride

Cadmium zinc telluride (variously denoted by CZT, CdZnTe, (Cd, Zn)Te, or $Cd_{(1-x)}Zn_xTe$) is a pseudo-ternary compound* that has been extensively studied at X-ray and gamma-ray wavelengths (see [62] for a review) and is probably the most widely used compound semiconductor. It was originally produced as a lattice-matched substrate for $Hg_{(1-x)}Cd_xTe$ epilayers for IR rather than X-ray or gamma ray applications. It has a cubic, zinc blende–type lattice with atomic numbers close to those of CdTe and a density ~3 times that of Si (5.8 g cm^{-3}). It is generally produced by the high pressure Bridgman (HPB) method, although low pressure Bridgman (LPB) material is becoming increasingly available and appears to offer good uniformity and comparable electronic properties to HPB CdZnTe. CdTe was originally the focus of experimental study in the 1960s, until it was discovered that the addition of a few percent of zinc to the melt results in an increased bandgap as well as the energy of defect formation [63] and the virtual elimination of polarization effects normally associated with CdTe [64]. This, in turn, increases bulk resistivities and reduces the dislocation density, resulting in lower leakage currents. Specifically, resistivities of CdZnTe are between one and two orders of magnitude greater than that for CdTe, and thus leakage currents are correspondingly lower. Alloying of CdTe with Zn or ZnTe has additional benefits, in that it strengthens the lat-

* Strictly speaking, it is an alloy of cadmium telluride and zinc telluride.

tice mechanically with a resulting lowering of defect densities. Depending on the zinc fraction, its bandgap is typically 1.45 to 1.65, making it suitable for room- or even elevated-temperature operation. In fact, Egarievwe et al. [65] have operated CZT detectors up to 70°C with little degradation in performance over that at room temperature.

The variation of bandgap with zinc fraction, x, can be described empirically [66] by

$$E_g(x) = 1.510 + 0.606x + 0.139x^2 \text{ (eV)} \tag{6.1}$$

While an increase in E_g increases Fano noise due to carrier generation statistics, it simultaneously reduces shot noise due thermal leakage currents. A trade-off can lead to a noise minimum at a given operating temperature and therefore an optimization of the spectral performance [67]. For example, a zinc fraction of 10% is optimum for operation at –30°C, whereas 70% is optimum at room temperature.

For 10% zinc fractions, typical energy resolutions at 59.54 keV are in the keV range at room temperature and decrease with decreasing temperature to a minimum resolution around –30°C. This is demonstrated in Figure 6.8,

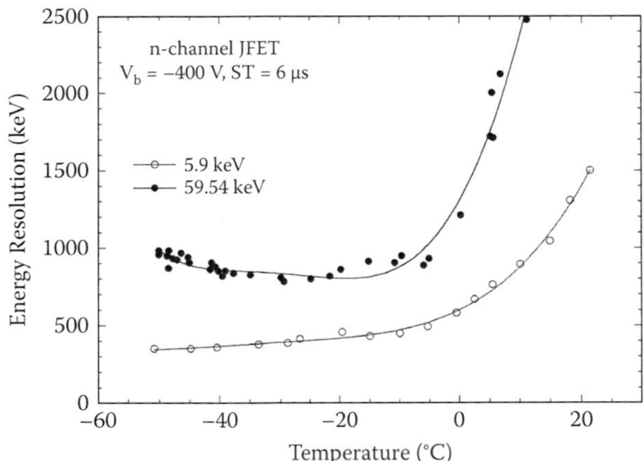

Figure 6.8 The temperature dependence of the FWHM energy resolutions measured with a 3.1-mm^2, 2.5-mm-thick CdZnTe at 5.9 keV and 59.54 keV under full-area illumination (from [68]). The solid lines are best-fit polynomials.

in which we show FWHM energy resolutions for a 3.1-mm^2, 2.5-mm-thick CdZnTe detector at 5.9 and 59.54 keV as a function of detector temperature [68]. Although the leakage currents are low enough to allow room-temperature operation, it can be seen that there is a marked improvement in both resolution functions with only a modest reduction in temperature. In fact, the energy resolution improves by a factor of three with a temperature reduction of only 20°C, compared with that measured at room temperature. At temperatures below –20°C, there is relatively little improvement, if any, in ΔE at both incident energies. We also note that, while the resolution at 5.9 keV steadily improves with decreasing detector temperature, the resolution at 59.54 keV shows a minimum near –30°C. This may be due to the holes freezing out, which would explain why resolution is only observable at the higher energies. At 60 keV, the drift lengths of holes are typically a few hundred microns, whereas at 6 keV, they are only a few microns. The optimum overall performance was found at a detector temperature of –37°C, at which the measured FWHM energy resolutions were 311 eV at 5.9 keV and 824 keV at 59.54 keV. The measured energy loss spectra are shown in Figure 6.9.

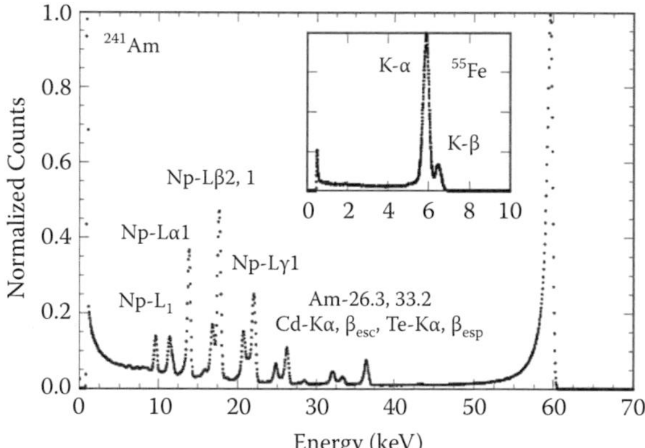

Figure 6.9 Composite of ^{241}Am and ^{55}Fe spectra taken with a 3.1-mm^2, 2.5-mm-thick CdZnTe detector (from [21]). The detector temperature was –37°C, and the applied bias +320 V. The measured FWHM energy resolutions were 311 eV at 5.9 keV and 824 eV at 59.54 keV. The corresponding pulser widths were 260 keV and 370 eV, respectively.

6.4.3 Cadmium Manganese Telluride

The pseudo-ternary compound cadmium manganese telluride (usually denoted by CMT, CdMnTe, (Cd, Mn)Te, or $Cd_{(1-x)}Mn_xTe$) is a magneto-optical material with spintronic properties [69] that has found application in the fabrication of Faraday rotators, optical isolators, and magnetic field sensors. It has traditionally been used in the production of LEDs, solar cells, and visible to mid-infrared tunable lasers [70] and has been proposed as a radiation detection medium [71]. Most recently it has attracted attention as an inexpensive alternative for CdZnTe, since it is easier to produce, because CdZnTe can only be reliably grown using high pressure Bridgman (HPB), which is complex and expensive. CdMnTe on the other hand, can be grown by the modified Bridgman technique, which is considerably less expensive than HPB. In addition, the segregation coefficient of Zn along the growth axis in CdZnTe is large (k = 1.35) and leads to substantial compositional variations along the growth direction, which in turn can lead to substantial variations in the performance in fabricated detectors. In contrast, the segregation coefficient of CdMnTe is almost unity, which means that large homogeneous crystals of the same composition as the starting liquid phase can be grown from melt. Finally, it is easier to engineer the bandgap in CdMnTe due to the large compositional influence of manganese. While the energy gap of CdZnTe increases only by 6.7 meV per atomic percent of Zn, the corresponding value for the Mn concentration in CdMnTe is 13 meV [72]. This means that less Mn is required than Zn to produce a specific bandgap, so for example, for $x = 0.05$, the bandgap of $Cd_{0.9}Mn_{0.05}Te$ is about 1.60 eV, which is the same as that of $Cd_{0.9}Zn_{0.1}Te$ (the standard composition for spectrometer-grade CdZnTe material). The room-temperature energy gap of $Cd_{1-x}Mn_xTe$ has been found to be linear with Mn fraction x [73],

$$E_g(x) = 1.526 + 1.316x \text{ (eV)} \tag{6.2}$$

In principle, it is possible to adjust the bandgap from 1.5 eV to 3.4 eV (in CdZnTe the range is 1.5 eV to 2.2 eV). Thus, CdMnTe could potentially operate over a wider temperature range than CdZnTe. The practical operation of CdZnTe detectors is limited to temperatures below ~70°C before irreversible damage sets in. At present, CdMnTe suffers from several major material problems. First, it is difficult to synthesize due to the high reactivity of Mn, which tends to bond with the residual oxides on the surface of the container. Second, compared to CdZnTe, the bond ionicity of CdMnTe is higher, resulting in a tendency to crystallize in the hexagonal form, or to produce a high degree of twinning in the zinc blende modification. Third, the resistivity of "as grown" CdMnTe crystals can be quite low ($<10^3$ Ω cm) due to a high concentration of cadmium vacancies which invariably result from Bridgman growth. These act as acceptors, and thus the material is *p*-type. However, the resistivity can be substantially increased ($>10^9$ Ω cm) by annealing the crystals in a Cd ambient, or more commonly, doping with a donor impurity, usually indium, chlorine, or vanadium. Mycielski et al. [74], in a critical study of the growth and preparation of CdMnTe for radiation detector applications, concluded that 6% atomic weight Mn content minimizes the density of twins and maximizes the grain size in "as-grown" crystals, and that annealing in Te vapor at the temperature gradient, followed by annealing in Cd vapor, minimizes impurity concentrations, Te precipitates, and cadmium vacancies. In an assessment of three contacting schemes they further conclude that amorphous layers of heavily doped semiconductors provide the best electrical contacts to semi-insulating (Cd,Mn)Te crystals.

Photoconduction measurements were carried out by Burger et al. [71] on a number of $Cd_{0.85}Mn_{0.15}Te$ and $Cd_{0.55}Mn_{0.45}Te$ crystals which were highly doped (~10^{19} cm^{-3}) with vanadium. The resistivity of the crystals exceeded 1×10^{10} Ω cm, but the mobility-lifetime product (μτ) was only 1×10^{-6} cm^2V^{-1}. A number of detectors were fabricated, and one device was found to be spectroscopic to photons. The device had an area of 1.8 mm^2 and a thickness 0.5 mm. A FWHM energy resolution of 40% at 59.54 keV was achieved for a detector bias of 500 V and a 1-μs peaking time. Parkin et al. [75] have produced detectors grown by the modified Bridgman technique. The detectors were found to be spectroscopic (~50% FWHM) to alpha particles but not to photons. Cui et al. [76] fabricated a number of $Cd_{0.94}Mn_{0.06}Te$ detectors from Bridgman-grown material. The "as-grown" crystals were annealed in a Cd ambient and doped with 5×10^{16} cm^{-3} of vanadium. One detector, of active area 28.3 mm^2 and thickness 1.8 mm, had a measured electron mu-tau product of 2.1×10^{-4} cm^{2}V^{-1}, over two orders of magnitude larger than the value achieved by Burger et al. [71]. From I/V measurements the resistivity was 3.2×10^{12} Ω cm. The detector was found to be spectroscopic to photons. When it was biased to collect electrons, the measured FWHM energy resolution was 27% at 60 keV. When biased to collect holes, a peak in the spectrum was also obtained, although poorly resolved. The hole mu-tau product could not be directly determined due to system noise. Using vertical Bridgman growth, Kim et al. [77] grew $Cd_{0.9}Mn_{0.1}Te$ crystals compensated with In (10^{17}cm^{-3}). The "as-grown" ingots were cut into 2-mm thick slices and mechanically and chemically polished down to a thickness of 1 mm. Detector blanks of area 2×2 mm^2 were diced from the wafers and contacted with Au electrodes deposited using thermal evaporation. Three detectors, of active areas of 0.25, 1, and 4 mm^2, were tested, and resistivities, determined from I/V measurements, were all found to be in the range 1 to 3×10^{10} Ω cm. The electron mobility-lifetime products were determined from a Hecht plot to be 1×10^{-3} cm^{2}V^{-1}, which is an order of magnitude higher than previous measurements. The detectors showed a spectroscopic response to both alpha particles and photons with room-temperature FWHM energy resolutions of 3% for 60 keV photons.

6.4.4 Cadmium Selenide

CdSe crystallizes into a hexagonal (wurtzite) closely packed structure at 1239°C, which contributes to its relatively high photon attenuation coefficients at hard X-ray wavelengths. These are comparable to CdTe despite a lower effective Z and density (5.8 g cm^{-3}). Its bandgap is 1.73 eV, which should be high enough to allow room-temperature operation, and its electron and hole mobilities (720 cm^{2}V^{-1}s^{-1} and 75 cm^{2}V^{-1}s^{-1}, respectively) are among the highest of II–VI compounds. CdSe was first explored as a room-temperature radiation detector by Burger et al. [78], who grew high-purity single crystals of CdSe by the vertical unseeded vapor growth method. The crystals were sliced in 250-μm disks. Although the as-grown crystals had a resistivity of only ~10 Ω cm, after annealing the resistivity of the disks increased considerably to 10^{12} Ω cm. The disks were polished and etched to a thickness of 200 μm, then washed and processed into detectors by evaporating 4 mm^2 Au contacts onto both sides. The finished detectors were found to be responsive to X-ray photons over the energy range 10 keV to 660 keV, although the energy resolutions were poor. For example, at 60 keV resolution was 45 keV (75%) FWHM.

Roth [79] produced high-resistivity (~ 10^9 Ω cm) CdSe crystals by the temperature gradient solution zoning technique. The crystals were sliced and fabricated into detectors of thicknesses ranging from 0.2 mm to 1.0 mm. The contacts were formed by painting Aquadag or evaporating Pd onto the crystal surfaces. A room-temperature energy resolution of 1.4 keV FWHM was obtained for the ^{55}Fe line at 5.9 keV and 8.5 keV FWHM for the ^{241}Am nuclear line at 59.54 keV.

Chen et al. [80] report measurements on undoped CdSe single crystals grown by an unspecified method. The measured resistivities were in the range 10^{10} Ω cm—comparable to that of CdZnTe grown by the high pressure Bridgman method. One crystal was cleaved into smaller samples having dimensions $4 \times 4 \times 1.2$ mm^3. The samples were polished, etched, and rinsed, leaving final detector thicknesses of 0.15, 0.3, and 1.2 mm. Au contacts were deposited by thermal evaporation immediately after chemical treatment to minimize surface oxidation. The devices were found to be spectroscopic to both alpha particles and photons. For example, at a photon energy of 60 keV, measured FWHM energy resolutions of 30% and 42 % were recorded for the 0.15-mm and 0.3-mm thick devices, respectively.

6.4.5 Cadmium Zinc Selenide

Burger et al. [81] also investigated the ternary system $Cd_{0.7}Zn_{0.3}Se$. It has a wurtzite crystal structure with a bandgap of 1.73 to 2.67 eV and a density of 5.4 to 5.8 g cm^{-3}, depending on the zinc fraction. As with the adding Zn to CdTe, the addition of Zn to the CdSe system also results in a higher bandgap, enabling the fabrication of detectors with lower leakage currents and sufficiently low noise to detect X-ray photons. In fact, leakage currents in a small $Cd_{0.7}Zn_{0.3}Se$ detector were measured to be approximately one order of magnitude lower than measured in a similar-sized detector of CdSe [79].

Burger et al. [81] grew single crystals by the temperature gradient solution zoning technique. The as-grown crystals had resistivities of the order of 10^6 Ω cm. The crystals were sliced into platelets with thickness varying from 0.1 mm to 1 mm. After post-growth treatment the measured resistivities increased to ~10^{10} Ω cm. Gold or carbon-based Aquadaq contacts were then applied. Measured FWHM energy resolutions of 1.8 keV at 5.9 keV and 4 keV at 27 keV were measured for a 0.45-mm-thick device. The 59.54 keV nuclear line from ^{241}Am exhibited poor resolution and the overlap of the Cd K_α and K_β escape peaks at 23 keV and 26 keV, respectively.

6.4.6 Cadmium Telluride Selenide

$Cd_xTe_{(1-x)}Se$ (or CdTeSe, Cd(Te, Se)) is seen as an attractive alternative to CdZnTe, because the binding energy of the CdSe system is 1.3 times higher than the corresponding CdTe system, and the lattice constant 0.9 times shorter. Thus, it is expected that CdSeTe crystals should have fewer Cd vacancies and increased hardness. In addition, the segregation coefficient (k) of Se in CdTe is ≤1, while that of Zn in CdTe is ≥1, which means that crystals of CdTeSe should be more uniform in terms of stoichiometry than CdZnTe.

Fiederle et al. [82] investigated the ternary systems (Cd,Zn)Te and Cd(Te,Se). Single crystals of CdTe, $Cd_{0.9}Zn_{0.1}Te$, and $CdTe_{0.9}Se_{0.1}$ were grown by the vertical high pressure Bridgman method. Simple monolithic detectors were then fabricated and tested. While spectroscopic results could be expected for CdTe and CdZnTe, $CdTe_{0.9}Se_{0.1}$ was also found to be spectroscopic to both alpha particles and photons. For a 1.34-mm-thick device, a FWHM energy resolution of ~10% was achieved for 60 keV photons [82]. Kim et al. [83] grew semi-insulating CdTeSe:Cl crystals using the vertical Bridgman method. Chlorine doping was used to obtain semi-insulating material by compensating for Cd vacancies. The ingot was sliced into 2-mm-thick wafers which were then chemically and mechanically polished down to a thickness of 1 mm. Detector blanks were diced out of the wafer. Based on I/V and Hall measurements, the resistivity of the material was found to be 5×10^9 Ω cm, and the electron and hole mobilities to be 59 and 33 $cm^{-2}V^{-1}s^{-1}$, respectively. The corresponding electron and hole mu-tau products were surprisingly high at 6.6×10^{-2} and 8.1×10^{-2} cm^2V^{-1}, respectively. Au electrodes of area 1×1 mm^2 were deposited onto the wafers by evaporation, and a bias of 100 V was applied. Although the detector material had high resistivity and mobility-lifetime products, the response of the detector to an ^{241}Am source yielded a FWHM energy resolution of only ~30% at 59.54 keV. The authors attribute this poor result to high leakage currents which limited the applied bias to 100 V, which in turn led to poor charge collection efficiencies.

6.4.7 Zinc Selenide

Zinc selenide crystallizes in the zinc blende configuration. It is a direct bandgap semiconductor with an energy gap of 2.7 eV at room temperature and a density of 5.3 g cm^{-3}. As-grown ZnSe is insulating due to self-compensation, but can be made semiconducting by annealing the crystal in molten zinc. It is an interesting material for several reasons. First, its lattice constant is 5.667 Å, which is almost lattice matched to GaAs. This makes epitaxial growth possible. It is also one of the few II–VI compounds for which both n- and *p*-type materials are available, which has opened up new avenues in optoelectronic applications, such as the production of blue LEDs and laser diodes. For radiation detector applications, ZnSe has been explored as a high-temperature alternative to CdTe and CdZnTe [84], because practical operation of CdTe and CdZnTe detectors is limited to ~70°C before irreversible damage sets in. Similarly, the maximum storage temperatures are around 100°C. ZnSe, by virtue of its large bandgap (2.7 eV as opposed to 1.47 eV for CdTe and 2.0 eV for CdZnTe), should operate to much higher temperatures, up to ~200°C.

Eissler and Lynn [84] grew ZnSe crystals using the high pressure Bridgman method. The crystals were sliced, and detectors were fabricated. Au or Pt electrodes were applied by sputtering to form a metal–semiconductor–metal (MSM) configuration. The electrode dimensions were 10 mm × 10 mm, and the detector was thickness 2 mm. The detectors were found to be spectroscopic to photons with room-temperature

FWHM energy resolutions of 25% at 22.1 keV. The device was found to function across the temperature range –70°C to +170°C. Up to ~100°C, very little change was seen in the spectral performance. Above this temperature, a steady degradation was observed. Although the peak width was more or less constant, the system gain systematically decreased, while simultaneously the noise floor increased. At 130°C, the FWHM energy resolution at 22.1 keV was 35%. Above 175°C, the system failed, although it was not clear exactly which component failed. However, these experiments demonstrate that ZnTe can function as a high-temperature spectroscopy medium.

6.5 Group III–VI Materials

Most of the III–VI compounds are chalcogenides, crystallizing in layer-type structures. The bonding is predominantly covalent within layers but much weaker van der Waals between layers. These materials are of interest because the behavior of electrons within the layers is quasi-two-dimensional (Q2D). Characteristic distances in the plane perpendicular to the layer are typically less than the de Broglie wavelength of an electron; hence the layer has the quantum properties of a lower dimension structure. While many of the features exhibited by these materials reflect those found in bulk materials, the anisotropy of the electron layer gives rise to additional properties and ordering, such as the quantum Hall effect resulting from planar magnetic order. (For a review of the electronic properties of Q2D systems, see Ando et al. [85].) Semiconducting examples of this group are at present limited to the gallium- and indium-based chalcogenides, GaS, GaSe, GaTe, InS, InSe, and InTe, of which GaSe and GaTe have shown responses to alpha and gamma radiation.

6.5.1 Gallium Selenide

Gallium selenide (GaSe) is a wide bandgap semiconductor (ε_g = 2 eV) with a density of 4.6 g cm^{-3}. Historically, GaSe was of interest because of its photoconducting and luminescence properties. However, recent research has focused on the generation and detection of broadband tunable terahertz (THz) radiation [86] by exploiting the highly anisotropic properties of its layered structure. GaSe was first investigated as a nuclear detector material in the early 1970s by Manfredotti et al. [87], who fabricated simple planar detectors by evaporating Au electrodes onto single crystals grown by the Bridgman method. A number of detectors were produced with thicknesses ranging from 50 to 150 μm and surface areas from 20 to 50 mm^2. The samples were generally n-type, with room-temperature resistivities in the range 10^8 to 10^9 Ω cm. The detectors were found to be sensitive to α-particles, with measured spectroscopic energy resolutions as low as 6.8% FWHM at 5.5 MeV. Sakai et al. [88] and later Nakatani et al. [89] reported on measurements taken from thin (~100 μm) detectors fabricated from platelets cleaved from ingots grown by the high pressure Bridgman method. Alpha particle resolutions of about 5% FWHM at 5.5 MeV were obtained. Yamazaki et al. [90] explored the properties of GaSe radiation detectors doped with Si, Ge, and Sn. They found that doping substantially decreased leakage currents, and they were able to realize FWHM energy resolutions as low as 4% for ^{241}Am 5.5 MeV alpha particles. In contrast, most undoped detectors did not function because of excessive leakage currents.

Castellano [91] explored the possibility of using GaSe for X-ray dosimetry. In this application, GaSe detectors were used as photoconduction devices by measuring the direct current induced by high fluxes of 130 and 170 kV X-rays. Mandal et al. [92] achieved X-ray detection with material grown by the modified vertical Bridgman technique. The ingots were cut, lapped, polished, and cleaved to produce planar detectors of area ~1 cm^2 and thickness 0.8 mm. After the deposition of 3-mm-diameter Au contacts, the resistivity was determined from the I/V characteristics to be in excess of 10^{10} Ω cm The mobility lifetime products were determined from a Hecht analysis to be ~1.4×10^{-5} cm^2V^{-1} for electrons and ~1.5×10^{-5} cm^2V^{-1} for holes. The devices showed a spectroscopic response to photons from an ^{241}Am source with a measured energy resolution of ~4% FWHM at 60 keV.

6.5.2 Gallium Telluride

Gallium telluride (GaTe) has attracted significant interest from the electronics industry, because unlike other compounds in this group, single crystals can be grown with low resistivity, making it possible to fabricate heterojunctions with negligible series resistance [93]. Recently, effort has been expended in exploring GaTe as a radiation detection medium, because its physical and mechanical properties are similar to GaSe and near ideal. It is nonhygroscopic, has a bandgap of 1.7 eV, a density of 5.44 g cm^{-3}, low melting point (824°C—low evaporation), and high-purity starting materials are readily available

Mandal [92] grew a large-diameter single crystal by the vertical Bridgman technique. Simple planar devices up to 1 cm^2 in area and 0.8 mm thick were fabricated and contacted with 3-mm-diameter Au pads. The devices also incorporated a guard ring to reduce leakage currents due to surface recombination. The resistivity of the devices, as determined from the I/V characteristics, was in excess of 10^9 Ω cm, and the measured mobility lifetime products of both electrons and holes were found to be similar to that of GaSe (~1.5×10^{-5} cm^2V^{-1}). The devices showed a spectroscopic response to 60 keV photons from an ^{241}Am source with a measured FWHM energy resolution of ~5% when operated with a guard ring [94]. The device was also found to be sensitive to 662 keV photons, although not spectroscopically. Without the guard ring, the devices showed no response.

6.6 Group n–VII Materials

Group n–VII (where n = II,II,IV) materials generally belong to the family of layered structured, heavy metal iodides and tellurides. The group VII anions form a hexagonal close packed arrangement, while the n cations fill all of the octahedral sites in alternate layers. The resultant structure is a layered lattice with the layers held in place by van der Waals forces and is typical for compounds of the form AB_2. The bonding within the layers is primarily covalent. The fact that layered compounds are strongly bound in two directions (by covalent bonding) and weakly bound in the third direction, along the c-axis, leads to an anisotropy of their structural and electronic properties. Materials in these groups tend to be mechanically soft, have low melting points and large dielectric constants, and show strong polarization effects. Large crystals may even deform under their own weight.

6.6.1 Mercuric Iodide

Mercuric iodide (HgI_2) has been investigated as a room-temperature X-ray and gamma ray detector since the early 1970s [95,96]. Its wide bandgap (2.1 eV) in conjunction with the high atomic numbers (80 and 53) of its constituent atoms makes it an attractive material for room-temperature X-ray and particularly gamma-ray spectrometers. In fact, because photoelectric absorption varies as Z^5, the specific sensitivity of HgI_2 is about 10 times greater than that of Ge for energies >100 keV. Additional advantages of this material are that it has been demonstrated to operate at elevated temperatures (55°C) with minimal impact on spectral performance [97], and it is extremely radiation hard, at least to a total proton fluence of ~3×10^{11} particles cm^{-2} [98,99]. However, HgI_2 suffers from relatively low transport properties ($\mu_e\tau_e = 3 \times 10^{-4}$ cm^2V^{-1} and $\mu_h\tau_h = 1 \times 10^{-5}$ cm^2V^{-1}) and severe material nonuniformity issues. For conventional planar contacts, these effectively limit detector thicknesses to ~3 mm thick in order to achieve acceptable spectroscopic results.

HgI_2 belongs to the family of layered structured, heavy metal iodides. It is a relatively soft material that forms a tetragonal lattice at temperatures below 130°C. This is known as the alpha phase, and the crystals appear red in color with a bandgap of 2.13 eV. At temperatures above 130°C, HgI_2 undergoes a phase transformation to an orthorhombic lattice (beta phase), appearing yellow in color with a bandgap of 2.5 eV. However, when cooled below 130°C, the material undergoes a destructive phase transition to alpha-HgI_2. This precludes melt growth. Spectroscopic-grade crystals were initially grown from solution growth [100], but now vapor phase is the preferred growth technique—specifically, the vertical

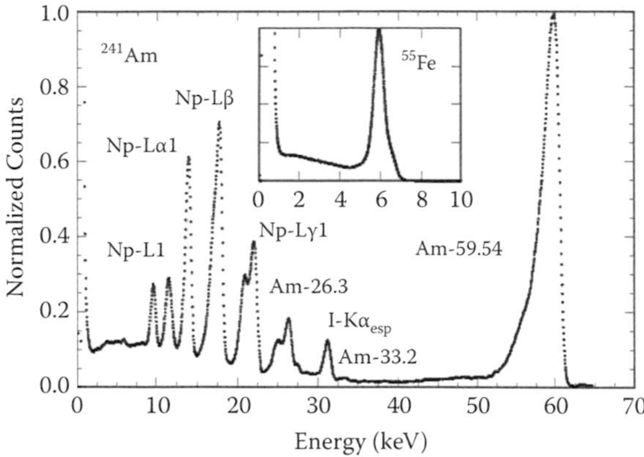

Figure 6.10 Composite ^{241}Am and ^{55}Fe spectra taken with a 7-mm^2, 0.5-mm-thick HgI_2 detector (from [21]). The detector temperature was +24°C, and the applied bias +800 V. The measured FWHM energy resolutions are 600 eV at 5.9 keV and 2.4 keV at 59.54 keV.

and horizontal ampoule methods [101]. Because of this, crystal growth is slow and the quality of large crystals inconsistent. Thus the cost of production is relatively high.

Detectors are generally prepared from bulk crystals by cleaving samples perpendicular to the crystallographic c-axis (i.e., parallel to the [011] planes). The bias is usually applied along the [001] direction, because hole mobilities are significantly higher in this direction. The fabrication of a p-n junction is usually not necessary, because dark currents are generally very low. HgI_2 is highly reactive with many metals; hence only a few materials can be used as electrical contacts, usually colloidal carbon (Aquadag) or Pd. Although reported energy resolutions are typically in the keV region [102,103], as shown in Figure 6.10, a series of small planar detectors developed for NASA's CRAF* space mission yielded near Fano-limited performances. For example, Iwanczyk et al. [104] reported an energy resolution of 198 eV FWHM at 5.9 keV, obtained with a 5-mm^2, 200-μm-thick detector operated at 0°C. These performances have not been equaled or surpassed in over two decades. In Figure 6.11 we show the

* Comet Rendezvous Asteroid Flyby (CRAF).

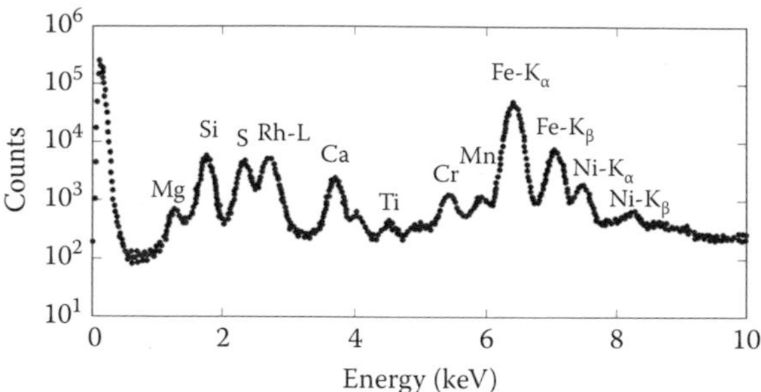

Figure 6.11 X-ray fluorescence spectrum of a sample of the Murchison meteorite taken with a 5-mm^2, 200-μm-thick HgI_2 detector operated at room temperature. The sample was excited using an X-ray tube source (from [105], © 1991 IEEE). The energy resolution is ~200 eV FWHM.

X-ray fluorescence spectrum of a sample of the Murchison meteorite* taken with such a detector at room temperature [105], illustrating the excellent energy resolution of the system. An X-ray tube with a rhodium anode was used for the excitation source. From the figure, we see that all major elements are clearly resolved.

6.6.2 Mercuric Bromoiodide

Mixed halides of mercury (HgXY) have been explored as possible alternatives to HgI_2. Although HgI_2 has many desirable properties (high density, wide bandgap, etc.) the presence of a low-temperature destructive solid–solid phase transition makes it difficult to grow large homogeneous crystals. HgXY compounds are stable at higher operating temperatures, have no phase transformations, and can be conveniently grown by the Bridgman–Stockbarger method. Of the available compounds, mercuric bromoiodide ($Hg(Br_xI_{1-x})_2$) appears the most promising. Its bandgap can be tuned by stoichiometry from 2.1 eV ($x = 0$) to 3.5 eV ($x = 1$) and for $x > 0.2$, the material is free from phase transitions. When coupled with its large density (6.2 g cm^{-3}) and high resistivity (~10^{12} Ω cm), it is an ideal candidate for room-temperature hard X-ray and gamma ray

* Named after the town of Murchison in Victoria, Australia, where the meteorite fell in 1969.

detector applications. The best measured electron and hole mobilities of 30 and 0.1 $cm^2V^{-1}s^{-1}$ [108], respectively, are comparable to those obtained with HgI_2.

Early work on this compound was carried out by Shah et al. [107], who investigated its use as a photodetector to readout scintillators, making use of the fact that the optical response can be tuned to match specific scintillators. Gospodinov et al. [108] fabricated simple planar detectors from plates of material cleaved along the [001] plane. The plates had an area of 16 mm^2 and thicknesses of 1 and 3 mm. Aquadag electrodes were then deposited directly on the upper and lower surfaces of these plates. The spectroscopic properties of the detector were assessed using ^{55}Fe and ^{241}Am radioactive sources. It was found that the best spectroscopic results were obtained with a 20% bromine fraction ($Hg(Br_{0.2}I_{0.8})_2$ [108]) and that crystals with slightly higher Br concentration gave significantly poorer energy resolutions. In Figure 6.12 we show the measured room-temperature spectra under full area illumination for 1-mm and 3-mm-thick devices fabricated from crystals with a 20% Br fraction at a detector bias of 500 V. The FWHM energy resolutions at 5.9 keV and 59.54 keV were 0.9 keV and 6.5 keV, respectively.

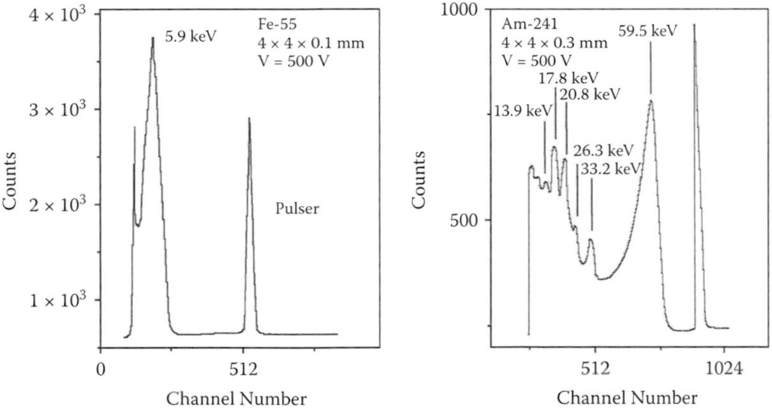

Figure 6.12 ^{55}Fe and ^{241}Am spectra, measured at room temperature with $Hg(Br_{0.2}I_{0.8})_2$ detectors (from [108]). The measured FWHM energy resolutions at 5.9 keV and 59.54 keV are 0.9 keV and 6.5 keV, respectively, under full area illumination.

6.6.3 Thallium Bromide

For room-temperature hard X-ray and gamma-ray applications, thallium bromide has emerged as a particularly interesting material in view of its wide bandgap (2.5 times that of Si) and high atomic numbers (Tl = 81, Br = 35) of its constituent atoms. In addition, its density (7.5 g cm^{-3}) is comparable to that of bismuth germanate (BGO), and thus it has excellent stopping power for hard X-rays and gamma rays. It was originally demonstrated as a radiation detector material by Hofstadter in 1949 [109], albeit with limited success due to purity and fabrication problems. Surprisingly, compared to other compound semiconductors, relatively little work has been carried out since this time [110–116]. The material has a CsCl-type simple cubic crystal structure and melts congruently at 480°C, with a single nondestructive phase transition below the melting point. Thus its physical properties are amenable to easy and rapid purification and growth using standard techniques. Its large bandgap of 2.68 eV suggests that detectors should operate at or above room temperature (RT) with low noise performance.

The relative softness* of TlBr (Knoop hardness 12 kg mm^{-2}) can cause major problems during detector fabrication because any mechanical processing, such as lapping and polishing, can create an inordinate amount of damage, principally in the form of dislocations oriented parallel to the surface being processed. The depth of the damage can be several microns, sufficient to have a significant effect on charge collection efficiencies. The effect is further exacerbated by the high dielectric constant of TlBr, which ensures a low energy of defect formation (1.1 eV [117]). Consequently, post-processing of the surface, such as etching with H_2O [118] is usually carried out prior to contacting.

Early TlBr detectors could only be operated at reduced temperatures because of stoichiometric and crystallographic imperfections mainly caused by interstitial impurities [109,110]. Shah et al. [111] found that zone refining the base material prior to crystal growth resulted in significantly

* About the same as refrigerated butter.

better performance, both in terms of resistivity and charge transport properties. At RT they measured FWHM energy resolutions of 1.5 keV, for the iron line at 5.9 keV, and 8 keV for the americium 59.54 keV nuclear line. Hitomi et al. [114,115] used multiple-pass zone refining and measured RT FWHM energy resolutions at 5.9 keV and 59.54 keV of 1.8 keV and 3.3 keV, respectively. The detectors they used had areas of 0.8 mm^2 and ~3 mm^2 and thicknesses of <100 μm and 150 μm. However, detector performances deteriorated with time, which was attributed to polarization effects* arising from two main sources: first, the modification of the internal electric field due to space charge effects caused by deep hole trapping, and second, a gradual increase in leakage currents caused by ionic conduction. TlBr is a mixed electronic–ionic conductor, with the ionic current significant even at room temperature. Above 250 K, the main conduction mechanism is due to Tl^+ ions [119]. Vaitkus et al. [120] proposed that ionic conductivity creates micro-inhomogeneities in the material which are activated by the electric field and by nonequilibrium carrier generation. The observed threshold-type effects are related to the growth of these structures and can be substantially reduced by lowering the temperature or increasing hydrostatic pressure. For example, Samaru [121] found that conductivities can vary by over a magnitude per decade change in temperature.

Owens et al. [116] carried out a series of experiments on prototype monolithic detectors of area ~8 mm^2 and thickness ~800 μm. Room temperature performances of 1.8 keV and 3.3 keV FWHM at 5.9 and 59.54 keV, respectively, were achieved. The measured energy loss spectra are shown in Figure 6.13. These detectors were operated at –22°C and showed stable and reproducible results over a time scale of 2 years. At higher energies, Hitomi et al. [122] demonstrated a room-temperature energy resolution of 1.3% FWHM at 662 keV using a 2 × 2 pixel array with depth correction. The pixel size was 1 × 1 mm^2, and the detector thickness was 4.2 mm. Without depth correction, the resolution degraded to 2%. Kim et al. [123]

* The term polarization effect is commonly used to refer to any change in the performance of a radiation detector over time that is not correlated with changes in operating parameters.

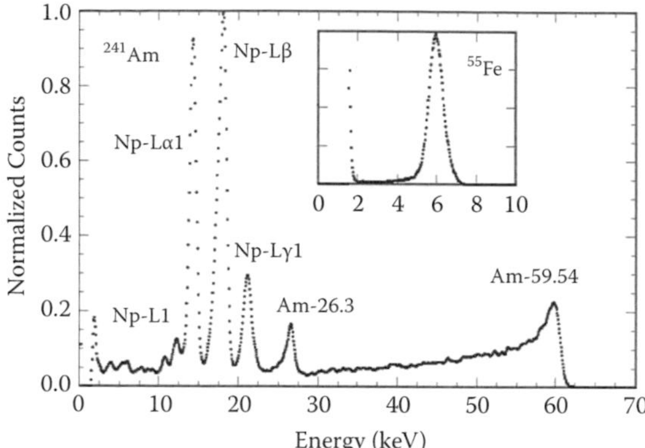

Figure 6.13 The response of an ~8-mm², 0.8-mm-thick TlBr detector to ^{241}Am and to ^{55}Fe (insert) using radioactive sources under full area illumination. The detector temperature was –22°C, and the pulser noise width 690 eV FWHM (from [21]).

fabricated a number of planar and pixel detectors Due to repeated and careful purification, electron μτ products as high as 3×10^{-3} cm^{2}V^{-1} were achieved, approaching those of CdTe. FWHM spectral resolutions of 5.3% at 122 keV and 1.7% at 662 keV were measured with a 3-mm-thick 2 × 2 pixel array (pixel size 1.3 × 1.3 mm^2). To reduce polarization effects, the detectors were operated at –18°C. For a larger 10 × 10 × 10 mm^3, 3 × 3 pixel array, a spectral resolution of 5.5% at 122 keV and 2.5% at 662 keV was achieved. The measured energy loss spectra are shown in Figure 6.14. The μτ values reported by Kim et al. [123] are somewhat surprising, because soft-lattice ionic compounds generally have inferior transport properties compared to covalent compounds, such as CdTe. Du [124] argued that the high μτ values are a consequence of the effective dielectric screening of charged defects and impurities, coupled with the electrically benign nature of the native defects. If true, the effect should be present in similar materials with high dielectric constants and may be used to prescreen materials.

6.6.4 Thallium Bromoiodide

While TlBr has many desirable properties (such as high density and wide bandgap), its mechanical softness does not lend itself

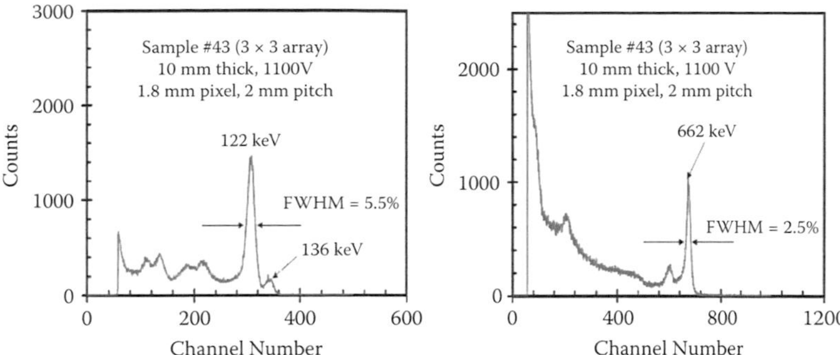

Figure 6.14 Room-temperature spectra of 122 keV (left) and 662 keV (right) gamma rays taken with a large 10 × 10 × 10 mm^3 pixel detector (from [123], © 2009 IEEE). The pixel size was 1.8 × 1.8 mm^2. The measured FWHM spectral resolutions at 122 keV and 662 keV were 5.5% and 2.5%, respectively.

to the growth of large homogeneous crystals, nor the mechanical processing required in the fabrication of detectors. In addition, the reproducibility of results is still a major problem due to stability and polarization issues. In view of these limitations, there has been recent interest in thallium salts as potential gamma ray detection media. Of the available compounds, thallium bromoiodide (Tl(Br, I), $TlBr_xI_{1-x}$) appears the most promising [125]. Its bandgap can be tuned by stoichiometry from 2.15 eV ($x = 0.3$) to 2.8 eV ($x = 1$), and for $x > 0.3$ the material is free from destructive phase transitions. When coupled with its large density (7.4 g cm^{-3}) and high resistivity (~10^{10} Ω cm), it is an ideal candidate for room-temperature hard X-ray and gamma ray detector applications. The best measured electron $\mu\tau$ values approach 10^{-3} cm^{2}V^{-1} [126], which are comparable to those obtained with TlBr. In addition, $TlBr_xI_{1-x}$ crystals are mechanically harder than TlBr, making them more tolerant to mechanical processing during device fabrication. In fact for $x = 0.35$, the Knoop hardness of thallium bromoiodide is about three times that of TlBr (~40 kg mm^{-2}).

Thallium bromoiodide, in the form $TlBr_{0.4}I_{0.6}$, has been used for several decades for infrared (IR) spectroscopy applications and is commonly known as KRS-5 [126], where the KRS prefix is an abbreviation of "Kristalle aus dem Schmelz-fluss" (crystals from the melt). It is principally used in the production of attenuated total reflection prisms, IR windows, and lenses, where transmission in the 0.6 μm–40 μm range is required.

For non-IR applications, thallium bromoiodide has been investigated as a photoconductor for the readout of scintillators [127], taking advantage of the fact that the bandgap can be tuned by composition to the peak emission spectrum of almost all common inorganic scintillators. Most recently, it has been investigated for gamma ray detection by Churilov et al. [125] who fabricated simple planar detectors from ingots grown by the travelled molten zone (TMZ) method. The detectors had compositions, x = 0.35, 0.5, and 0.65. The measured resistivities were found to be >10^{10} Ω cm, which ensured that leakage currents were sufficiently low to allow room-temperature operation. One device, a 0.6-mm-thick $TlBr_{0.35}I_{0.65}$ planar detector, showed a spectroscopic response to a ^{109}Cd source. This device had 2-mm-diameter chromium and gold electrodes and was operated at 40-V bias and 12-μs shaping time. The measured FWHM energy resolution was ~40% at 22 keV. The $(\mu\tau)_e$ product was determined from the Hecht relationship to be ~10^{-3}cm^{2}V^{-1}. However, while this device was stable, other devices fabricated from the same ingot degraded quickly.

6.6.5 Lead Iodide

Lead iodide (PbI_2) is a layered compound crystallizing in a hexagonal close packed lattice whose structure displays a large degree of polytypism. Although over 32 polytypes are known, it generally solidifies into the 2H form. It has been considered as a suitable material for X-ray and gamma-ray detection since the 1970s in view of its high density (6.2 g cm^{-3}) and wide bandgap (2.3 to 2.5 eV), which should allow detectors to operate at, or even above, room temperature. It has several advantages over HgI_2. For example, it is environmentally very stable (low vapor pressure) and, unlike HgI_2, does not undergo a phase transformation below the melting point (403°C). This makes it possible to grow lead iodide monocrystals directly from the melt or to use sublimation near the melting point for purification and film deposition. Unfortunately, at the present time carrier mobility-lifetime products are poor—on the order $\mu_e\tau_e = 1 \times 10^{-5}$ cm^{2}V^{-1} and $\mu_h\tau_h = 3 \times 10^{-7}$ cm^{2}V^{-1}. This is roughly an order of magnitude lower than HgI_2, which effectively precludes the fabrication of thick detectors if spectral performance is to be maintained. However, the high atomic number of its elements (Z_{Pb} = 82, Z_I = 53) ensures good stopping power well into the hard X-ray region, and so detector thickness can be minimized for a given detection efficiency. For example, at 100 keV, the detector needs only be ~1 mm thick to absorb 90% of the incident radiation.

Detectors produced to date give reasonable results but only for thicknesses <200 μm (a direct consequence of the poor μτ products). For example, Lund et al. [128] fabricated radiation detectors from melt-grown crystals. The detectors exhibited good energy resolution (915 eV FWHM at 5.9 keV at 20°C). Results also indicated they were more stable than HgI_2 detectors and capable of operating at temperatures over 100°C. Similar results were found by Deich and Roth [129], who fabricated detectors from boules produced by both the horizontal traveled molten zone (TMZ) and Bridgman methods. The detectors had a planar geometry with painted Aquadag* electrodes of area 3 mm^2. For a 107-μm-thick detector, a spectroscopic energy resolution of 712 eV (12%) FWHM was be obtained for 5.9 keV X-rays and 1.8 keV (3%) FWHM for 60 keV gamma rays. Shah et al. [130] used the Bridgman technique to produce 1-mm^2, 150-μm-thick detectors. By careful attention to detector fabrication techniques and preamplifier electronics design, they were able to achieve room-temperature energy resolutions of 415 eV FWHM and 1.38 keV FWHM at 5.9 keV and 60 keV, respectively.

6.6.6 Bismuth Triiodide

Bismuth triiodide (BiI_3) is a high-density (5.8 g cm^{-3}), direct bandgap semiconductor with a bandgap of 1.7 eV. Although several polymorphs exist, below 408°C BiI_3 crystallizes into a mechanically soft, layered, hexagonal form, similar to tetragonal HgI_2. BiI_3 is unusual in that it belongs to a family of iodides (with AsI_3 and SbI_3) that retain their semiconducting properties even in the liquid state [131] and have found applications

* A colloidal suspension of ~18% of submicron-sized graphite powder in a water carrier that dries to form an adherent, conductive film (<300 Ω/□) on virtually any surface, including glass and flexible materials.

in photography and holography. Because BiI_3 is materially similar to α-HgI_2, it has been proposed as a gamma ray detection medium [132]. Nason and Keller [132] grew single crystals by physical vapor transport from seed. A detector was fabricated by applying colloidal graphite electrodes and palladium wire leads to an as-grown $1.2 \times 1.2 \times 0.4$ cm^3 crystal. Electrical conductivity measurements indicated a resistivity of 2×10^9 Ω cm, although no alpha-particle or gamma-ray response was detected. Similarly, Dmitriev et al. [133] fabricated detectors from polycrystals grown by the vertical Bridgman method. Although resistivities were of the order of 1 GΩ cm, and the measured electron mobility-lifetime products were ~10^5 cm^2V^{-1}, no radiation response was detected. Matsumoto et al. [134] grew BiI_3 crystals by the vertical Bridgman method using commercially available powder. The crystals were fabricated into radiation detectors by cleaving into wafers of thickness approximately 100 μm and depositing Pd electrodes by vacuum on both cleaved surfaces. Electrical signals were extracted using thin Pd wires, bonded to the electrodes using silver epoxy. The measured resistivities were estimated to be 2×10^{10} Ω cm, which is about one order of magnitude higher than previously reported. The detector was found to be spectroscopic to alpha particles, with a measured FWHM energy resolution of 2.2 MeV at 5.5 MeV. Fornaro et al. [135] produced BiI_3 monocrystals by the travelled molten zone method, which yielded detector material, but not of spectrometric grade. The measured resistivities were up to 2×10^{12} Ω cm, and unlike previous work, small detectors of thickness ranging from 50 to 80 μm showed a response to 60 keV gamma rays.

6.7 Ternary Compounds

Other than pseudo-ternary alloys of the form AB_xC_{1-x} (e.g., $Cd_{1-x}Zn_xTe$), very little work has been carried out on true ternary materials, like $A^{II}B^{IV}C_2^{V}$ or $A_2^{II}B^{V}C^{VII}$, mainly because of the difficulties in maintaining compositional homogeneity during growth. Consequently, these materials are generally grown by MBE or MOCVD methods, which provide finer control over stoichiometry than other techniques. As a rule, ternary compounds based on the heavier metallic elements (such as Hg and Bi) are of lower mechanical strength, poorer phase and chemical stability than their binary derivatives.

6.7.1 Thallium Lead Iodide

The ternary compound, $TlPbI_3$ has been investigated as a room-temperature X-ray and gamma ray detection medium, in view of its high Z, high density (6.6 g cm^{-3}), and bandgap of 2.3 eV. Its low melting point (346°C), low vapor pressure, and lack of a destructive phase transition between room temperature and its melting point facilitate purification and crystal growth directly from the melt. It crystallizes in a rhombohedral, Perovskite-like (ABX_3), structure.

Kocsis [136] synthesized $TlPbI_3$ from the melt and sliced a 0.8-mm-thick disk from the ingot. A detector was fabricated by applying colloidal graphite on the top and bottom surfaces of the disk. The measured resistivity was 2.5×10^{12} Ω cm. While the device gave measureable photoconductivity response when exposed to an X-ray tube, no measurements were reported using alpha-particle or gamma-ray sources. Hitomi et al. [137], grew $TlPbI_3$ crystals using the vertical Bridgman method. Several detectors were fabricated by sawing the as-grown crystals into several 0.24-mm-thick wafers and applying 0.1-mm-diameter Au electrodes by vacuum evaporation. Pd contact wires were then bonded to the electrodes using silver epoxy. The resistivities, as evaluated from their current-voltage characteristics, were typically greater than 10^{11} Ω cm. The detectors showed a clear peak when exposed to 5.5 MeV alpha-particles from an ^{241}Am source. However, when exposed to a ^{137}Cs source, the 662 keV photon peak was not resolved, although an increase in counts above the noise spectrum was observed. It was also reported that the performance of the detectors was unstable with time.

6.7.2 Thallium Chalcohalides

Recently Johnsen et al. [138,139] have proposed investigating a number of thallium chalcohalide–based semiconductors, namely $TlGaSe_2$, Tl_6SeI_4, and Tl_2Au_4Se. They point out that the heavy metal halides semiconductors tend to have large bandgaps (>2.6 eV) and be mechanically soft, whereas in the metal chalcogenides the band gaps are too small to allow room temperature operation but are mechanically robust. By combining binary halides (MX_n, where M is a heavy metal X is a halogen) and binary chalcogenides (M_xQ_y, where Q is a chalcogen), it is possible to form hybrid chalcohalide compounds ($M_xQ_yX_z$) that have energy gaps which lie between the corresponding end members of the binary chalcogenides and binary halides (that is, 1.6 eV to 2.0 eV) while still retaining the mechanical properties of the chalcogenides. Their main advantages over the heavy metal halides, such as HgI_2, TlBr, PbI_2, and BiI_3 is that they tend to have higher densities, larger mu-tau products, and are mechanically stronger (typically 3 to 6 times harder). The latter quality is particularly important for fabricating detectors since any mechanical processing introduces defects in soft materials. Two compounds in particular have shown promising results, $TlGaSe_2$ and Tl_6I_4Se.

6.7.2.1 *Thallium Gallium Selenide*

Thallium gallium selenide ($TlGaSe_2$) melts congruently at 350°C and crystallizes in a layered type structure. The bonding is strongly covalent within layers and much weaker between layers. It is an indirect bandgap material (ε_g = 1.95 eV) and has a density of 6.4 gcm^{-3}. Johnsen et al. [138] synthesised $TlGaSe_2$ from a stohchiometric combination of TlSe, Ga, and Se and grew single crystalline material using a modified vertical Bridgman method. The ingot was cut into wafers from which single crystals were cleaved. A detector $3 \times 5 \times 0.87$ mm^3 in size was fabricated from one crystal and contacted with an evaporation of Ti/Au. From I/V measurements, the resistivity was estimated to be ~ 10^9 Ω cm. The mobility-lifetime products were determined from photoconductivity measurements to be $\mu\tau_e = 6 \times 10^{-5}$ cm^2V^{-1} for electrons and $\mu\tau_h = 9 \times 10^{-6}$ cm^2V^{-1} for holes. The X-ray response of the device was investigated using a Ag X-ray tube. The measurements were carried out at room temperature with an operating bias of 190V. Although the device was spectroscopic, resolving the characteristic Kα line at 22.2 keV, the Kβ line at 24.9 keV was not resolved.

6.7.2.2 *Thallium Iodide Selenide*

Thallium iodide selenide (Tl_6I_4Se) is a direct gap material with a bandgap of 1.86 eV. It melts congruently at 432 °C and crystallizes into a dense (7.4 g cm^{-3}) tetragonal structure. Johnsen et al. [139] grew single crystalline material by a modified vertical Bridgman method. The resulting sample shows single-crystalline domains from which wafers were cut perpendicular to the growth direction. A detector was fabricated from a 6 × 4 × 2 mm^3 single crystal diced from a wafer and Ti/Au contacts evaporated on to the top and bottom faces. From I/V measurements, the resistivity was found to be 4 × 10^{12} Ω cm along the ⟨001⟩ crystallographic direction. The mobility-lifetime products were determined from photoconductivity measurements and found to comparable to CdZnTe for electrons ($\mu\tau_e$ = 7 × 10^{-3} cm^{2}V^{-1}) and an order of magnitude larger for holes ($\mu\tau_h$ = 6 × 10^{-4} cm^{2}V^{-1}).

Figure 6.15 shows pulse height spectra recorded using a ^{57}Co radioactive source from which we see that the principal line emissions at 14.4, 122.1, and 136.5 keV are clearly resolved. The operating bias was 290V and the measurement carried out at room temperature. The measured FWHM energy resolution

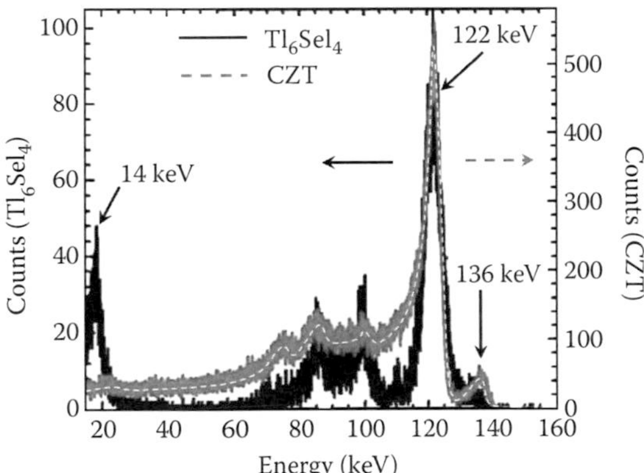

Figure 6.15 Recorded pulse height spectrum from gamma radiation from a ^{57}Co-57 source using a 6 × 4 × 2 mm^3 ⟨001⟩ Tl_6I_4Se wafer (dark solid line) (from [139], © 2011 American Chemical Society). For comparison, the measured pulse height spectrum from a commercial 5 × 5 × 5 mm^3 CZT detector is also shown (dashed line; SPEAR detector manufactured by EI Detection & Imaging Systems). The measurements were carried out at 295 K.

was 5.7 keV (4.7%) at 122.1 keV. For comparison, Figure 6.15 also shows the measured spectrum obtained with a commercial $5 \times 5 \times 5$ mm^3 CdZnTe detector. In this case, the measured FWHM energy resolution is 5.54 keV (4.5%) at 122.1 keV.

6.8 Other Inorganic Compounds

Numerous other compounds have been proposed and tested (e.g., AlSb [140], Bi_2S_3 [141], $PbBr_2$ [142], and $HgBr_2$ [143]). None has shown a response to alpha particles.

6.9 Organic Compounds

So far we have concentrated on inorganic semiconductors for radiation detection. However, there is no reason why organic (polymer) semiconductors may not also serve the same function (see [144]). These compounds belong to groups I–IV, I–V–VI, and other organic (polymer, oligomers) derivatives. They are the so-called "plastic" semiconductors, although the term is often widely used to encompass other semiconducting organic materials. They are polymers with a delocalized π-electron system along the polymer backbone This results in the creation of alternating single and double bonds between the carbons by weak *pz–pz* bonding (–C=C–C=C–), which in turn results in the creation of a bandgap of ~2.5 eV. The great advantages of organic semiconductors are their diversity and the relative ease of changing their properties to specific applications. The number of known organic semiconductors is huge and includes plant and animal chelates such as carotene, chlorophyll, and even blood pigments.

At present, these materials are being exploited for use in low-cost flexible displays and low-end data storage media. Such materials offer numerous advantages in terms of easy processing (spin coating as opposed to epitaxial growth), and good compatibility with a wide variety of substrates. However, as pointed out by Beckerle and Strobele [145], there is no reason why polymer semiconductors may not be used for charged particle detection. They demonstrated that stretched polyacetylene foils were sensitive to 5.5 MeV alpha particles. The foils had dimensions of 1 cm × 1 cm and 1 cm × 2 cm with thicknesses of 10, 50, and 100 μm. The electrical contacts were formed by sputtering 200-μm thick gold electrodes onto the foils. Without stretching, the mobility of the free carriers is ~10^7 to 10^9 times smaller than in silicon, leading to very low drift velocities, which in turn limit the use of such films to low count rate experiments. However, stretching the film creates a degree of alignment of the polymer chains in the stretched direction in which both drift velocities and electron mobilities increase markedly with stretch factor. This is because the alignment of the polymer chains with respect to the electric field should reduce the number of hopping processes per unit length. In an unstretched foil, the efficiency for detecting 5.5 MeV alpha particles in a 10-μm film is around 35% for drift lengths <3 mm. Stretching the foil by a factor of three increases the detection efficiency to ~70%. However, drift velocities are still very low, and long-term material stability and radiation damage issues may prohibit the use of polymer foils for most applications.

Blakesley et al. [146] investigated the potential for using polymeric semiconductors in medical X-ray imaging applications. They found that polymer photodiodes coupled to phosphor screens show a response to X-ray radiation with a good efficiency. The photodiodes were fabricated on glass substrates with indium–tin oxide (ITO) bottom contacts. A layer of poly(3,4 ethylenedioxythiothene) poly(styrenesulfonate) (PEDOT-PSS) is spin coated onto the substrate and dried. Finally, aluminum contacts are evaporated onto the top surface.

6.10 Discussion

Although compound semiconductor radiation detectors have been under development for almost four decades, progress has been incremental. In all cases, this can be directly traced back to a single material issue, namely the difficulty of producing crystallographically perfect crystals of high purity and exact stoichiometry. For III–V materials the problems usually occur during the growth process (for example, defects introduced due to lattice mismatched substrates), while for II–VI materials detector handling and fabrication techniques can play a significant role, in view of the soft or layered nature of some materials. In Table 6.3 we summarize some of the best reported spectral resolutions at soft and hard X-ray wavelengths for each of the above materials.

6.11 Neutron Detection

Since neutrons are uncharged, the only practical detection method is through the observation of the reaction products following neutron capture. It is noteworthy that very few elements have high enough capture cross sections to be used in practical detection systems. These are listed in Table 5.1 along with their principal capture processes (e.g., (n,γ), (n,p), (n,α), (n,β) or (n,fission) reactions).

6.11.1 Indirect Neutron Detection

Until recently, work has concentrated on so-called indirect techniques, which normally involve depositing a thin layer of a highly neutron absorbing material (such as Li, B, or Gd) on top of an active detector—usually a planar device. The neutrons are converted into charged particles in this layer following neutron capture, and these secondary products are detected in the detector. One such implementation has been studied by McGregor et al. [149]—a boron-coated GaAs diode (see Figure 6.16). Thermal neutrons are absorbed in the ^{10}B film by neutron capture. The excited nucleus then de-excites with the ejection of an energetic ^{7}Li ion (840 keV) and an alpha particle (1.47 MeV) in the opposite direction. The diode detects one or both of the reaction products. The advantage of such techniques is its relative simplicity; the disadvantage is that the boron coating is limited to a thickness of a few microns due to self-absorption of the alpha particles. In practice, this limits the detection efficiency to about 4%. It is worth noting that most nuclear materials emit 10 or more times as many gamma rays as neutrons. Thus, the

TABLE 6.3
The Best Energy Resolutions Achieved with Simple Planar Detectors

Material	Detector Size Area, Thickness	ΔE @ 5.9 KeV (eV)	ΔE @ 59.5 KeV (eV)	Reference
Si	0.8 mm^2, 500 μm	245 at –15°C	524	Owens et al. [21]
		750 at +15°C	800	
GaAs	0.8 mm^2, 40 μm	450 at –22°C	670	Owens et al. [21]
		572 at RT	780	
GaAs pixel	250×250×40 μm^3	219 at –30°C	468	Owens et al. [21]
		266 at RT	487	
SiC-4H pixel	0.03 mm^2, 25 μm	196 at +30°C	not measured	Bertuccio et al. [12]
		233 at +100°C	not measured	
SiC-4H	0.79 mm^2	Not measured	550 at RT	Phlips et al. [11]
InP	3.142 mm^2, 200 μm	911 at –170°C	3050	Owens et al. [21]
		2480 at –60°C	9200	
CdTe	16 mm^2, 1.2 mm	310 at –60°C	600	Loupilov et al. [147]
CdZnTe	4 mm^2, 2 mm	240 at –40°C	1200 at –30°C	Niemela & Silipa [148]
	3.142 mm^2, 2.5 mm	1508 at RT	2900	Owens et al. [68]
HgI_2	5 mm^2, 200 μm	198 at 0°C	650	Iwanczyk et al. [104]

(continued)

TABLE 6.3 (CONTINUED)
The Best Energy Resolutions Achieved with Simple Planar Detectors

Material	Detector Size Area, Thickness	ΔE @ 5.9 KeV (eV)	ΔE @ 59.5 KeV (eV)	Reference
HgBrI	16 mm^2, 1 mm	860 eV at RT	not measured	Gospodinov et al. [108]
	16 mm^2, 3 mm	not measured	6500 eV at RT	Gospodinov et al. [108]
PbI_2	1 mm^2, 50 μm	415 at RT	1380	Shah et al. [130]
TlBr	3.142 mm^2, 800 μm	800 at –30°C	2300	Owens et al. [116]
		1800 at RT	3300	

Note: The figures are quoted for the collection of both carriers. Spectral enhancement techniques involving single carrier collection have not been employed. The measurements were carried out under uniform illumination using ^{55}Fe and ^{241}Am radioactive sources. For completeness, we also list the resolutions at room temperature (RT) where possible, because there are many applications in which resolving power is not a primary requirement.

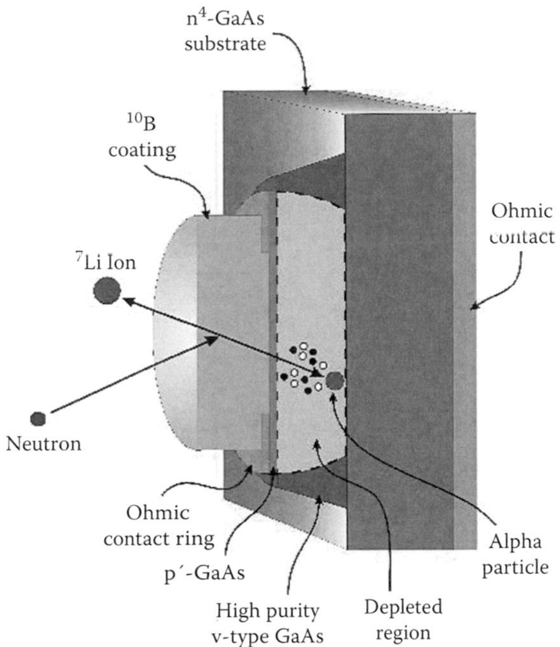

Figure 6.16 Cross-section depiction of a self-biased GaAs neutron detector (from [150]). The internal potential formed at the p+/ν-type junction is sufficient to deplete the high-purity GaAs material. The device is coated with pure ^{10}B. Neutrons absorbed in the ^{10}B layer discharge energetic ions (^{7}Li ions and α-particles) in opposite directions. Charges excited within the depleted region are drifted by the internal potential to form measurable pulses.

wgamma ray sensitivity of a neutron detector can be an important criterion in material selection. This may preclude the use of the heavier compounds in some applications because their gamma ray response may overwhelm their neutron response. For, example, for ^{113}Cd and ^{157}Gd, the resulting low-energy conversion electrons and gamma ray emissions following neutron capture reactions can be difficult to distinguish from background gamma rays in a high radiation field.

6.11.2 Direct Neutron Detection

Direct neutron detection in a highly absorbent semiconducting material offers one major advantage—namely high detection efficiency, because the device acts as both absorber and detector. A direct neutron detection system offers additional advantages over indirect methods. For example, because much charge is generated per incident event, detection systems can operate without bias. This attribute can be further exploited for power generation, potentially leading to an alternate to radioisotope thermoelectric generators for outer planetary missions.

Attempts at direct neutron detection date back to the 1960s. Johnson [151] reported the detection of induced radioactivity produced by fast neutrons in a CdS crystal operating as both neutron absorber and detector. Following irradiation, changes in the post-irradiation resistivity were observed. This was attributed to self-ionization of the sample by radioactive decay of transmuted atoms produced during the irradiation, specifically from the decays of ^{115}Cd and ^{32}P.

Fasasi et al. [152] exposed two cadmium telluride detectors to a known thermal neutron beam in an attempt to detect decay gamma rays from ^{113}Cd following neutron capture. ^{113}Cd has a large thermal neutron cross section of 20,000 barns. Because the isotopic abundance of ^{113}Cd is 12.6%, this leads to an effective absorption cross section of 2450 barns for natural cadmium. Thermal neutrons react with ^{113}Cd via the reaction ^{113}Cd(n,γ)^{114}Cd. Almost 500 gamma ray transitions are possible, of which the most dominant produce prompt emission lines at 559 keV (100%), 651 keV (19%), 806 keV (7%), 1209 keV (6%), and 1364 keV (6%) [153]. Satisfactory agreement was observed between calculations and experimental data, and spectral line features were detected at 96 keV and 560 keV. Fasasi et al. [152] point out that, because the atomic density of CdTe is much higher than in gas detectors, its efficiency for small detection volumes can be comparable to that of a ^{3}He tube, when the entire pulse-height spectrum is used. The authors report an efficiency of about 5%, when using the counts in the entire spectrum, 0.5% when only the 96 keV gamma-ray line is used, and 0.38% when the 558 keV line is measured with a 2-mm-thick detector.

McGregor et al. [154] have proposed using an active CdZnTe detector to detect thermal neutrons. In the exposure of a $10 \times 10 \times 3$ mm^3 CdZnTe detector to a thermal neutron source, they found clear signatures of the 586-keV and 651-keV gamma-ray lines. The detection sensitivity is, however, dependent on the gamma ray absorption efficiency, which in turn depends on the detector active volume. In this case, the detection sensitivity using the 586-keV line is ~4%. Also, as the authors point out, because CdZnTe is an efficient X-ray and gamma ray detection medium, a weak thermal neutron source may easily be masked by a high gamma ray background.

Beyerle and Hull [155] and Melamud et al. [156] have pointed out that mercuric iodide can act as an efficient neutron detector via the reaction $^{199}Hg(n,\gamma)^{200}Hg$, which has a capture cross section of 2150 barns. Taking into account the isotopic abundance of ^{199}Hg, the effective cross section of natural Hg is 374 barns. The main gamma ray lines emitted are prompt of energies, 368 keV (81%) and 1694 keV (14%). However neutron absorption can also proceed via the production of radioactive ^{128}I, which produces a strong continuum with a half-life of 25 mins. As such, the buildup of an induced background can swamp the prompt neutron response of Hg. As with CdZnTe, HgI_2 is also a very efficient gamma ray detection medium, and so its operation as a thermal neutron detector in a high gamma background environment will be compromised. Bell et al. [157] proposed improving the sensitivity by coating the surfaces of an HgI_2 detector with boron. The boron film acts as a converter by capturing neutrons entering the HgI_2 crystal and generating 478 keV gamma rays via the transition to the first excited state of the $^{10}B(n,\alpha)^{7}Li^{*}$ reaction. This occurs in 94% of thermal captures on boron and is the only gamma ray emitted in this reaction. An HgI_2 detector a few millimeters thick would be expected to have good photopeak efficiency at this energy. The presence of both the 368-keV and 478-keV gamma rays in the HgI_2 detector with the correct ratio results in an improvement of the neutron capture signature. Bell et al. [158] extended this analysis and demonstrated that the information in both lines could be used to estimate the average energy of incident neutrons and also to distinguish between unmoderated radioactive and fission sources.

It should be noted that, since the energies of these gamma rays are lower than the main lines from Cd observed in a CdZnTe detector, the probability of gamma ray detection is higher in an HgI_2-based device. For a 25 mm × 25 mm × 2.6 mm HgI_2 crystal, Bell et al. [158] estimate the thermal neutron capture efficiency of a boron-clad detector to be 4%.

6.11.3 Choice of Compound

The ideal candidate for direct solid state thermal neutron detection should satisfy the following criteria.

1. It should have a high probability of interaction with a thermal neutron, and the reaction should give an easily identifiable and unambiguous result.
2. There should be a low probability of interference due to other radiation.
3. The characteristics of either the element or one of its compounds should be semiconductor-like.
4. It should be in sufficient abundance to make it affordable for the application.
5. The detector should have chemical and physical compatibility with a suitable substrate.
6. It should also have chemical and electrical compatibility with a contacting system.

Of the available compounds, those based on boron are the most promising in view of its high thermal neutron absorption cross section, zero gamma ray emission, low gamma ray absorption cross section, and a large number of boron compounds available. The electron deficiency of boron together with its small atomic size allow boron atoms to coalesce in three-dimensional atomic networks of icosahedral boron linked together to form the various polymorphs. Boron bonds with the p-block elements of the periodic table to form a variety of both binary and ternary covalently bonded compounds like BN (both hexagonal and cubic), BP (cubic and rhombohedral), B_xC, BAs, BC_xO_y, and similar compounds. Thermal neutrons are detected via the capture reaction $^{10}B(n,\alpha)^{7}Li^{*}$ which has a cross section of 3840 barns for meV neutron energies. There are actually two decay modes, to the first excited state and directly to the ground state:

$$^{10}\text{B} + \text{n} \Rightarrow {}^{7}\text{Li}\ (0.84\ \text{MeV}) + {}^{4}\text{He}\ (1.47\ \text{MeV}) + \gamma\ (0.48\ \text{MeV}) \quad Q = 2.31\ \text{MeV}$$

or

$$^{10}\text{B} + \text{n} \Rightarrow {}^{7}\text{Li}\ (1.02\ \text{MeV}) + {}^{4}\text{He}\ (1.78\ \text{MeV}) \quad Q = 2.79\ \text{MeV} \tag{6.3}$$

with 94% and 6% probability, respectively. Because the Q value for the dominant decay mode is 2.31 MeV, over 10^5 electron–hole pairs are generated for each captured neutron. Thus, given that the range of the reaction products is negligible (~4 μm for the alpha particle and ~2 μm for the Li recoil), a solid-state device which acts as both absorber and detector would be highly compact and efficient. Because the densities of boron compounds are low, the background from gamma rays, which invariably accompany neutron production, will also be low. A schematic of such a detector is given in Figure 6.17. In principle, we require a depleted thickness of around 200 μm for 100% detection efficiency. For reference, the diodes described here would have detection efficiencies of 1.5% to 4%, depending on thickness.

For neutron detection, BP, BN, B_4C, and B are particularly interesting since they are refractory materials and

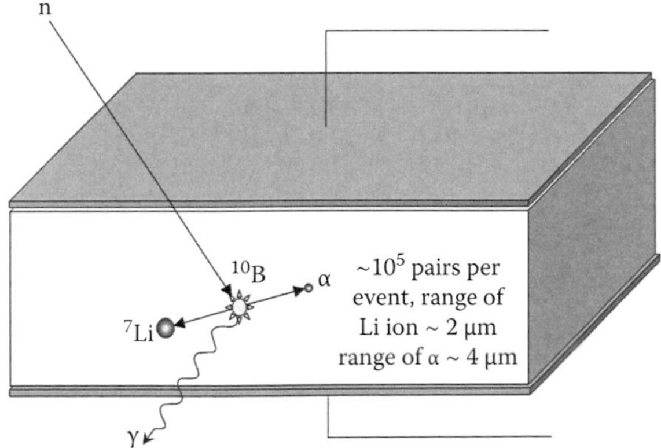

Figure 6.17 Schematic of the principal interaction mechanism in a direct neutron detection device based on a boron compound.

chemically inert, and are already being actively developed for high-temperature/high-power electronic applications [e.g., 159]. The phosphide has a cubic zinc blende structure, while the nitride and carbide are available in three phases—cubic, hexagonal, and amorphous. Until recently, impurities have prevented the production of detector-quality monocrystalline material. However, Ananthanarayanan et al. [160] have demonstrated that small 6.4-mm-diameter, 1-mm-thick polycrystalline BP and BN detectors are responsive to thermal neutron fluxes of 10^8 cm^{-2}s^{-1}. Kumashiro et al. [161] grew single crystal (10 mm × 20 mm × 300 μm thick) wafers of ^{10}BP using chemical vapor deposition on Si substrates. In exposure to thermal neutrons of fluence ~10^4 n cm^{-2}s^{-1}, no changes in electrical properties were found for the wafers grown on Si(100) substrates, and only a slight change for those grown on Si(111) substrates.

Recently, there has been a breakthrough in the production of monocrystalline material for all three boron compounds. Zhang et al. [162] have reported the production of cubic BN films, Kumashiro [163] has reported the production of single-crystalline BP wafers, and Robertson et al. [164] have reported direct neutron detection with icosahedral B_5C. At the present time it is unclear whether the neutron signal originates in the B_5C bulk or in a boronated Si layer at the B_5C–Si substrate interface [165]. In fact, two known compounds of boron and silicon exist, B_6Si and B_4Si, which are both semiconductors with bandgaps close to 0.5 eV.

Lund et al. [166] fabricated radiation detectors from BP films grown by the chemical vapor deposition technique on (100)-oriented *n*-type Si substrates. The BP films were typically 1 to 10 mm thick. The detectors showed a spectroscopic response to 5.5 MeV alpha particles, but were unresponsive to thermal neutrons. Kaneko et al. [167] fabricated radiation detectors using single-crystal and polycrystalline cubic boron nitride (cBN) crystals synthesized using a high-pressure and high-temperature method. Surprisingly, while they found that the single-crystal device did not detect alpha particles due to high leakage currents, the polycrystalline device was responsive. Although, its response to alpha particles was not spectroscopic, it did show a direct sensitivity to neutrons.

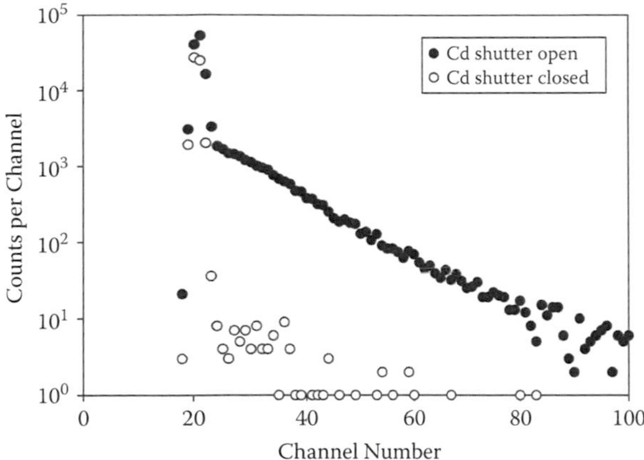

Figure 6.18 Pulse height spectra from a p-BN detector exposed to a thermal neutron beam from a reactor demonstrating a clear response (from [168]).

McGregor et al. [168] reported a series of experiments on commercially obtained pyrolitic* boron nitride (p-BN) material. The devices were constructed from 5 mm × 5 mm × 1 mm thick samples of p-BN. Circular contacts of 2.5 mm diameter were applied to both sides of the samples by evaporating 600 Å of Ti followed by a further evaporation of 1000 Å Au through a shadow mask. The devices were mounted on sapphire substrates and operated as traditional planar semiconductor detectors. The I/V characteristics show nonrectifying behavior with a bulk resistivity of ~10^{12} Ω cm. At a nominal bias of 400 V the devices showed a clear response to thermal neutrons from a reactor (see Figure 6.18). The thermal neutron counting efficiency was found to vary from 1.2% to 7.2% between samples—the variation between attributed to the fact that the bulk of the response is confined to a few channels of the spectrum above the noise and is therefore extremely sensitive to threshold and polarization effects. Based on the ^{10}B content of the samples, the expected efficiencies are of the order of ~98%, indicating poor charge transport probably caused by nonuniform electric fields.

It should be noted that boron is itself a semiconductor, and so it should be possible to fabricate a detector from epitaxially

* Pyrolytic in this case meaning an ordered solid of individual crystallites showing strong anisotropic properties.

grown boron films, thus greatly simplifying the growth and fabrication process. Boron is the only element in group IIIb which possesses semiconducting properties with a bandgap of 1.50 or 1.56 eV. It exists in several allotropes: amorphous boron is a brown powder, while crystalline boron is black, hard (~3300 kgf mm^{-2} on the Knoop microhardness scale) and a weak conductor at room temperature. Growth difficulties have so far prevented the use of boron as a semiconductor, and so it is hardly surprising that little work has been reported in the literature. The electron and hole mobilities are reported to be 0.7 and 2 $cm^2V^{-1}s^{-1}$, which are very low. The carrier lifetimes are unknown and, if poor, would severely limit the thickness of an active boron layer—thus capping the achievable detection efficiency.

Tomov et al. [169] fabricated detectors from crystalline boron films grown on Si substrates. The film thicknesses ranged from 500 nm to 3.8 μm. All devices showed strongly rectifying behavior and a spectroscopic response to alpha particles ($\Delta E/E$ ~5%). They also show a response to neutrons. This is illustrated in Figure 6.19. For the neutron response, we can identify the individual components of the spectrum. Because the diodes were fabricated from natural boron, efficiencies are expected to be low because of the low isotropic concentration

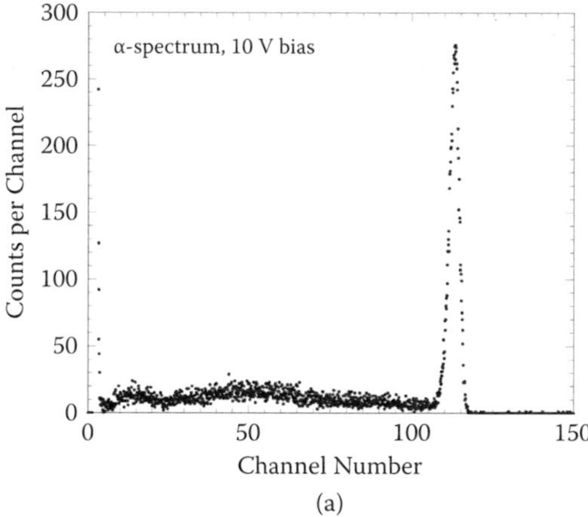

Figure 6.19 (a) The response of a 12.6-mm^2, 1.7-μm-thick diode to 5.5 MeV alpha particles. The bias was 10 V. (b) The response to thermal neutrons from the Delft 2 MW nuclear reactor. Even under zero bias the diode shows a response.

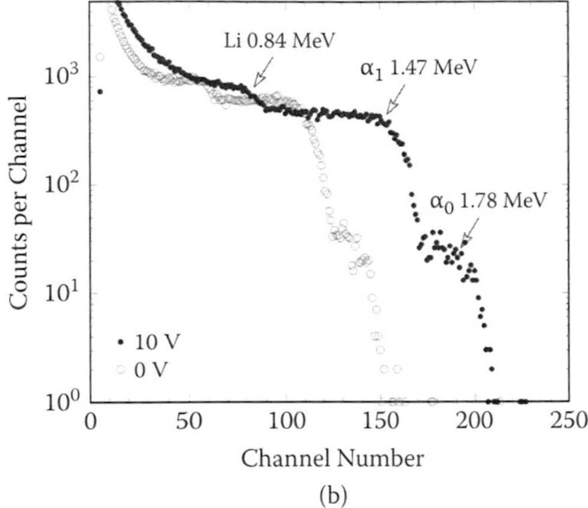

Figure 6.19 (Continued)

of ^{10}B. In fact, the measured efficiency of a 2.5-μm-thick diode was 0.5%. Coupled with the relatively thin boron layers, it is not clear if the response is due to an active boron layer, a boronated Si p-n junction, or both. The latest generation of devices now have low enough leakage currents (~100 nA) that they also show a response to 60-keV gamma rays. This is illustrated in Figure 6.20, in which we show the measured

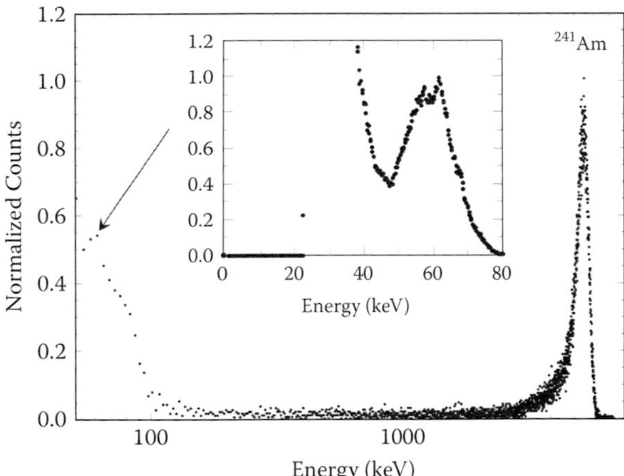

Figure 6.20 Alpha and photon response of a 16-mm^2, 2.8-μm-thick boron diode to an ^{241}Am radioactive source. The inset shows an expansion of the spectrum in the vicinity of 60 keV, showing that the 60 keV gamma-ray line is clearly resolved. The detector bias was 6 V, and shaping time 2 μs.

energy loss spectrum of a 15-mm^2, 2.8-micron-thick device when exposed to an ^{241}Am radioactive source. A 60-keV peak is clearly resolved (shown in the inset) with an energy resolution of ~30% FWHM. The FWHM energy resolution of the alpha peak at 5.5 MeV is 12%. Note that the source used is "covered" in that the active material is covered with a thin protective film. The width of the 5.5-MeV line is largely due to straggling in this film.

In Table 6.4 we list the advantages and disadvantages of the compounds listed above which mostly relate to growth. For completeness, we also list boron-doped diamond, for which neutron detection has also been claimed [170,171]. However, because the boron is a dopant it will constitute a small fraction of the overall detector volume, with corresponding impacts on detection efficiency.

TABLE 6.4
A Comparison of the Pros and Cons of Potential Neutron Detection Materials

Material	Advantages	Disadvantages
Boron (B)	Simple, 100% ^{10}B composition possible	Difficult to grow, slow growth, no obvious substrate
Boron nitride (BN)	Benign process	Difficult to grow, slow growth, diamond substrate required
Boron phosphide (BP, $B_{12}P_2$)	Simple CVD, reasonable growth rates, cubic and rhombohedral forms look promising	Phosphine gas precursor, no obvious substrate, high-temperature growth
Boron carbide (B_4C, B_5C)	Low-temperature deposition, single, solid precursor, no lattice-matching required, neutron detection claimed	Expensive precursor, many polytypes, heterostructure may be required
Doped diamond C:B	Neutron detection claimed	Expensive, single uninterested vendor, low detection efficiency

References

1. G. Armantrout, S. Swierkowski, J. Sherohman, J. Yee, "What Can Be Expected from High Z Semiconductor Detectors," *IEEE Trans. Nucl. Sci.*, Vol. NS-24 (1977) pp. 121–125.
2. R.N. Ghosh, R. Loloee, T. Isaacs-Smith, J.R. Williams, "High Temperature Reliability of SiC n-MOS Devices up to 630°C," *Mat. Sci. Forum, Silicon Carbide and Related Materials*, Vols. 527–529 (2006) pp. 1039–1042.
3. A.R. Verma, P. Krishna, *Polymorphism and Polytypism in Crystals*, John Wiley & Sons Inc. New York (1966).
4. R.S. Ramsdell, "Studies on Silicon Carbide," *Am. Mineral.*, Vol. 32 (1947) pp. 64–82.
5. R.V. Babcock, S.L. Ruby, F.D. Schupp, K.H. Sun, *Miniature Neutron Detectors*, Westinghouse Elec. Corp., Materials Engineering Report No. 5711-6600-A, November (1957).
6. R.V. Babcock, H.C. Chang, "SiC Neutron Detectors for High Temperature Operation, Neutron Dosimetry," *Proceedings of the Symposium on Neutron Detection, Dosimetry and Standardization*, Vol. 1, International Atomic Energy Agency (IAEA), Vienna, December (1962) pp. 613–622.
7. R.R. Ferber, G.N. Hamilton, *Silicon Carbide High Temperature Neutron Detectors for Reactor Instrumentation*, Westinghouse Research and Development Center, Document No. 65-1 C2-RDFCT-P3, June (1965).
8. G. Bertuccio, R. Casiraghi, F. Nava, "Epitaxial Silicon Carbide for X-Ray Detection," *IEEE Trans. Nucl. Sci.*, Vol. NS-48 (2001) pp. 232–233.
9. G. Bertuccio, R. Casiraghi, E. Gatti, D. Maiocchi, F. Nava, C. Canali, A. Cetronio, C. Lanzieri, "SiC X-Ray Detectors for Spectroscopy and Imaging in a Wide Temperature Range," *Materials Science Forum*, Vols. 433–436 (2003) pp. 941–944.
10. G. Bertuccio, R. Casiraghi, A. Cetronio, C. Lanzieri, F. Nava, "Silicon Carbide for High Resolution X-Ray Detectors Operating up to 100°C," *Nucl. Instr. and Meth.*, Vol. A522 (2004) pp. 413–419.
11. B.F. Phlips, K.D. Hobart, F.J. Kub, R.E. Stahlbush, M.K. Das, B.A. Hull, G. De Geronimo, P. O'Connor, "Silicon Carbide PiN Diodes as Radiation Detectors," *Materials Science Forum*, Vols. 527–529 (2006) pp. 1465–1468.

12. G. Bertuccio, R. Casiraghi, A. Cetronio, C. Lanzieri, F. Nava, "Advances in Silicon Carbide X-Ray Detectors," *Nucl. Instr. and Meth.*, Vol. A652 (2010) pp. 193–196.
13. P. Bergonzo, A. Brambilla, D. Tromson, C. Mer, B. Guizard, R. Marshall, F. Foulon, "CVD Diamond for Nuclear Detection Applications," *Nucl. Instr. and Meth.*, Vol. A476 (2002) pp. 694–700.
14. A. Mainwood, "CVD Diamond Particle Detectors," *Diamond and Related Materials*, Vol. 7 (1998) pp. 504–509.
15. F. Nava, C. Canali, M. Artuso, E. Gatti, P.F. Manfredi, S.F. Kozlov, "Transport Properties of Natural Diamond Used As Nuclear Particle Detector for a Wide Temperature Range," *IEEE Trans. Nucl. Sci.*, Vol. NS-26 (1979) pp. 308–315.
16. J.H. Kaneko, T. Tanaka, T. Imai, Y. Tanimura, M. Katagiri, T. Nishitani, H. Takeuchi, T. Sawamura, T. Iida, "Radiation Detector Made of a Diamond Single Crystal Grown by a Chemical Vapor Deposition Method," *Nucl. Instr. and Meth.*, Vol. A505 (2003) pp. 187–190.
17. W.R. Harding, C. Hilsum, M.E. Moncaster, D.C. Northrop, O. Simpson, "Gallium Arsenide for γ-Ray Spectroscopy," *Nature*, Vol. 187 (1960) pp. 405–405.
18. J.E. Eberhardt, R.D. Ryan, A.J. Tavendale, "High Resolution Radiation Detectors from Epitaxial n-GaAs," *Appl. Phys. Lett.*, Vol. 17 (1970) pp. 427–429.
19. J.E. Eberhardt, R.D. Ryan, A.J. Tavendale, "Evaluation of Epitaxial n-GaAs for Nuclear Radiation Detection," *Nucl. Instr. and Meth.*, Vol. 94 (1971) pp. 463–476.
20. D.S. McGregor, H. Hermon, "Room-Temperature Compound Semiconductor Radiation Detectors," *Nucl. Instr. and Meth.*, Vol. 395, (1997) pp. 101–124.
21. A. Owens, A. Peacock, M. Bavdaz, "Progress in Compound Semiconductors," *Proc. of the SPIE*, Vol. 4851 (2003) pp. 1059–1070.
22. A. Owens, M. Bavdaz, A. Peacock, A. Poelaert, H. Andersson, S. Nenonen, L. Tröger, G. Bertuccio, "High Resolution X-ray Spectroscopy Using GaAs Arrays," *J. App. Phys.*, Vol. 90 (2001) pp. 5367–5381.
23. A. Owens, H. Andersson, M. Campbell, D. Lumb, S. Nenonen, L. Tlustos, "GaAs Arrays for X-Ray Spectroscopy," *Proc. of the SPIE*, Vol. 5501 (2004) pp. 241–248.
24. M. Campbell, H.M. Heijne, G. Meddeler, E. Pernigotti, W. Snoeys, "Readout for a 64×64 Pixel Matrix with 15-Bit Single Photon Counting," *IEEE Trans. Nucl. Sci.*, Vol. 45 (3) (1998) pp. 751–753.

25. L.R. Weisberg, B. Goldstein, "GaAs and GaP for Room Temperature Gamma-Ray Counters." In *Nucleonics in Aerospace*, P. Polishuk (ed.), Plenum Press, New York (1968) pp. 182–186.
26. P. Litovchenko, D. Bisello, A. Litovchenko, S. Kanevskyj, V. Opilat, M. Pinkovska, V. Tartachnyk, R. Rando, P. Giubilato, V. Khomenkov, "Some Features of Current-Voltage Characteristics of Irradiated GaP Light Diodes," *Nucl. Instr. and Meth.*, Vol. A552 (2005) pp. 93–97.
27. A. Owens, S. Andersson, R. den Hartog, F. Quarati, A. Webb, E. Welter, "Hard X-ray detection with a GaP Schottky diode," *Nucl. Instr. and Meth.*, Vol. A581 (2007) pp. 709–712.
28. J. Vaitkus, W. Cunningham, E. Gaubas, M. Rahman, S. Sakai, K.M. Smith, T. Wang, "Semi-insulating GaN and Its Evaluation for a Particle Detection," *Nucl. Instr. and Meth.*, Vol. 509 (2003) pp. 60–64.
29. A. Owens, A. Barnes, R.A. Farley, M. Germain, P.J. Sellin, "GaN detector development for particle and X-ray detection," *Nucl. Instr. and Meth.*, Section A (2011), doi: 10.1016/j.nima.2011.11.02.
30. F.J. Leonberger, P.F. Moulton, "High-Speed InP Optoelectronic Switch," *Appl. Phys. Lett.*, Vol. 35 (1979) pp. 712–714.
31. A.G. Foyt, F.J. Leonberger, R.C. Wiamson, "Picosecond InP Optoelectronic Switches," *Appl. Phys. Lett.*, Vol. 40 (1982) pp. 447–449.
32. J. Lund, F. Olscher, F. Sinclair, M.R. Squillante, "Indium Phosphide Particle Detectors for Low Energy Solar Neutrino Spectroscopy," *Nucl. Instr. and Meth.*, Vol. A272 (1988) pp. 885–888.
33. T.F. Deutsch, F.J. Leonberger, A.G. Foyt, D. Mills, "High-Speed Ultraviolet and X-Ray-Sensitive InP Photoconductive Detectors," *Appl. Phys. Lett.*, Vol. 41 (1982) pp. 403–405.
34. D. Kania, R. Bartlett, R. Wagner, R. Hammond, "Pulsed Soft X-Ray Response of InP:Fe Photoconductors," *Appl. Phys. Lett.*, Vol. 44 (1984) pp. 1059–1061.
35. Y. Suzuki, Y. Fukuda, Y. Nagashima, "An Indium Phosphide Solid State Detector," *Nucl. Instr. and Meth.*, Vol. A275 (1989) pp. 142–148.
36. F. Olschner, J.C. Lund, M.R. Squillante, D.L. Kelly, "Indium Phosphide Particle Detectors," *IEEE Trans. Nucl. Sci.*, Vol. NS-36 (1989) pp. 210–212.
37. P. Jayavel, S. Ghosh, A. Jhingan, D.K. Avasthi, K. Asokan, J. Kumar, "Study on the Performance of SI–GaAs and SI–InP Surface Barrier Detectors for Alpha and Gamma Detection," *Nucl. Instr. and Meth.*, Vol. A454 (2000) pp. 252–256.

38. F. Dubecký, B. Zaťko, V. Nečas, M. Sekáčová, R. Fornari, E. Gombia, P. Boháček, M. Krempaský, P.G. Pelfer, "Recent Improvements in Detection Performances of Radiation Detectors Based on Bulk Semi-Insulating InP," *Nucl. Instr. and Meth.*, Vol. A487 (2002) pp. 27–32.
39. A. Owens, M. Bavdaz, V. Gostilo, D. Gryaznov, A. Loupilov, A. Peacock, H. Sipila, "The X-Ray Response of InP," *Nucl. Instr. and Meth.*, Vol. A487 (2002) pp. 435–440.
40. V. Gorodynskyy, K. Zdansky, L. Pekarek, V. Malina, S. Vackova, "Ti and Mn Co-Doped Semi-Insulating InP Particle Detectors Operating at Room Temperature," *Nucl. Instr. and Meth.*, Vol. 555 (2005) pp. 288–293.
41. R. Yatskiv, K. Zdansky, L. Pekarek, "Room-Temperature Particle Detectors with Guard Rings Based on Semi-Insulating InP Co-Doped with Ti and Zn," *Nucl. Instr. and Meth.*, Vol. 598 (2009) pp. 759–763.
42. P.G Pelfer, F. Dubecky, R. Fornari M. Pikna, E. Gombia, J. Darmo, M. Krempaský, M. Sekácová, "Present Status and Perspectives of the Radiation Detectors Based on InP Materials," *Nucl. Instr. Meth.*, Vol. A458 (2001) pp. 400–405.
43. K. Zdansky, L. Pekarek, P. Kacerovsky, "Evaluation of Semi-Insulating Ti-Doped and Mn-Doped InP for Radiation Detection," *Semi. Sci. and Tech.*, Vol. 16, Issue 12 (2001) pp. 1002–1007.
44. M.R. Squillante, C. Zhou, J. Zhang, L.P. Moy, K.S. Shah, "InI Nuclear Radiation Detectors," *IEEE Trans. Nucl. Sci.*, Vol. 40 (1993) pp. 364–366.
45. T. Onodera, K. Hitomi, T. Shoji, "Fabrication of Indium Iodide X- and Gamma-Ray Detectors," *IEEE Trans. Nucl. Sci.*, Vol. 52 (2006) pp. 2056–2059.
46. P. Bhattacharya, M. Groza, Y. Cui, D. Caudel, T. Wrenn, A. Nwankwo, A. Burger, G. Slack, A.G. Ostrogorsky, "Growth of InI Single Crystals for Nuclear Detection Applications," *J. Cryst. Growth*, Vol. 312, Issue 8 (2010) pp. 1228–1232.
47. A.W. Hoffman, E. Corrales, P. Love, "2K × 2K InSb for Astronomy," *Proc. of the SPIE*, Vol. 5499 (2004) pp. 59–67.
48. A. Säynätjoki, P. Kostamo, J. Sormunen, J. Riikonen, A. Lankinen, H. Lipsanen, H. Andersson, K. Banzuzi, S. Nenonen, H. Sipilä, S. Vaijärvi, D. Lumb, "InAs Pixel Matrix Detectors Fabricated by Diffusion of Zn Utilising Metal-Organic Vapour Phase Epitaxy," *Nucl. Instr. and Meth.*, Vol. A563 (2006) pp. 24–26.

49. W.C. Harris, "InSb as a γ-Ray Detector," *Nucl. Instr. and Meth.*, Vol. 242 (1986) pp. 373–375.
50. I. Kanno, F. Yoshihara, R. Nouchi, O. Sugiura, T. Nakamura, M. Katagiri, "Cryogenic InSb Detector for Radiation Measurement," *Rev. Sci. Instr.*, Vol. 73 (2002) pp. 2533–2536.
51. I. Kanno, S. Hishiki, O. Sugiura, R. Xiang, T. Nakamura, M. Katagiri, "InSb cryogenic radiation detectors," *Nucl. Instr. and Meth.*, Vol. A568 (2006) pp. 416–420.
52. Y. Sato, Y. Morita, T. Harai, I. Kanno, "Photopeak detection by an InSb radiation detector made of liquid phase epitaxially grown crystals," *Nucl. Instr. and Meth.*, Vol. A621 (2010) pp. 383–386.
53. Y. Sato, K. Watanabe, A. Yamazaki, I. Kanno, "Charge Collection Process of a Liquid-Phase Epitaxially Grown InSb Detector," *Jpn. J. Appl. Phys.*, Vol. 50 (2011) pp. 096401–096405.
54. R.M. Park, M.B. Trofer, C.M. Rouleau, J.M. Depuydt, M.A. Haase, "P-Type ZnSe by Nitrogen Atom Beam Doping During Molecular Beam Epitaxial Growth," *Appl. Phys. Lett.*, Vol. 57 (1990) pp. 2127–2129.
55. W. Akutagawa, K. Zanio, J. Mayer, "CdTe as a Gamma-Detector," *Nucl. lnstr. and Meth.*, Vol. 55 (1967) pp. 383–385.
56. T. Takahashi, S. Watanabe, "Recent Progress in CdTe and CdZnTe Detectors," *IEEE Trans Nucl. Sci.*, Vol. 48 (2000) pp. 950–959.
57. T. Takahashi, T. Mitani, Y. Kobayashi, M. Kouda, G. Sato, S. Watanabe, K. Nakazawa, Y. Okada, M. Funaki, R. Ohno, K. Mori, "High-Resolution Schottky CdTe Diode Detector," *IEEE Trans. Nucl. Sci.*, Vol. 49 (2002) pp. 1297–1303.
58. A. Niemela, H. Sipila, V.I. Ivanov, "High-Resolution p–i–n CdTe and CdZnTe X-Ray Detectors with Cooling and Rise-Time Discrimination," *IEEE Trans. Nucl. Sci.*, Vol. 43 (1996) pp. 1476–1480.
59. A. Khusainov, R. Arlt, P. Siffert, "Performance of a High Resolution CdTe and CdZnTe P–I–N Detectors," *Nucl. Instr. and Meth.*, Vol. A380 (1996) pp. 245–251.
60. A. Khusainov, J.S. Iwanczyk, B.E. Patt, A.M. Pirogov, D.T. Voa, P.A. Russo, "Approaching Cryogenic Ge Performance with Peltier Cooled CdTe," *Proc. SPIE*, Vol. 4507 (2001) pp. 50–56.
61. M. Mahdavi, K.L. Giboni, S. Vajda, J.S. Schweitzer, J.A. Truax, "First Year PIDDP Report On Gamma-Ray and X-Ray Spectroscopy X-Ray Remote Sensing and in Situ Spectroscopy for Planetary Exploration Missions and Gamma-Ray Remote Sensing and in Situ Spectroscopy for Planetary Exploration Missions," NASA document ID 19950009501 (1994).

62. R.B. James, T.E. Schlesinger, J. Lund, M. Schieber, "$Cd_{1-x}Zn_xTe$ Spectrometers for Gamma and X-Ray Applications." In *Semiconductors for Room Temperature Nuclear Detection Applications*, eds. T.E. Schlesinger, R.B. James, Academic Press, New York (1995) pp.335–384.
63. A.W. Webb, S.B. Quadri, E.R. Carpenter, E.F. Skelton, "Effects of Pressure on $Cd_{1-x}Zn_xTe$ Alloys ($0 \le x < 0.5$)," *J. Appl. Phys.*, 61 (1987) pp. 2492–2494.
64. J.F. Butler, C.L. Lingren, F.P. Doty, "$Cd_{1-x}Zn_xTe$ Gamma Ray Detectors," *IEEE Trans. Nucl. Sci.*, Vol. 39 (1992) pp. 605–609.
65. U. Egarievwe, L. Salary, K.T. Chen, A. Burger, R.B. James, "Performances of CdTe and $Cd_{1-x}Zn_xTe$ Gamma-Ray Detectors at Elevated Temperatures," *Proc. SPIE*, Vol. 2305 (1994) pp. 167–173.
66. D. Olega, J. Faurie, S. Sivananthan, P. Raccah, "Optoelectronic Properties of $Cd_{1-x}Zn_xTe$ Films Grown by Molecular Beam Epitaxy on GaAs Substrates," *Appl. Phys. Lett.*, Vol. 47 (1985) pp.1172–1174.
67. J.E. Toney, T.E. Schlesinger, R.B. James, "Optimal Bandgap Variants of $Cd_{1-x}Zn_xTe$ for High-Resolution X-Ray and Gamma-Ray Spectroscopy," *Nucl. Inst. and Meth.*, Vol. A428 (1999) pp. 14–24.
68. A. Owens, M. Bavdaz, H. Andersson, T, Gagliardi, M. Krumrey, S. Nenonen, A. Peacock, I. Taylor, "The X-Ray Response of CdZnTe," *Nucl. Instr. and Meth.*, Vol. A484 (2002) pp. 242–250.
69. J. Frey, R. Frey, C. Flytzanis, R. Triboulet, "Theoretical and Experimental Investigation of Nonlinear Faraday Processes in Diluted Magnetic Semiconductors," *J. Opt. Soc. Am. B*, Vol. 9, No. 1 (1992) pp. 132–142.
70. V.V. Fedorov, W. Mallory, S.B. Mirov, U. Hőmmerich, S.B. Trivedi, W. Palosz, "Iron-Doped $Cd_xMn_{1-x}Te$ Crystals for Mid-IR Room-Temperature Lasers," *J. of Cryst. Growth*, Vol. 310 (2008) pp. 4438–4442.
71. A. Burger, K. Chattopadhyay, H. Chen, J.O. Ndap, X. Ma, S. Trivedi, S.W. Kutcher, R. Chen, R.D. Rosemeier, "Crystal Growth, Fabrication and Evaluation of Cadmium Manganese Telluride Gamma Ray Detectors," *J. of Cryst. Growth*, Vol. 198/199 (1999) pp. 872–876.
72. A. Mycielski, A. Burger, M. Sowinska, M. Groza, A. Szadkowski, P. Wojnar, B. Witkowska, W. Kaliszek, P. Siffert, "Is the (Cd,Mn)Te Crystal a Prospective Material for X-Ray and γ-Ray Detectors?" *Phys. Stat. Sol. (c)*, Vol. 2, No. 5, (2005) pp. 1578–1585.

73. R. Triboulet, A. Heurtel, J. Rioux, "Twin-Free (Cd, Mn)Te Substrates," *J. Cryst. Growth*, Vol. 101 (1990) pp. 131–134.
74. A. Mycielski, D. Kochanowska, M. Witkowska, R.J. Baran, A. Szadkowski, B. Witkowska, W. Kaliszek, B. Kowalski, A. Reszka, P. Łach, K. Izdebska, A. Suchocki, R. Jakieła, V. Domukhovski, T. Wojtowicz, M. Wiater, M. Węgrzycki, Ł. Kilański, "Studies of (Cd,Mn)Te Crystals as a Material for X- and Gamma Ray Detectors: Where We Are?" invited paper, IEEE NSS/MIC and 17th RTSD workshop, Oct. 30–Nov. 6, Knoxville (2011).
75. J. Parkin, P.J. Sellin, A.W. Davies, A. Lohstroh, M.E. Őzsan, P. Seller, "α Particle Response of Undoped CdMnTe," *Nucl. Instr. and Meth.*, Vol. A573 (2007) pp. 220–223.
76. Y. Cui, A. Bolotnikov, A. Hossain, G. Camarda, A. Mycielski, G. Yang, D. Kochanowska, M. Witkowska-Baran, R. James, "CdMnTe in X-Ray and Gamma-Ray Detection: Potential Applications," *Proc. of the SPIE*, Vol. 7079, SPIE (2008) pp. 70790N-1–70790N-9.
77. K. Kim, S. Cho, J. Suh, J. Hong, S. Kim, "Gamma-Ray Response of Semi-Insulating CdMnTe Crystals," *IEEE Trans. Nucl. Sci.*, Vol. 56, issue 3, no. 2 (2009) pp. 858–862.
78. A. Burger, I. Shilo, M. Schieber, "Cadmium Selenide: a Promising Novel Room Temperature Radiation Detector," *IEEE Trans. Nuc. Sci.*, Vol. NS-30 (1983) pp. 368–370.
79. M. Roth, "Advantages and Limitations of Cadmium Selenide Room Temperature Gamma-Ray Ray Detectors," *Nucl. Instr. and Meth.*, Vol. A283 (1989) pp. 291–298.
80. H. Chen, M. Hayes, X. Ma, Y.-F. Chen, S.U. Egarievwe, J.O. Ndap, K. Chattopadhyay, A. Burger, J. Leist, "Physical Properties and Evaluation of Spectrometer Grade CdSe Single Crystal," *Proc. of the SPIE*, Vol. 3446 (1998) pp. 17–28.
81. A. Burger, M. Roth, M. Schieber, "The Ternary $Cd_{0.7}Zn_{0.3}Se$ Compound, a Novel Room Temperature X-Ray Detector," *IEEE Trans. Nuc. Sci.*, Vol. NS-32 (1985) pp. 556–558.
82. M. Fiederle, D. Ebling, C. Eiche, D.M. Hofmann, M. Salk, W. Stadler, K. Benz, B.K. Meyer, "Comparison of CdTe, $Cd_{0.9}Zn_{0.1}Te$ and $CdTe_{0.9}Se_{0.1}$ Crystals: Application for γ- and X-Ray Detectors," *J. Cryst. Growth*, Vol. 138 (1994) pp. 529–533.
83. K. Kim, J. Hong, S.U. Kim, "Electrical Properties of Semi-Insulating $CdTe_{0:9}Se_{0:1}$:Cl Crystal and Its Surface Preparation," *J. Cryst. Growth*, Vol. 310, Issue 1 (2008) pp. 91–95.
84. E.E. Eissler, K.G. Lynn, "Properties of Melt-Grown ZnSe Solid-State Radiation Detectors," *IEEE Trans. Nucl. Sci.*, Vol. 42 (1995) pp. 663–667.

85. T. Ando, A.B. Fowler, F. Stern, "Electronic Properties of Two-Dimensional Systems," *Rev. Mod. Phys.*, Vol. 54 (1982) pp. 437–672.
86. W. Shi, Y.J. Ding, N. Fernelius, K. Vodopyanov, "Efficient, Tunable and Coherent 0.18–5.27-THz Source Based on GaSe Crystal," *Opt. Lett.*, Vol. 27 (2002) pp. 1454–1456.
87. C. Manfredotti, R. Murri, L. Vasanelli, "GaSe as Nuclear Particle Detector," *Nucl. Instr. and Meth.*, Vol. 115 (1974) pp. 349–353.
88. E. Sakai, H. Nakatani, C. Tatsuyama, F. Takeda, "Average Energy Needed to Produce an Electron-Hole Pair in GaSe Nuclear Particle Detectors," *IEEE Trans. Nucl. Sci.*, Vol. 35, No. 1 (1988) pp. 85–88.
89. H. Nakatani, E. Sakai, C. Tatsuyama, F. Takeda, "GaSe Nuclear Particle Detectors," *Nucl. Instr. and Meth.*, Vol. A283 (1989) pp. 303–309.
90. T. Yamazaki, H. Nakatani, N. Ikeda, "Characteristics of Impurity-Doped GaSe Radiation Detectors," *Jap. J. of Appl. Phys.*, Vol. 32, Issue 4R (1993) pp. 1857–1858.
91. A. Castellano, "GaSe Detectors for X-Ray Beams," *Appl. Phys. Lett.*, Vol. 48 (1996) pp. 298–299.
92. K.C. Mandal, M. Choi, S.H. Kang, R.D. Rauh, J. Wei, H. Zhang, L. Zheng, Y. Cui, M. Groza, A. Burger, "GaSe and GaTe Anisotropic Layered Semiconductors for Radiation Detectors," *Proc. SPIE*, Vol. 6706 (2007) 67060E1–67060E10, doi:10.1117/12.739399.
93. V.N. Katerinchuk, M.Z. Kovalyuk, "Gallium Telluride Heterojunctions," *Tech. Phys. Letts.*, Vol. 25, no.1 (2007) pp. 54–55.
94. K.C. Mandal, private communication.
95. W.R. Willig, "Mercury Iodide As a Gamma-Ray Spectrometer," *Nucl. Instr. and Meth.*, Vol. 96 (1971) pp. 615–616.
96. H.L. Malm, "A Mercuric Iodide Gamma-Ray Spectrometer," *IEEE Trans. Nucl. Sci.*, Vol. 19 (1972) pp. 263–265.
97. L. van den Berg, A.E. Proctor, K.R. Pohl, "Spectral Performance of Mercuric Iodide Gamma-Ray Detectors at Elevated Temperatures," *Proc. of the SPIE*, Vol. 5198 (2004) pp. 144–149.
98. A. Owens, L. Alha, H. Andersson, M. Bavdaz, G. Brammertz, K. Helariutta, A. Peacock, V. Lämsä, S. Nenonen, "The Effects of Proton-Induced Radiation Damage on Compound-Semiconductor X-Ray Detectors," *Proc. of the SPIE*, Vol. 5501 (2004) pp. 403–411.
99. B.E. Patt, R.C. Dolin, T.M. Devore, J.M. Markakis, J.S. Iwanczyk, N. Dorri, "Radiation Damage Resistance in Mercuric Iodide X-Ray Detectors," *Nucl. Instr. and Meth.*, Vol. A299 (1990) pp. 176–181.

100. I.F. Nicolau, J.P. Joly, "Solution Growth of Sparingly Soluble Single Crystals from Soluble Complexes—III. Growth of α-HgI_2 Single Crystals from Dimethylsulfoxide Complexes," *J. Crystal Growth*, Vol. 48 (1980) pp. 61–73.
101. M. Schieber, W.F. Schnepple, L. van den Berg, "Vapor Growth of HgI_2 by Periodic Source or Crystal Temperature Oscillation," *J. Crystal Growth*, Vol. 33 (1976) pp. 125–135.
102. A.M. Gerrish, L. van den Berg, "Perspectives on Mercuric Iodide As a Radiation Detector Material for Space Measurements," Conference on the High Energy Radiation Background in Space, Cherbs, The IEEE Nuclear and Plasma Sciences Society and The Institute of Electrical and Electronic Engineers, Inc. (1998) pp. 94–98.
103. A. Owens, M. Bavdaz, G. Brammertz, M. Krumrey, D. Martin, A. Peacock, L. Tröger, "The Hard X-ray Response of HgI_2," *Nucl. Instr. and Meth.*, Vol. A479 (2002) pp. 535–547.
104. J.S. Iwanczyk, Y.J. Yang, J.G. Bradley, J.M. Conley, A.L. Albee, T.E. Economou, "Performance and Durability of HgI_2 X-Ray Detectors for Space Missions," *IEEE Trans. Nucl. Sci.*, Vol. NS-36 (1989) pp. 841–845.
105. J.S. Iwanczyk, Y.J. Wang, N. Dorri, A.-J. Dabrowski, T.E. Economou, A.L. Turkevich, "Use Of Mercuric Iodide X-ray Detectors With Alpha Backscattering Spectrometers For Space Applications," *IEEE Trans. Nucl., Sci.*, Vol. NS-38 (1991) pp. 574–579.
106. V. Marinova, I. Yanchev, M. Daviti, K. Kyritsi, A.N. Anagnostopoulos, "Electron- and Hole-Mobility of $Hg(Br_xI_{1-x})_2$ Crystals (x = 0.25, 0.50, 0.75)," *Mat. Res. Bull.*, Vol. 37 (2002) pp. 1991–1995.
107. K.S. Shah, L. Moy, J. Zhang, F. Olschner, J.C. Lund, M.R. Squillante, "$HgBr_xI_{2-x}$ Photodetectors for Use in Scintillation Spectroscopy," *Nucl. Instr. Meth. Phys.*, Vol. A322 (1992) pp. 509–513.
108. M.M. Gospodinov, D. Petrova, I.Y. Yanchev, M. Daviti, M. Manolopoulou, K.M. Paraskevopoulos, A.N. Anagnostopoulos, E.K. Polychroniadis, "Growth of Single Crystals of $Hg(Br_xI_{1-x})_2$ and Their Detection Capability," *Journal of Alloys and Compounds*, Vol. 400 (2005) pp. 249–251.
109. R. Hofstadter, "Crystal Counters I," *Nucleonics*, April (1949) pp. 2–27.
110. I.U. Rahman, W.A. Fisher, R. Hofstadter, J. Shen, "Behavior of Thallium Bromide Conduction Counters," *Nucl. Instr. and Meth.*, Vol. A261, (1987) pp. 427–439.
111. K.S. Shah, F. Olschner, L.P. Moy, J.C. Lund, W.R. Squillante, "Characterization of Thallium Bromide Nuclear Detectors," *Nucl. Inst. and Meth.*, Vol. A299 (1990) pp. 57–59.

112. F. Olscher, K. Shah, J. Lund, J. Zhang, K. Daley, S. Medrick, W.R. Squillante, "Thallium Bromide Semiconductor X-Ray and γ-Ray Detectors," *Nucl. Instr. and Meth.*, Vol. A322 (1992) pp. 504–508.
113. K. Shah, J. Lund, F. Olschner, L. Moy, M. Squillante, "Thallium Bromide Radiation Detectors," *IEEE Trans. Nucl. Sci.*, Vol. 36 (1989) pp. 199–202.
114. K. Hitomi, T. Murayama, T. Shoji, T. Suehiro, Y. Hiratate, "Improved Spectrometric Characteristics of Thallium Bromide Nuclear Radiation Detectors," *Nucl. Instr. and Meth.*, Vol. A428 (1999) pp. 372–378.
115. K. Hitomi, O. Muroi, T. Shoji, T. Suehiro, Y. Hiratate, "Room Temperature X- and Gamma-Ray Detectors Using Thallium Bromide Crystals," *Nucl. Instr. and Meth.*, Vol. A436 (1999) pp. 160–164.
116. A. Owens, M. Bavdaz, G. Brammertz, G. Gostilo, H. Graafsma, A. Kozorezov, M. Krumrey, I. Lisjutin, A. Peacock, A. Puig, H. Sipila, S. Zatoloka, "The X-Ray Response of TlBr," *Nucl. Instr. and Meth.*, Vol. A497 (2003) pp. 370–380.
117. A.K. Shukla, S. Radmas, C.N.R. Rao, "Formation Energies of Schottky and Frenkel Defects in Thallium Halides," *J. Phys. Chem. Solids*, Vol. 34(4) (1973) pp. 761–764.
118. L.F. Voss, A.M. Conway, R.T. Graff, P.R. Beck, R J. Nikolic, A.J. Nelson, S.A.Payne, H. Kim, L. Cirignano, K. Shah, "Surface processing of TlBr for improved gamma spectroscopy," *IEEE Nucl. Sci. Symp. Conf. Rec.*, NSS/MIC (2010) pp. 3746–3748.
119. J. Vaitkus, J. Banys, V. Gostilo, S. Zatoloka, A. Mekys, J. Storasta, A. Žindulis, "Influence of Electronic and Ionic Processes on Electrical Properties of TlBr Crystals," *Nucl. Instr. and Meth.*, Vol. 546 (2005) pp. 188–191.
120. J. Vaitkus, V. Gostilo, R. Jasinskaite, A. Mekys, A. Owens, S. Zataloka, A. Zindulis, "Investigation of Degradation of Electrical and Photoelectrical Properties in TlBr Crystals," *Nucl. Instr. and Meth.*, A531 (2004) pp. 192–196.
121. G.A. Samara, "Pressure and Temperature Dependences of the Ionic Conductivities of the Thallous Halides TlCl, TlBr, and TlI," *Phys. Rev. B*, Vol. 23, no. 2 (1981) pp. 575–586.
122. K. Hitomi, T. Onodera, T. Shoji, Z. He, "Investigation of Pixellated TlBr Gamma-Ray Spectrometers with the Depth Sensing Technique," *Nucl. Instr. and Meth.*, Vol. A591 (2008) pp. 276–278.
123. H. Kim, L. Cirignamo, A. Churliov, G. Ciampi, W. Higgins, F. Olschner, K. Shah, "Developing Larger TlBr Detectors—Detector Performance," *IEEE Trans. Nucl. Sci.*, Vol. 56, issue 3, no. 2 (2009) pp. 819–823.

124. M.-H. Du, "First-Principles Study of Native Defects in TlBr: Carrier Trapping, Compensation, and Polarization Phenomenon," *J. Appl. Phys.*, Vol 108 (2010) pp. 053506-1–053506-4.
125. A.V. Churilov, G. Ciampi, H. Kim, W.M. Higgins, L.J. Cirignano, F. Olschner, V. Biteman, M. Minchello, K.S. Shah, "TlBr and $TlBr_xI_{1-x}$ Crystals for γ-Ray Detectors," *J. Cryst. Growth*, Vol. 312 (2010) pp. 1221–1227.
126. W.J. Tropf, "Cubic Thallium(I) Halides." In *Handbook of Optical Constants of Solids,* Vol. III, Ed. E.D. Palik, Elsevier Science (1997).
127. K.S. Shah, J.C. Lund, F. Olschner, J. Zhang, L.P. Moy, M.R. Squillante, W.W. Moses, and S.E. Derenzo, "$TlBr_xI_{1-x}$ Photo Detectors for Scintillation Spectroscopy," *IEEE Trans. Nucl. Sci.*, Vol. NS-41 (1994) pp. 2715–2718.
128. J.C. Lund, K.S. Shah, M.R. Squillante, L.P. Moy, F. Sinclair, G. Entine, "Properties of Lead Iodide Semiconductor Radiation Detectors," *Nucl. Instr. and Meth.*, Vol. A283 (1989) pp. 299–302.
129. V. Deich, M. Roth, "Improved Performance Lead Iodide Nuclear Radiation Detectors," *Nucl. Instr. and Meth.*, Vol. A380 (1996) pp. 169–172.
130. K.S. Shah, F. Olschner, L.P. Moy, P. Bennett, M. Misra, J. Zhang, M.R. Squillante, J.C. Lund, "Lead Iodide X-Ray Detection Systems," *Nucl. Instr. and Meth.*, Vol. A380 (1996) pp. 266–270.
131. G. Fischer, "The Electrical Resistivity of Solid and Liquid Tri-iodides of Antimony and Bismuth," *Helv. Phys. Acta*, Vol. 34 (1961) pp. 827–833.
132. D. Nason, L. Keller, "The Growth and Crystallography of Bismuth Tri-Iodide Crystals Grown by Vapor Transport," *J. Cryst. Growth*, Vol. 156, No. 3 (1995) pp. 221–226.
133. Y.N. Dmitriev, P.R. Beimett, L.J. Cirignano, M. Kiugerman, K.S. Shah, "Bismuth Iodide Crystals as a Detector Material: Some Optical and Electrical Properties," *Proc. of the SPIE*, Vol. 3768 (1999) pp. 520–529.
134. M. Matsumoto, K. Hitomi, T. Shoji, Y. Hiratate, "Bismuth Tri-Iodide Crystal for Nuclear Radiation Detectors," *IEEE Trans. Nucl. Sci.*, Vol. 49 (2002) pp. 2517–2520.
135. L. Fornaro, A. Cuña, A. Noguera, M. Pérez, L. Mussio, "Growth of Bismuth Tri-Iodide Platelets for Room Temperature X-ray Detection," *IEEE Trans. Nucl. Sci.*, Vol. 51 (2004) pp. 2461–2465.
136. M. Kocsis, "Proposal for a New Room Temperature X-Ray Detector-Thallium Lead Iodide," *IEEE Trans. Nuc. Sci.*, Vol. 47 (2000) pp. 1945–1947.

137. K. Hitomi, T. Onodero, T. Shoji, Y. Hiratate, "Thallium Lead Iodide Radiation Detectors," *IEEE Trans. Nucl. Sci.*, Vol. 50 (2003) pp. 1039–1042.
138. S. Johnsen, Z. Liu, J.A. Peters, J.-H. Song, S.C. Peter, Christos, D. Malliakas, N.K. Cho, H. Jin, A.J. Freeman, B.W. Wessels, M.G. Kanatzidis, "Thallium Chalcogenide-Based Wide-Band-Gap Semiconductors: $TlGaSe_2$ for Radiation Detectors," *Chem. Mater.*, Vol. 23, no. 12 (2011) pp. 3120–3128.
139. S. Johnsen, Z. Liu, J.A. Peters, J.-H. Song, S. Nguyen, C.D. Malliakas, H. Jin, A.J. Freeman, B.W. Wessels, M.G. Kanatzidis, "Thallium Chalcohalides for X-ray and γ-ray Detection," *J. Am. Chem. Soc.*, Vol. 133, no. 26 (2011) pp. 10030–10033.
140. V.E. Kutny, A.V. Rybka, A.S. Abyzov, L.N. Davydov, V.K. Komar, M.S. Rowland, C.F. Smith, "AlSb Single-Crystal Grown by HPBM," *Nucl. Instr. and Meth.*, Vol. A458 (2001) pp. 448–454.
141. F.V. Wald, J. Bullitt, R.O. Bell, "Bi_2S_3 as a High-Z Material for γ-Ray Detectors," *IEEE Trans. Nucl. Sci.*, Vol. NS-22 (1975) pp. 246–250.
142. M. Giles, A. Cuna, N. Sasen, M. Llorente, L. Fornaro, "Growth of Lead Bromide Polycrystalline Films," *Cryst. Res. Technol.*, Vol. 39 (2004) pp. 906–911.
143. L. Fornaro, N. Sasen, M. Pérez, A. Noguera, I. Aguiar, "Comparison of Mercuric Bromide and Lead Bromide Layers as Photoconductors for Direct X-Ray Imaging Applications," *IEEE Trans. Nucl. Sci.*, Conf. Record, (2006) pp. 3750–3754.
144. D. Natali and M. Sampietro, "Detectors Based on Organic Materials: Status and Perspectives," *Nucl. Instr. and Meth.*, Vol. A512 (2003) pp. 419–426.
145. P. Beckerle, H. Strobele, "Charged Particle Detection in Organic Semiconductors," *Nucl. Instr. and Meth.*, Vol. A449 (2000) pp. 302–310.
146. J.C. Blakesley, P.E. Keivanidis, M. Campoy-Quiles, C.R. Newman, Y. Jin, R. Speller, H. Sirringhaus, N.C. Greenham, J. Nelson, P. Stavrinou, "Organic Semiconductor Devices for X-Ray Imaging," *Nucl. Instr. and Meth.*, Vol. A580 (2007) pp. 774–777.
147. L. Loupilov, A. Sokolov, V. Gostilo, "X-Ray Peltier Cooled Detectors for X-Ray Fluorescence Analysis," *Rad. Phys. and Chem.*, Vol. 61 (2001) pp. 463–464.
148. A. Niemela, H. Sipila, "Evaluation of CdZnTe Detectors for Soft X-Ray Applications," *IEEE Trans. Nucl. Sci.*, Vol. 41 (1994) pp. 1054–1057.

149. D. McGregor, R.T. Klann, H.K. Gersch, Y.-H. Yang, "Thin-Film-Coated Bulk GaAs Detectors for Thermal and Fast Neutron Measurements," *Nucl. Instr. and Meth.*, Vol. A466 (2001) pp. 126–141.
150. D. McGregor, "Self-Biased ^{10}B-Coated High Purity Epitaxial GaAs Neutron Detectors," DOE, Phase II SBIR with SPIRE Corporation report, University of Michigan, Ann Arbor (2002).
151. R.T. Johnson, Jr., "Fast-Neutron Irradiation Effects in CdS," *J. Appl. Phys.*, Vol. 39 (1968) pp. 3517–3526.
152. M. Fasasi, M. Jung, P. Siffert, C. Teissier, "Thermal Neutron Dosimetry with Cadmium Telluride Detectors," *Radiat. Prot. Dosim.*, Vol. 23 (1988) pp. 429–431.
153. J.K. Tuli, "Thermal Neutron Capture Gamma Rays," BNLNCS-5 1647 Report, UC-34-C, Brookhaven National Laboratory (1983).
154. D.S. McGregor, J.T. Lindsay, R.W. Olsen, "Thermal Neutron Detection with $Cadmium_{1-x}$ $Zinc_x$ Telluride, Semiconductor Detectors," *Nucl. Instr. and Meth.*, Vol. A381 (1996) pp. 498–501.
155. A.G. Beyerle, K.L. Hull, "Neutron Detection with Mercuric Iodide Detectors," *Nucl. Instr. and Meth.*, Vol. A256 (1980) pp. 377–380.
156. M. Melamud, Z. Burshtein, A. Levi, M.M. Schieber, "New thermal-Neutron Solid State Electronic Based on HgI_2 Single Crystals," *Appl. Phys. Lett.*, Vol. 43 (1983) pp. 275–277.
157. Z.W. Bell, K.R. Pohl, L. van den Berg, "Neutron Detection with Mercuric Iodide," *IEEE Trans. Nucl. Sci.*, Vol. 51 (2004) pp. 1163–1165.
158. Z.W. Bell, W.G. West, K.R. Pohl, L. van den Berg, "Monte Carlo Analysis of a Mercuric Iodide Neutron/Gamma Detector," *IEEE Trans. Nucl. Sci.*, Vol. 52 (2005) pp. 2030– 2034.
159. R. Kirschman, ed., *High-Temperature Electronics*, John Wiley & Sons, Ltd. (1998).
160. K.P. Ananthanarayanan, P.J. Gielisse, A. Choudry, "Boron Compounds for Thermal Neutron Detection," *Nucl. Instr. and Meth.*, Vol. 118 (1974) pp. 45–48.
161. Y. Kumashiro, K. Kudo, K. Matsumoto, Y. Okado, T. Koshiro, "Thermal Neutron Irradiation Experiments on ^{10}BP Single-Crystal Wafers," *J. Less-Common Metals*, Vol. 143 (1988) 71–75.
162. W. Zhang, H.-G. Boyen, N. Deyneka, P. Ziemann, F. Banhart, M. Schreck, "Epitaxy of Cubic Boron Nitride on (001)-Oriented Diamond," *Nature Materials*, Vol. 2 (2003) pp. 312–315.
163. Y. Kumashiro, "Refractory Semiconductor of Boron Phosphide," *J. Mat. Res.*, Vol. 5 (1996) pp. 2933–2947.

164. B.W. Robertson, S. Adenwalla, A. Harken, P. Welsch, J.I. Brand, P.A. Dowben, J.P. Claassen, "A Class of Boron Based Solid State Neutron Detectors," *Appl. Phys. Letts.*, Vol. 80 (2002) pp. 3644–3646.
165. D.S. McGregor, J.K. Shultis, "Spectral Identification of Thin-Film-Coated and Solid-Form Semiconductor Neutron Detectors," *Nucl. Instr. and Meth.*, A517 (2004) pp. 180–188.
166. J.C. Lund, F. Olschner, F. Ahmed, K.S. Shah, "Boron Phosphide on Silicon for Radiation Detectors," *Proc. Mat. Res. Soc. Symp.*, Vol. 162 (1990) pp. 601–604.
167. J.H. Kaneko, T. Taniguchi, S. Kawamura, K. Satou, F. Fujita, A. Homma, M. Furusaka, "Development of a Radiation Detector Made of a Cubic Boron Nitride Polycrystal," *Nucl. Instr. and Meth.*, Vol. A576 (2007) pp. 417–421.
168. D.S. McGregor, T.C. Unruh, W.J. McNeil, "Thermal Neutron Detection with Pyrolytic Boron Nitride," *Nucl. Instr. and Meth.*, Vol. A591 (2008) pp. 530–533.
169. R. Tomov, R. Venn, A. Owens, A. Peacock, "The Development of a Portable Thermal Neutron Detector Based on a Boron Rich Heterodiode," *Proc. of the SPIE*, Vol. 7119 (2008) pp. 71190H-1–71190H-9.
170. A.J. Whitehead, "Diamond Radiation Detector," United States Patent application 6952016 (2005).
171. C. Mer, M. Pomorski, P. Bergonzo, D. Tromson, M. Rebisz, T. Domenech, J.-C. Vuillemin, F. Foulon, M. Nesladek, O.A. Williams, R.B. Jackman, "An Insight into Neutron Detection from Polycrystalline CVD Diamond Films," *Diamond and Related Materials*, Vol. 13, Issues 4–8 (2004) pp. 791–795.

7
Improving Performance

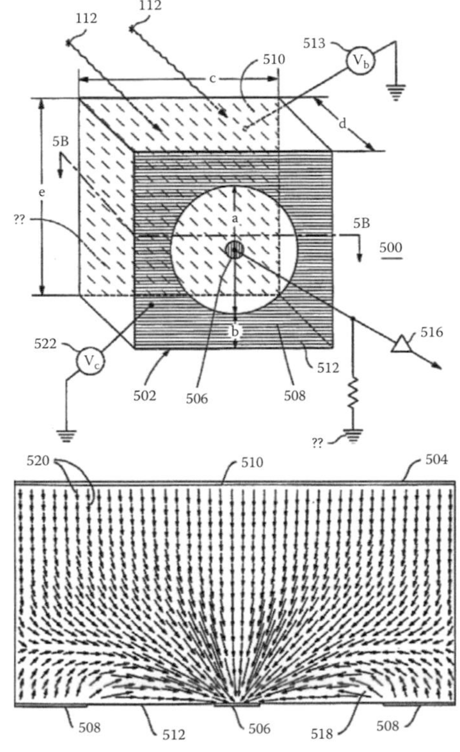

Frontispiece Lingren et al., "Semiconductor Radiation Detector with Enhanced Charge Collection," United States Patent No. US2002/0079456A1 (2002).

For conventional radiation detectors fabricated from compound semiconductors, the wide disparity between the transport properties of the electrons and holes ensures that detector performances are limited by the carrier with the poorer mobility-lifetime product ($\mu\tau$). In CdZnTe, for example, there is an order of magnitude difference between its electron and hole mobilities, 1350 $cm^2V^{-1}s^{-1}$ for electrons and 120 cm^2V^{-1} s^{-1} for holes. Coupled with the fact that the mean free drift times are 5 times smaller for holes, the mu-tau product of holes is thus 50 times worse than electrons. The resultant reduced drift lengths introduce an energy-dependent depth term into the charge collection process, which effectively limits maximum detector thicknesses to a few millimeters for spectroscopy applications and about a centimeter for counter applications. Because mobility is a fundamental material property, the only practical way of improving the $\mu\tau$ product is to increase the carrier lifetime, which in turn depends greatly on detector material quality and stoichiometry. Until the specific traps and defects can be identified and corrected, stack geometries, single carrier pulse processing (such as rise time compensation), or sensing techniques (for example coplanar grids) offer the only practical means of alleviating the problem, allowing detectors with relatively large active volumes to be constructed. In this chapter we will review carrier collection in compound semiconductor radiation detectors and examine various approaches to single carrier correction and collection techniques.

7.1 Single Carrier Collection and Correction Techniques

Single polarity sensing and correction techniques have been widely applied to overcome poor hole transport, particularly to CdZnTe detectors. For conventional detection systems, the electrons are almost fully collected for interactions occurring throughout the detector volume, but not the holes. In fact, spectral tailing can be attributed almost exclusively to holes. Part of the problem can be attributed to the much lower drift mobilities of holes, resulting in inordinately long transit

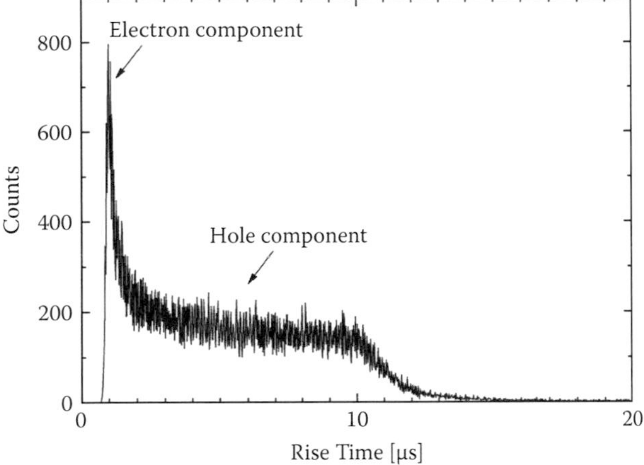

Figure 7.1 Distribution of preamplifier output pulse rise times from a 5-mm^2, 0.5-mm-thick TlBr planar detector illustrating how the wide disparity in carrier mobilities leads to their clear separation in the time domain (from [1]). For complete charge collection, a trade-off is required between an optimum shaping time to minimize electronic noise and a longer shaping time needed to collect the entire signal.

times, and partly to preferential hole trapping compounded by the long transit times, which results in the removal of holes from the signal before they can be collected. An example of the first is illustrated in Figure 7.1, in which we show the temporal distribution of rise times obtained from a 5 × 5-mm^2, 0.5-mm-thick TlBr detector [1]. We see clearly how the wide disparity in carrier mobilities leads to their separation in the time domain. The figure also illustrates that there are constraints on amplifier shaping times. If they are too long, the signal will be degraded by electronic noise,* and if they are too short, ballistic deficiency effects will occur,† also leading to a degradation of the signal. Significant improvement in spectral acuity can be achieved by either discarding the carrier with the poorer transport properties, which is the most common approach, or correcting for it. These two approaches are described below.

* Specifically parallel noise due to the leakage current which increases as the shaping time increases.

† That is, not all the signal will be collected, since in theory the shaping time should be infinitely long.

7.1.1 Rise Time Discrimination

Rise time discrimination (RTD) is a relatively simple method [2,3] to improve spectral shape and relies on the fact that, for widely different transport properties, the rise times of the electron and hole current pulses are quite different. With reference to Figure 7.2, the times taken for an electron and a hole

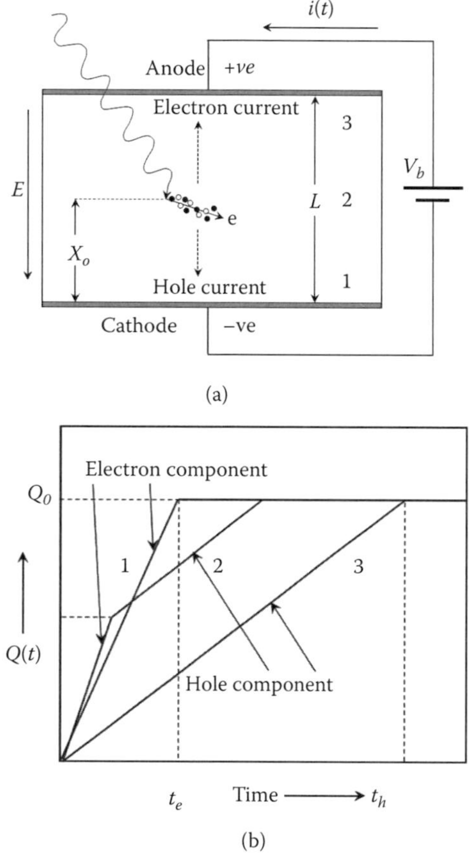

Figure 7.2 (a) Schematic of a simple planar detector showing a photoelectric interaction. Ionizing radiation absorbed in the sensitive volume generates electron–hole pairs in direct proportion to the energy deposited. These are subsequently swept toward the appropriate electrode by the electric field, E, induced by the bias voltage, V_b. (b) The time dependence of the induced signal for three different interaction sites in the detector in the absence of trapping. The fast-rising part of the signal is due to the electron component, while the slower-rising part is caused by the holes (from [4]).

to traverse the detector width L are:

$$t_e = \frac{L}{\mu_e E} \text{ and } t_h = \frac{L}{\mu_h E} \tag{7.1}$$

These are illustrated by positions 1 and 3 in Figure 7.2b. Position 1 illustrates the case where charge is produced very close to the cathode, and so the entire induced signal is due to the motion of the electrons, while in position 3 the charge is produced very close to the anode, and the signal is now due entirely to the motion of the holes. In the absence of trapping, the induced signal will build up linearly to the value of the initial charge Q_o, at a rate dependent on the hole drift velocity. The current pulse begins when the carriers induce charge on the electrodes as prescribed by the Shockley–Ramo theorem [5,6]. Given the difference in carrier mobilities, there will be two distinct current pulses, one from holes and one from electrons

$$I_{ho} = Q_o\left(\frac{\mu_h E}{L}\right), \quad I_{eo} = Q_o\left(\frac{\mu_e E}{L}\right) \tag{7.2}$$

We note from Equations 7.1 and 7.2 that, for the case when $\mu_e >> \mu_h$ (which is generally true), the current pulse induced by the electrons will have a much larger amplitude and shorter duration than that induced by the holes, as evident in Figure 7.1. For the general case, shown by position 2, the induced signal will be a composite of electron and hole components whose relative strengths will depend on the depth of the interaction.

In rise time discrimination methods, all pulses whose rise time exceeds a preset threshold are rejected, specifically all those events that would normally lie in the tail (see Figure 7.1). The net effect of rise time discrimination (RTD) is demonstrated in Figure 7.3, in which we show the measured response of a 3 × 3 × 2 mm^3 CdZnTe detector [7]. While the resolution is improved from 1 keV FWHM (full width at half maximum) at 59.54 keV to 700 eV using RTD, the efficiency is still lower than would be expected from the physical dimensions of the detector, because many counts are rejected. This is also demonstrated in Figure 7.3, in which we can see that the amplitude of the lowest peak at ~9 keV is essentially the same

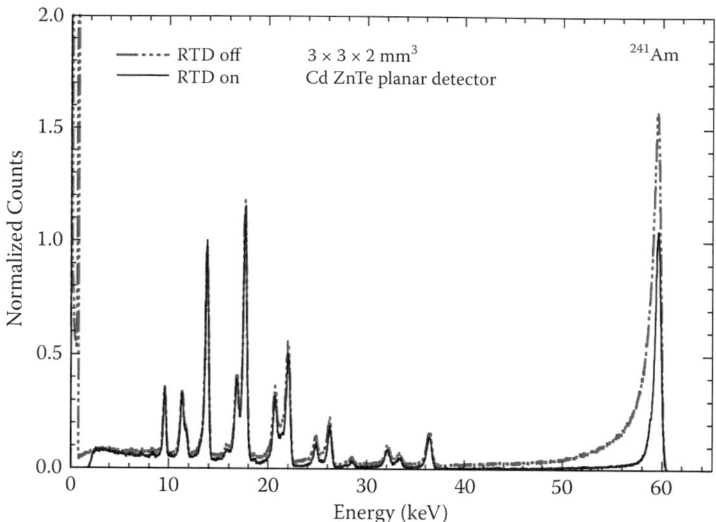

Figure 7.3 Two ^{241}Am spectra taken with a $3 \times 3 \times 2$ mm^3 CdZnTe detector, illustrating the effectiveness of rise time discrimination (RTD). From the figure we see that hole tailing is substantially reduced when RTD is employed, and in fact the FWHM energy resolution improves from 1 keV to 700 eV at 59.54 keV. However, an energy-dependent decrease in efficiency is also apparent (from [7]).

in both spectra, because the entire signal is due to the electrons. However, there is an increasing loss of counts in subsequent peaks of the RTD-on spectrum as the energy increases. In fact, the ratio of counts in the 60 keV photopeak (i.e., RTD-on/RTD-off) is ~0.35, and the ratio of peak heights is ~0.65.

7.1.2 Bi-Parametric Techniques

Since both rise time and signal amplitude depend on the depth of interaction, one can use correlative measurements to correct the measured signal for the charge lost on the way to the collecting electrode, thus improving spectroscopic performance. Such methods are termed bi-parametric. In practice, this is achieved by plotting rise time as a function of pulse amplitude and energy, from which one can generate a set of correction factors. The measured charge is then multiplied by the appropriate correction factor, yielding a corrected pulse amplitude [8]. The technique permits high resolution with little loss of sensitive volume. Verger et al., [9] report a room temperature FWHM energy resolution of 6.5% at 122 keV using $4 \times 4 \times 6\ mm^3$ CdZnTe detector with standard planar electrodes. The efficiency was > 80%.

Tada et al. [1] applied a digital pulse processing technique to a simple planar $5 \times 5\ mm^2$, 0.5-mm-thick TlBr detector to improve spectrometric performance. By applying two shaping times—one optimized to minimize electronic noise and the other to minimize ballistic deficiency effects, it was possible to derive an interaction depth correction (basically the ratio of the photopeak pulse heights from the fast and slow shapers) and use it to renormalize the charge collected with optimized system noise shaping. When operated as a single planar detector with optimum shaping, the measured FWHM energy resolution was 5.8%. This improved to 4.3% FWHM with the correction.

Bi-parametric techniques, while intuitively simple, suffer from several disadvantages. First, the electronics is relatively complicated, and second, because the pulse height deficit varies nonlinearly with rise time, the technique is only easily applied over a limited detector depth.

7.1.3 Stack Geometries

One simple way to increase active volumes and improve carrier collection is the use of stack geometries. The basic principle is illustrated in Figure 7.4. An array of single planar detectors is

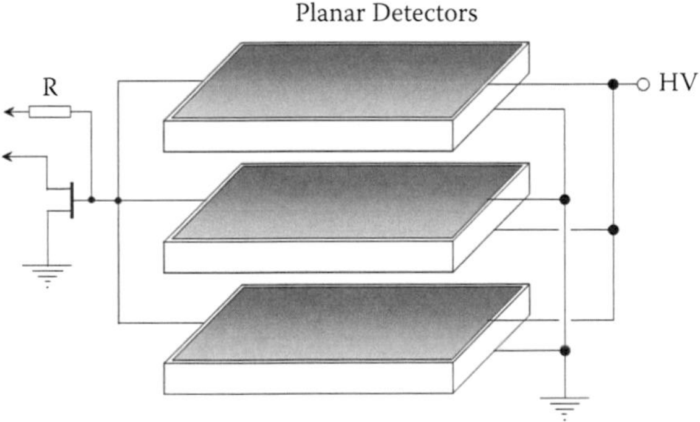

Figure 7.4 Stack detector concept. The signal path of each element is essentially or-ed. The bias is common between adjacent planes and alternate between cathodes and anodes.

stacked on top of each other. Each individual element is thin enough to allow good charge transport, but with the outputs summed so that the full volume of the stack is used. Charge collection efficiency is thus much higher than in an equivalently sized, planar detector because the travel distance to the collecting electrode can be made much less than the trapping lengths of the carriers. To date, this technique has only been implemented for CdTe because the hole mu-tau products are an order of magnitude higher than in other materials, and this particular technique relies on the efficient collection of both carriers.

Watanabe et al. [10] constructed a CdTe stacked detector with 10 large, thin, CdTe diodes, each with an area of 21.5 × 21.5 mm² and a thickness of 0.5 mm. A measured FWHM energy resolution of 7.9 keV at 662 keV was achieved at a detector operating temperature of –20°C. Redus et al. [11] describe two "stack" detectors based on three and five CdTe detector elements, each 1 mm thick and 25 mm² in area. In Figure 7.5, we show ^{137}Cs spectra taken with both detectors along with a spectrum from a single-element planar device, from which we can see a significant improvement in efficiency. For example, the three- and five-element stacks had 3.8 and 8 times as many counts in the photopeak as the single-element detector. The measured energy resolutions of the three- and five-element devices were 2.5% FWHM at 662 keV.

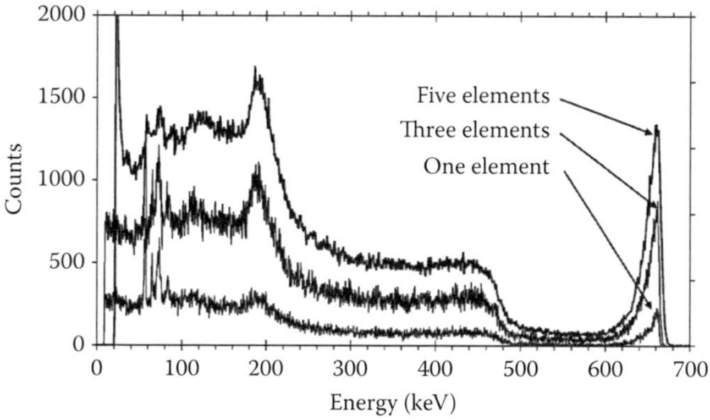

Figure 7.5 ^{137}Cs spectra obtained using the first generation stack detectors, with one, three, and five elements. The improvement in both the resolution and efficiency with the number of elements is clearly evident (from [11], © 2004 IEEE).

7.1.4 Hemispherical Detectors

Improvements in spectral acuity can also be effected by careful electrode design. One way this can be achieved is by designing a detector's geometry such that the part of the detector volume from which the electrons are collected is much larger than that of the holes. The simplest implementation is the so-called hemispherical configuration, in which the cathode is an extended electrode, of radius r_c, which surrounds a much smaller anode electrode [12]. Because the electric field varies as r^{-2}, a high field exists close to the anode and a low field near the cathode. This is illustrated in Figure 7.6a. Thus the drift length of an electron is now a function of its radial distance from the anode. Carriers generated near the cathode, which encompasses the bulk of the active volume for geometric reasons, have to traverse a predominantly low field, with the holes migrating toward the cathode and the electrons toward the anode. Because the lifetime for the electrons is typically far larger than that of the holes, the electrons have a high probability of traversing this region and arriving in the high

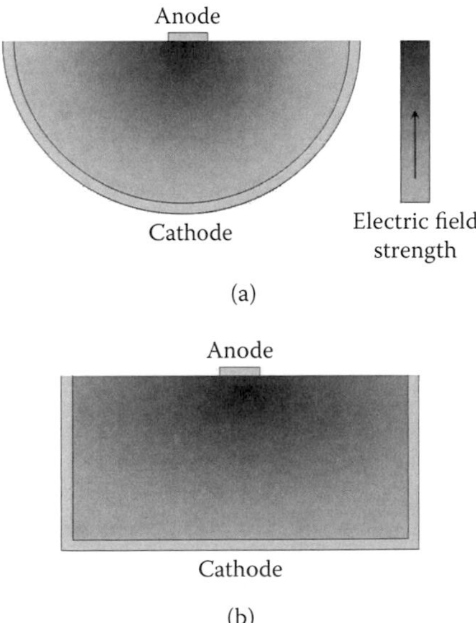

Figure 7.6 Distribution of the equipotential field lines in (a) hemispherical and (b) pseudo-hemisphere detector geometries.

field region. Once in the high field, the electrons will induce a charge dQ on the anode given by

$$dQ = (q/V)\,E(r)\,dr \tag{7.3}$$

in the absence of trapped charge. Here $E(r)$ is the electric field at r, and V is the applied bias voltage. In contrast, due to their low mobility and short lifetime, holes induce negligible charge on the cathode. For simplicity, detailed calculations usually proceed by assuming a spherical geometry [12].

Although conceptually straightforward, hemispherically shaped detectors are difficult to fabricate. However, the shape can be approximated by a cubic detector, in which the cathode covers five sides and the anode is a small pad located in the center of the sixth side (see Figure 7.6b). In Figure 7.7 we show a simulated comparison of ^{137}Cs spectra in a 0.5 cm^3 planar detector and a 0.5 cm^3 quasi-hemispherical CdZnTe detector (taken from [13]) which illustrates the effectiveness of the technique. Typically, quasi-hemispherical detector volumes range between 0.01 cm^3 to 1 cm^3. FWHM energy resolutions of ≤3% at 662 keV have been achieved with 0.5 cm^3 devices [13,14].

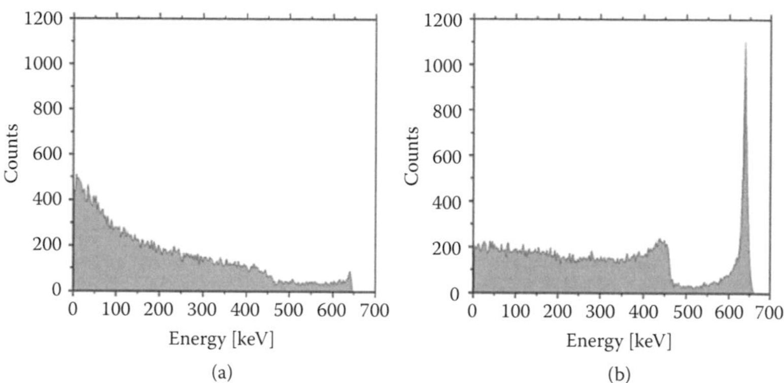

Figure 7.7 Comparison between the spectroscopic capabilities of a simple planar detector and a quasi hemispheric CdZnTe detector of the same size. Left: A simulated ^{137}Cs spectrum for a $10 \times 10 \times 5$ mm^3 planar detector. Right: simulated ^{137}Cs spectrum for a $10 \times 10 \times 5$ mm^3 quasi-hemispherical detector (from [13]).

7.1.5 Coaxial Geometries

In a variation on this theme, Lund et al. [15] constructed a CdZnTe detector in a coaxial configuration, similar to that commonly used for HPGe detectors. This can be thought of as a two-dimensional implementation of a hemispherical detector and, as such, should have similar advantages. While planar detectors offer high resolution for incident energies up to a few hundred keV, they lack sensitivity at higher energies, primarily due to the drop in electric field strength with detector thickness. In addition, large area devices suffer from poorer energy resolution, due to the increased capacitance of the collecting electrode. Coaxial configurations, on the other hand, offer a marked increase in field strength near the central collecting electrode, while maintaining a relatively low capacitance, resulting in better high-energy performance. However, near the outer surface, the electric field strength is much weaker, resulting in a relatively poor low-energy performance due to trapping.

Lund et al. [15] fabricated a 0.625 cm diameter × 0.7 cm long coaxial CdZnTe device of active volume 0.66 cm^3. However, when exposed to a ^{137}Cs radiation source, the response was found to be no better than that obtained with an equivalently sized planar detector. The response of the device could be predicted with good precision by the formulation of Sakai [16], originally developed for coaxial Ge detectors:

$$Q(V,r_o) = Q_o\left\{\int_{r_1}^{r_o}\frac{1}{r}\left[-\frac{\left(r_o^2-r^2\right)}{2\mu_e\tau_e V}\right]dr + \int_{r_o}^{r_2}\frac{1}{r}\left[-\frac{\left(r^2-r_o^2\right)}{2\mu_h\tau_h V}\right]dr\right\} \quad (7.4)$$

Here, $Q(V,r_o)$ is the charge collected as a function of radial position of the interaction, r_o, in the detector, Q_o is the total charge created by the interaction, $\mu_e\tau_e$ and $\mu_h\tau_h$ are the electron and hole mobility-lifetime products, respectively, and V is the potential difference between the contacts at r_1 and r_2. Using the measured transport parameters in Equation (7.4), Lund el al. [15] were able to show theoretically that a substantial improvement in performance could be obtained by altering the geometry of the device or increasing the lifetime of the holes.

7.2 Electrode Design and the Near-Field Effect

Both hemispherical and coaxial geometries achieve improved charge collection by shaping the electric field so that the induced charge is generated in a small high-field region centered on the collecting electrode, within which the signal is nearly independent of where the carrier is. This is the so-called near-field effect and can be understood in the context of the Shockley–Ramo theorem [5,6], which states that, regardless of bias or the presence of space charge, the change in the induced charge Q at an electrode caused by a charge q moving from x_i to x_f is given by

$$\Delta Q = \int_{x_i}^{x_f} q\vec{E}_w \cdot d\vec{x} = -q[\Phi_w(x_f) - \Phi_w(x_i)] = -q\Delta\Phi \tag{7.5}$$

where Φ_w and $\vec{E}_w$ are known as the weighting potential and weighting field, respectively. These are not the actual potential or permanent electric field, E, established by the bias,* but representations in which the dependences on factors other than charge induction have been removed. To calculate the weighting potential, one must solve Laplace's equation, $\nabla^2\Phi = 0$, for the geometry of the detector but with artificial boundary conditions. These are (a) the voltage on the electrode on which the induced charge is to be calculated is set to unity, (b) the voltages on all other electrodes are set to zero, and (c) the trapped charge within the detector volume is ignored (otherwise a formal solution of the Poisson equation would be required). The weighting field is then given by the gradient of the weighting potential, –grad Φ_w. For simple symmetric geometries, such as planar, hemispherical, or coaxial, analytical solutions may be easily deduced. For more complicated geometries, numerical solutions are usually obtained using 3-D electrostatic codes (see, for example [17]).

* For two-electrode configurations, the electric field and the weighting field have the same form.

We see from Equation (7.5) that the induced charge on the electrode is dependent only on weighting potentials. The greater the change in weighting potential the carrier undergoes on its journey to the collecting electrode, the larger the charge induced on that electrode. The instantaneous current, i, induced on the electrode is given by

$$i(t) = \frac{dQ}{dt} = q\vec{v}\vec{E}_w = q\mu\, E\, \vec{E}_w. \tag{7.6}$$

The utility of Equation (7.6) is that the field terms have now been separated into two, a static field term, E, which determines the path and the drift velocity of the carrier and a weighting field term, $\vec{E}_w$, which determines how the motion of the charge is coupled to a particular electrode. The important point to note is that these terms can be altered to improve charge induction. However, while one can shape the internal electric field, it usually affects adversely the leakage current and charge collection efficiency. The same is not true for the weighting field, which can be manipulated by careful electrode design to improve charge induction without affecting the applied field and carrier dynamics. By simulating weighting potentials for a given electrode design and detector geometry, it may be possible to find a self-consistent set of solutions in which the carrier with the poorest transport properties must traverse a region with weak signal induction and be subject to enhanced trapping, while the carrier with the best transport properties must travel through a high-field region with high signal induction in which it is less likely to be trapped. The net result is improved spectral performance, and a number of detector concepts, such as Frisch grid/ring [18], coplanar grids [19], strip [20], and small-pixel geometries [21], have evolved into practical systems by specifically exploiting the near-field effect.

7.2.1 Frisch Grid/Ring Detectors

Frisch grid/ring detectors achieve single carrier sensing in a manner emulating the Frisch grid scheme used in gas detectors to reduce the positional dependence of charge collection.

The idea has been pursued by McGregor and coworkers in a series of papers (e.g., see [18,22,23]). In a gas detector, the electron mobility is much higher than the positive ion mobility, and hence the extraction times of the electrons are considerably shorter than those for the ions. For typical microsecond integration times, the measured pulse amplitude becomes dependent on the location of the initial interaction in the chamber, and wide variations in pulse amplitude are possible. The effect was significantly reduced by Frisch [24] who incorporated a conductive grid in the chamber near the anode. The grid effectively separates the chamber into two regions—a large region between the grid and cathode where the majority of gamma ray interactions occur and a much smaller region between the grid and anode in which charge induction takes place. Gamma rays interacting in the large volume release electron–ion pairs which drift in opposite directions when an external electric field is applied. The electrons drift through the grid and into the measurement region of the device, while the slower-moving ions drift way from the grid. By appropriate design of the Frisch grid system, $\vec{E}_w$ can be optimized. As a consequence, an induced charge on the anode will result only from charge carriers moving between the conductive grid and the anode, and not from carrier motion between the cathode and the grid, thus greatly reducing the pulse shape dependence on the gamma ray interaction position. Simply stated, for semiconductors this is achieved by maximizing $\Delta\Phi_w$ for electrons, while minimizing it for holes.

While it is impractical to construct a grid inside a semiconductor detector, the concept has been successfully demonstrated using an external pair of "pseudo" grids [18]. In essence, this is an additional electrode system which is patterned onto the detector bulk just above the anode as illustrated in Figure 7.8 (left). The effectiveness of the technique is demonstrated in Figure 7.9 (left), in which is shown the response of a $5 \times 2 \times 5$ mm^3 CdZnTe device to a ^{137}Cs radioactive source [18]. With the grid disconnected, no full energy peak is observed. However, with the grid active, a full energy peak at 662 keV is clearly observed with a measured energy resolution of 6% FWHM. The major disadvantages of this design are, first, it adds complexity to the fabrication, and second,

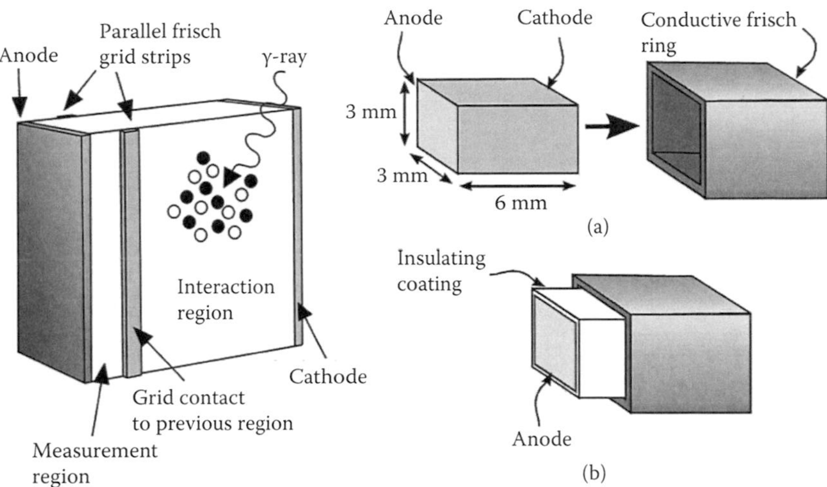

Figure 7.8 **(See color insert.)** Two implementations of the Frisch grid concept. Left: a prototype semiconductor design (from [18]). Two parallel contact strips are fabricated on the sides of the detector between the anode and cathode planes and act as a pseudo-Frisch grid. Charge carriers are excited in the interaction region, and the electrons are drifted through the parallel grids by an applied electric field. Right: schematic of a capacitively coupled Frisch grid detector, consisting of (a) a bar-shaped detector placed inside an isolated but conductive ring (b) (from [23]). This implementation effectively eliminates leakage current between the grid and the anode, because the ring is not actually connected to the detector.

the grid system is difficult to optimize in terms of noise and uniformity of response. At low energies, the system noise is dominated by the grid system itself, which introduces additional noise from leakage currents flowing between the anode and the grid. At higher energies, border effects are induced by the grid, which lead to nonuniformities in the response. These effects can be largely eliminated while still maintaining single carrier collection using so-called "noncontacting" Frisch grid schemes. For example, Montemont et al. [25] and McNeil et al. [23] proposed a capacitively coupled grid which used a metal collar as a conduction screen with which to modify charge induction. The screen is grounded, but isolated from the sides of the detector, so that it modifies only the transient electric behavior of the detector given by the weighting field, without perturbing the static applied field. A schematic

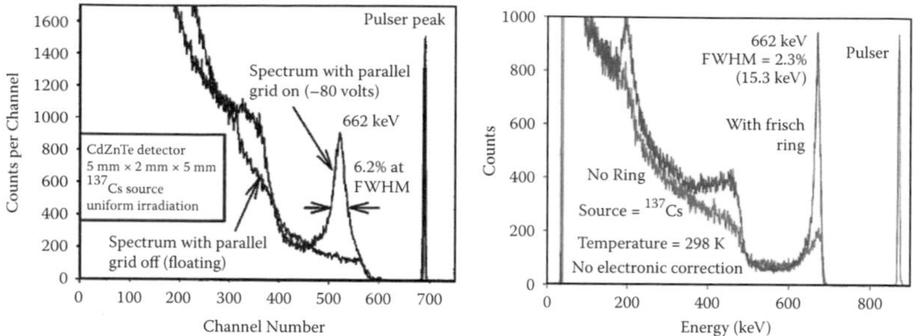

Figure 7.9 **(See color insert.)** Left: response of a $5 \times 2 \times 5$ mm^3 CdZnTe patterned Frisch grid detector to a ^{137}Cs radioactive source under full area illumination (from [18]). Shown are spectra taken with the grid turned off and on. No full energy peak is apparent when the parallel grid is off; however a full energy peak of 6.2% FWHM at 662 keV becomes obvious when the grid is activated. Right: ^{137}Cs spectra from a $3 \times 3 \times 6$ mm^3 CdZnTe device with a 5 mm insulated capacitively coupled Frisch ring (from [26]). The thin-lined blue spectrum shows the response of the device with the ring connected, and the thick-lined red spectrum with it disconnected. With the ring connected, the FWHM energy resolution is 2.3% at 662 keV.

of such a device is shown in Figure 7.8 (right). The conductive ring, when correctly biased, confines the largest change in the weighting potential to the vicinity of the anode while effectively screening induction from charge motion elsewhere. As well as being easier to fabricate than a patterned grid system, the design results in much reduced leakage currents. In Figure 7.9 (right) we show the response of a $3 \times 3 \times 6$ mm^3 CdZnTe device with a 5-mm insulated Frisch ring to a ^{137}Cs radioactive source [25]. The thick-lined red spectrum shows the results without a Frisch ring, and the thin-lined blue spectrum shows the results with the ring active. With the grid active, the measured spectral resolution was 2.3% FWHM at 662 keV. While this design eliminates ambient grid anode leakage currents at low energies, its major limitation is noise introduced by interelectrode capacity.

7.2.2 Small-Pixel Effect Detectors

For pixel detectors, Barrett et al. [21] have shown that the deleterious effects of hole trapping can be greatly reduced if the pixel dimension, w, is made small relative to the detector

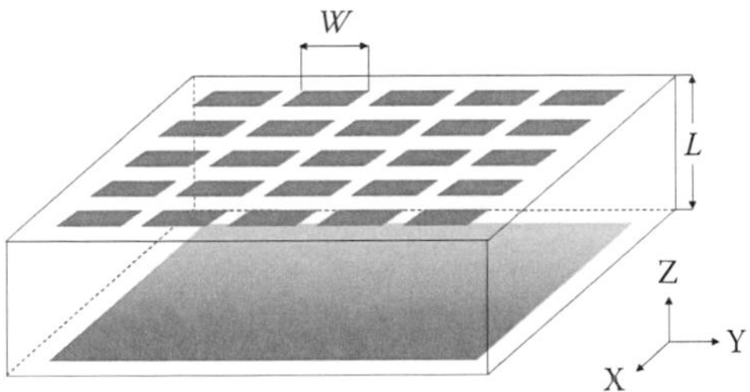

Figure 7.10 Small-pixel geometry. The detector consists of a pixelated array with pixels of size w, located above a planar cathode. The detector thickness is L.

thickness, L, as illustrated in Figure 7.10. This is generally referred to as the small-pixel effect and makes use of highly nonuniform weighting potentials that can be generated within the pixel by the electrode geometry. Eskin et al. [27] describe three methods for calculating the induced signal on a pixel. These comprise a formal solution of the three-dimensional Laplace equation, a solution based on Green's functions, and a solution based on the Shockley–Ramo theorem and weighting potentials. It is found that the size of the near-field region depends on the ratio w/L and is roughly hemispherical in shape with the characteristic size of the lateral pixel dimension. Because charge induction takes place within this region, the ratio w/L also determines the relative contribution of electrons and holes. For large values of w/L, the pixel behaves as a planar device, and charge is collected from the entire detector volume, resulting in poor spectroscopy because of a large hole component. For very small values of w/L, the near-field region is so small that spectra acuity is lost due to charge sharing with other pixels. Clearly an optimum ratio exists in which significant charge is registered only when the electrons move into the near-field close to the anode and the holes move into an almost exclusively weak field region and contribute little to the induced signal. The near-field region can be visualized in Figure 7.11, for two pixel geometries, one with $w/L = 2$ and the other with $w/L = 0.25$. In the first case (planar), the field is uniform over most of the detector, and so charge induction takes place over a large area; in fact 90% of the charge originates from 80% of the detector thickness. In the second case

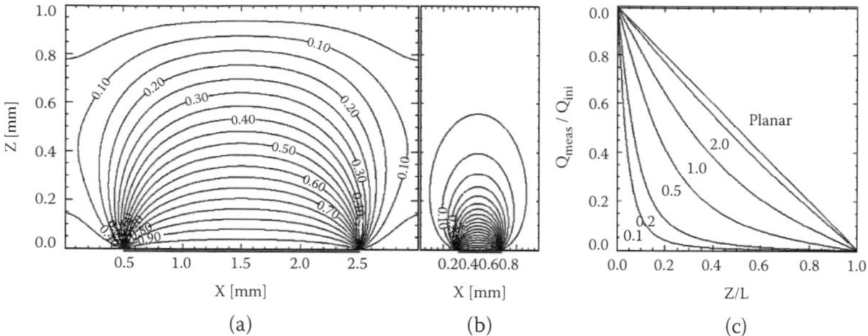

Figure 7.11 (a) and (b). Illustration of the weighting potentials $\Phi_w(x)$ for two ratios of pixel size to detector thickness, namely, $w/L = 2.0$ and 0.38 (from [28]). The contour lines in (a) and (b) are equivalent with the fraction of the total charge generated at $z = 1$ that are detected when the electrons are drawn toward the pixel at $z = 0$ mm and all the holes are lost. Thick black lines indicate the location of the pixel electrodes. In (c) the fraction of the generated charge measured as a function of photon absorption depth is shown for different pixel geometries, characterized by w/L.

the near field is much more tightly confined to the anode, and 90% of charge induction now occurs within 40% of the detector thickness from the anode.

The effect of the weighting potential on charge collection can be seen in Figure 7.11c, where the accumulation of induced charge is plotted as a function of height above the electrode for different pixel geometries. The curves can be understood as follows. Compared to the planar case ($w/L = 1$) in which the induced charge is directly proportional to the distance of the carriers from the anode, the other curves become increasingly nonlinear near the anode as $w/L \to 0$. For interactions close to the cathode ($z = 1$), both carriers move though a region in which the weighting potential changes little and so contributes little to the induced signal. The holes continue to move in decreasing potential and eventually terminate on the cathode. The electrons, on the other hand, approach the anode and eventually enter the near-field region. At this point the weighting potentials increase rapidly, as does the induced signal on the anode, which ceases when the electrons reach the anode. The measured energy-loss spectrum consists of a narrow photopeak, which becomes narrower for smaller pixels. However, the relative contributions of the electron and hole signals to the measured energy-loss spectrum depend not only on the ratio w/L but also on the trapping length of the electrons, and this ensures that there is a minimum pixel size. Less than this, the peak will again broaden, since an increasingly proportion of electrons are lost to trapping in the low-field region of the detector, before they reach the near-field region in which the signal is generated.

To demonstrate the spectral improvement that can be made by exploiting the small-pixel effect, Figure 7.12a shows a comparison between two ^{241}Am spectra. The first spectrum was obtained with a prototype 3 × 3 TlBr pixel array fabricated on monocrystalline material of size 2.7 × 2.7 × 0.8 mm^3. The pixel size was 350 × 350 microns^2, pitch 450 μm, and $w/L = 0.35$ (broken line). The second was obtained from a planar detector fabricated from the same wafer material and using the same contacting technology [28]. In this case, $w/L = 3.4$ (solid line). The energy scale and energy-loss spectrum obtained with the planar detector have been renormalized to those of the pixel detector for the neptunium Lα line at 13.64 keV. At this energy, it is assumed that X-rays interact so close to the cathode that both devices are 100% efficient. From the figure, it is clear that the gain is different for the planar and pixel

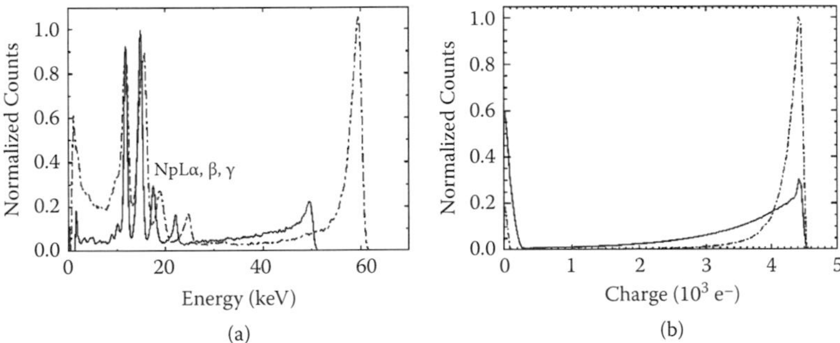

Figure 7.12 (a) Demonstration of the small-pixel effect in TlBr, in which its spectral properties are greatly influenced by pixel geometry and specifically the ratio of the pixel dimension, w, to its thickness, L. We show two ^{241}Am spectra (from [28]), one recorded by a pixel detector with a ratio $w/L = 0.35$ (broken line) and the other by a 2.7 × 2.7 × 0.8 mm^3 planar detector with a ratio $w/L = 3.4$ (solid line). (b) Modeled 60 keV spectra for the two detectors for the same operating conditions. Both spectra have been convolved with a Gaussian of FWHM 690 eV to simulate the electronic noise of the system.

detectors, and therefore their charge collection efficiencies (CCEs) are different, because the apparent measured energy losses become increasingly divergent at higher energies. At 59.54 keV, the CCE of the pixel detector is 22% larger than that of the planar device. Clearly, the tailing due to hole trapping is greatly reduced, and the amplitude of the nuclear line is increased. Note that the bias applied to the planar detector was larger than that applied to the pixel detectors (250 V as opposed to 170 V). This accounts for its superior spectral resolution at low energies.

In Figure 7.12b we show the calculated spectra for the planar and pixel detectors for the same operating conditions as the experimental values. The photo-absorption sites were distributed according to the expected absorption characteristics of TlBr. The values for the mobility and lifetimes of electrons and holes used in these simulations were $\mu_e = 20\ cm^2V^{-1}s^{-1}$; $\tau_e = 30\ \mu s$; $\mu_h = 1\ cm^2V^{-1}s^{-1}$; $\tau_h = 1\ \mu s$. These values reflect the measured $\mu\tau$ products for electrons and holes. Both spectra were convolved with a Gaussian of FWHM 690 eV to account for the additional line broadening due to electronic noise. Although we did not attempt to model the Np lines, nor the difference in gain between the two detectors, it is clear that the simulated spectra bear credible similarity to experimental measurement.

7.2.3 Drift-Strip Detectors

A variation on electrode design and bi-parametric themes is the drift-strip detector proposed by van Pamelen and Budtz-Jørgensen [20]. This device consists of a number of drift detectors with a single anode readout strip. The detector geometry is illustrated in Figure 7.13. The drift-strip electrodes are biased in such a way that the electrons move toward the anode strip. The electrodes also provide an electrostatic shield so that the movement of the positive charge carriers will induce only a small signal at the anode strip, thus reducing the sensitivity to holes. However, the signal from the planar cathode electrode is strongly influenced by the holes. From the ratio, $r_{pa} = q_{planar}/q_{anode}$, that is, the charge induced on the planar cathode divided by the charge induced on the anode strip, it is possible to correct for the residual contribution of the holes at the anode. The corrections are

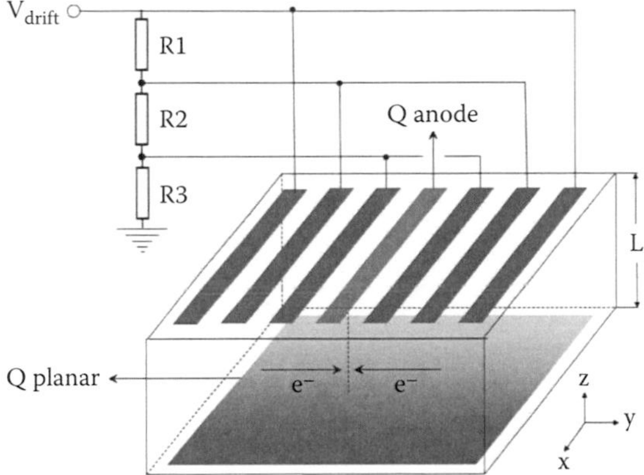

Figure 7.13 Drift-strip detector geometry. The drift-strip electrodes are biased in such a way that the electrons move to the anode strips.

obtained from the bi-parametric distribution of q_{anode} and r_{pa} as demonstrated by van Pamelen and Budtz-Jørgensen [20]. Energy resolutions of 0.6% FWHM have been obtained at 662 keV [29] using a $10 \times 10 \times 3$ mm^3 discriminator grade* CdZnTe crystal.

7.2.4 Coplanar Grid Detectors

The coplanar grid technique suggested by Luke [19] is essentially another variant of the classical Frisch grid and is again easily understood using the Shockley–Ramo theorem [5,6]. In the coplanar grid scheme [19,30], the anode electrodes take the form of inter-digitated grids that are connected to separate charge-sensitive preamplifiers (see Figure 7.14). An electric field is established in the detector bulk by applying bias to the cathode, which is a full-area planar contact located on the side opposite the grid electrodes. The two grid preamplifiers are connected to a subtraction circuit to produce a difference signal. Bias is applied between the grid electrodes, so that one of the grids preferentially collects charge. Charge motion within the bulk of the detector is sensed equally by the grid

* Commercial CdZnTe planar crystals are usually classified as counter, discriminator, or spectrometer grade, based on their FWHM energy resolution at 60 keV. Nominally, resolution should be better than 25% for counter grade, 15% for discriminator grade, and 10% for spectrometer grade.

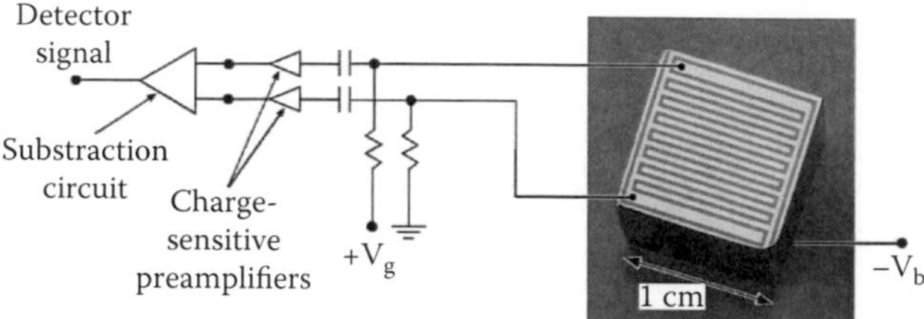

Figure 7.14 Schematic illustrating the essential components of a coplanar grid detector (from [31]).

electrodes, so that the difference signal is essentially zero. However, when charge approaches the anode, the grid signals begin to differ due to the difference in grid potentials, and a signal is registered at the output of the difference circuit. In a well-designed detector and in the absence of electron trapping, the magnitude of the difference signal is the same no matter where the charge is generated in the device. This results in a large improvement in performance for gamma-ray spectroscopy when compared to conventional planar device technology. In fact, operating the grids as a simple planar electrode, (i.e., at the same bias voltage) results in no spectroscopic signal, but when operated in the coplanar geometry the results are quite dramatic, as is illustrated in Figure 7.15.

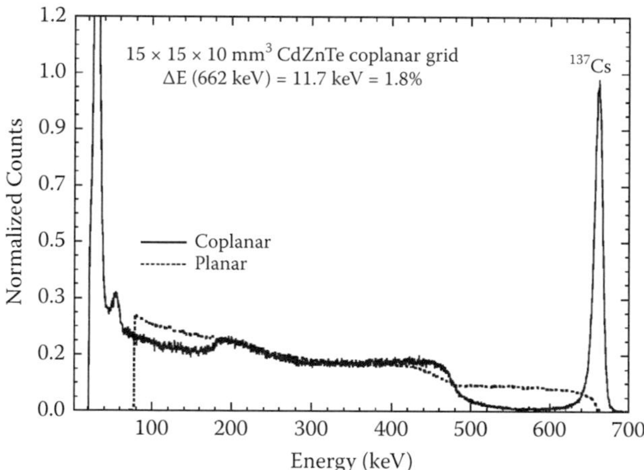

Figure 7.15 Comparison of pulse height spectra measured with a 2.25-cm^3 device (15 × 15 × 10 mm^3), when operated as a planar detector and as a coplanar grid detector [32].

The principal strengths of coplanar grid systems are that they offer large detection volumes with good energy resolution. For example, a 2.25 cm^3 coplanar grid detector fabricated out of CdZnTe had a measured FWHM energy resolution of 1.8% at 662 keV [32]. Recently, Gostilo et al. [33] reported fabricating a 10 cm^3 detector. While the performance of this device was disappointing due to crystal imperfections, spectral resolutions of 8% FWHM at 662 keV were still achieved. Although coplanar grid structures offer superlative performances at gamma-ray energies, they do require complicated lithography, material processing, and large single crystals.

7.2.5 Ring-Drift Detectors

Most recently a new type of geometry has been introduced—the so-called ring-drift detector [34]. In silicon technology, which is far more advanced, a considerable reduction in noise can be made by shaping the electric field and channeling charge to a central, small, readout anode with a capacitance which is much smaller than that of the active surface area. In a drift device [35], this is achieved using a number of concentric ring electrodes designed and biased so that the potential gradient induces a transverse electric field, pushing electrons toward the central anode. A so-called ring-drift detector follows a similar approach for compound semiconductors. However, unlike Si, the motivation is not centered on achieving low readout noise, but on channeling electrons directly to the readout node, while incurring minimum charge loss. At the same time, holes drift to their collecting electrodes further away from the near-field region of the central anode. This results in their contribution to the detector response being greatly diminished, making the detector essentially a single carrier sensing device. The device is illustrated schematically in Figure 7.16. An interesting feature of this design is that, by the proper adjustment of the ring potentials V_1 and V_2, the detector may be operated in two modes of charge collection—pseudo-hemispherical and drift modes. Measurements have been carried out on a small prototype detector fabricated on a CdZnTe crystal of size $5 \times 5 \times 1$ mm^3. On the top face, the crystal is patterned by evaporating gold electrodes consisting

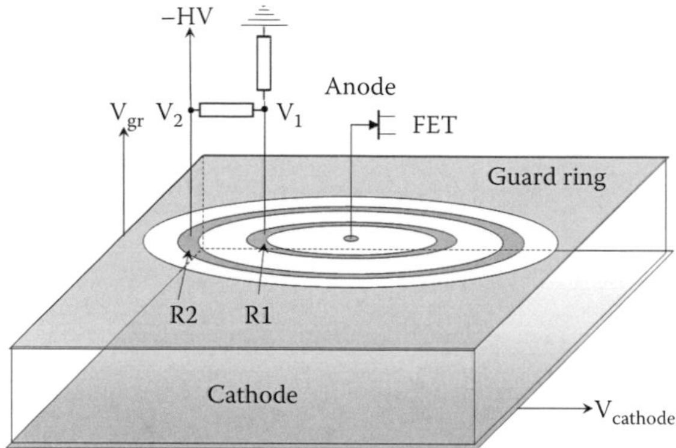

Figure 7.16 **(See color insert.)** Schematic image of the prototype ring-drift detector. The crystal has dimensions 5 × 5 × 1 mm^3. The inner anode has a diameter of 80 μm, and the centers of the other two ring electrodes, R1 and R2, are 0.19 and 0.39 mm, away from the center of the anode. The guard ring extends beyond a radius of 0.59 mm.

of a small circular anode, a double ring electrode structure, and finally a guard ring. A planar cathode is deposited on the back side. The electrode geometry is illustrated in Figure 7.16.

The device was tested in both modes using highly monochromatic X-ray pencil beams across the energy range 10.5 keV to 100 keV. The results showed that it gave simultaneously excellent energy resolution and a wide dynamic range, which makes it particularly attractive in X-ray fluorescence (XRF), electron microprobe analysis systems, and nuclear medicine applications. For, example, the measured FWHM energy resolutions under pencil beam illumination were 6.3% (drift mode) and 8.1% (hemispherical mode) at 10.5 keV, falling to 1.4% (drift mode) and 2.0% (hemispherical mode) at 60 keV. Using full-area illumination, the measured FWHMs are slightly worse. At 60 keV they are 2.0% (drift mode) and 2.5% (hemispherical mode). The detector was also found to maintain good spectral resolution up to gamma ray energies. Specifically, at 662 keV, the FWHM energy resolutions were 0.73% (drift mode) and 0.79% (hemispherical mode) [35]. The use of a small readout node ensures that electronic noise due to anode capacitance is low and independent of the active detector area, which implies that bigger detection areas can be fabricated with little loss of performance.

7.2.6 Other Implementations

If energy resolution is not critical, simpler designs, such as two-terminal* [36] and three-electrode geometries [37] may be used to advantage. In essence, these designs generally use a combination of the small-pixel effect and steering electrodes to enhance single carrier collection.

7.2.7 Combinations of Techniques

As mentioned previously, significant improvements in spectral acuity can be achieved by applying a combination of techniques. For example, Verger et al. [38] show that, using a mixed electrode scheme which combined a noncontacting Frisch grid and the pixel field effect with bi-parametric techniques, an energy resolution of 0.7% at 662 keV could be achieved with a large 1 cm^3 CdZnTe crystal. Similarly, Zhang et al. [39] achieved an energy resolution of 0.8% at 662 keV with a large 2.25 cm^3 pixelated detector by merging steering, or focusing, grids and 3D depth correction.

7.3 Discussion and Conclusions

At the present time, problems of hole trapping and material uniformity limit the useful thickness of simple detectors from about 0.2 to a few millimeters and hence their high energy performance. Until the particular defects can be identified, single carrier pulse processing and sensing techniques offer the only practical means of obviating the problem. A summary and compilation of the various single carrier sensing and correction schemes is given in Table 7.1. The performance of a particular technique is largely driven by the costs of production and ancillary support equipment, which in turn have impacts on the practicality of the device and its ease of use. We see from Table 7.1 that most techniques work reasonably well for volumes < 0.5 cm^{-3}. However above this value only

* Essentially a hemispherical detector with two asymmetric electrodes.

TABLE 7.1
Summary of the Salient Properties of the Various Single Charge Sensing/Correction Schemes (adapted from [4])

Technique	Resolution at 662 KeV, Size	Advantages	Disadvantages
Planar [40]	8%, $18 \times 18 \times 2$ mm^3	Simple	Thickness limited to a few millimeters Resolution heavily dependent on carrier transport properties
RTD [40]	1.4%, $18 \times 18 \times 2$ mm^3	Good resolution	Loss of counts in photopeaks (~60%) Thickness limited to a few millimeters Improvement limited to particular energy window Additional electronics
Bi-parametric [41]	1%, $4 \times 4 \times 5$ mm^3	High resolution Little loss of photopeak events	Complicated electronics Complicated calibration Depth correction limited to ~1 cm Expensive
Stack [11]	1%, $5 \times 5 \times 2.25$ mm^3	Relatively simple Good resolution	Volume limited by complexity of stack fabrication
Hemispherical [13]	1.6%, $10 \times 10 \times 5$ mm^3	Simple, good resolution Similar performance to planar + RTD	Fabrication difficulties Volume limited to ~ 1 cm^3
Drift-strip [29]	0.6%, $10 \times 10 \times 3$ mm^3	Very high resolution	Thickness limited to ~few millimeters

Small pixel [42]	1.2%, 20 × 20 × 10.5 mm³ 2.46 mm pixel size	Simple High resolution	Large number of pixels to achieve large area leading to complex electronics
+ steering grids and 3D sensing [39]	0.8%, 15 × 15 × 10 mm³, 11 × 11 pixels	Very high resolution	More complicated electronics than above
Virtual Frisch-grid [43]	1.4%, 5 × 5 × 14 mm³	Simple construction High resolution Large thickness possible	Essentially pixel device Small area without applying other techniques
+ "mixed" electrode and bi-parametric [38]	0.7%, 10 × 10 × 10 mm³	Very high resolution	More complicated electronics than above
Coplanar grid [32]	1.8%, 15 × 15 × 10 mm³	Good performance Large volumes ~few cubic centimeters	Complicated photolithography Max volume limited by crystal availability
Ring-drift [34]	0.7%, 5 × 5 × 1 mm³	Simple Very high resolution	Complicated spatial response Only small volumes demonstrated

Note: For the purposes of comparison, we only list only room-temperature results for CdZnTe detectors. Better energy resolution can generally be achieved with cooling. Improvements can also be achieved by employing a combination of techniques, for which two examples are given.

coplanar grid techniques can offer a substantial increase in volume while still maintaining high spectral acuity. In an analysis of a number of techniques, Luke [44] has shown that, with the exception of coplanar grid designs, electrode structures need to be specifically designed to match material characteristics and detector operating conditions, if optimum spectral performances are to be achieved. Coplanar grid detectors, on the other hand, can be optimized by a simple gain adjustment between the collecting and noncollecting grids. However, while coplanar grid detectors would seem to be the obvious candidates for all gamma-ray applications, Kozorezov et al. [45] have shown that the energy resolution of coplanar grid detectors is limited by broadening due to an intrinsic lateral inhomogeneity inherent in the design. For common designs, this means a limiting energy resolution of about 2% FWHM at 662 keV.

To achieve even higher resolutions a combination of techniques can be employed [38,39], but at the expense of increased complexity, which again emphasizes the fact that, while novel designs and processing techniques can improve energy resolution, carrier transport still limits the maximum detection volume. Thus, ultimately the quality of the material has to be improved if a substantial increase in detector volume is to be achieved, sufficient to produce an effective MeV gamma ray detector. However, as Luke [44] has pointed out, with the exception of simple planar geometries, improving hole collection efficiency may actually degrade spectral performance as single carrier collection techniques become less effective.

7.4 The Future

The requirements of gamma-ray spectroscopy should drive the future development of compound semiconductors, because material properties are critical for thick detectors but not for thin detectors. By contrast, for X-ray detection, lifetime-mobility products need only be 10^{-4} cm^{2}V^{-1} or less to ensure the efficient collection of carriers and energy resolutions near the Fano limit, provided detector thicknesses are kept below

200 μm. Thus, X-ray applications are beneficial in driving the development of the front-end electronics, and it should be borne in mind that detectors up to 200 microns thick still have nearly 100% quantum efficiency for energies up to 20 keV. For gamma-ray applications, the objective should be to produce a detector that will operate at or above room temperature, with a FWHM energy resolution of 1% or less at 500 keV and a usable active volume of 10 cm^3 or more.

7.4.1 General Requirements on Detector Material

In terms of general requirements, *Z* should be greater than 40 to yield a high stopping power. Structurally, the lattice should have a close packed geometry (such as a face centered cubic structure) to optimize density. The material should have a low dielectric constant, to ensure low capacitance and therefore system noise. For practical systems, preference should be given to binary or pseudo-alloyed binary systems, as opposed to ternary and higher order compounds, to reduce the multiplication of stoichiometry errors. Additionally, such a restriction would also clean up the response function by reducing the number of unwanted absorption and emission features in the measured energy-loss spectra. The selection criteria can be further extended to exclude most II–VI compounds in view of their propensity for toxicity, deep trapping, and polarization effects, coupled with their low melting points. The latter makes it difficult to perform the required thermal annealing to activate implanted dopants.

From an electronic point of view, the material should have an indirect bandgap to limit radiative recombination processes. The bandgap energy should be greater than 0.14 eV, so that there is no thermal generation of carriers at room temperature, but less than ~2.2 eV, based on the fact that carrier mobilities tend to drop rapidly with increasing bandgap due to polar lattice scattering. Impurity carrier scattering can also decrease appreciably mobilities and effective masses, and therefore impurity concentrations should be kept to a level of 10^{15} cm^{-3} or less (see [46,47]).

Material resistivity should be greater than 10^8 Ω cm to allow large biases to be applied, resulting in faster drift

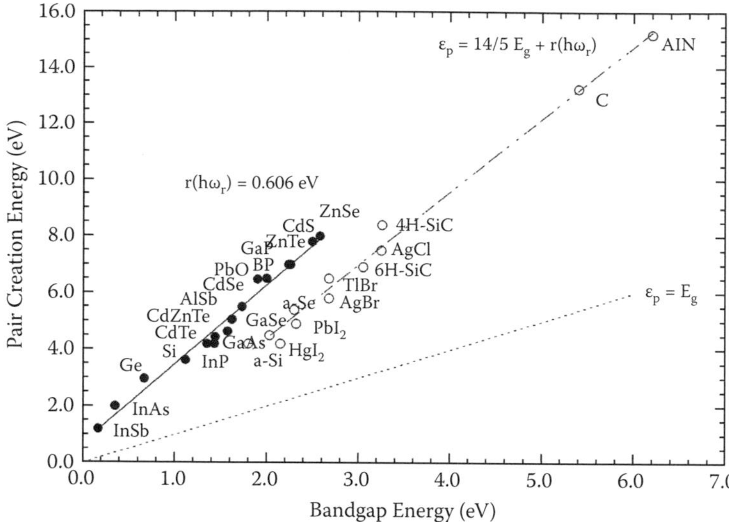

Figure 7.17 The average energy to create an electron–hole pair as a function of bandgap energy for a selection of semiconductors (from [48]). Two main bands are evident—the main branch found by Klein [49] (solid line) and the *n*-VIIB branch (dashed-dotted line). The dotted line denotes the limiting case when $\varepsilon_p = E_g$. The difference between this curve and the measured curves is due to optical phonon losses and the residual kinetic energy left over from impact ionization thresholding effects. Note the solid lines through the two branches are best-fit Klein functions of the form $14/5\, E_g + a_1$, in which a_1 is a free parameter. In order to obtain good fits to both the main and secondary branches, both AlN and diamond were fit as part of the secondary branch because they are clearly displaced from the main branch. Note also that, while the parameter $a_1 = 0.6$ for the main branch is reasonable, in that it should lie in the range $0.5 \leq a_1 \leq 1.0$ [50], the fitted value for the secondary branch is unphysical (i.e., $a_1 = -4.8$).

velocities and deeper depletion depths. For the highest Z materials, Fano noise can also be substantially reduced and hence energy resolution improved, by choosing compounds from groups n(period 6)–VII, where n = II, III, or IV (e.g., HgI_2). This is illustrated in Figure 7.17, in which we plot the bandgap energy as a function of electron–hole pair creation energy. From the graph we see that the compounds HgI_2, PbI_2, and TlBr are clearly displaced from the line describing the bulk of the semiconductors, giving a ~30% reduction in the mean pair creation energy for a given bandgap compared to the main branch. However, heavy compounds in this category tend to be mechanically soft or layered, making them difficult to handle. Consequently, they do not lend themselves to standard processing techniques, particularly array replication.

In terms of carrier transport properties, the majority carrier effective mass should be low (<0.1 m_o) to ensure high-speed operation. Note that effective mass and bandgap are related, and that smaller bandgap materials tend to have lower effective masses. So for high-speed operation and room-temperature operation bandgap energies should be close to 1.3 eV. The electron and hole mobility-lifetime products should be greater than 10^{-3} and 10^{-4} cm^2V^{-1}, respectively, to ensure good carrier transport and therefore spectral performance. This in turn places a limit on the density of typical trapping centers of $<5 \times 10^{12}$ cm^{-3}. In a study of high-performance detector operation, Luke and Amman [51] have suggested that carrier mobilities should have values of at least 500 $cm^2V^{-1}s^{-1}$ in order to ensure that the amount of charge induced on the electrodes and collected is reasonably flat with energy. Any nonuniformities will result in signal variations which in turn will degrade the detector's spectral resolution. The requirement on mu-tau products places a lower limit on carrier lifetimes of ~20 μs. Unfortunately, carrier lifetimes are not an intrinsic property of the material but are determined by the concentration of carrier-trapping defects and their capture cross sections. Therefore, while we may select compounds with the appropriate mobilities, the resultant mu-tau products are largely unpredictable. An alternative approach is to pursue a single carrier sensing and "kill off" the carrier with poor transport properties, because it will dominate the spectral response. However, the mu-tau product for this carrier cannot be made too small, because it will lead to polarization effects at high count rates and/or large energy depositions.

In Table 7.2, we summarize desirable material properties in the form of a "wish list." Because of the uncertainty in end applications and compatibility between requirements, some of the recommendations are by necessity subjective. Applying the above criteria reduces the number of potential compounds to about three (HgI_2, CdTe, and CdZnTe).

TABLE 7.2
Summary of Desirable Material Parameters for the Development of the Next Generation of Compound Semiconductor Detectors Based on Minimum Material and/or Preferred Values

Parameter	Minimum or Recommended Value	Reason/Comments
Composition	Binary (or pseudo binary)	To minimize stoichiometry errors.
Structure	Cubic (close packed)	To optimize density.
Density	>5 g cm^{-3}	To ensure good stopping power.
Growth technique	Epitaxial, FZ	Allows the possibility of integrated electronic structures.
Contact barrier height	<0.4 eV	To ensure the contacts “look” Ohmic (i.e., with a resistivity $< 10^{-3}\ \Omega\ m^2$).
Effective atomic number, Z_{eff}	>40	For high stopping power.
Resistivity	$>10^8\ \Omega$ cm	To allow high biases to be applied.
Hardness	>500 kgf mm^{-2} (Knoop scale)	Chosen to be high enough to allow the use of a range of mechanical processing and bonding technologies.
Bandgap	Indirect	To limit radiative recombination.
Bandgap energy	$1.4 < \varepsilon_g < 2.2$ eV	Lower limit for room temperature operation. Upper limit imposed by mobility losses due to polar lattice scattering.
Static dielectric constant, ε (0)	<5	To ensure low capacitance.
Ionicity	<0.3 (Phillips scale)	Ionicity should be low to prevent problems with ionic conductivity and polarization.

Majority carrier trapping center density	$< 5 \times 10^{12}$ cm^{-3}	To ensure good charge collection.
Impurity concentration	$<10^{15}$ cm^{-3}	Can adversely affect resistivity, mobilities and effective mass.
Mobility electrons	>500 cm^{2}V^{-1}s^{-1}	Charge induction considerations.
Mobility holes	10 or 500 cm^{2}V^{-1}s^{-1}	Lower value if single carrier sensing is used, upper value if not.
Majority carrier lifetime	>20 µs	To ensure reasonably high mu-tau products.
Electron mobility lifetime ($\mu\tau_e$)	10^{-3} cm^{2}V^{-1}	Minimum value really depends on application, X-rays or γ-rays.
Hole mobility lifetime ($\mu\tau_h$)	$<10^{-4}$ or 10^{-3} cm^{2}V^{-1}	($\mu\tau_h$) should be < 0.1 ($\mu\tau_e$) if single carrier sensing is employed.
Majority carrier effective mass	0.1 m_o	To ensure high speed operation.

Note: Because of the uncertainty in end applications and compatibility between requirements, some of the recommendations are by necessity subjective.

7.4.2 The Longer Term

At present, material issues limit the effective exploitation of compound semiconductors. Until these are resolved, the future lies in the controlled and directed manipulation of charge, whether through the development of heterostructures and inserted interface layers to overcome contacting problems, single carrier sensing and quantum cages to reduce trapping, or quantum heterostructures and superlattices to facilitate low-noise readout. In the preceding chapters we have concentrated purely on material developments. In the longer term, material improvements must be paralleled by corresponding developments in heterostructure and quantum heterostructure technology, provided of course that strain is not a limiting factor. Doping a semiconductor like Si or GaAs provides control over the sign and density of the charge carriers, but by combining different semiconductors in heterostructures, one gains control over far more parameters, including the bandgap, refractive index, carrier mass and mobility, and other fundamental quantities. For example, work on heterostructures may resolve the problem of contacting by the buildup of a series of semiconductor layers until it is possible to satisfy the relationship that the work function of the metal contact is less that that of the semiconductor (i.e., $\phi_m < \phi_s$) for n-type material. An alternative approach is to use a so-called “interface control layer” at the junction between the contact and the semiconductor. It has been shown that such a layer can lower the overall Schottky barrier height (SBH), allowing better contacting to wide-bandgap semiconductors. The reduction in barrier height is accomplished largely through the breakdown of one large (Schottky) barrier height into two smaller ones (a SBH and a heterojunction band-offset). HgCdTe contact layers have been used on CdTe detectors to form a heterojunction HgCdTe/CdTe/HgCdTe p-i-n detector [52].

In principal, quantum heterostructures could also facilitate ultra-low-noise operation by constructing a series of quantum valleys, which have the dimensions of the order of 0.1 nm at the contact side and millimeters or even centimeters on the intrinsic-layer side. As such, the readout would have approximately the dimensions of the readout node of a

Si CCD, which is directly responsible for their ultra-low-noise operation. The basis of operation is based on the work of Esaki and Tsu [53], who suggested that it would be possible to grow alternating layers of GaAs and AlGaAs in a periodic array to form a superlattice, which would have remarkably different electronic properties from those of bulk GaAs or AlGaAs. When a layer of GaAs is sandwiched between two "infinite" layers of AlGaAs, the carriers in the GaAs are trapped in the GaAs layer along the growth direction. In this structure, the energy levels in the well are raised in the conduction band for the electrons and lowered in the valence band for the holes. This leads to the confinement of electrons along the growth direction and characterizes the well as a structure that has a one-dimensional confinement for charged carriers. The composite of many such layers would form a bi-periodic array of rectangular quantum wires. At the present time, the first purpose-built heterostructures for X-ray applications are being reported in the literature. For example, Silenas et al. [54] have constructed an n-GaAs–p-AlGaAs graded-gap X-ray detector in which the AlGaAs layer functions as the classical absorption and detection layer and the n-GaAs layer as a carrier multiplication zone. Early results show the device is sensitive to alpha particles, and gains of up to 100 can be achieved. Lees et al. [55] tried a simpler approach and report the construction of an $Al_{0.8}Ga_{0.2}As$ p^+–p^-–n^+ diode of diameter 200 μm and thickness ~3 μm. The detector was found to be spectroscopic to low-energy X-rays with a measured FWHM energy resolution of 1.5 keV at 5.9 keV at room temperature. At elevated reverse bias levels (>20 V), avalanche multiplication was observed. In fact, gains of 3 to 4 could be achieved which, while it improved the signal-to-noise ratio, simultaneously degraded the spectroscopic performance.

Perhaps even more speculative, we note reports of porosity in III–V compounds, particularly, GaAs, GaP, and InP [56]. While one can envision photonic crystal applications, depending on cytotoxicity this also leads to the possibility of animal cell semiconductor interfacing as in the case of nanostructured Si [57]. One idea is to translate the electrical activity of neurons directly into measurable photonic signals. In this case, the porous semiconductor serves both as the adhesion substrate for the cells and as the transducer. Preliminary work has demonstrated that mammalian cells can be cultured on nanostructured Si and remain viable over time, both in terms of respiration and membrane integrity [58].

References

1. T. Tada, K. Hitomi, T. Tanaka, Y. Wu, S.Y. Kim, H. Yamazaki, K. Ishii, "Digital Pulse Processing and Electronic Noise Analysis for Improving Energy Resolution in Planar TlBr Detectors," *Nucl. Instr. and Meth.*, Vol. A638 (2011) pp. 92–95.
2. V.T. Jordanov, J.A. Pantazis, A. Huber, "Compact Circuit for Pulse Rise-Time Discrimination," *Nucl. Instr. and Meth.*, Vol. A380 (1996) pp. 353–357.
3. R. Redus, M.R. Squillante, J. Lund, *"Electronics for High Resolution Spectroscopy with Compound Semiconductors," Nucl. Instr. and Meth.*, Vol. A380 (1996) pp. 312–317.
4. A. Owens, A.G. Kozorezov, "Single Carrier Sensing in Compound Semiconductor Detectors," *Nucl. Instr. and Meth.*, Vol. A563 (2006) pp. 31–36.
5. W. Shockley, "Currents to Conductors Induced by a Moving Point Charge," *J. Appl. Phys.*, Vol. 9 (1938) pp. 635–636.
6. S. Ramo, "Currents Induced by Electron Motion," *Proc. IRE*, Vol. 27 (1939) pp. 584–585.
7. A. Owens, T. Buslaps, C. Erd, H. Graafsma, D. Lumb, E. Welter, "Hard X- and γ-Ray Measurements with a $3 \times 3 \times 3$ mm^3 CdZnTe Detector," *Nucl. Instr. and Meth.*, Vol. A563 (2005) pp. 268–273.
8. M. Richter, P. Siffert, "High Resolution Gamma Ray Spectroscopy," *Nucl. Instr. and Meth.*, Vol. A322 (1992) pp. 529–537.
9. L. Verger, J.P. Bonnefoy, F. Glasser, P. Ouvrier-Buffet, "New Developments in CdTe and CdZnTe Detectors for X- and Gamma-Ray Applications," *J. Electron., Mater.*, Vol. 26 (1997) pp. 738–744.
10. S. Watanabe, T. Takahashi, Y. Okada, G. Sato, M. Kouda, T. Mitani, Y. Kobayashi, K. Nakazawa, Y. Kuroda, M. Onishi, "CdTe Stacked Detectors for Gamma-Ray Detection," *IEEE Trans. Nucl. Sci.*, Vol. 49 (2001) pp. 1292–1296.
11. R. Redus, A. Huber, J. Pantazis, T. Pantazis, T. Takahashi, S. Woolf, "Multielement CdTe Stack Detectors for Gamma-Ray Spectroscopy," *IEEE Trans. Nucl. Sci.*, Vol. 51 (2004) pp. 2386–2394.

12. H.L. Malm, C. Canali, J.M. Mayer, M-A. Nicolet, K.R. Zanio, W. Akutagawa, "Gamma–Ray Spectroscopy with Single-Carrier Collection in High-Resistivity Semiconductors," *App. Phys. Letts.*, Vol. 26 (1975) pp. 344–346.
13. D.S. Bale, C. Szeles, "Design of High Performance CdZnTe Quasi-Hemispherical Gamma-Ray CAPture™ Plus Detectors," *Proc. of the SPIE*, Vol. 6319 (2006) pp. 1–11.
14. http://www.eurorad.com/PDF/BR_hemisph.pdf
15. J.C. Lund, R.W. Olsen, R.B. James, J.M. Van Scyoc, E.E. Eissler, M.M. Blakeley, J.B. Glick, C.J. Johnson, "Performance of a Coaxial $Cd_{1-x}Zn_xTe$ Detector," *Nucl. Instr. and Meth.*, Vol. A377 (1996) pp. 479–483.
16. E. Sakai, "Charge Collection in Coaxial Ge(Li) Detectors," *IEEE Trans. Nucl. Sci.*, Vol. NS-15 (1968) pp. 310–320.
17. Z. He, G.F. Knoll, D.K.Wehe, Y.F. Du, "Coplanar Grid Patterns and Their Effect on Energy Resolution of CdZnTe Detectors," *Nucl. Instr. Meth.*, Vol. A411 (1998) pp. 107–113.
18. D.S. McGregor, Z. He, H.A. Seifert, D.K. Wehe, R.A. Rojeski, "Single Charge Carrier Type Sensing with a Parallel Strip Pseudo-Frisch-Grid CdZnTe Semiconductor Radiation Detector," *Appl. Phys. Letts.*, Vol. 72, no. 7 (1998) pp. 792–795.
19. P.N. Luke, "Single-Polarity Charge Sensing in Ionization Detectors Using Coplanar Electrodes," *Appl. Phys. Lett.*, Vol. 65 (1994) pp. 2884–2886.
20. M. van Pamelen, C. Budtz-Jørgensen, "CdZnTe Drift Detector with Correction for Hole Trapping," *Nucl. Instr. and Meth.*, Vol. A411 (1998) pp. 197–200.
21. H. Barrett, J. Eskin, H. Barber, "Charge Transport in Arrays of Semiconductor Gamma-Ray Detectors," *Phys. Rev. Letts.*, Vol. 75 (1995) pp. 156–159.
22. D.S. McGregor, R.A. Rojeski, Z. He, D.K. Wehe, M. Driver, M. Blakely, "Geometrically Weighted Semiconductor Frisch Grid Radiation Spectrometers," *Nucl. Instr. and Meth.*, Vol. A422 (1999) pp. 164–168.
23. W.J. McNeil, D.S. McGregor, A.E. Bolotnikov, G.W. Wright, R.B. James, "Single-Charge-Carrier-Type Sensing with an Insulated Frisch Ring CdZnTe Semiconductor Radiation Detector," *Appl. Phys. Lett.*, Vol. 84 (2004) pp. 1988–1991, doi:10.1063/1.1668332.
24. O. Frisch, "Isotope Analysis of Uranium Samples by Means of Their α-Ray Groups," British Atomic Energy Report, BR-49 (1944).
25. G. Montemont, M. Arques, L. Verger, J. Rustique, "A Capacitive Frisch Grid Structure for CdZnTe Detectors," *IEEE Trans. Nucl. Sci.*, Vol. 48, No. 3 (2001) pp. 278–281.

26. D.S. McGregor, D. Schinstock, M. Harrison, A. Kargar, W. McNeil, A.E. Bolotnikov, G.W. Wright, R.B. James, "Semiconductor Radiation Detectors with Frisch Collars and Collimators for Gamma Ray Spectroscopy and Imaging," DOE NEER Grant Number 031D14498, Progress Report for year 1 covering the period June, 2003–May, 2004.
27. J. Eskin, H. Barrett, H. Barber, "Signals Induced in Semiconductor Gamma-Ray Imaging Detectors," *J. Appl. Phys.*, Vol. 85 (1999) pp. 647–659.
28. R. den Hartog, A. Owens, A.G. Kozorezov, M. Bavdaz, A. Peacock, V. Gostilo, I. Lisjutin, S. Zatoloka, "Optimization of Array Design for TlBr Imaging Detectors," *Proc. of the SPIE*, 4851 (2003) pp. 922–932.
29. I. Kuvvetli, C. Budtz-Jørgensen, L. Gerward, C.M. Stahle, "Response of CZT Drift-Strip Detector to X- and Gamma Rays," *Rad. Phys. and Chem.*, Vol. 61 (2001) pp. 457–460.
30. P.N. Luke, "Unipolar Charge Sensing with Coplanar Electrodes—Application to Semiconductor Detectors," *IEEE Trans. Nuc. Sci.*, Vol. NS-42 (1995) pp. 207–213.
31. P.N. Luke, M. Amman, C. Tindall, J.S. Lee, "Recent Developments in Semiconductor Gamma-Ray Detectors," *J. Radioanal. Nucl. Chem.*, Vol. 264 (2005) pp.145–153.
32. A. Owens, T. Buslaps, V. Gostilo, H. Graafsma, R. Hijmering, A. Kozorezov, A. Loupilov, D. Lumb, E. Welter, "Hard X- and Gamma-Ray Measurements with a Large Volume Coplanar Grid CdZnTe Detector," *Nucl. Instr. and Meth.*, Vol. A563 (2006) pp. 242–248.
33. V. Gostilo, Z. He, V. Ivanov, L. Li, A. Loupilov, I. Tsirkova, "Preliminary Results of Large Volume Multi-Pair Coplanar Grid CdZnTe Detector Fabrication," *IEEE Trans. Nucl. Sci.*, Vol. NS-35 (2005) pp. 1402–1407.
34. A. Owens, R. den Hartog, F. Quarati, V. Gostilo, V. Kondratjev, A. Loupilov, A.G. Kozorezov, J.K. Wigmore, A. Webb, E. Welter, "The Hard X-Ray Response of a CdZnTe Ring-Drift Detector," *J. Appl. Phys. Letts.*, Vol. 102 (2007) pp. 054505-1–054505-9.
35. P. Lechner, S. Eckbauer, R. Hartmann, S. Krisch, D. Hauff, R. Richter, H. Soltau, L. Strüder, C. Fiorini, E. Gatti, A. Longoni, M. Sampietro, "Silicon Drift Detectors for High Resolution Room Temperature X-Ray Spectroscopy," *Nucl. Instr. and Meth.*, Vol. A377 (1996) 346–351.
36. K. Parnham, C. Szeles, K. Lynn, R. Tjossem, "Performance Improvement of CdZnTe Detectors Using Modified Two-Terminal Electrode Geometry," *Proc. of the SPIE*, Vol. 3768 (1999) pp. 49–54.

37. H. Kim, L. Cirignano, K. Shah, M. Squillante, P. Wong, "Investigation of the Energy Resolution and Charge Collection Efficiency of Cd(Zn)Te Detectors with Three Electrodes," *IEEE Trans. Nucl. Sci.*, Vol. 51, no. 2 (2004) 1229–1234.
38. L. Verger, P. Ouvrier-Buffet, F. Mathy, G. Montemont, M. Picone, J. Rustique, C. Riffard, "Performance of a New CdZnTe Portable Spectrometric System for High Energy Applications," *IEEE Trans. Nucl. Sci.*, Vol. 52 (2005) pp. 1733–1738.
39. F. Zhang, Z. He, G.F. Knoll, D.K. Wehe, J.E. Berry, "3-D Position Sensitive CdZnTe Spectrometer Performance Using Third Generation VAS/TAT Readout Electronics," *IEEE Trans. Nucl. Sci.*, Vol. NS-52 (2005) pp. 2009–2016.
40. J.C. Lund, J.M. Van Scyoc, R.B. James, D.S. McGregor, R.W. Olsen, "Large Volume Room Temperature Gamma-Ray Spectrometers from CZT," *Nucl. Instr. and Meth.*, Vol. A380 (1996) pp. 256–261.
41. M. Mangun Panitra, A. Uritani, J. Kawarabayshi, T. Iguchi, H. Sakai, "Pulse Shape Analysis on Mixed Beta Particle and Gamma-Ray Source Measured by CdZnTe Semiconductor Detector by Means of Digital-Analog Hybrid Signal Processing Method," *J. Nucl. Sci. and Tech.*, Vol. 38 (2001) pp. 306–311.
42. H. Chen, S.A. Awadalla, K. Iniewski, P.H. Lu, F. Harris, J. Mackenzie, T. Hasanen, W. Chen, R. Redden, G. Bindley, I. Kuvvetli, C. Budtz-Jørgensen, P. Luke, M. Amman, J.S. Lee, A.E. Bolotnikov, G.S. Camarda, Y. Cui, A. Hossain, R.B. James, "Characterization of Large Cadmium Zinc Telluride Crystals Grown by Traveled Heater Method," *J. Appl. Phys.*, Vol. 103 (2008) pp. 014903-1–014903-5.
43. H. Chen, S.A. Awadalla, J. Mackenzie, R. Redden, G. Bindley, A.E. Bolotnikov, G.S. Camarda, G. Carini, R.B. James, "Characterization of Traveled Heater Method (THM) Grown $Cd_{0:9}Zn_{0:1}Te$ Crystals," *IEEE Trans. Nucl. Sci.*, Vol. 54, no 4 (2007) pp. 811–816.
44. P.N. Luke, "Electrode Configuration and Energy Resolution in Gamma-Ray Detectors," *Nucl. Instr. and Meth.*, Vol. A380 (1996) pp. 232–237.
45. A.G. Kozorezov, A. Owens, A. Peacock, J.K. Wigmore, "Carrier Dynamic and Resolution of Co-Planar Grid Radiation Detection," *Nucl. Instr. and Meth.*, Vol. A563 (2006) pp. 37–40.
46. H.E. Ruda, "A Theoretical Analysis of Electron Transport in ZnSe," *J. Appl. Phys.*, Vol. 59 (1986) pp. 1220–1231.
47. M. Lundstrom, *Fundamentals of Carrier Transport*, Cambridge University Press, 2nd ed. (2002).

48. A. Owens, A. Peacock, "Compound Semiconductor Radiation Detectors," *Nucl. Instr. and Meth.*, Vol. A531 (2004) pp. 18–37.
49. C.A. Klein, "Bandgap Dependence and Related Features of Radiation Ionization Energies in Semiconductors," *J. Appl. Phys.*, Vol. 4 (1968) pp. 2029–2033.
50. W. Que, J.A. Rowlands, "X-Ray Photogeneration in Amorphous Selenium: Geminate Versus Columnar Recombination," *Phys. Rev. B*, Vol. 51 (1995) pp. 10500–10507.
51. P.N. Luke, M. Amman, "Room-Temperature Replacement for Ge Detectors—Are We There Yet?" *IEEE Trans. Nucl. Sci.*, Vol. 54, no. 4 (2007) pp. 834–842.
52. F.J. Ryan, S.H. Shin, D.D. Edwall, J.G. Pasko, M. Khoshnevisan, C.I. Westmark, C. Fuller, "Gamma Ray Detectors with HgCdTe Contact Layers," *Appl. Phys. Lett.*, Vol. 46 (1985) pp. 274–276.
53. L. Esaki, R. Tsu, "Superlattice and Negative Differential Conductivity in Semi-conductors," *J. Res. Develop.*, Vol. 24 (1970) pp. 61–65.
54. A. Silenas, K. Pozela, L. Dapkus, V. Jasutis, V. Juciene, J. Pozela, K.M. Smith, "Graded-Gap $Al_xGa_{1-x}As$ X-Ray Detector with Collected Charge Multiplication," *Nucl. Instr. and Meth.*, Vol. 509 (2003) pp. 30–33.
55. J.E. Lees, D.J. Bassford, J.S. Ng, C.H. Tan, J.P.R. David, "AlGaAs Diodes for X-Ray Spectroscopy," *Nucl. Instr. and Meth.*, Vol. 594 (2003) pp. 202–205.
56. H. Főll, J. Carstensen, S. Langa, M. Christophersen, I.M. Tiginyanu, "Porous III-V Compound Semiconductors: Formation, Properties, and Comparison to Silicon," *Phys. Stat. Sol. A*, Vol. 197 (2003) pp. 61–70.
57. S.C. Bayliss, P.J. Harris, L.D. Buckberry, C. Rousseau, "Phosphate and Cell Growth on Nanostructured Semiconductors," *J. Mat. Sci. Lett.*, Vol. 17 (1997) pp. 737–740.
58. S.C. Bayliss, R. Heald, D.I. Fletcher, L.D. Buckberry, "The Culture of Mammalian Cells on Nanostructured Silicon," *Advanced Materials*, Vol. 11, Issue 4 (1999) pp. 318–321.

Appendix A: Table of Physical Constants

Quantity	Symbol	Value	Units
speed of light in vacuum	c	2.997 924 58	10^{8} m s^{-1}
permeability of vacuum	μ_O	1.256 637 061 4	10^{-6} H m^{-1}
permittivity of vacuum	ε_O	8.854 187 817	10^{-12} F m^{-1}
Newtonian constant of gravitation	G	6.672 59	10^{-11} $Nmkg^{-2}$
Planck constant,	h	6.626 075 5	10^{-34} Js
in electron volts		4.135 669 2	10^{-15} eV s
h/2π,	ħ	1.054 572 66	10^{-34} Js
in electron volts		6.682 122 0	10^{-16} eV s
Avogadro constant	N_A	6.022 136 7	10^{23} mol^{-1}
atomic mass unit	amu	1.660 540 2	10^{-27} kg
Faraday constant	F	9.648 530 9	C mol^{-1}
Boltzmann constant,	k	1.380 658	10^{-23} JK^{-1}
kT (300K)	kT	2.585	10^{-2} eV
Stefan–Boltzmann constant, $(\pi^2/60)k^4/\hbar^3c^2$	σ	5.670 51	10^{-8} $Jm^{-2}s^{-1}$ K^{-4}
Wien displacement law constant, $b = \lambda_{max}T$	b	2.897 756	10^{-3} m K
standard atmosphere	atm	1.013 25	10^{2} kPa

(continued)

Quantity	Symbol	Value	Units
acceleration of free fall, standard (Sèvres, France),	g	9.806 65	m s^{-2}
local—US datum	g(CB)	9.801 043	m s^{-2}
local—UK datum	g(BFS)	9.811 818	m s^{-2}
electronic charge	e	1.602 177 33	10^{-19} C
Bohr magneton	μ_B	9.274 015 4	10^{-21} erg G^{-1}
nuclear magneton	μ_N	5.050 786 6	10^{-27} JT^{-1}
fine-structure constant, $\mu_O ce^2/2h$	α	7.297 353 08	10^{-3}
inverse fine-structure constant	α^{-1}	137.035 989 5	
Rydberg constant, $m_e c\alpha^2/2h$	R_∞	1.097 373 153 4	10^7 m
in hertz		3.289 841 949 9	10^{15} Hz
in ergs		2.179 874 1	10^{-11} erg
in electron volts		13.605 698 1	eV
Bohr radius, $\alpha/4\pi R_\infty$	a_O	5.291 772 49	10^{-11} m
electron mass	m_e	9.109 389 7	10^{-31} kg
		5.485 799 03	10^{-4} amu
		0.510 999 06	MeV/c^2
proton mass	m_p	1.672 623 1	10^{-27} kg
		1.007 276 470	amu
		938.272 31	MeV/c^2
neutron mass	m_n	1.674 928 6	10^{-27} kg
		1.008 664 904	amu
		939.565 63	MeV/c^2
muon mass	m_μ	1.883 532 7	10^{-28} kg
		0.113 428 913	amu
		105.658 389	MeV/c^2

Quantity	Symbol	Value	Units
proton-electron mass ratio	m_p/m_e	1.836 152 701	10^3
electron-proton mass ratio	m_e/m_p	5.446 170 13	10^{-4}
neutron-electron mass ratio	m_n/m_e	1.838 683 662	10^3
electron specific charge	$-e/m_e$	−1.758 819 62	10^{11} C kg^{-1}
proton specific charge	e/m_p	9.578 830 9	10^7 C kg^{-1}
Compton wavelength	λ_C	2.426 310 58	10^{-12} m
proton Compton wavelength, h/m_pc	$\lambda_{C,p}$	1.321 410 02	10^{-15} m
electron magnetic moment	μ_e	9.284 770 1	10^{-3} JT^{-1}
proton magnetic moment	μ_p	1.410 607 61	10 JT^{-1}
classical electron radius, $\alpha^2 a_O$	r_e	2.817 940 92	10^{-15} m
Thomson cross-section, $(8\pi/3)(r_e)^2$	σ_e	6.652 461 6	10^{-29} m

A comprehensive and up to date listing of fundamental constants can be found in: P.J. Mohr, B.N. Taylor, "CODATA Recommended Values of the Fundamental Physical Constants: 2006," *Rev. Mod. Phys.*, 80 (2008) pp. 633–730. The data are also available on the WWW at http://physics.nist.gov/cuu/Constants/index.html

Appendix B: Units and Conversions

Parameter	Unit	Unit
Length		
	1 meter = $1.000\ 00 \times 10^{2}$	centimeter
	1 light year = $9.460\ 53 \times 10^{15}$	meter
	1 parsec = $3.085\ 68 \times 10^{16}$	meter
	1 ångstrom = $1.000\ 01 \times 10^{-10}$	meter
	1 ångstrom = $1.000\ 01 \times 10^{-8}$	centimeter
	1 micron = $1.000\ 00 \times 10^{-6}$	meter
	1 fermi = $1.000\ 00 \times 10^{-15}$	meter
	1 nautical mile = $1.852\ 00 \times 10^{3}$	meter
	1 statute mile = $1.609\ 34 \times 10^{3}$	meter
	1 astrono unit (AU) = $1.495\ 99 \times 10^{11}$	meter
	1 parsec = 3.086×10^{16}	meter
	1 solar radius = $6.959\ 90 \times 10^{8}$	meter
	1 inch = $2.540\ 00 \times 10^{-2}$	meter
Mass		
	1 kilogram = $1.000\ 00 \times 10^{3}$	gram
	1 atomic mass unit (amu) = $1.660\ 54 \times 10^{-24}$	gram

(continued)

Parameter	Unit	Unit
	1 atomic mass unit (amu) = $1.660\ 54 \times 10^{-27}$	kilogram
	1 solar mass = $1.989\ 10 \times 10^{33}$	gram
	1 solar mass = $1.989\ 10 \times 10^{30}$	kilogram
	1 gram = $6.022\ 14 \times 10^{23}$	atomic mass unit (amu)
	1 gram = $5.027\ 40 \times 10^{-34}$	solar mass
	1 kilogram = $6.022\ 14 \times 10^{26}$	atomic mass unit (amu)
	1 kilogram = $5.027\ 40 \times 10^{-31}$	solar mass
	1 kilogram = 2.204 62	pound
	1 kilogram = $3.527\ 40 \times 10$	ounce
	1 pound = $4.535\ 92 \times 10^{-1}$	kilogram
	1 pound = $1.600\ 00 \times 10$	ounce
	1 ounce = $2.834\ 95 \times 10$	gram
	1 gram = $3.527\ 40 \times 10^{-2}$	ounce
Energy		
	1 joule = $1.000\ 00 \times 10^{7}$	erg
	1 joule = $6.241\ 51 \times 10^{18}$	electron volt (eV)
	1 erg = $1.000\ 00 \times 10^{-7}$	joule
	1 erg = $6.241\ 51 \times 10^{11}$	electron volt (eV)
	1 electron volt = $1.602\ 18 \times 10^{-12}$	erg
	1 amu × c^2 = $9.314\ 95 \times 10^{8}$	electron volt (eV)
	1 g × c^2 = $5.609\ 59 \times 10^{32}$	electron volt (eV)
	1 calorie = $4.184\ 00 \times 10$	joule
Force		
	1 newton = $1.000\ 00 \times 10^{5}$	dyne
	1 dyne = $1.000\ 00 \times 10^{-5}$	newton
Pressure		
	1 pascal = $1.000\ 00 \times 10$	newton m^{-2}
	1 pascal = 1.019716×10^{-7}	kg force mm^{-2}
	1 Knoop = 9.80665×10^{6}	pascal
	1 bar = $1.000\ 00 \times 10^{6}$	dyne cm^{-2}
	1 bar = $9.869\ 23 \times 10^{-1}$	atmosphere
	1 torr = $1.333\ 22 \times 10^{-3}$	bar

Parameter	Unit	Unit
Power		
	1 watt = $1.000\,00 \times 10^{7}$	erg s^{-1}
	1 horsepower = $7.457\,00 \times 10^{2}$	watt
	1 BTU s^{-1} = $1.055\,80 \times 10^{3}$	watt
Time		
	1 minute = 6.000 00	second
	1 hour = $3.600\,00 \times 10^{3}$	second
	day = $8.640\,00 \times 10^{4}$	second
	1 tropical year = $3.155\,69 \times 10^{7}$	second
	1 tropical year = $3.652\,42 \times 10^{2}$	day
	1 second = $3.168\,88 \times 10^{-8}$	tropical year
	1 sidereal second = $9.972\,70 \times 10^{-1}$	second
	1 sidereal year = $3.652\,56 \times 10^{2}$	day
Temperature		
	T kelvin = T – 273.15	Celsius
	T kelvin = (9/5) × (T – 273.15) + 32	Fahrenheit
	T Celsius = T + 273.15	kelvin
	T Fahrenheit = (5/9) × (T – 32) + 273.15	kelvin
	T Celsius = (9/5) × T + 32	Fahrenheit
	T Fahrenheit = (5/9) × (T – 32)	Celsius
	1 electron volt = $1.160\,48 \times 10^{4}$	kelvin
	1 kelvin = $8.617\,12 \times 10^{-5}$	electron volt
Electricity and Magnetism		
	1 coulomb = $2.997\,92 \times 10^{9}$	statcoulomb
	1 coulomb m^{-3} = $2.997\,92 \times 10^{3}$	statcoul cm^{-3}
	1 ampere = 1	coulomb s^{-1}
	1 ampere = $2.997\,92 \times 10^{9}$	statampere

(*continued*)

Parameter	Unit	Unit
	1 ampere m^{-2} = $2.997\ 92 \times 10^{5}$	statamp cm^{-2}
	1 volt m^{-1} = $3.335\ 65 \times 10^{-5}$	statvolt cm^{-1}
	1 volt = $3.335\ 65 \times 10^{-3}$	statvolt
	1 ohm = $1.112\ 65 \times 10^{-12}$	statohm
	1 farad = $8.987\ 52 \times 10^{11}$	cm
	1 weber = $1.000\ 00 \times 10^{8}$	gauss cm^{2}
	1 tesla = $1.000\ 00 \times 10^{4}$	gauss
	1 ampere-turn m^{-1} = $1.256\ 64 \times 10^{-2}$	oersted
Miscellaneous		
	1 curie = $3.700\ 00 \times 10^{10}$	disintegrations s^{-1}
	1 rayleigh = $7.957\ 75 \times 10^{4}$	ph cm^{-2} s^{-1} sr^{-1}
	1 fu or jansky = $1.000\ 00 \times 10^{-26}$	watt m^{-2} Hz^{-1}
	1 jansky = $1.000\ 00 \times 10^{-8}$	erg cm^{-2} s^{-1} Hz^{-1}
	1 jansky = $2.420\ 00 \times 10^{-9}$	erg cm^{-2} s^{-1} keV^{-1}
	1 jansky = $1.509\ 00 \times 10$	keV cm^{-2} s^{-1} keV^{-1}
	1 eV = $1.239\ 85 \times 10^{4}$	ångstrom
	1 eV = $2.417\ 97 \times 10^{14}$	Hz
	1 ångstrom = $1.239\ 85 \times 10^{4}$	eV
	1 arcsec = $4.848\ 14 \times 10^{-6}$	radian
	1 arcmin = $2.908\ 88 \times 10^{-4}$	radian
	1 degree = $1.745\ 33 \times 10^{-2}$	radian
	1 $arcsec^{2}$ = $2.350\ 40 \times 10^{-11}$	steradian
	1 $arcmin^{2}$ = $8.461\ 70 \times 10^{-8}$	steradian
	1 deg^{2} = $3.046\ 20 \times 10^{-4}$	steradian

Appendix C: Periodic Table of the Elements

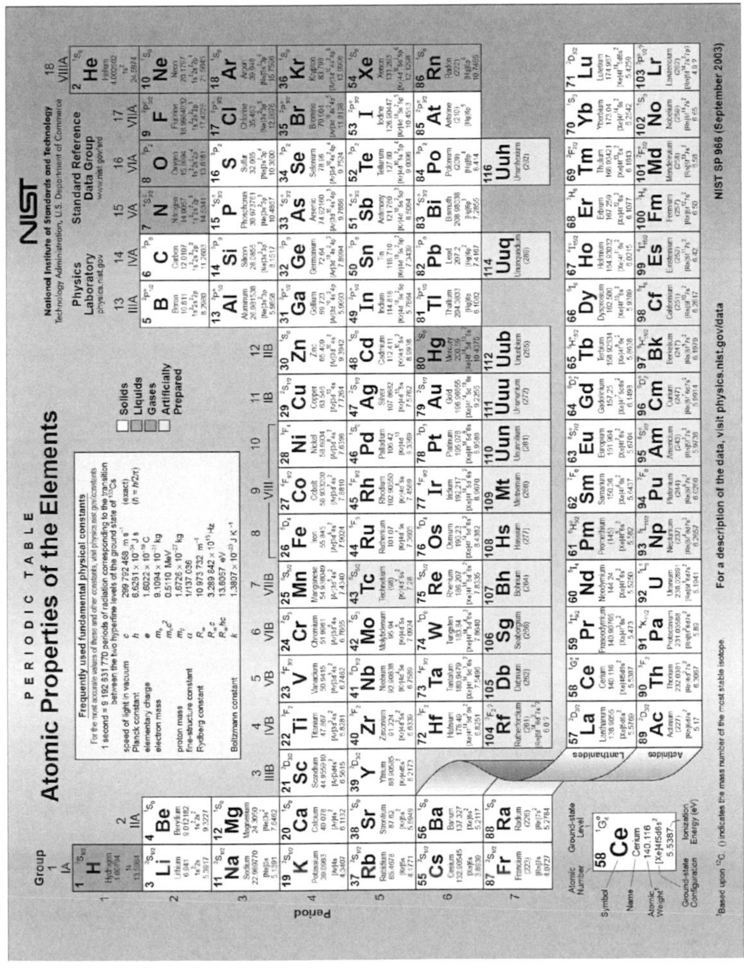

Image courtesy NIST.

Appendix D: Properties of the Elements

The following table lists the atomic weights, densities, melting and boiling points, first ionization potentials, and specific heats of the elements. The table was reproduced from the *X-Ray Data Booklet* [1], courtesy of the Lawrence Berkeley National Laboratory. Atomic weights apply to elements as they exist naturally on earth or, in the cases of radium, actinium, thorium, protactinium, and neptunium, to the isotopes with the longest half-lives. Values in parentheses are the mass numbers for the longest-lived isotopes. Specific heats are given for the elements at 25°C. Densities for solids and liquids are given at 20°C unless otherwise indicated by a superscript temperature (in °C); densities for the gaseous elements are for the liquids at their boiling points.

References

1. D. Vaughan, Ed., *X-Ray Data Booklet*, Lawrence Berkeley National Laboratory PUB-940, University of California, Berkeley, CA (1986).

Atomic Number (Z)	Element	Atomic Weight	Density (g/cm³)	Melting Point (°C)	Boiling Point (°C)	Ionization Potential (eV)	Specific Heat (cal/g·K)
1	Hydrogen	1.00794	0.0708	–259.14	–252.87	13.598	3.41
2	Helium	4.00260	1.122	–272.2	–268.934	24.587	1.24
3	Lithium	6.941	0.533	180.54	1342	5.392	0.834
4	Beryllium	9.01218	1.845	1278	2970	9.322	0.436
5	Boron	10.81	2.34	2079	2550[c]	8.298	0.245
6	Carbon	12.011	2.26	3550	3367[c]	11.260	0.170
7	Nitrogen	14.0067	0.81	–209.86	–195.8	14.534	0.249
8	Oxygen	15.9994	1.14	–218.4	–182.962	13.618	0.219
9	Fluorine	18.998403	1.108	–219.62	–188.14	17.422	0.197
10	Neon	20.179	1.207	–248.67	–246.048	21.564	0.246
11	Sodium	22.98977	0.969	97.81	882.9	5.139	0.292
12	Magnesium	24.305	1.735	648.8	1090	7.646	0.245
13	Aluminum	26.98154	2.6941	660.37	2467	5.986	0.215
14	Silicon	28.0855	2.32^{25}	1410	2355	8.151	0.168
15	Phosphorus	30.97376	1.82	44.1	280	10.486	0.181
16	Sulfur	32.06	2.07	112.8	444.674	10.360	0.175
17	Chlorine	35.453	1.56	–100.98	–34.6	12.967	0.114
18	Argon	39.948	1.40	–189.2	–185.7	15.759	0.124

19	Potassium	39.0983	0.860	63.25	760	4.341	0.180
20	Calcium	40.08	1.55	839	1484	6.113	0.155
21	Scandium	44.9559	2.980^{25}	1541	2831	6.54	0.1173
22	Titanium	47.88	4.53	1660	3287	6.82	0.1248
23	Vanadium	50.9415	$6.10^{18.7}$	1890	3380	6.74	0.116
24	Chromium	51.996	7.18	1857	2672	6.766	0.107
25	Manganese	54.9380	7.43	1244	1962	7.435	0.114
26	Iron	55.847	7.860	1535	2750	7.870	0.1075
27	Cobalt	58.9332	8.9	1495	2870	7.86	0.107
28	Nickel	58.69	8.876^{25}	1453	2732	7.635	0.1061
29	Copper	63.546	8.94	1083.4	2567	7.726	0.0924
30	Zinc	65.38	7.112^{25}	419.58	907	9.394	0.0922
31	Gallium	69.72	$5.877^{29.6}$	29.78	2403	5.999	0.088
32	Germanium	72.59	5.307^{25}	937.4	2830	7.899	0.077
33	Arsenic	74.9216	5.72	$817^{28\ \mathrm{atm}}$	613[c]	9.81	0.0785
34	Selenium	78.96	4.78	217	684.9	9.752	0.0767
35	Bromine	79.904	3.11	–7.2	58.78	11.814	0.0537
36	Krypton	83.80	2.6	–156.6	–152.30	13.999	0.059
37	Rubidium	85.4678	1.529	38.89	686	4.177	0.0860
38	Strontium	87.62	2.54	769	1384	5.695	0.0719

(continued)

Atomic Number (Z)	Element	Atomic Weight	Density (g/cm3)	Melting Point (°C)	Boiling Point (°C)	Ionization Potential (eV)	Specific Heat (cal/g·K)
39	Yttrium	88.9059	4.456[25]	1522	3338	6.38	0.0713
40	Zirconium	91.22	6.494	1852	4377	6.84	0.0660
41	Niobium	92.9064	8.55	2468	4742	6.88	0.0663
42	Molybdenum	95.94	10.20	2617	4612	7.099	0.0597
43	Technetium	(98)	11.48[a]	2172	4877	7.28	0.058
44	Ruthenium	101.07	12.39	2310	3900	7.37	0.0569
45	Rhodium	102.9055	12.39	1966	3727	7.46	0.0580
46	Palladium	106.42	12.00	1554	2970	8.34	0.0583
47	Silver	107.8682	10.48	961.93	2212	7.576	0.0562
48	Cadmium	112.41	8.63	320.9	765	8.993	0.0552
49	Indium	114.82	7.30	156.61	2080	5.786	0.0556
50	Tin	118.69	7.30	231.9681	2270	7.344	0.0519
51	Antimony	121.75	6.679	630.74	1950	8.641	0.0495
52	Tellurium	127.60	6.23	449.5	989.8	9.009	0.0481
53	Iodine	126.9045	4.92	113.5	184.35	10.451	0.102
54	Xenon	131.29	3.52	–111.9	–107.1	12.130	0.0378
55	Cesium	132.9054	1.870	28.40	669.3	3.894	0.0575
56	Barium	137.33	3.5	725	1640	5.212	0.0362

57	Lanthanum	138.9055	6.127^{25}	921	3457	5.577	0.0479
58	Cerium	140.12	6.637^{25}	799	3426	5.47	0.0459
59	Praseodymium	140.9077	6.761	931	3512	5.42	0.0467
60	Neodymium	144.24	6.994	1021	3068	5.49	0.0453
61	Promethium	(145)	7.20^{25}	1168	2460	5.55	0.0442
62	Samarium	150.36	7.51	1077	1791	5.63	0.0469
63	Europium	151.96	5.228^{25}	822	1597	5.67	0.0326
64	Gadolinium	157.25	7.8772^{25}	1313	3266	6.14	0.056
65	Terbium	158.9254	8.214	1356	3123	5.85	0.0435
66	Dysprosium	162.50	8.525^{25}	1412	2562	5.93	0.0414
67	Holmium	164.9304	8.769^{25}	1474	2695	6.02	0.0394
68	Erbium	167.26	9.039^{25}	159	2863	6.10	0.0401
69	Thulium	168.9342	9.294^{25}	1545	1947	6.18	0.0382
70	Ytterbium	173.04	6.953	819	1194	6.254	0.0287
71	Lutetium	174.967	9.811^{25}	1663	3395	5.426	0.0285
72	Hafnium	178.49	13.29	2227	4602	7.0	0.028
73	Tantalum	180.9479	16.624	2996	5425	7.98	0.0334
74	Tungsten	183.85	19.3	3410	5660	7.98	0.0322
75	Rhenium	186.207	20.98	3180	5627[b]	7.88	0.0330
76	Osmium	190.2	22.53	3045	5027	8.7	0.0310

(continued)

Atomic Number (Z)	Element	Atomic Weight	Density (g/cm^3)	Melting Point (°C)	Boiling Point (°C)	Ionization Potential (eV)	Specific Heat (cal/g·K)
77	Iridium	192.22	22.39[17]	2410	4130	9.1	0.0312
78	Platinum	195.08	21.41	1772	3827	9.0	0.0317
79	Gold	196.9665	18.85	1064.43	3080	9.225	0.0308
80	Mercury	200.59	13.522	–38.842	356.58	10.437	0.0333
81	Thallium	204.383	11.83	303.5	1457	6.108	0.0307
82	Lead	207.2	11.33	327.502	1740	7.416	0.0305
83	Bismuth	208.9804	9.730	271.3	1560	7.289	0.0238
84	Polonium	(209)	9.3	254	962	8.42	0.030
85	Astatine	(210)	—	302	337[b]	—	—
86	Radon	(222)	4.4	–71	–61.8	10.748	0.0224
87	Francium	(223)	—	27	677	—	—
88	Radium	226.0254	5	700	1140	5.279	0.0288
89	Actinium	227.0278	10.05[a]	1050	3200[b]	6.9	—
90	Thorium	232.0381	11.70	1750	4790	—	0.0281
91	Protactinium	231.0359	15.34[a]	<1600	—	—	0.029
92	Uranium	238.0289	18.92	1132.3	3818	—	0.0278
93	Neptunium	237.0482	20.21	640	3902[b]	—	—
94	Plutonium	(244)	19.80	641	3232	5.8	—

95	Americium	(243)	13.64	994	2607	6.0	
96	Curium	(247)	13.49[a]	1340	—	—	—
97	Berkelium	(247)	14[b]	—	—	—	—
98	Californium	(251)	—	—	—	—	—
99	Einsteinium	(252)	—	—	—	—	—
100	Fermium	(257)	—	—	—	—	—
101	Mendelevium	(258)	—	—	—	—	—
102	Nobelium	(259)	—	—	—	—	—
103	Lawrencium	(260)	—	—	—	—	—

[a] Calculated.
[b] Estimated.
[c] Sublimes.

Appendix E: General Properties of Semiconducting Materials

The data compiled in these tables have been gathered from many sources. Particularly useful compilations can be found in references [1–9]. Note that published values can vary widely, particularly for transport parameters. Indeed, no values exist for many materials, and so the relevant entries have been left blank. Where wide discrepancies in published data exist, average values have been used, or a judicious choice has been made. Needless to say, values quoted in these tables should be used with caution.

E.1 Table Headings, Nomenclature and Explanation

Material: The semiconductor is denoted by its chemical symbol (e.g., GaAs). We consider only the stable solid form. Where more than one form exists, we consider its thermodynamically favored allotrope at ambient temperature.

Crystal structure: The structure of a crystal is usually defined in terms of lattice points, which mark the positions of the atoms forming the basic unit cell of the crystal. Cullity [10] defines a lattice point as "*An array of points in space so arranged that each point has (statistically) identical surroundings.*" The word "statistically" is introduced to allow for solid solutions, where fractional atoms would otherwise result. The following abbreviations have been used. Amorphous material is denoted by **a**, crystalline material by **c**. For cubic lattices: **dia** denotes a diamond type structure (i.e., two intersecting face centered cubic lattices); **NaCl** denotes a sodium chloride type lattice structure—also referred to as "rock salt" structure (face centered cubic); **CsCl** a caesium chloride type structure (body centered cubic), and **ZB** for a zinc blende structure (diamond structure for binary compounds where the two atom types form two interpenetrating face centered cubic lattices). For hexagonal lattices—**W** denotes a wurtzite structure (an AB binary system of alternating tetrahedrally coordinated atoms stacked in an ABABAB pattern), and **H**, a pure hexagonal structure (containing six atoms per unit cell)—essentially the hexagonal analog of zinc blende; **ortho** refers to orthorhombic, **rhomb** to rhombohedral, **trig** to trigonal, **tetra** to tetragonal, and **layered** to a layered structure (i.e., cubic or hexagonal structured layers separated by van der Waal forces). Most elemental semiconductors crystallize in a diamond structure. Most compound semiconductors crystallize in either the **ZB** or **W** forms (see Figure E.1).

Pearson symbols, space groups: Parameters which define a crystal's structure and its underlying symmetry. While the Bravias lattice designations identify crystal types, they cannot uniquely identify particular crystals. There are several systems for classifying crystal structure, most of which are based on assigning a specific letter to each of the Bravais lattices. However, with the exception of the Pearson classification scheme, none is self-defining, and even the Pearson scheme does not uniquely define a particular crystal structure. The Pearson notation [11] is a simple and convenient scheme

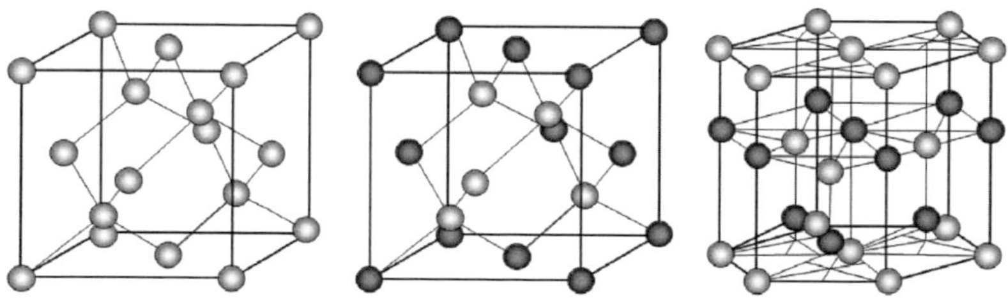

Figure E.1 The most common semiconductor lattice structures. Left to right—diamond, zinc blende, and wurtzite (adapted from [2]).

and is based on the so-called Pearson symbols, of which there are three. The first symbol is a lowercase letter designating the crystal type (i.e., a = triclinic, m = monoclinic, o = othorhombic, t = tetragonal, h = hexagonal and rhombohedral, c = cubic). The second symbol is a capital letter which designates the lattice centering (i.e., P = primitive, C = side face centered, F = all face centered, I = body centered, R = rhombohedral). Thus, the 14 unique Bravais lattices can be characterized by a two-letter mnemonic as summarized in Table 2.1. The third Pearson symbol is a number which designates the number of atoms in the conventional unit cell. Therefore, a diamond structure, which is cubic, face centered, and has 8 atoms in its unit cell, is represented by *cF8*. To use the Pearson system effectively, however, we need to know a structure, or prototype, which is the classic example of that particular structure. For example, both GaAs and MgSe have a Pearson designation cF8, but GaAs has a classic zinc blende (i.e., "ZnS") structure, and MgSe a "NaCl" type structure. While both structures are formed by two interpenetrating face-centered cubic lattices, they differ in how the two lattices are positioned relative to one another.

While Pearson symbols categorize crystal structures into particular patterns and are conceptually simple and easy to use, not every structure is uniquely defined. The space group notation, also known as the International or Hermann–Maguin system [12], is a mathematical description of the symmetry inherent in a crystal structure and is also represented by a set of numbers and symbols. The space groups in three dimensions are made from combinations of the 32 crystallographic point groups (given in parentheses in the tables) with the 14 Bravais lattices which belong to one of the 7 basic crystal systems. This results in a space group being some combination of the translational symmetry of a unit cell including lattice centering and the point group symmetry operations of reflection, rotation, and improper rotation. The combination of all these symmetry operations results in a total of 230 *unique* space groups describing all possible crystal symmetries. The relationship between the basic crystal systems, the Bravais lattices, and the point and space groups is illustrated in Figure E.2. The International Union of Crystallography publishes comprehensive tables [13] of all space groups and assigns each a unique number.

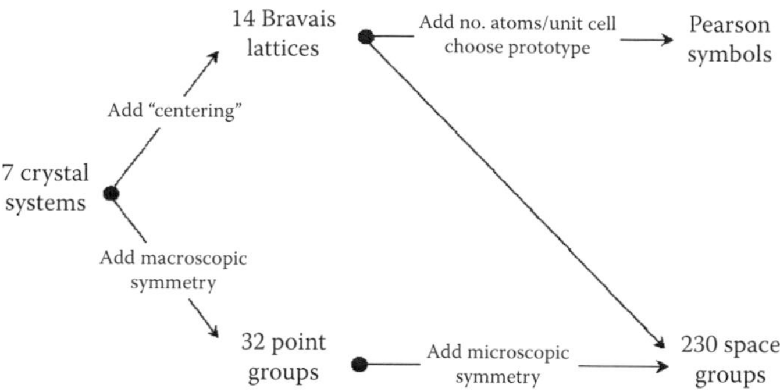

Figure E.2 The interrelationship between the basic crystal systems, the Bravais lattices, point groups, and the Pearson symbols and space groups. The macroscopic symmetry elements are those operations (e.g., reflection and translation) which take place over unit cell dimensions, whereas the microscopic symmetry elements add small translations (less than a unit cell vector) to the macroscopic symmetry operations. A point group is a representation of the ways that the macroscopic symmetry elements (operations) can be self-consistently arranged around a single geometric point. There are 32 unique ways in which this can be done.

Lattice constants: These are the lattice parameters in units of nanometers, generally denoted by the letters a, b, and c. For cubic crystal structures, the unit cell sizes are equal, and we only refer to a. Similarly, for hexagonal crystal structures, the a and b lengths are equal, and so we only give the a and c values. Orthorhombic structures are rectangular prisms so that the sides have unequal lengths. Therefore, the a, b, and c lengths need to be specified. As expected, the lattice constant tends to be smaller when a compound consists of smaller-diameter atoms than when it consists of larger atoms.

Atomic No(s): The atomic number(s) of the constituent element(s). The atomic number corresponds to the number of protons in the nucleus of an atom of that element.

Molecular mass (molecular weight) is the mass of one molecule of a substance and is expressed as a unified atomic mass unit (amu). One amu is equal to 1/12 the mass of one atom of carbon-12.

Density: The density (mass per unit volume) of the material in grams per cubic centimeter. Data given here refer to the solid. Density is temperature dependent, and different allotropes possess different densities. Values are given for the thermodynamically most favored allotrope at ambient temperature.

Melting point: Perhaps the most important thermophysical parameter. Melting point is the temperature at which a material changes from the solid to liquid state and at which both phases exist in equilibrium. Values are given in units of K at normal pressure unless otherwise stated. Materials that sublime, that is, transit directly from the solid to the gaseous phase, are noted.

$\boldsymbol{\varepsilon_r(0)}$ is the static dielectric constant, or more correctly, the relative static permittivity of a material, which is a measure of its ability to concentrate electrostatic lines of flux. It is defined as the ratio of the amount of electrical energy stored in a material when a potential is applied, relative to that stored in a vacuum. In reality, if a material with a high relative permittivity is placed in an electric field, the magnitude of that field will be measurably reduced within the volume of the dielectric, by an amount proportional to the relative dielectric constant. Because a high dielectric constant is generally associated with a high dielectric strength, dielectric breakdown will occur at higher electric fields. This can have some advantage in detection systems, in that higher biases can be applied to these materials, leading to better charge collection. However, a highly dielectric medium can also lead to polarization effects, which introduce a time dependence in their detection properties. Numerically, the relative static permittivity is the same as the relative permittivity evaluated for a frequency of zero, $\varepsilon(0)$. The static dielectric constant is important in that it forms the constant of proportionality between the potential and the charge density in Poisson's equation. It is a key parameter for several scattering mechanisms, while both the zero-frequency and high-frequency dielectric constants are employed in the description of polar optical scattering. Note for zinc blende structures, ε_r values are the same along each crystallographic axis, and only one value is quoted, while for wurtzite materials the values along the a- and c-axes are different, and so two values are usually given: $\varepsilon(0)$ perpendicular ($\perp$) to the c-axis and $\varepsilon(0)$ parallel ($\parallel$) to the c-axis. In some cases, however, only the c-axis value is quoted for wurtzite materials.

Ionicity f_i is an important physical concept and generally exists only in chemical bonds between different atoms. It allows us to form a quantitative description of chemical bonding and predict structural phase stability. In essence, the ionicity of a semiconductor is a measure of the partial charges created due to the asymmetric distribution of electrons in chemical bonds. This charge is a property only of zones within the distribution, and not the assemblage as a whole. For example, when an electrically neutral atom bonds chemically to another neutral atom that is more electronegative, its electrons are partially drawn away. This leaves the region about that atom's nucleus with a partial positive charge, which in turn creates a partial negative charge on the atom to which it is bonded. The ionicity of a bond can be defined as the fraction of ionic or heteropolar part of the bond, f_i, compared with the fraction of covalent or homopolar bonding, f_h. By definition, $f_i + f_h = 1$. Values are quoted for the Phillips [14] ionicity scale unless otherwise stated.* For the elemental semiconductors, such as Si, bonding is entirely covalent and $f_i = 0$, whereas for some of the alkali halides the bond is more than 90% ionic and $f_i \rightarrow 1$. Generally, compounds with ionicities less than 0.1 are classified as covalent, compounds with ionicities in the range 0.2 to 0.7 are considered partially ionic, and compounds with ionicities greater than 0.7 are considered ionic. For ionic semiconductors, electrical conductivity is due primarily to the movement of ions rather than electrons and holes.

Phillips [14] demonstrated that structural properties such as the cohesive energy and heats of formation depend linearly on f_i. Structurally, covalent bonding gives rise to structures with small coordination numbers, while ionic bonding appears in high-symmetry structures. A value of $f_i = 0.785$ is found to mark the transition between tetrahedral (fourfold) and octahedral (sixfold) coordination, meaning that covalent bonding is not sufficiently strong to stabilize tetrahedrally bonded structures [16] above this value.

* In some cases the Tubbs values are given. The Tubbs ionicity scale [15] was devised to order the many compounds intermediate between the alkali halides and the group III–V semiconductors, for which conventional ionicity definitions based on the heats of formation (e.g., [14]) are of limited value.

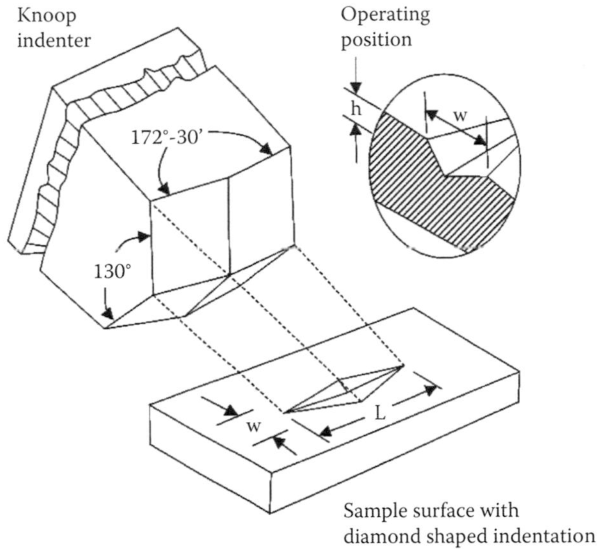

Figure E.3 Diagram showing the mechanics of the Knoop indenter and hardness measurement. *L* is the long diagonal of the diamond-shaped indentation, measured in microns (from [18]).

KH denotes the hardness of the material and is defined in terms of the Knoop micro-hardness [17] named after Frederick Knoop of the US National Bureau of Standards. The Knoop hardness (KH) number of a substance is determined by measuring the area of a "diamond-shaped" indentation made by a particular set of facets of a diamond pressed into the substance at a set pressure for a set time. Specifically, it is given by the ratio of the load applied to the indenter, *P* to the unrecovered projected area *A* illustrated in Figure E.3.

$$\mathrm{KH} = \mathrm{P/A} = \mathrm{P/CL^2} \tag{E.1}$$

where L = measured length of long diagonal of indentation in mm, and C is a constant relating the shape of the indenter to L (= 0.07028). KH values are usually given in units of kgf·mm^{-2}, although pascals are also sometimes used (1 kgf·mm^{-2} = 9.80665 MPa). Typical KH values lie in the range from 10 (refrigerated butter) to 10,000 (diamond) and vary with temperature—materials are softer at higher temperatures (see [19]). For example, hardness values for GaAs decrease by a factor of ~25 between room temperature and 300°C. Other measurement protocols are the Brinell hardness scale [20], Vickers hardness test [21], and the Mohs* scale [22]. Note that Brinell and Vickers hardness measurements can be accurately related, but only over a limited range of values due to the different indenter morphologies. For example, the Brinell 3000-kgf, 10-mm steel ball test and the Vickers diamond pyramid test are essentially the same for values between 100 and 500 kg mm^{-2}. The Knoop and Vickers scales are linearly related ($HV \approx 0.2\ KH$), however the Mohs scale is not simply related to any other test, because it is benchmarked on a set of standard minerals. The scale runs from 1 (talc), the softest, to 10 (diamond), the hardest, such that any mineral assigned a greater value can scratch a mineral of a lesser value. As a consequence, the Mohs scale is nonlinear, since the procedure for assigning values cannot easily cope with noninteger values. In Figure E.4 we show a conversion for the reference minerals used to define the Mohs scale, which can be used to estimate KH values when only Mohs values are available.†

Type: Denotes whether the semiconductor is a direct (D) or indirect (I) bandgap material, as illustrated in Figure E.5. In a direct (D) bandgap material, the minimum energy of the conduction band lies directly above the maximum energy of the valence band in momentum (k) space, such that the electrons at the conduction band minimum can combine directly with holes at the valence band maximum, while conserving momentum. In an indirect (I) bandgap material, the minimum energy in the conduction band is shifted in $\boldsymbol{k}$ space relative to the valence band. An electron must therefore undergo a significant change in momentum to move from the bottom of the conduction band to the top of the valence band, which can only be achieved with the mediation of a third body, such as a phonon or crystallographic defect. Examples of direct

* Named after the German mineralogist Friedrich Mohs (b. 1773; d. 1839) who devised the scale in 1812.

† Generally given for soft materials for which the Knoop indenter may not leave a clear and undistorted impression.

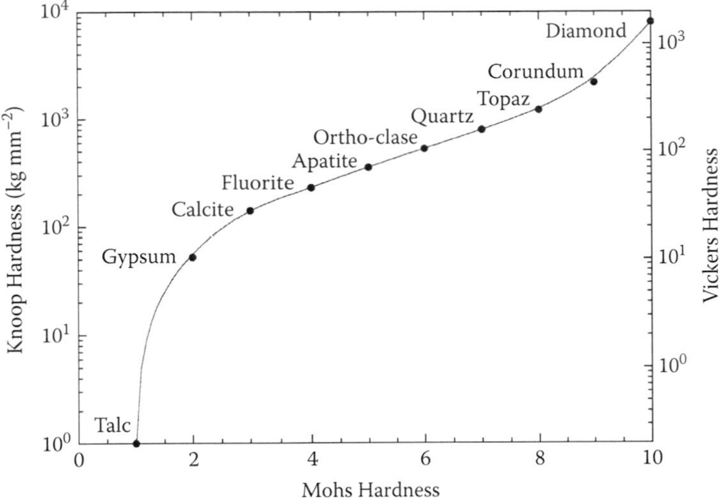

Figure E.4 A plot of the Knoop hardness scale versus the Mohs scale for the standard minerals used to define the scale. The right-hand ordinate gives the corresponding values on the Vickers hardness (HV) scale. Note: by convention HV is usually quoted without units.

bandgap materials are GaAs, InP, and CdTe; examples of indirect bandgap materials are Si, Ge, and GaP. SM signifies that the material is a semimetal. Semimetals have indirect bandgaps, but unlike other indirect materials, the top of the valence band is at a higher energy than the bottom of the conduction band, and consequently bandgap energies are frequently negative.

Bandgap: The bandgap energy in units of eV. The bandgap is defined as the energy difference between the top of the valence band and the bottom of the conduction band (see Figure E.5). The bandgap energy is a weak function of temperature. Values are given at 300 K unless otherwise stated.

Pair energy: The energy consumed to create an electron–hole pair in units of eV. The pair energy is roughly three times larger than the bandgap energy—the difference due to the energy lost to phonons and plasmons when crossing the bandgap.

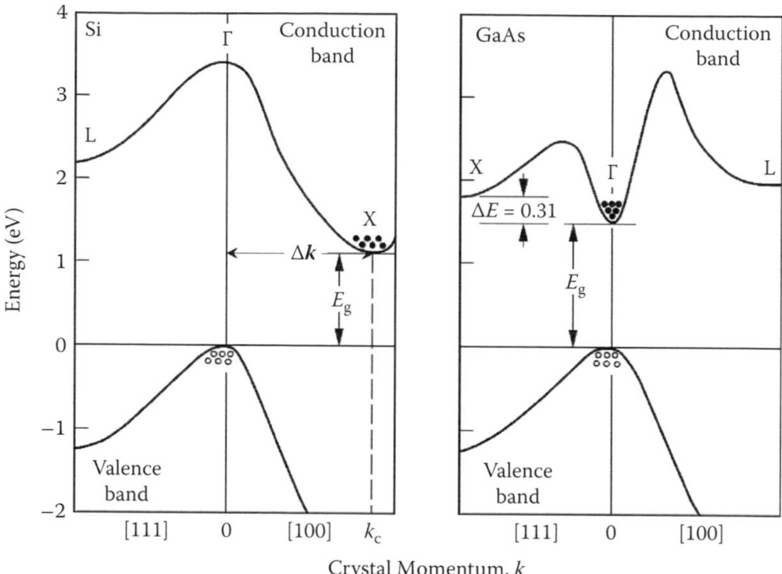

Figure E.5 Illustration of direct (left) and indirect (right) bandgaps in which we show the band structure in energy and momentum space, $E,\mathbf{k}$ (from [23]). Here, E_g is the bandgap energy. The illustrated examples are Si and GaAs. Circles (o) indicate holes in the valence bands, and dots (•) indicate electrons in the conduction bands. The important difference is that, for a direct bandgap material, an electron can transit from the lowest potential in the conduction band to the highest potential in the valence band without a change in momentum, $\Delta\mathbf{k}$, whereas for an indirect bandgap material it cannot transit without the mediation of a third body (e.g., phonon) to conserve momentum.

Resistivity: The electrical resistivity, or specific resistance, ρ, is the resistance between the opposite faces of a meter cube of a material. It is a property of a semiconductor that depends on the free electron and hole densities and their respective mobilities. Resistivity has units of Ω-cm and is the reciprocal of conductivity. It is strongly dependent on temperature. For hexagonal or layered materials, resistivity can also depend upon the direction along which the measurement is made.

Electron mobility, μ_e: Defined as the ratio of the electron velocity (cm s^{-1}) to the electric field (V cm^{-1}) through which it is moving. Mobility has units of $cm^2V^{-1}s^{-1}$ and is usually determined using one of two methods—Hall or drift. Hall mobilities, μ_H, are derived from Hall effect measurements [24] and drift or conductivity mobilities, μ_d, are derived using the Haynes–Shockley technique [25]. It is important to note that they can be different. The ratio μ_H/μ_d is usually close to unity for direct bandgap semiconductors, but can be greater than unity for indirect bandgap semiconductors. Unless otherwise stated, the values quoted here are drift mobilities at a temperature of 300 K. Note that for hexagonal and wurtzite structures, the mobilities can be different along the a and c crystallographic axes. Whenever possible, two values are given: μ perpendicular (⊥) to the c-axis and μ parallel (||) to the c-axis.

Hole mobility, μ_h: Defined as the ratio of the hole velocity (cm s^{-1}) to the electric field (V cm^{-1}) through which it is moving. Mobility has units of $cm^2V^{-1}s^{-1}$. All remarks in the previous paragraph are equally valid for hole mobilities.

$\mu_e\tau_e$: This is the electron mobility-lifetime product expressed in units of cm^2V^{-1}. Values are quoted at a temperature of 300 K unless otherwise stated.

$\mu_h\tau_h$: This is the hole mobility-lifetime product expressed in units of cm^2V^{-1}. Values are quoted at a temperature of 300 K unless otherwise stated.

E.2 Accompanying Notes for the Tables

The following tables list the physical and electronic properties of the various groupings of the elemental and compound semiconductors. These are groups III, IV, III–V, IV–IV, II–VI, I–VII, their binary alloys and ternary compounds. For detector applications, there are numerically approximately 16 group III–V compounds of interest, 19 group II–VI compounds, 6 group

Column / Period	I	II	III	IV	V	VI	VII
2			5 B	6 C	7 N	8 O	
3		12 Mg	13 Al	14 Si	15 P	16 S	17 Cl
4	29 Cu	30 Zn	31 Ga	32 Ge	33 As	34 Se	35 Br
5	47 Ag	48 Cd	49 In	50 Sn	51 Sb	52 Te	53 I
6		80 Hg	81 Tl	82 Pb	83 Bi		

Transition metal | Alkaline earth | Metalloid | Other metals | Non metal

Figure E.6 Section of the periodic table from which the bulk of the semiconductor groups are derived. Most compounds derived from this table will form tetrahedral diamond structures which satisfy the Mooser–Pearson [26] rules which are a simple set of tests to determine whether a compound is likely to display semiconducting properties.

I–VII, and a virtually infinite number of ternary and quaternary compounds. Figure E.6 shows the section of the periodic table from which the bulk of the semiconductor groups are derived. The main groups and how they are formed are described in some detail below. For completeness, we include the elemental semiconductors.

Group VI (B) elements. These are the classical elemental semiconductors, Si, Ge, C (diamond), and gray tin (α-Sn) which crystallize in the diamond structure and are unique in the periodic table in that their outer shells are exactly half filled. Consequently they bond exclusively covalently. An examination of properties of group IV elements shows that bandgap energies, hardness, and melting points all decrease with increasing Z, while charge carrier mobilities, densities, and lattice constants generally increase. These trends may be attributed to the progressive “metallization” of the elements with increasing Z within the group. One can also combine two different group IV semiconductors to obtain compounds such as SiC and SiGe whose physical and electronic properties are intermediate between both species.

Group III–V compounds. These are compounds which combine an anion* from group V (nitrogen on down) and a cation† from group III (usually Al, Ga, or In). Each group III atom is bound to four group V atoms and vice versa—thus each atom has a filled (8 electron) valence band. Although bonding would appear to be entirely covalent, the shift of valence charge from the group V atoms to the group III atoms induces a component of ionic bonding to the crystal [14]. This ionicity causes significant changes in semiconducting properties. For example, it increases both the Coulomb attraction between the ions and the forbidden bandgap energy. When grown epitaxially (MBE, MOCVD, and variants), III–V materials usually assume a zinc blende (ZB) structure—so in their basic electronic and crystal structures they are completely analogous to the group IV elements. The stable bulk allotrope often has a wurtzite structure. Both the zinc blende and wurtzite lattices are tetrahedrally bonded. They differ only by the orientation of the nearest-neighbor tetrahedrons. The zinc blende form differs from a diamond lattice only in that the two interpenetrating face centered lattices are occupied by different atoms. Representative III–V compounds are InSb, InAs, GaAs, GaP, AlAs, and AlP. In terms of bandgap, indium compounds have the smallest energy gap, followed by gallium, boron, and aluminum compounds. Likewise within these groups the antimonides have the smallest bandgaps, followed by the arsenides, phosphides, and finally the nitrides—reflecting the decreasing size of the atomic nuclei (r = 145 pm (Sb), 115 pm (As), 100 pm (P), 65 pm (N), where 1 pm = 10^{-12} m). Similarly, because the bond lengths are reducing with decreasing z, (going from the antimonides to the nitrides) ionicity and hardness correspondingly increase. Electronically, group III–V materials tend to have larger electron mobilities at low electric fields than the elemental semiconductors, which make them attractive candidates for high-speed applications. For zinc blende structures there is a transition from a direct to an indirect bandgap

* The electronegative component of a compound, e.g., As in GaAs, Te in CdTe.
† The electropositive component of a compound, e.g., Ga in GaAs, Cd in CdTe.

somewhere between GaAs (1.4 eV direct) and AlSb (1.6 eV indirect).

Ternary III–V alloys have the general form $(A_{1x}, A_{21-x})B$ with two group III atoms used to fill the group III atoms in the lattice, or $A(B_{1x}, B_{21-x})$ using two group V atoms in the group V atomic positions in the lattice. Here A and B represent elements from groups III and V, respectively, and x is the mole fraction in the range 0 to 1. The quaternary semiconductors use two group III and two group V elements, yielding a general form $(A_{1x}, A_{21-x})(B_{1y}, B_{21-y})$ for $0 < x < 1$; $0 < y < 1$ (e.g., $Ga_{0.12}In_{0.88}As_{0.23}P_{0.77}$). The lattice constants of ternary and quaternary compounds can be calculated with good precision using Vegard's law [27], which gives a value equal to the weighted average of all of the four possible constituent binaries. For example, the lattice constant of the quaternary compound $A_{1-x}B_xC_yD_{1-y}$ is given by

$$a(x,y) = x\,y\,a_{BC} + x\,(1-y)\,a_{BD} + (1-x)\,y\,a_{AC} + (1-x)(1-y)\,a_{AD} \tag{E.2}$$

Except for binary alloys, a similar expression relating the weighted bandgap energy to the bandgap energies of the constituent elements does not really exist, due the presence of multiple minima in the conduction band which, in theory, should be taken into account in the weighting. In fact, for the higher-order alloys, the bandgap can actually change type (direct to indirect and vice versa), depending on composition and whether the semiconductor is strained or not. Generally, however, a so-called one valley bandgap fit can give reasonable agreement over a limited range of x and y.

Group II–VI compounds. These are compounds which combine a group IIb metal (for example, Zn, Cd, and Hg in periods 3, 4, and 5, respectively) with a group VIa cation. The latter is usually S, Se, or Te. Structurally, it forms when atomic elements from one type bond to the four neighbors of the other type, as shown in Figure 2.6c. Three crystal structures dominate II–VI compounds. These are zinc blende, wurtzite, and the rock salt (NaCl) structure. A major motivation for developing II–VI semiconductors is their broad range of bandgaps (from 0.15 eV for HgTe to 4.4 eV for MgS), high effective Z, and a demonstrated capability for making MBE- and MOCVD-grown heterostructures, as for III–V systems. Additionally, all II–VI binaries have direct bandgaps which make them particularly attractive for optoelectronic applications. In terms of bandgap, mercury compounds have the smallest energy gap, followed by cadmium, zinc, and magnesium compounds. Likewise, within these groups the tellurides have the smallest bandgaps, followed by the selenides and the sulfides—again reflecting the decreasing size of the atomic nuclei. In this case, the atomic radii are 140 pm (Te), 115 pm (Se) and 100 pm (S), respectively. Compounds generally crystallize naturally in a hexagonal or NaCl structure. Representative compounds are CdTe and HgTe. Pseudo-binary alloys with Zn, Se, Mn, or Cd are also common, particularly for radiation detector and optoelectronic applications (for example, $Cd_{(1-x)}Zn_xTe$, $Cd_{(1-x)}Mn_xTe$, and $Hg_{(1-x)}Cd_xTe$). Group II–VI compounds typically exhibit a larger degree of ionic bonding than III–V materials, because their constituent elements differ more in electron affinity due to their location in the periodic table. A major limitation of II–VI compounds is the difficulty in forming n-type and p-type material of the same compound. Also, it is difficult to control the defect state density within the bandgap due to self-compensation. Group II–VI semiconductors can be created in ternary and quaternary forms, although these are less common than III–V varieties.

Group III–VI compounds. Most of the III–VI compounds crystallize in layer type structures. The bonding is predominantly covalent within the layers and much weaker van der Waals type between layers. Because of this, the behavior of electrons within the layers is quasi-two-dimensional. Ionicities tend to be higher than in III–V materials but lower than in II–VI compounds. Examples of this group are at present limited to the Ga-based compounds GaS, GaSe, and GaTe, which are being studied primarily for photoconductivity and luminescence applications.

Group I–VII compounds. In comparison to other semiconductors groups, all the I–VII group of compounds are very similar, displaying little variation in their mechanical and electrical properties (such as melting point, density, hardness, and bandgap). They bond ionically and are consequently characterized by high ionicities. They have wide energy gaps which are considerably larger than in many III–V materials. In, fact there is a clear tendency for an increase in the energy gap and melting point with increasing ionic bonding, going from $A^{IV}A^{IV}$ to $A^{III}B^{V}$ to $A^{II}B^{VI}$ through to $A^{I}B^{VII}$ systems. Specific examples of this group include the silver halogenides, AgCl and AgBr, which were some of the first compound semiconductors demonstrated to be sensitive to ionizing radiation. Under normal conditions, they crystallize in a rock salt (NaCl) form, while the other main members of the group, the copper halogenides (CuCl, CuBr, and CuI), crystallize in a zinc blende configuration.

Group IV–VI compounds. Stable IV–VI compounds (also known as the group IV-chalcogenides*) exist in various stoichiometric compositions and generally have very narrow bandgaps. In fact, most are aligned with the IR waveband. They have large ionicities, six-fold coordination, high mobilities, and are electronically highly polarizable. Perhaps the most interesting are the lead chalcogenides, which are unique in that their energy gaps increase with increasing temperature, as opposed to other semiconductors which have negative temperature coefficients. PbSe is a particularly interesting material, in that excess Pb atoms in PbSe act as electron donors and excess Se atoms act as electron acceptors. Thus by changing the relative concentrations of these elements, it is possible to choose the nature of the material, n-type or p-type. The main application of IV–VI compounds is in the production of light-emitting devices and IR detectors.

Group n–VII compounds. Group *n*–VII (where *n* = II,II,IV) materials generally belong to the family of layered structured, heavy metal iodides and tellurides. The group VII anions form a hexagonal close packed arrangement while the group *n* cations fill all the octahedral sites in alternate layers. The resultant structure is a layered lattice with the layers held in place by van der Waals forces and is typical for compounds of the form AB_2. The bonding within the layers is mainly covalent. Physically, materials in these groups

* Derived from the Greek words *chalcos* and *gen* meaning "ore-forming." These are elements from group VIB of the periodic table (i.e., O, S, Se, and Te).

tend to be mechanically soft with easy cleavage planes, have low melting points and large dielectric constants, and show strong polarization effects. The fact that layered compounds are strongly bound in two directions and weakly bound in the third direction (i.e., along the c-axis), leads to an anisotropy of their structural and electronic properties. Because of this, electrons can display quasi-two-dimensional behavior within layers, which can potentially be exploited in transistor-like structures. In addition, because only van der Waals forces act between the layers, it is possible to introduce a variety of foreign atoms and organic molecules between the layers forming intercalation compounds, which can significantly modify the original physical and electronic properties. The reversibility of intercalation processes can also be exploited, for example, in the production of high energy density rechargeable batteries in which the intercalation layers form the cathodes and the cell reaction powers the reverse intercalation [28].

Group I–II–V compounds. These are the filled tetrahedral semiconductors of the form $A^{I}B^{II}C^{V}$. Structurally the lattices can be viewed as "zinc blende" derivatives partially filled with group I interstitials. An example would be LiZnX, where X = N, P, and As. However, in complete analogs of III–V zinc blende materials, the interstitials have the effect of preserving both a direct and a wide bandgap. The threshold between direct and indirect bandgaps in III–V zinc blende materials lies somewhere between GaAs (1.4 eV direct) and AlSb (1.6 eV indirect). Consequently, these semiconductors are currently being investigated for optoelectronic applications.

Group I–III–VI_2 compounds. Chalcopyrites,* such as $CuAlS_2$, $CuGaS_2$, and $CuInSe_2$. Other, non-Cu-based, members of the I–III–VI group are also usually referred to as chalcopyrites, because they adopt a similar crystal structure.† This is a ternary-compound equivalent of the diamond structure, in which every atom is bonded to four first neighbors in a tetrahedral structure. The atomic bonds are mainly covalent.

* Named after the mineral chalcopyrite, $CuFeS_2$.
† Also adopted by a number of II–IV–V2 compounds.

Instead of bonding to four group II elements as in a group II–VI semiconductor, the group VI element bonds to two group I and two group III elements in the I–III–VI_2 ternary system. Cu-In-Se systems are the most studied variants at this time, especially the alloy $CuIn_{1-x}Ga_xSe_2$ ($Cu(In,Ga)Se_2$). Group I–III–VI systems offer direct gap semiconductors over a broad range of lattice constants and bandgaps and are presently being investigated for exploitation as photovoltaic materials. In fact, the Cu-chalcopyrite family of semiconductors produces some of the best thin-film solar cell absorbers with power conversion efficiencies up to ~20% [29].

Group I–IV, I–V–VI and other organic (polymer, oligomers) derivatives. These are the so-called "plastic" semiconductors. They are polymers with a delocalized π-electron system along the polymer backbone. This gives rise to the creation of alternating single and double bonds by weak *pz–pz* bonding, which in turn results in the creation of a bandgap of energy ~2.5 eV. Such materials offer numerous advantages in terms of easy processing (spin or dip coating as opposed to epitaxial growth) and good compatibility with a wide variety of substrates. At present, organic materials are being exploited for use in low-cost flexible displays, low-end data storage media, and inexpensive solar cells. For the latter, efficiencies of up to 2.5% have been achieved [30]. They are also of interest for medical applications because they are "tissue-equivalent" materials, in that the low atomic numbers of their constituent atoms closely match those of biological tissue.

Ternary compounds, such as $A^{II}B^{IV}C_2^{V}$ or $A_2^{II}B^{V}C^{VII}$ and combinations of binary alloys (e.g., $A_xB_{1-x}C$, where x is the fractional concentration of A), offer a much wider choice of physical parameters, such as effective atomic number and bandgap energy. Whereas the bandgap of ternary compounds is fixed, alloying gives access to a continuous range of bandgaps. Alloys of binary compounds from the same column of the periodic table can also be alloyed to arbitrary composition such that their physical properties, such as bandgap width, change smoothly with x. However, it is difficult to achieve a high degree of compositional homogeneity and crystal quality.

Consequently, these materials are generally grown by MBE or MOCVD, which provide better control over stoichiometry than other techniques. Most properties, such as effective mass, vary quadratically and monotonically with alloy fraction. Alloy scattering is largest near a 50%, mix and transport properties tend not to vary monotonically. As a rule, ternary compounds based on the heavier metallic elements (e.g., Hg and Bi) are of lower mechanical strength, poorer phase and chemical stability than $A^{III}B^{V}$, $A^{II}B^{VI}$, or $A^{I}B^{VII}$ compounds. Also, because they do not possess four valence electrons per atom and are not full analogs of diamond, it is more difficult to synthesize these materials to the required purity and stoichiometry or even to predict their properties.

E.2.1 General Properties of the Groups

Based on the above summaries and the accompanying tables, we can make several generalizations about compound semiconductors. To begin with, hexagonal materials are generally harder than cubic materials. However, when a material can crystallize into say, stable wurtzite and zinc blende forms, the zinc blende structure is generally stronger. Following on, heavy materials tend to be mechanically soft. This is borne out in Figure E.7 (left) in which we plot density versus Knoop microhardness (KH) for a range of semiconductors. Note, "*Others*" here refers to semiconductors not in groups IV, IV–IV, II–V, II–VI, and I–VII (e.g., groups IV–VI). The data show a power law correlation, with the heaviest materials having lower KH values than light materials. In fact, on average the KH decreases by an order of magnitude for every ~2 g cm^{-3} increase in density. Similarly, we might then expect heavier compounds to have lower melting points because they are likely to be softer. In this case (see Figure E.7 right), the correlation is much weaker, but does show the expected trend and appears to be independent of semiconductor group.

The most obvious material variations are those with lattice constant. For $A^{N}B^{8-N}$ semiconductors a plot of lattice constant,

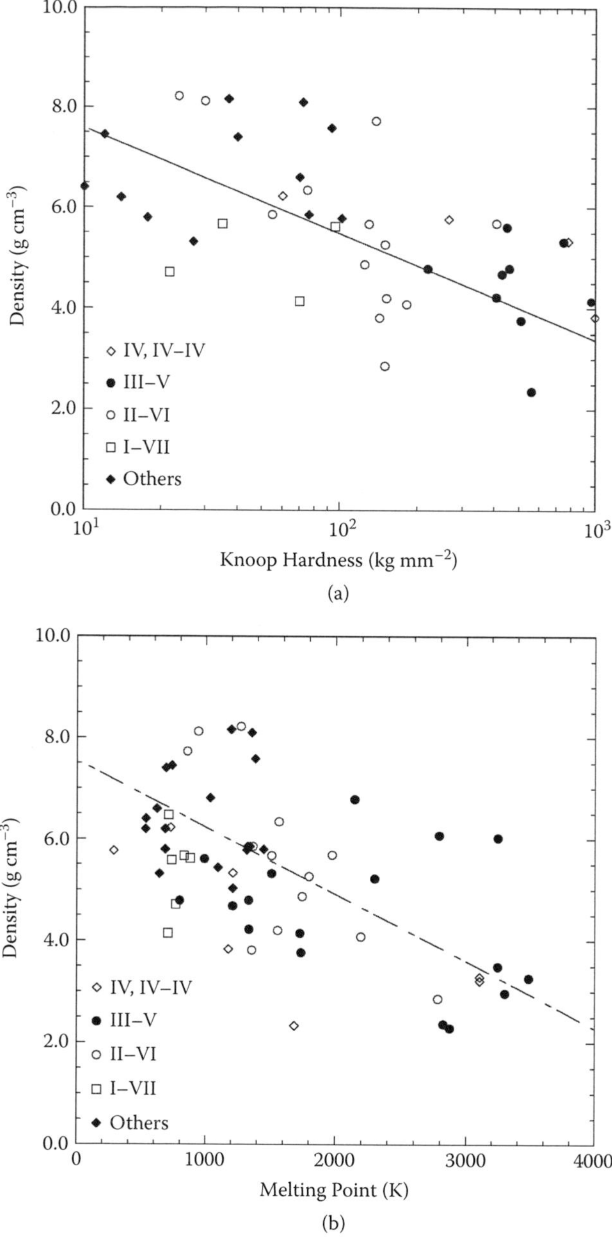

Figure E.7 Correlation between (a) density and hardness and (b) density and melting point. The solid lines are drawn to guide the eye. Others here refer to semiconductors not in groups IV, IV–IV, II–V, II–VI and I–VII (e.g., groups IV–VI).

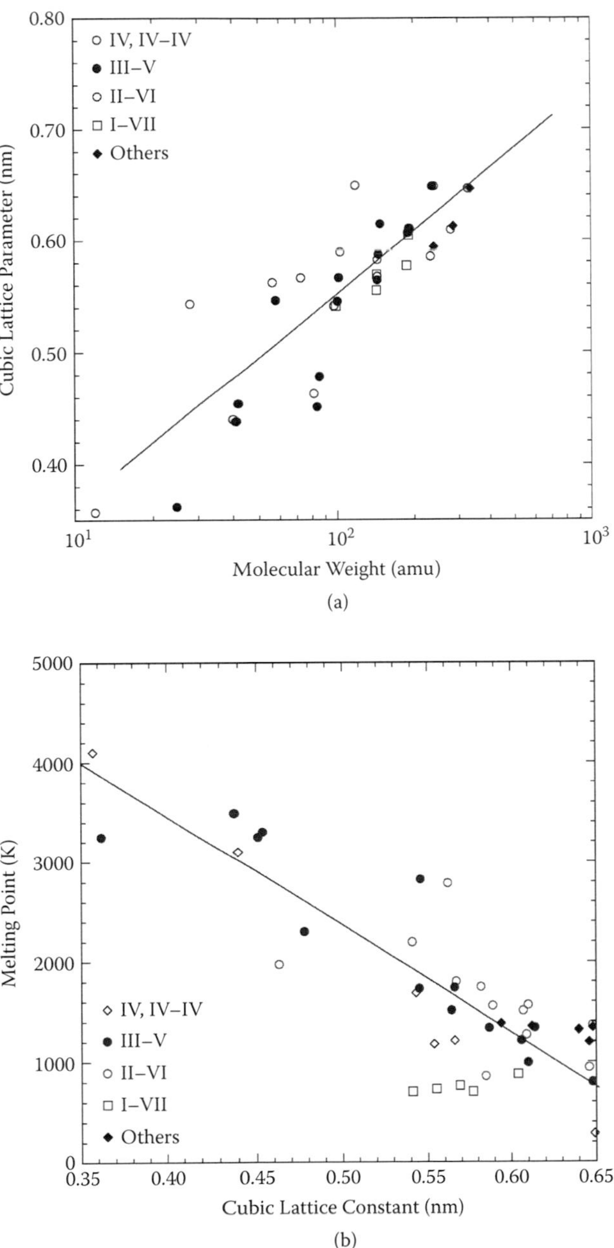

Figure E.8 (a) Molecular weight and (b) melting point as a function of lattice parameter for a range of semiconductors. The solid lines are drawn to guide the eye.

a, versus molecular weight* (see Figure E.8a) can be approximated by a linear function of the form [4],

$$a(\text{nm}) = 0.08 + 0.23 \ln \text{MW(amu)} \tag{E.3}$$

where MW = M_A+M_B. In Figure E.8b, we note there is a clear tendency for the melting point to increase with decreasing interatomic bond length by roughly 1000 degrees K per ångstrom. It can be approximated by,

$$T_m\,(\text{K}) = 7743 - 10723\,a(\text{nm}) \tag{E.4}$$

Note that for I–VII materials, the melting point is independent of lattice constant. For hexagonal semiconductors, a should be replaced by an effective lattice constant, a_{eff}, given by

$$a_{\text{eff}} = (\sqrt{3}a^2c)^{1/3} \tag{E.5}$$

where c is the interatomic bond length along the c crystallographic direction.

In Figure E.9a we plot Knoop microhardness as a function of lattice parameter for a number of materials, from which we can see the larger the lattice parameter, the softer or more fragile the material. Interestingly, groups III–V and II–VI materials lie on different curves, but with the same slope, and can be approximated by,

$$\text{KH} = a_1 \exp(-12.13a_o(\text{nm}))\ \text{kg mm}^{-2} \tag{E.6}$$

Here $a_1 = 5.884 \times 10^5$ kg mm^{-2} for IV, IV–IV, and III–V materials and 1.556×10^5 kg mm^{-2} for II–VI and "other" materials.

Note that KH values also decrease with increasing temperature. For example, GaAs values decrease by a factor of ~25 between room temperature and ~300°C.

* Note: the molecular weight for an A^NB^{8-N} (N≠4) compound is simply given by the sum of the atomic weights of atoms A and B. For an elemental semiconductor (N = 4), it is given as the atomic weight of the elemental atom. The molecular weight M of alloy semiconductors can be obtained by linear interpolation.

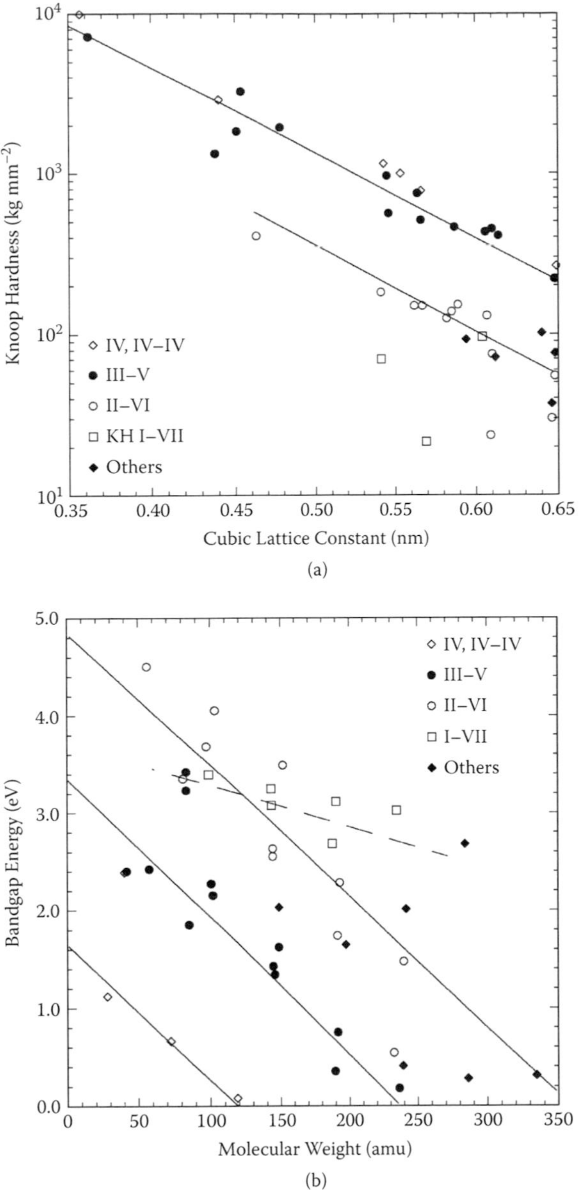

Figure E.9 (a) Knoop microhardness as a function of lattice parameter for a range of semiconductors. Notice that the Knoop hardness data seems to separate into two distinct distributions: groups IV, IV–IV and III–V; and groups II–VI and "other" materials. (b) The variation of bandgap energy with molecular weight for groups IV, III–V, II–VI and I–VII semiconductors.

In Figure E.9b we plot the bandgap energy as a function of molecular weight (MW), from which we note that the different groups lie on separate curves. In fact, for groups IV, IV–IV, III–V, and II–VI, the curves are very nearly parallel to each other and can be reasonably well approximated by the function

$$\varepsilon_g = a_1 - 0.014 \text{ MW (eV)} \tag{E.7}$$

where $a_1 = 1.64$ eV for IV, IV–IV materials, 3.3 for III–V materials, and 4.8 for II–VI materials. However, it is also clear from Figure E.9b that I–VII materials lie on a quite different slope:

$$\varepsilon_g = 3.63 - 0.0032287 \text{ MW (eV)} \tag{E.8}$$

In Figure E.10a we see there is also a clear tendency for the bandgap energy, ε_g, to decrease with increasing lattice constant. In fact, ε_g changes by roughly a factor of 1.4 per angstrom, except for group I–VII materials, which show a much less pronounced variation.

In Figure E.10a, we note that the static dielectric constant varies exponentially with bandgap energy. Smaller bandgap energies have larger dielectric constants; this is particularly well correlated for groups III–V materials. A reasonably good approximation is,

$$\varepsilon(0) = 14.9 \exp(-0.289\varepsilon_g) \tag{E.9}$$

Thus, for room-temperature operation we should expect to have a dielectric constant of ~12.

Electronically, we note that mobilities tend to be larger in direct bandgap materials than in indirect bandgap materials and lower in heavy compounds. In Figure E.11a we plot both electron and hole mobilities as a function of bandgap energy. The electron mobilities show a weak correlation with bandgap energy for bandgaps less than 3 eV, which appear to increase with deceasing bandgap. Above 3 eV they are essentially constant. The hole mobilities show no obvious variation with bandgap energy. We also note that, while alloying can result in an increase in resistivity, it is usually accompanied

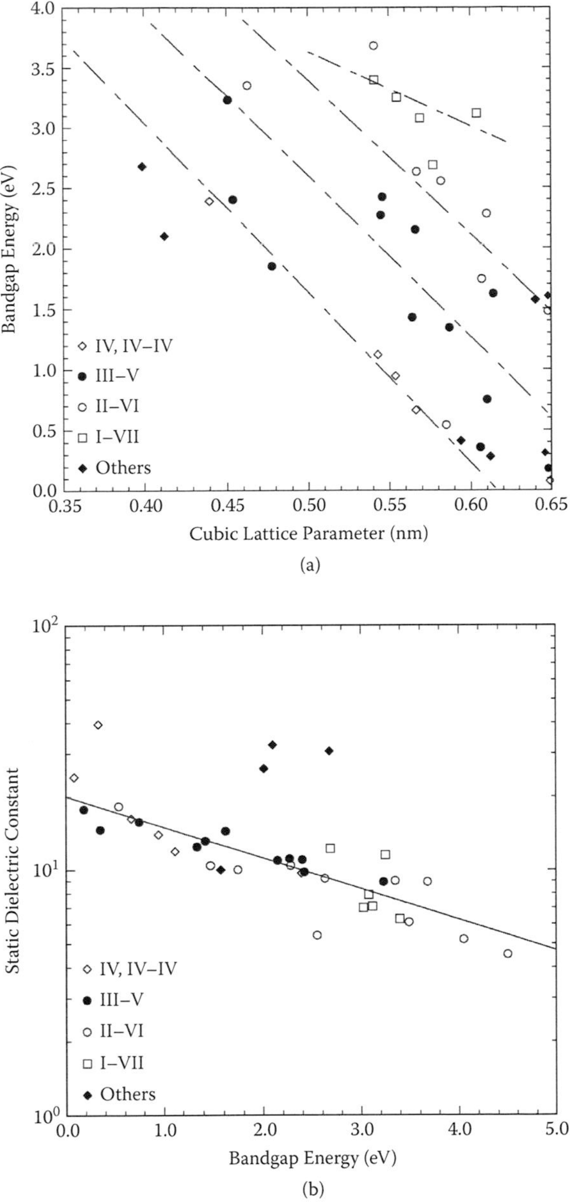

Figure E.10 (a) Bandgap energy as a function of lattice parameter for a range of semiconductors. Notice that the data seems to separate into distinct distributions for each semiconductor family. (b) The variation of bandgap energy with static dielectric constant for groups IV, IV–IV, III–V, II–VI, I–VII, and other semiconductors.

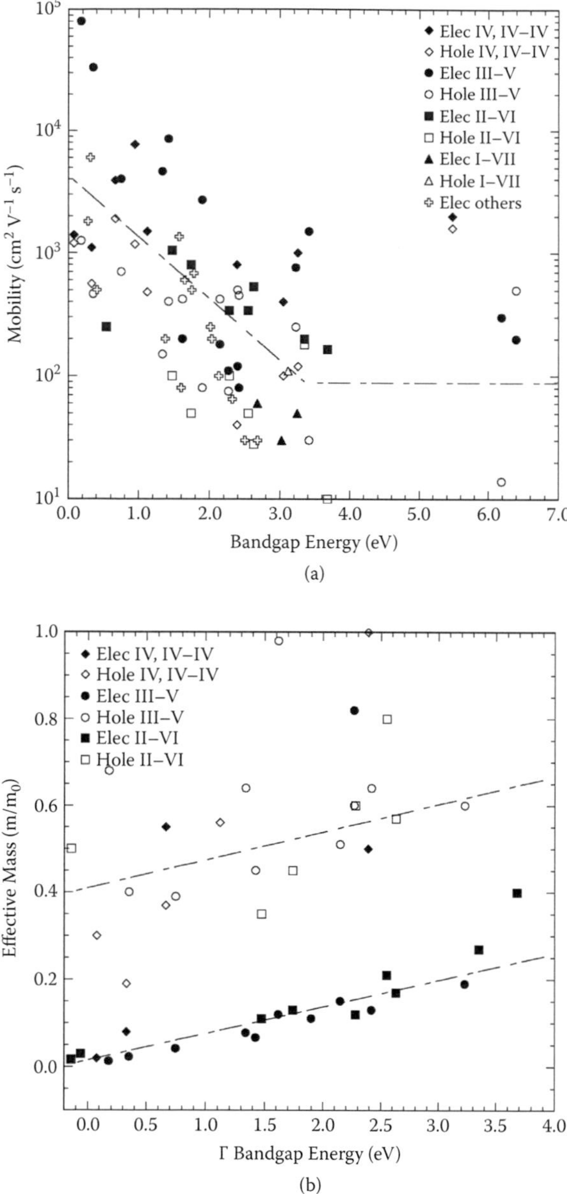

Figure E.11 (a) Electron and hole mobilities plotted as a function of bandgap for the various semiconductor groups. The lines are drawn to guide the eye. (b) Variation of effective mass with direct bandgap energy. For indirect bandgap materials this corresponds to the Γ valley.

by a fall in the mu-tau products, due to impurity scattering [31,32].

In Figure E.11a we plot the electron and hole effective masses as a function of the direct bandgap energy for a range of semiconductors. In the case of indirect bandgap materials, such as Ge and Si, we have used Γ energy gap (see Figure 1.7) at $\boldsymbol{k} = 0$. From the figure we see there is a good correlation between the electron effective mass and the bandgap. The data can be reasonably represented by

$$m_e{}^* = 0.0625\ \varepsilon_{g\Gamma} \tag{E.10}$$

where $m_e{}^* = m^*/m_o$ is the effective mass in units of electron mass. The correlation between the hole effective mass and the bandgap energy is much weaker than that of the electrons; however, we note that the slope of the curve is similar to that for the electrons and may be represented by

$$m_h{}^* = 0.41 + m_e{}^* \tag{E.11}$$

Thus,

$$m_h{}^* = 0.0625\ \varepsilon_{g\Gamma} + 0.41 \tag{E.12}$$

Because effective mass and mobility are inversely related, Equation (E.10) implies that, for applications where high speed is required, one should choose a semiconductor with a small bandgap.

The interrelationship between semiconductor parameters is summarized in Table E.1. In order to explore the differences between semiconductor groupings, we compare the average values of a number of key mechanical and electronic parameters. The results are tabulated in Table E.2. From an inspection of the table we see that densities, bandgaps, and ionicities tend to increase going from groups IV to III–V to II–VI and finally to I–VII, while melting point, hardness, and dielectric constant tend to decrease.

TABLE E.1
A Summary of the Interrelationships between Semiconductor Parameters

Increasing Parameter		ρ	MW	KH	MP	A_o	ε_G	ε(0)	μ	M*	R
Density,	ρ↑	ρ↑	MW↑	KH↓	MP↓	none	none	none	none	none	none
Molecular weight,	MW↑	ρ↑	MW↑	KH↓	MP↓	a_o↑	ε_g↓	ε(0)↑	none	m*↓e	none
Hardness,	KH↑	ρ↓	MW↓	KH↑	MP↑	a_o↓	none	none	none	m*↑e	R↑
Melting point,	MP↑	ρ↓	MW↓	KH↑	MP↑	a_o↓	none	none	none	m*↑	none
Lattice constant,	a_o↑	none	MW↑	KH↓	MP↓	a_o↑	ε_g↓	none	none	none	none
Bandgap energy,	ε_g↑	none	MW↓	none	none	a_o↓	ε_g↑	ε(0)↓	μ↓	m*↑	none
Dielectric constant,	ε(0)↑	none	MW↑	none	none	none	ε_g↓	ε(0)↑	none	m*↓e	none
Electron mobility,	μ↑	none	none	none	none	none	ε_g↓	none	μ↑	m*↓	none
Effective mass,	m*↑	none	MW↓e	KH↑e	MP↑	none	ε_g↑	ε(0)↓e	μ↓	m*↑	none
Resistivity,	R↑	none	none	KH↑	none	none	none	none	none	none	R↑

Note: Where no apparent correlation between parameters is found, the corresponding entry in the table is marked "none." Here *MW* is the molecular weight, *R* is the resistivity, and ρ the density. The other symbols have their usual meanings. The suffix "e" on effective mass refers to electron.

TABLE E.2
Average Values of the Main Mechanical and Electronic Characteristics of the Main Semiconductor Groupings (Groups IV and IV–IV, Groups III–V, Groups II–VI, and Groups I–VII)

Parameter Groups	Density (g cm^{-3})	Melting Point (K)	Lattice Parameter (nm)	Bandgap (eV)	Knoop Hardness (kg mm^{-2})	Ionicity (Phillips) f_I	Dielectric Constant $\varepsilon(0)$	Electron Mobility (cm^{2}V^{-1})	Hole Mobility (cm^{2}V^{-1})
IV, IV–IV	4.0 ± 0.5	2054 ± 441	0.52 ± 0.04	1.9 ± 0.6	2515 ± 1031	0.06 ± 0.03	17 ± 4	2200 ± 765	800 ± 230
III–V	4.5 ± 0.3	2167 ± 213	0.53 ± 0.02	2.5 ± 0.5	1435 ± 406	0.32 ± 0.04	12 ± 1	9080 ± 5508	380 ± 82
II–VI	5.6 ± 0.5	1611 ± 143	0.58 ± 0.01	2.8 ± 0.4	136 ± 27	0.68 ± 0.02	11 ± 2	4868 ± 3000	97 ± 37
I–VII	5.4 ± 0.3	769 ± 29	0.57 ± 0.01	3.1 ± 0.1	40 ± 15	0.77 ± 0.03	9 ± 1	47 ± 9	56 ± 54

Note: Because of the wide dispersion in values with groups, we quote the average error on the measurement, i.e., standard deviation/number of data points.

References

1. O. Madelung, M. Schulz, H. Weiss, Springer-Verlag, Eds., *Landolt-Börnstein: Numerical Data and Functional Relationships in Science and Technology*, *Springer-Verlag*, New York, Vol. 17 (1982).
2. L.I. Berger, *Semiconductor Materials*, CRC Press, Boca Raton, (1997).
3. O. Madelung, *Semiconductors: Data Handbook*, 3rd edition, Springer Verlag (2004).
4. S.Adachi, *Properties of Group-IV, III-V and II-VI Semiconductors*, John Wiley & Sons (2005).
5. S. Adachi, *Properties of Semiconductor Alloys: Group-IV, III-V and II-VI Semiconductors*, John Wiley & Sons (2009).
6. *CRC Handbook of Chemistry & Physics*, 82nd edition, "Properties of Semiconductors," CRC Press, Boca Raton, FL (2001) pp. 12-97–12-106.
7. "NSM Archive - Physical Properties of Semiconductors," http://www.ioffe.ru/SVA/NSM/Semicond/.
8. S.M. Sze, *Physics of Semiconductor Devices,* Wiley, John & Sons, 2nd edition (1981) ISBN-13: 9780471056614.
9. R.C. Alig, S. Bloom, C.W. Struck, "Scattering by Ionization and Phonon Emission in Semiconductors," *Physical Review B*, Vol. 27, no. 12 (1980) pp. 5565–5582.
10. B.D. Cullity, *Elements of X-Ray Diffraction*, Addison-Wesley Pub. Co., Reading, Mass. (1956). Revised edition: B.D. Cullity, S.R. Stock, *Elements of X-Ray Diffraction*, 2nd Ed., Prentice Hall: New Jersey (2001).
11. W.B. Pearson, *A Handbook of Lattice Spacings and Structures of Metals and Alloys*, Vol. 2, Pergamon Press, Oxford (1967).
12. C. Hermann, Ed., *Internationale Tabellen zur Bestimmung von Kristallstrukturen*, Gebruder Borntraeger, Berlin, Vols. I and II (1935).
13. T. Hahn, *International Tables for Crystallography, Volume A: Space Group Symmetry,* 5th ed., Springer-Verlag, Berlin, (2002).
14. J.C. Phillips, "Ionicity of the Chemical Bond in Crystals," *Rev. Mod. Phys.*, Vol. 42 (1970) pp. 317–356.
15. M.R. Tubbs, "A Spectroscopic Interpretation of Crystalline Ionicity," *Phys. Stat. Sol.*, Vol. 41 (1970) K61–K64.

16. N.E. Christensen, S. Satpathy, Z. Pawlowska, "Bonding and Ionicity in Semiconductors," *Phys. Rev.*, Vol. B36 (1987) pp. 1032–1050.
17. F. Knoop, C.G. Peters, W.B. Emerson, "A Sensitive Pyramidal-Diamond Tool for Indentation Measurements," *Journal of Research of the National Bureau of Standards*, Vol. 23 (1939) pp. 39–61.
18. A. Banerjee, M. Sherriff, E.A.M. Kidd, T.F. Watson, "A Confocal Microscopic Study Relating the Autofluorescence of Carious Dentine to Its Microhardness," *British Dental Journal*, Vol. 187, no. 4 (1999) pp. 206–210.
19. I. Yonenaga "High-Temperature Strength of III–V Nitride Crystals," *J. Phys. Condens. Mat.*, Vol. 14 (2002) pp. 12947–12951.
20. H. Chandler, *Hardness Testing*, ASM International; 2nd edition (1999).
21. R.L. Smith, G.E. Sandland, "An Accurate Method of Determining the Hardness of Metals, with Particular Reference to Those of a High Degree of Hardness," *Proceedings of the Institution of Mechanical Engineers*, Vol. I (1922) pp. 623–641.
22. F. Mohs, *Grundriss der Mineralogie*, 2 volumes, Dresden (as Prismatisches Scheel-Erz) (1822).
23. S.M. Sze, *Semiconductor Devices: Physics and Technology*, John Wiley & Sons, Hoboken, N.J., 2nd edition, (2002).
24. "Appendix A: Hall Effect Measurements," Lake Shore 7500/9500 Series Hall System User's Manual, Lake Shore Cryotronics, Inc., Westerville, OH.
25. J.R. Haynes, W. Shockley, "The Mobility and Life of Injected Holes and Electrons in Germanium," *Phys. Rev.*, Vol. 81, Issue 5 (1951) pp. 835–843.
26. E. Mooser, W.B. Pearson, "The Chemical Bond in Semiconductors," *J. Electron.*, Vol. 1 (1956) pp. 629–645.
27. L. Vegard, "Die Konstitution der Mischkristalle und die Raumfüllung der Atome," *Z. Phys.*, Vol. 5 (1921) pp. 17–26.
28. C. Julien, G.A. Nazri, *Solid State Batteries: Materials Design and Optimization*, The Springer International Series in Engineering and Computer Science, Vol. 271 (1994).
29. I. Repins, M.A. Contreras, B. Egaas, C. DeHart, J. Scharf, C.L. Perkines, B. To, R. Noufi, "19·9%-Efficient ZnO/CdS/CuInGaSe^2 Solar Cell with 81·2% Fill Factor," *Progress in Photovoltaics: Research and Applications*, Vol. 16, Issue 3 (2008) pp. 235–239.

30. S.E. Shaheen, C.J. Brabec, N.S. Sariciftci, F. Padinger, T. Fromherz, J.C. Hummelen, "2.5% Efficient Organic Plastic Solar Cells," *Appl. Phys. Lett.*, Vol. 78, no. 6 (2001) pp. 841–843.
31. H.E. Ruda, "A Theoretical Analysis of Electron Transport in ZnSe," *J. Appl. Phys.*, Vol. 59 (1986) pp. 1220–1231.
32. M. Lundstrom, *Fundamentals of Carrier Transport*, Cambridge University Press, 2nd ed. (2002).
33. J. Wu, W. Walukiewicz, K.M. Yu, J.W. Ager, E.E. Haller, H. Lu, W.J. Schaff, Y. Saito, Y. Nanishi, "Unusual Properties of the Fundamental Band Gap of InN," *Appl. Phys. Letts.*, Vol. 80, no. 21 (2002) pp. 3967–3969.
34. C. Wu, T. Li, L. Lei, S. Hu, Y. Liu, Y. Xie, "Indium Nitride from Indium Iodide at Low Temperatures: Synthesis and Their Optical Properties," *New J. Chem.*, Vol. 29 (2005) pp. 1610–1615.
35. D.J. Stuckel, "Electronic Structure and Optical Spectrum of Boron Arsenide," *Phys. Rev. B*, Vol. 1, issue 8 (1970) pp. 3458–3463.
36. R.M. Chrenko, "Ultraviolet and infrared spectra of cubic boron nitride," *Solid State Commun.*, Vol. 14 (1974) pp. 511–515.

The Elemental Semiconductors

Physical Properties—Elemental Semiconductors

Material	Group	Crystal Structure	Pearson Symbols/ Space/Point Group	Lattice Constants *a* or *a, c* (nm)	Atomic Number (Z)	Av. Mol. Weight	Density (g cm^{-3})	Melting Point (K)	Iconicity f_i	KH (kg mm^{-2})
β-B	III	rhomb	hR105,R$\bar{3}$m	1.0194, 2.381	5	10.8	2.33	2365	0	2110–2580
Grey-Sn (α)	IV	dia	cF8, Fd$\bar{3}$m(O_h^7)	0.649	50	118.7	5.77	286	0	265
Ge	IV	dia	cF8, Fd$\bar{3}$m(O_h^7)	0.566	32	72.6	5.33	1210	0	780
Si	IV	dia	cF8, Fd$\bar{3}$m(O_h^7)	0.543	14	28.1	2.33	1687	0	1150
a-Si	IV	a	—	0.543	14	28.1	2.25	1420	0	1183
C (Diamond)	IV	dia	cF8, Fd$\bar{3}$m (O_h^7)	0.357	6	12.0	3.52	4100	0	5700–10400
a-Se	VI	a	—	0.435, 0.495	34	79.0	4.28	494[a]	0	45
Te	VI	trig	hP, Im3(T_h^6)	0.445, 0.591	52	127.6	6.23	723	0	2.25[b]

[a] Transforms to gray Se at 360 K.
[b] Mohs scale.

Electronic Properties—Elemental Semiconductors

Material	Type	Bandgap (eV)	Pair Energy (eV)	Dielectric Constant $\varepsilon_r(0)$	Resistivity (Ω-cm)	Electron Mobility ($cm^2V^{-1}s^{-1}$)	Hole Mobility ($cm^2V^{-1}s^{-1}$)	$\mu_e\tau_e$ (cm^2V^{-1})	$\mu_h\tau_h$ (cm^2V^{-1})
β-B	I	1.5		10.6	10^6	10–300	2		
Grey-Sn (α)	SM	0.08		24	2×10^{-4}	1400	1200		
Ge	I	0.664	2.96[a]	16.2	53	3900	1900	>1	>1
Si	I	1.12	3.62	11.9	up to 10^4	1500	480	>1	~1
a-Si	D[b]	1.8	4.2	11.7	10^{12}	1	0.05	2×10^{-7}	3×10^{-8}
C (Diamond)	I	5.48	13.25	5.7	$>10^{12}$	2000	1600	2×10^{-5}	$<1.6 \times 10^{-5}$
a-Se	I	2.3	5.4[c]	6.6	10^{12}	0.005	0.14	6×10^{-6}	2×10^{-5}
Te	D	0.33		39.5	0.44	2380 (‖)	1260 (‖)[d]		
						1150 (⊥)	650 (⊥)[e]		

[a] 77 K.
[b] Pseudo direct.
[c] Strongly dependent on field strength.
[d] (‖) = Parallel to crystallographic c axis.
[e] (⊥) = Perpendicular to the crystallographic c axis.

Group IV–IV Materials

Physical Properties—Group IV–IV Materials

Material	Crystal Structure	Pearson Symbols/Space/ Point Group	Lattice Constants *a* or *a, c* (nm)	Atomic Number (Z)	Av. Mol. Weight	Density (g cm^{-3})	Melting Point (K)	Ionicity f_i	KH (kg mm^{-2})
2H-SiC	H	hP4,$C4_{6v}$-$P6_3mc$	0.3073, 0.5048	14, 6	40.1	3.219	1670[a]		2460
3C-SiC (β)	ZB	cF8,$F\bar{4}3m(T_d^2)$	0.4359	14, 6	40.1	3.215	3103[b]	0.177	2900
4H-SiC	H[c]	hP8,$P6_3mc(C_{6v}^4)$	0.3073, 1.01153	14, 6	40.1	3.290	3103[b]	0.12	3980
6H-SiC	H[c]	hP12,$P6_3mc(C_{6v}^4)$	0.3081, 1.51173	14, 6	40.1	3.215	3103[b]	0.177	2172–2755
15R-SiC	rhomb[c]	hR30,$R3m(C_{3v})$	0.3079, 3.778	14, 6	40.1		2810	0.177	3059
$Si_{1-x}Ge_x$	dia	cF8,$Fd3m(O_h^7)$	$0.531 - 0.02x + 0.003x^2$	14, 32	–	$2.329 + 3.493x - 0.499x^2$	1412 $x = 0$[d] 937 $x = 1$	0.06[e]	$1150 - 350x$
C_xSi_{1-x}			$0.543 - 0.24x + 0.057x^2$	6,14	–		$1700 + 2400x$	0.0	$1020 - 1532x + 9180x^2$

[a] Undergoes solid-state phase transformation to the 3C structure.
[b] At 35 atms (sublimes at 2370 K at 1 atm).
[c] Mixed cubic–hexagonal character.
[d] Liquidus surface $Tl(x) = 937 + 916x - 442x^2$.
[e] Shows little variation with x.

Electronic Properties—Group IV–IV Materials

Material	Type	Bandgap (eV)	Pair Energy (eV)	Dielectric Constant $\varepsilon_r(0)$[a]	Resistivity (Ω-cm)	Electron Mobility ($cm^2V^{-1}s^{-1}$)	Hole Mobility ($cm^2V^{-1}s^{-1}$)	$\mu_e\tau_e$ (cm^2V^{-1})	$\mu_h\tau_h$ (cm^2V^{-1})
2H-SiC	I	3.23			10^2–10^3	1000[b]			
3C-SiC (α)	I	2.39		9.72	150	800	40		
4H-SiC	I	3.26	7.8	10.0(⊥), 9.8(‖))	$>10^{12}$	1000	120	4×10^{-4}	8×10^{-5}
6H-SiC (β)	I	3.05	6.9	10.0(⊥), 9.8(‖)	0.0015–10^3	400	100		
15R-SiC	I	2.99		9.7(⊥), 6.5(‖)		500			
$Si_{1-x}Ge_x$	I	$1.12 - 0.41x + 0.008x^2$ ($x<0.85$) $1.8 - 1.2x$ ($x>0.85$)		$11.7 + 4.5x$	(0.68–1.2) $\times 10^5$	$1450–4325x$[c]	$450 - 865x$[c]	4.5–13.5	2.06–1.72
C_xSi_{1-x}	I/D								

[a] (‖) = Parallel to c axis, (⊥) = perpendicular to c axis.
[b] Calculated.
[c] For $0 < x < 0.3$.

Group III–V Materials

Physical Properties—Group III–V Materials

Material	Crystal Structure	Pearson Symbols/ Space/Point Group	Lattice Constants a or a, c (nm)	Atomic Number (Z)	Av. Mol. Weight	Density (g cm^{-3})	Melting Point (K)	Ionicity f_i	KH (kg mm^{-2})
InSb	ZB	cF8, $F\bar{4}3m(T_d^2)$	0.648	49, 51	236.6	5.78	797	0.321	220
InAs	ZB	cF8, $F\bar{4}3m(T_d^2)$	0.606	49, 33	189.7	4.68	1210	0.357	430
InP	ZB	cF8, $F\bar{4}3m(T_d^2)$	0.587	49, 15	145.8	4.79	1335	0.421	460
InN	W	hP4, $P6_3mc(C_{6v}^4)$	0.354, 0.870	49, 7	128.8	6.81	2146	0.578	1140
GaSb	ZB	cF8, $F\bar{4}3m(T_d^2)$	0.610	31, 51	191.5	5.61	991	0.261	450
GaAs	ZB	cF8, $F\bar{4}3m(T_d^2)$	0.564	31, 33	144.6	5.32	1513	0.310	750
GaP	ZB	cF8, $F\bar{4}3m(T_d^2)$	0.545	31, 15	100.7	4.14	1730	0.327	964
β-GaN	ZB	cF8, $F\bar{4}3m(T_d^2)$	0.451	31, 7	83.7	6.15	3246[a]	0.500	1830
α-GaN	W	hP4, $P6_3mc(C_{6v}^4)$	0.319, 0.518	31, 7	83.7	6.07	2791	0.500	1200–1700
BAs	ZB	cF8, $F\bar{4}3m(T_d^2)$	0.478	5, 33	85.7	5.22	2027	0.044	1937
BP	ZB	cF8, $F\bar{4}3m(T_d^2)$	0.4538	5, 15	41.8	2.97	3300	0.032	3263
β-BN	ZB	cF8, $F\bar{4}3m(T_d^2)$	0.362	5, 7	24.8	3.48	3246	0.256	6730–7648
α-BN	H	hP4, $P6_3/mmc(D_{6h})$	0.250, 0.666	5, 7	24.8	3.28	2873[2]	0.221	1489

(continued)

Physical Properties—Group III–V Materials

Material	Crystal Structure	Pearson Symbols/ Space/Point Group	Lattice Constants *a* or *a*, *c* (nm)	Atomic Number (Z)	Av. Mol. Weight	Density (g cm^{-3})	Melting Point (K)	Ionicity f_i	KH (kg mm^{-2})
AlSb	ZB	cF8, $F\bar{4}3m(T_d^2)$	0.614	13, 51	148.7	4.22	1338	0.250	408
AlAs	ZB	cF8, $F\bar{4}3m(T_d^2)$	0.566	13, 33	101.9	3.76	1710	0.274	510
AlP	ZB	cF8, $F\bar{4}3m(T_d^2)$	0.546	13, 15	58.0	2.36	2823	0.307	561
AlN	W	hP4, $P6_3mc(C_{6v}^4)$	0.311, 0.498	13, 7	41.0	3.26	3487	0.449	1020–1427

[a] Sublimes/decomposes.

Electronic Properties—Group III–V Materials

Material	Type	Bandgap (eV)	Pair Energy (eV)	Dielectric Constant $\varepsilon_r(0)$[a]	Resistivity (Ω-cm)	Electron Mobility ($cm^2V^{-1}s^{-1}$)	Hole Mobility ($cm^2V^{-1}s^{-1}$)	$\mu_e\tau_e$ (cm^2V^{-1})	$\mu_h\tau_h$ (cm^2V^{-1})
InSb	D	0.18	1.2	17.7	~100 @ 4 K	80000	1250	7×10^{-5}	
InAs	D	0.354	2.0	15.1	0.03	33000	460		
InP	D	1.34	4.2	12.4	10^8	4600	150	2×10^{-5}	1×10^{-5}
InN	D	1.89[b]		13.1($\perp$), 14.4($\parallel$)		3200	<80		
GaSb	D	0.75		15.7	0.04	4000	700	1.4×10^{-5}	
GaAs	D	1.425	4.35	13.1	10^{10}	8500	400	1×10^{-4}	4×10^{-6}
GaP	I	2.27	6.5	11.1	10^8–10^{11}	110	75	1.4×10^{-6}	
β-GaN	D	3.23		8.9	10^6	760	250		
α-GaN	D	3.42		10.4($\perp$), 9.5($\parallel$)	$>10^{10}$	1500	30		
BAs	I	1.85[c]		9.9					
BP	I	2.4	6.5	11.0	20	30–120	285–500		
BN (c)	I	6.4		7.1	10^5	<200	<500		
BN (h)	I	5.2		7.04($\perp$), 5.1($\parallel$)	10^{14}				
AlSb	I	1.62	6.58	14.4	0.04–10^7	200	420		
AlAs	I	2.15		10.9	0.2	200	100		
AlP	I	2.42		9.8	1×10^{-5}	60	450		
AlN (h)	D	6.19		8.3($\perp$), 8.9()	10^{13}	300	14		

[a] ($\parallel$) = Parallel to c axis, ($\perp$) = perpendicular to c axis.

[b] Early studies showed 1.9–2.05. A recent study claims a value of 0.7–0.8 [33] and is supported by measurement [34].

[c] Calculated value [35]. Optical absorption measurements suggest a bandgap energy of 0.67 eV [36].

Group II–VI Materials

Physical Properties—Group II–VI Materials

Material	Crystal Structure	Pearson Symbols/Space/ Point Group	Lattice Constants *a* or *a, c* (nm)	Atomic Number (Z)	Av. Mol. Weight	Density (g cm^{-3})	Melting Point (K)	Ionicity f_i	KH (kg mm^{-2})
HgTe	ZB	cF8, $F\bar{4}3m(T_d^2)$	0.646	80, 52	328.2	8.12	943	0.650	23.5–37.7
HgSe	ZB	cF8, $F\bar{4}3m(T_d^2)$	0.609	80, 34	279.6	8.22	1270	0.680	23.5
α-HgS	W	hP6,$P6_3mc(C_{6v}^4)$	0.415, 0.950	80, 16	232.7	8.19	1093	0.790	91
β-HgS	ZB	cF8, $F\bar{4}3m(T_d^2)$	0.585	80, 16	232.7	7.73	857	0.790	138[a]
CdTe	ZB	cF8, $F\bar{4}3m(T_d^2)$	0.648	48, 52	240.0	5.85	1366	0.675	45.9–61.2
CdSe (c)	ZB	cF8, $F\bar{4}3m(T_d^2)$	0.607	48, 34	191.4	5.67	1512	0.699	130
CdSe	W	hP4,$P6_3mc(C_{6v}^4)$	0.430, 0.701	48, 34	191.4	5.81	2023	0.699	91.8
CdS (c)	ZB	cF8, $F\bar{4}3m(T_d^2)$	0.582	48, 16	144.5	4.87	1748	0.685	125
CdS (h)	W	hp4, $P6_3mc(C_{6v}^4)$	0.414, 0.671	48, 16	144.5	4.83	1748	0.685	123–235
ZnTe	ZB	cF8, $F\bar{4}3m(T_d^2)$	0.610	30, 52	193.0	6.34	1568	0.609	61.2–91.8
ZnSe (c)	ZB	cF8, $F\bar{4}3m(T_d^2)$	0.567	30, 34	144.3	5.26	1799	0.630	139.7–186.6
β-ZnS	ZB	cF8, $F\bar{4}3m(T_d^2)$	0.541	30, 16	97.5	4.08	1991[b]	0.623	178
α-ZnS	W	hP4, $P6_3mc(C_{6v}^4)$	0.382, 0.626	30, 16	97.5	4.08	2196	0.623	178
ZnO	ZB	cF8, $F\bar{4}3m(T_d^2)$	0.463	30, 8	81.4	5.68	2248	0.616	5.0 (M[c])

ZnO	W	hP4,$P6_3mc(C_{6v}^4)$	0.3250, 0.5204	30, 8	81.4	5.61	1975[b]	0.616	407.9[d]
β-MgTe	W	hp4, $P6_3mc(C_{6v}^4)$	0.4530, 0.7406	25, 52	151.9	3.81	1360	0.554	143
β-MgSe	NaCl	cF8, $Fm\bar{3}m(O_h^5)$	0.589	12, 34	103.3	4.20	1560	0.790	152
β-MgS	NaCl	cF8, $Fm3m(O_h^5)$	0.520	12, 16	56.4	2.86	2783	0.786	3.5 (M[c])[e]

[a] Estimated.
[b] Decomposes/sublimes.
[c] Mohs scale.
[d] Hardest of the II–VI–VI compounds.
[e] Value given for its mineral form, niningerite.

Electronic Properties—Group II–VI Materials

Material	Type	Bandgap (eV)	Pair Energy (eV)	Dielectric Constant $\varepsilon_r(0)$[a]	Resistivity (Ω-cm)	Electron Mobility ($cm^2V^{-1}s^{-1}$)	Hole Mobility ($cm^2V^{-1}s^{-1}$)	$\mu_e\tau_e$ (cm^2V^{-1})	$\mu_h\tau_h$ (cm^2V^{-1})
HgTe	SM	–0.14		20.8	10^{-2}–10^{-3}	25000	350		
HgSe	SM	–0.06		25.6		20000	2		
α-HgS	D	2.03		23.5(‖), 18.2(⊥)	3450 (‖), 11080 (⊥)	30(‖), 10(⊥)			
β-HgS	SM	0.54		18.2		250			
CdTe	D	1.48	4.43	10.4	10^9	1050	100	3.3×10^{-3}	2×10^{-4}
CdSe (c)	D	1.74	5.5	10.0	10^5	800	50	7.2×10^{-4}	7.5×10^{-5}
CdSe (h)	D	1.75	5.5	9.3(⊥), 10.2(‖)	10^8	840	75	6×10^{-5}	7.5×10^{-5}
CdS (c)	D	2.42	6.3	5.4		340	50		
CdS (h)	D	2.51	7.8	8.3(⊥), 8.7(‖)	10^{10}	330	48	10^{-7}	
ZnTe	D	2.28	7.0	10.4	10^{10}	340	100	1.4×10^{-6}	7×10^{-5}
ZnSe (c)	D	2.63	8.0	9.2	2×10^9	540	28	4×10^{-5}	
β-ZnS (c)	D	3.58	8.2	8.9	10^{10}	165	10		
α-ZnS (w)	D	3.91		8.6(⊥), 8.4(‖)	10^8–10^{12}	280(‖), 165(⊥)	100–800		
ZnO (c)	D	3.35	7.5	9.0		200	180		
ZnO (w)	D	3.37		7.8(⊥), 8.8(‖)	10^9	250(‖), 280(⊥)	125		
β-MgTe	D	3.49		6.1					

[a] (‖) = Parallel to c axis, (⊥) = perpendicular to c axis.

Group I–VII

Physical Properties—Group I–VII Materials

Material	Crystal Structure	Pearson Symbols/ Space/Point Group	Lattice Constants a or a, c (nm)	Atomic Number (Z)	Av. Mol. Weight	Density (g cm^{-3})	Melting Point (K)	Ionicity f_i	KH (kg mm^{-2})
β-MgSe	D	4.05		5.2					
β-MgS	D	4.5 (77 K)		4.5					
AgBr	NaCl	cF8, $Fm\bar{3}m(O_h^5)$	0.577	47, 35	187.8	6.47	705	0.850	7
β-AgI	W	hP4, $P6_3mc(C_{6v}^4)$	0.458, 0.749	47, 53	243.8	5.67	831	0.770	24–52[a]
γ-CuBr	ZB	cF8, $F\bar{4}3m(T_d^2)$	0.569	29, 35	143.5	4.71	765	0.735	21.6
γ-CuI	ZB	cF8, $F\bar{4}3m(T_d^2)$	0.604	29, 53	190.5	5.62	879	0.692	96[a]
AgCl	NaCl	cF8, $Fm\bar{3}m(O_h^5)$	0.555	47, 17	143.3	5.59	730	0.856	9.5
γ-CuCl	ZB	cF8, $F\bar{4}3m(T_d^2)$	0.541	29, 17	99.0	4.14	703	0.746	52–96[a]

[a] Estimated.

Electronic Properties—Group I–VII Materials

Material	Type	Bandgap (eV)	Pair Energy (eV)	Dielectric Constant $\varepsilon_r(0)$	Resistivity (Ω-cm)	Electron Mobility ($cm^2V^{-1}s^{-1}$)	Hole Mobility ($cm^2V^{-1}s^{-1}$)	$\mu_e\tau_e$ (cm^2V^{-1})	$\mu_h\tau_h$ (cm^2V^{-1})
AgBr	I	2.684	5.8	12.2		60	2	2	
β-AgI	D	3.024		7.0		30			
γ-CuBr	D	3.077		7.9					
γ-CuI	D	3.115		7.1	2×10^6		110		
AgCl	I	3.249	7.5	11.5	4×10^8	50			
γ-CuCl	D	3.395		6.3	1×10^8				

Other Groups: II–V, IV–VI, V–VI, and n–VII Materials

Physical Properties—Other Groups: II–V, III–VI, IV–VI, V–VI, and n–VII Materials

Material	Group	Crystal Structure	Pearson Symbols/Space/Point Group	Lattice Constants *a* or *a, c* (nm)	Atomic Numbers	Av. Mol. Weight	Density (g cm^{-3})	Melting Point (K)	Iconicity f_i	KH (kg mm^{-2})
CdSb	II–V	ortho	oP16,Pbca(D_{2h}^{15})	0.647, 0.8248, 0.8531	48, 51	234.2	6.98	729[a]	0.699	
CdSb	II–V	W	hP4	0.4298, 0.7002	48, 51	234.2	5.67		0.699	91.8–132.6
α-HgI_2	II–VII	tetra,layered	P41/nmc(D_{3d}^{5})	0.4361, 1.245	80, 53	454.4	6.4	532	0.67	<10
B_4C	III–IV	rhomb	$R\overline{3}m(D_{3d}^{5})$	0.560, 1.207	5, 6	55.3	2.52	2623	0.45	2900–3580
GaTe	III–VI	H, layered	hP8, C2lm	0.402, 1.671	31,52	197.3	5.44	1097[a]	0.26	35 (HB)[b]
GaSe	III–VI	H, layered	hP8,P6m2	0.374, 1.589	31, 34	148.7	5.03	1211	0.37	7.8
GaS	III–VI	H, layered	hP8,P6_3mmc	0.359, 1.549	31, 16	101.8	3.86	1233[a]	0.43	29 (HB)[b]

(continued)

Physical Properties—Other Groups: II–V, III–VI, IV–VI, V–VI, and n–VII Materials (Continued)

Material	Group	Crystal Structure	Pearson Symbols/Space/Point Group	Lattice Constants *a* or *a, c* (nm)	Atomic Numbers	Av. Mol. Weight	Density (g cm^{-3})	Melting Point (K)	Iconicity f_i	KH (kg mm^{-2})
InI	III–VII	ortho	CmCm(D2)	0.476, 1.278, 0.4912	49, 53	241.7	5.32	637	0.80	27
TlBr	III–VII	CsCl	cP2,Pm$\underline{3}$m	0.399	81, 35	284.3	7.45	733[a]	0.81	12
SnS	IV–VI	ortho	oP8,Pnma(D_{2h}^{16})	0.4349, 1.1202, 0.3988	50, 16	150.8	5.08	1153	0.76	2 (M)[c]
PbTe	IV–VI	NaCl	cF8,Pnma(D_{2h}^{16})	0.646	82, 52	334.8	8.16	1197	0.74	37
PbSe	IV–VI	NaCl	cF8,Pnma(D_{2h}^{16})	0.612	82, 34	286.2	8.1	1355	0.77	72
PbS	IV–VI	NaCl	cF8,Pnma(D_{2h}^{16})	0.594	82, 16	239.3	7.59	1383	0.79	93
PbO	IV–VI	tetra	tP4,P4/nmm(D_{4h}^{7})	0.3976, 0.5023	82, 8	223.2	9.53	1163		2 (M)[c]
PbI_2	IV–VII	H, layered	hP3,P-3m1	0.4557, 0.6979	82, 53	461.0	6.2	681	0.8	<10
Bi_2S_3	V–VI	ortho, layered	oP20, Pnma(D_{2h})	1.111, 1.125, 0.397	83, 16	514.2	6.81	1036[a]		2(M)[c]

BiI_3	V–VII	rhomb,layered	hR24,$R\overline{3}h(D_{3d}^5)$	0.770, 2.073	83, 53	589.7	5.78	681[a]	0.41 (T[d])	17.7[e]
Bi_2Te_3	V–VI	trig, layered	hR15, R-3m,	0.438, 3.036	83, 52	800.8	7.86	858		155

[a] Melts congruently.
[b] Brinell hardness.
[c] Mohs scale.
[d] Tubbs ionicity [15].
[e] Vickers scale.

Electronic Properties—Other Groups: II–V, IV–VI, V–VI, and n–VII Materials

Material	Type	Bandgap (eV)	Pair Energy (eV)	Dielectric Constant $\varepsilon_r(0)$[a]	Resistivity (Ω-cm)	Electron Mobility ($cm^2V^{-1}s^{-1}$)	Hole Mobility ($cm^2V^{-1}s^{-1}$)	$\mu_e\tau_e$ (cm^2V^{-1})	$\mu_h\tau_h$ (cm^2V^{-1})
CdSb (O)	I	0.49		16.4	2	300–700	100–660		
CdSb (W)	D	1.72				900	50		
α-HgI_2	D	2.13	4.2	25.9(⊥), 8.5(‖)	10^{13}	100	4	1×10^{-4}	4×10^{-5}
B_4C	D	1.5		10.0	0.1–10	0.7	2		
GaTe	D[b]	1.65		10.6(⊥), 9.7(‖)	10^9	40–600	25–40 (Hall)	1.5×10^{-5}	1.5×10^{-5}
GaSe	D	2.03	4.5	10.6(⊥), 6.2(‖)	10^{10}	(‖) 80, (⊥) 300	45	3.5×10^{-5}	1×10^{-5}
GaS	I	2.5		10.0(⊥), 5.9(‖)	7×10^8	12 (Hall)	80		
InI	D	2.01[c]		26	5×10^{11}	250		5×10^{-6}	7×10^{-5}
TlBr	I	2.68	6.5	30.6	10^{12}	30	4	3×10^{-4}	6×10^{-5}
SnS	D	0.094 (2.0)		24	894	13	90 (⊥)		
PbTe	D	0.31		412	0.01	1900	900		
PbSe	D	0.28		250	0.1	1800	930		
PbS	I	0.41		174	0.005	500	600		
PbO	I	1.9	6.36	25.9	10^{11}–10^{13}	100 (Hall)	80		
PbI_2	D	2.32	4.9	26.4(⊥), 9.3(‖)	10^{13}	65	20	1×10^{-5}	1×10^{-6}

Bi_2S_3	D	1.38	54(⊥), 8.6(‖)	10^9	200	1100	2×10^{-8}	1×10^{-7}
BiI_3	I[d]	1.78	54(⊥), 8.6(‖)	2×10^{12}	680	20	1×10^{-5}	$<1 \times 10^{-7}$
Bi_2Te_3	I	0.16			1140	680		

[a] (‖) = Parallel to c axis, (⊥) = perpendicular to c axis.

[b] The exact electronic band structure of GaTe is not known, but has been established to be a direct semiconductor.

[c] Controversial.

[d] Hexagonal form has a direct bandgap.

Binary Alloys and Ternary Compounds

Electronic Properties—Binary Alloys and Ternary Compounds

Material	Crystal Structure	Pearson Symbols/ Space/Point Group	Lattice Constants *a* or *a*, *c* (nm)	Atomic Number (Z)	Density (g cm^{-3})	Melting Point (K)	Diclectric Constant $\varepsilon_r(0)$	Ionicity f_i	KH (kg mm^{-2})
$Hg_{1-x}Cd_xTe$	ZB	cF8	$0.6461 + (8.4x + 11.68x^2 - 5.7x^3)10^{-4}$	80, 48, 52	$8.05 - 2.3x$	1223	19.5 (x = 0.2) T = 4.2 K		70 max at x = 0.75
$In_xAs_{1-x}Sb$	ZB	cF8	0.636 for x=0.35	49, 13, 51	$5.68 + 0.09x$		$15.15 + 1.65x$		
$In_xGa_{1-x}As$[a]	ZB	cF8	0.5869 for x=0.53	49, 31, 33					
$In_xAl_{1-x}P$[b]	ZB	cF8	0.5653 for x=0.5	49, 13, 15					
$Ga_xIn_{1-x}Sb$	ZB	cF8		31, 49, 51	$5.77 - 0.16x$		$16.8 - 1.1x$		265–459
$Ga_xAs_{1-x}Sb$	ZB	cF8		31, 33, 51	$5.32 + 0.29x$		$12.90 + 2.8x$		450–750
$Ga_xIn_{1-x}As$	ZB	cF8, $F\bar{4}3m$(Td2)	$0.60583 - 0.0405x$	31, 49, 33	$5.68 - 0.37x$	~1373	$15.1 - 2.87x + 0.67x^2$		~560

$Ga_xIn_{1-x}P$	ZB	cF8,$F\bar{4}3m(T_d^2)$	0.5868 – 0.00418x	31, 49, 15	4.8 – 0.67x		12.5 – 1.4x		1122–530
$Ga_xIn_{1-x}N$	ZB	cF8							
$Al_xGa_{1-x}Sb$	ZB	cF8		13, 31, 51					
$Al_xIn_{1-x}Sb$	ZB	cF8		13, 49, 33					
$Al_xIn_{1-x}As$[a]	ZB	cF8		13, 49, 33					
$Al_xGa_{1-x}As$	ZB	cF8/$F\bar{4}3m(T_d^2)$	0.56533 + 0.00078x	13, 31, 33	5.32 – 1.56x	12401082x + 582x^2[c]	12.90 – 2.84x	0.31 – 0.36x	730–510
$Cd_{1-x}Zn_xTe$	ZB	cF8	0.6477 – 0.03772x	48, 30, 52	5.85	1365–1568	10.4 (x=0) 9.7 (x=1)		102 (x = 4%)
$CdTe_{1-x}Se_x$	ZB	cF8	0.6481 – 0.411x	48, 52, 34	5.85–6.3	1323			
$Cd_xZn_{1-x}Se$	ZB/W[d]	cF8/$P6_3$	0.567 + 0.04x	48, 30, 34	5.4–5.8	1512–1793			
$Cd_{1-x}Mn_xTe$	ZB[e]	cF8	0.6482 – 0.0150x	48, 25, 52	5.3–5.7	1343, x=0.45			76 (x = 45%)
$HgBr_xI_{2-x}$[f]	ortho,layered	oP, C_{2v}^{12}	0.772, 1.3101, 0.604	80, 35, 53	6.2	502–532			14

(continued)

Electronic Properties—Binary Alloys and Ternary Compounds (Continued)

Material	Crystal Structure	Pearson Symbols/ Space/Point Group	Lattice Constants *a* or *a*, *c* (nm)	Atomic Numbers	Density (g cm^{-3})	Melting Point (K)	$\varepsilon_r(0)$	Iconicity f_i	KH (kg mm^{-2})
$TlBr_xI_{1-x}$[g]	CsCl[h]	cP2	0.4125	81, 35, 53	7.4	688	32.5[i]		40[i]
$TlPbI_3$	ortho	oP	0.462, 1.49, 1.19	81, 82, 53	6.6	619			70
Tl_6SeI_4	H, layered	P4/mnc	0.918, 0.968	81, 79, 53	7.4	710			72
$TlGaSe_2$	monoclinic	C2/c	1.077, 1.563	81, 32, 79	6.4				
$TlAu_4S_3$	ortho	Pmmm	0.751, 1.1919, 0.469	81, 79, 16	10.2				

[a] Can be lattice matched to InP
[b] Can be latticed matched to GaAs.
[c] Liquidus surface.
[d] Zinc blende structure observed for $0.7 < x < 1.0$, wurtzite structure for $x<0.68$.
[e] Zinc blende structure observed for $0 < x < 0.77$.
[f] Forms solid solutions for $0.2 < x < 1.0$.
[g] Also known as KRS-5 for $x = 0.4$.
[h] For $x>0.3$. At $x = 0.3$, bulk $TlBr_xI_{1-x}$ transforms from the cubic to the orthorhombic phase when cooled to LN_2 temperatures.
[i] Quoted for KRS-5.

Electronic Properties—Binary Alloys and Ternary Compounds

Material	Type	Bandgap (eV)	Pair Energy (eV)	Resistivity (Ω-cm)	Electron Mobility ($cm^2V^{-1}s^{-1}$)	Hole Mobility ($cm^2V^{-1}s^{-1}$)	$\mu_e\tau_e$ (cm^2V^{-1})	$\mu_h\tau_h$ (cm^2V^{-1})
$Hg_{1-x}Cd_xTe$	D (for $x >$ 0.15)	–0.3 ($x = 0$)–1.6 ($x =$ 1)[a]	$3.4\varepsilon_g + 0.95$	100	$10^4/(8.754x - 1.044)$[b]	40–80		
$In_xAs_{1-x}Sb$	D	0.1 for $x = 0.35$			5×105 for $x =$ 0.35	<500		
$In_xGa_{1-x}As$	D	0.75 for $x = 0.53$		2430	10000 $x = 0.53$	150		
$In_xAl_{1-x}P$	D/I	2.35 for $x = 0.5$						
$Ga_xIn_{1-x}Sb$	D	$0.172 + 0.165x + 0.43x^2$			5000 – 8000 $x = 0.08$–0.14	7000–80000		
$Ga_xAs_{1-x}Sb$	D	$1.42 - 1.9x + 1.2x^2$ (for $0 < x < 0.3$)						
$Ga_xIn_{1-x}As$	D	$0.36 + 0.63x + 0.43x^2$			$(40 - 80.7x + 49.2x^2)$	450–300	700–7000	
$Ga_xIn_{1-x}P$	I / D	$1.34 + 0.511x + 0.604x^2$ (for $0.49 < x < 0.55$)			6500–1000	180–140		
$Ga_xIn_{1-x}N$	D	$3.21 + 1.969x$			1200–200			
$Al_xGa_{1-x}Sb$	D / I	$0.73 + 1.10x + 0.47x^2$			12000–1000	1000–100		

(continued)

Electronic Properties—Binary Alloys and Ternary Compounds (Continued)

Material	Type	Bandgap (eV)	Pair Energy (eV)	Resistivity (Ω-cm)	Electron Mobility ($cm^2V^{-1}s^{-1}$)	Hole Mobility ($cm^2V^{-1}s^{-1}$)	$\mu_e\tau_e$ (cm^2V^{-1})	$\mu_h\tau_h$ (cm^2V^{-1})
$Al_xIn_{1-x}Sb$	D / I	$0.172 + 1.621x + 0.43x^2$			76000–1000	1100–400		
$Al_xIn_{1-x}As$	D / I	$0.36 + 2.35x + 0.24x^2$			34000–1000	450–100		
$Al_xGa_{1-x}As$	D ($x <$ 0.45) I ($x > 0.45$)	$1.424 + 1.087x + 0.438x^2$ $1.9 + 0.125x + 0.1438x^2$			$8{\cdot}10^3 - 2.2{\cdot}10^4x + 10^4{\cdot}x^2$	$370 - 970x + 740x^2$		
$Cd_{1-x}Zn_xTe$	D	$1.51 + 0.606x + 0.139x^2$	5.0	$>10^{10}$	1350	120	1×10^{-2}	2×10^{-4}
$CdTe_{1-x}Se_x$	D	$1.48(1 - x) + 1.74x - 0.88x(1 - x)$		10^9	59	33	10^{-2}	10^{-2}
$Cd_xZn_{1-x}Se$	D	$2.63x + 1.74(1 - x) - 0.73x(1 - x)$		10^{11}	~500		10^{-4}	
$Cd_{1-x}Mn_xTe$	D	$1.526 + 1.316x$	5.0, $x = 0.1$	3×10^{10}	10–80	40	1×10^{-3}	
$HgBr_xI_{2-x}$	D	2.1–3.4		5×10^{13}	30 ($x = 0.25$), 0.45 ($x = 0.5$)	0.1 ($x = 0.25$), 0.01 ($x = 0.5$)	7.1×10^{-6}	1×10^{-5}

$TlBr_xI_{1-x}$ [c]	I	$0.983x + 1.82$	3×10^{10} ($x = 0.35$)			1×10^{-3} [d]	
			5×10^{11} ($x = 1$)				
$TlPbI_3$		2.3	10^{11}				
Tl_6SeI_4	D	1.86	4×10^{12}			7×10^{-3}	6×10^{-4}
$TlGaSe_2$	I	1.95	10^{10} (⊥),10^4(‖)	23-99	61-65	1×10^{-4}	3×10^{-5}
$TiAu_4S_3$	D	1.63	6×10^7			1×10^{-4}	1×10^{-6}

[a] $E_g(\text{eV}) = -0.313 + 1787x + 0.444x^2 - 1.237x^3 + 0.932x^4 + (6.67 \times 10^{-4} - 1.714 \times 10^{-3}x + 7.6 \times 10^{-4}x^2).T(\text{K})$, for T>70 K.
[b] $0.18 \le x \le 0.25$.
[c] Also known as KRS-5 for x = 0.4.
[d] for x = 0.35.

Appendix F: Table of Radioactive Calibration Sources

The activity of a radioactive source is defined as its rate of decay and is given by

$$\left.\frac{dn}{dt}\right|_{decay} = -\lambda n \tag{F.1}$$

where n is the number of radioactive nuclei, and λ is defined as the decay constant. Historically the unit of activity has been the curie (Ci), which is defined as 3.7×10^{10} disintegrations/sec and initially arose as the best estimate for the total activity of 1 g of pure ^{226}Ra. For laboratory measurements involving calibrating standards, the mCi (10^{-3} Ci) and the μCi (10^{-6} Ci) are most commonly used. Other commonly used units are the roentgen and the rem. The roentgen is named after the discoverer of X-rays. It is defined as that amount of radiation required to produce 0.001293 grams of air ions carrying one electrostatic unit of electricity of either sign. This unit is only used with X-rays and gamma rays. The common submultiple is the milliroentgen or mR. For ionizing radiations we use the rem (roentgen equivalent for man) which includes a factor to take into account the abilities of the different radiations to produce biological injury. The common submultiple is the millirem or mrem. The amount of radiation received by a particular body is known as the dose and is defined as the amount of energy imparted to matter by ionizing particles, per unit mass of irradiated material, at the place of interest. Radiation doses are usually quantified in terms of the rem and the rad. The rad is a measure of the amount of the energy absorbed in a certain amount of material

(100 ergs per gram). The SI unit of absorbed dose is the gray (Gy) where 1 Gy = 100 rads. Because the rad does not specify the medium, a medium should be stated unless clearly implied. For example, the term "tissue rad" should be used in the case of exposure of soft tissue. The rem and the rad are related by the relative biological effectiveness (RBE) which expresses the effectiveness of a particular type of radiation in producing the same biological response as X-rays or gamma radiation. For applied radiation protection the quality factor (QF) is the preferred term under most practical standards.

F.2 Radiation Quantities and Units

Unit of activity = **curie**:
1 Ci = 3.7×10^{10} disintegration s^{-1}

SI unit of activity = **becquerel**:
1 Bq = 1 disintegration s^{-1} = 2.703×10^{-11} Ci

Unit of exposure dose for X- and γ-radiation = **roentgen**:
1 R = 1 esu cm^{-3} = 87.8 erg g^{-1} (5.49×10^{7} MeV g^{-1}) of air

Unit of absorbed dose = **rad**:
1 rad = 100 erg g^{-1} (6.25×10^{-7} MeV g^{-1}) in any material
1 Gy = 100 rad

Unit dose equivalent (for protection): **rem**
1 rem (roentgen equivalent for man) = 1 rad × QF
where QF is the quality factor which depends upon the type of radiation and other factors. For X- and gamma rays, QF ≅ 1; for beta rays and electrons, QF=1; for thermal neutrons, QF ≅ 2.5; for protons and fast neutrons, QF=10; for alpha particles, QF=10; and for heavy ions, QF ranges up to 20
1 sievert = 100 rem

TABLE F.1
Radiation Exposure (mrem yr^{-1}) of a Typical Person in the United States*

Natural Sources	Exposure (mrem yr^{-1})	Artificial Sources	Exposure (mrem yr^{-1})
Cosmic radiation	50	Environmental	8
Terrestrial radiation	50	Medical	92
Internal isotopes	25	Occupational	1
Total natural	125	Nuclear power	0.3
		Miscellaneous	5
Total dosage range (geographical and other factors)		75–50,00 mrem yr^{-1}	

* This table has been adapted from *"Particle Properties Data Booklet,"* Lawrence Berkeley National Laboratory, Berkeley, CA (1980) and A.C. Upton, "The Biological Effects of Low-Level Ionizing Radiation," *Sci. Amer.*, Vol. 246 (1982) pp. 41–49.

Maximum permissible occupational dose for the whole body:
5 rem yr^{-1} (or ~100 mrem wk^{-1})

Fluxes (per cm^2) to liberate 1 rad in carbon:
3.5×10^7 minimum ionizing singly charged particles
1.0×10^9 photons of 1 MeV energy
(These fluxes are correct to within a factor of 2 for all materials.)

F.3 Table of Radionuclides

In the following table, we list over 50 radionuclides commonly used as alpha, beta, X- and gamma-ray calibration standards. Sources may be obtained from Isotope Products Laboratories (1800 N. Keystone St., Burbank, CA, 91504), Amersham Corporation (2636 Clearbrook Dr., Arlington Heights, IL, 60005), or the National Institute of Standards & Technology (US Dept. of Commerce, NIST, Bldg. 202, Rm. 204,

Gaithersburg, MD, 20899) and are generally available with strengths ranging from μCi to mCi and calibrated activities from 1 to 20%. The first column of the table, lists the parent nuclide. Some daughter nuclides may be in equilibrium with the parent nuclide when source is supplied. In cases where this may occur, the transition probabilities for the daughters relate to the disintegrations of each daughter. This is stated in the tables. Daughters with half-lives greater than the parent nuclide have not been listed because they would be present only in insignificant amounts. The second column gives the type of decay (*i.e.*, α, β, X, or γ), and the third lists the corresponding decay energies in units of MeV. For β-emission, the end-point energy is quoted. The fourth column gives the transition probabilities for each mode of the primary decay. They are expressed as percentages of the total number of nuclear transformations of the relevant nuclides. The last two columns give the photon energies and branching ratios for electromagnetic transitions.

Abbreviations:
Half-lives

y—years
d—days
h—hours
m—minutes
s—seconds
ms—milliseconds
μs—microseconds

Type of decay

e.c.—electron capture
i.t.—isometric transition
s.f.—spontaneous fission

Photons emitted

IC—photons of the stated energy are ~100% internally converted

Nuclide Index

Nuclide and Half-Life	Type of Decay	Particle Energies and Transition Probabilities		Electromagnetic Transitions	
		Energy MeV	Transition Probability	Photon Energy MeV	Photons Emitted
Americium-241	α	5.387	1.6%	0.026	2.5%
(433 y)		5.442	12.5%	0.033	0.1%
		5.484	85.2%	0.043	0.1%
		5.511	0.20%	0.0595	35.9%
		5.543	0.34%	0.099	0.02%
		others	low	0.103	0.02%
				0.125	0.004%
				others	low
				Np LX-rays	~40%
				(0.012 –0.022)	
Antimony-124	β–	0.21	9%	0.603	98.0%
(60.2 d)		0.61	52%	0.646	7.2%
		0.86	4%	0.709	1.4%
		0.94	2%	0.714	2.3%
		1.57	5%	0.723	11.2%
		1.65	3%	0.791	0.7%
		2.30	23%	0.968	1.9%
		others	low	1.045	1.9%
				1.325	1.5%
				1.355	0.9%
				1.368	2.5%
				1.437	1.1%
				1.691	50.4%
				2.091	6.1%
				others	<0.5% each

(continued)

Nuclide and Half-Life	Type of Decay	Particle Energies and Transition Probabilities		Electromagnetic Transitions	
		Energy MeV	Transition Probability	Photon Energy MeV	Photons Emitted
Antimony-125	β–	0.094	13.5%	0.035	4.5%
(2.77 y)		0.124	5.7%	0.176	6.8%
		0.130	18.1%	0.321	0.5%
		0.241	1.5%	0.381	1.5%
		0.302	40.2%	0.428	29.8%
		0.332	0.3%	0.463	10.4%
		0.445	7.2%	0.601	17.8%
		0.621	13.5%	0.607	4.9%
				0.636	11.4%
				0.671	1.7%
				others	<0.5% each
				0.027 –0.031	~50%(Te KX-rays)
	Daughter ^{125m}Te(58 d)			(23% of ^{125}Sb decays via ^{125m}Te)	
	i.t. 100%			0.035	7%
				0.109	0.3%
				0.027 –0.032	~110%(Te KX-rays)
Barium-133	e.c.		100%	0.053	2.2%
(10.8 y)				0.080	2.6%
				0.081	33.9%
				0.161	0.7%
				0.223	0.4%
				0.276	7.1%
				0.303	18.4%
				0.356	62.2%
				0.384	8.9%

Nuclide and Half-Life	Type of Decay	Particle Energies and Transition Probabilities		Electromagnetic Transitions	
		Energy MeV	Transition Probability	Photon Energy MeV	Photons Emitted
				Cs KX-rays (0.030–0.036)	~120%
Barium-140	β-	0.468	24%	0.014	1.3%
(12.80 d)		0.582	10%	0.030	14%
		0.886	2.6%	0.163	6.2%
		1.005	46%	0.305	4.5%
		1.019	17%	0.424	3.2%
		others	0.4%	0.438	2.1%
				0.537	23.8%
				0.602	0.6%
				0.661	0.7%
				others	low intensity
	Daughter ^{140}La				
Beryllium-7	e.c.		100%	0.478	10.4%
(53.3 d)					
Bromine-82	β-	0.263	1.7%	0.221	2.3%
(35.3 h)		0.444	98.3%	0.554	72%
				0.606	1.0%
				0.619	43%
				0.698	27%
				0.777	83%
				0.828	24%
				1.008	1.7%
				1.044	29%

(*continued*)

Nuclide and Half-Life	Type of Decay	Particle Energies and Transition Probabilities		Electromagnetic Transitions	
		Energy MeV	Transition Probability	Photon Energy MeV	Photons Emitted
				1.317	28%
				1.475	17%
				1.651	0.9%
				others	
				up to 1.96	<1% each
Cadmium-109	e.c		100%	Ag KX-rays	67.7%
(462 d)				(0.022 –0.026)	
	via ^{109m}Ag(40 s)			0.088	3.8%
				Ag KX-rays	~34.5%
				(0.022 –0.026)	
Calcium-45	β–	0.257	100%		
(164 d)					
Calcium-47	β-	0.69	82%	0.489	6.8%
(4.54 d)		1.22	0.1%	0.530	0.1%
		1.99	17.9%	0.767	0.2%
				0.808	6.8%
				1.297	75.1%
				others	low intensity
	Daughter ^{47}Sc(3.48 d)				
		0.44	70%	0.159	69.7%
		0.60	30%		
Carbon-14	β–	0.156	100%		
(5730 y)					

Nuclide and Half-Life	Type of Decay	Particle Energies and Transition Probabilities		Electromagnetic Transitions	
		Energy MeV	Transition Probability	Photon Energy MeV	Photons Emitted
Cerium-139	e.c.		100%	0.166	79.9%
(137.5 d)				0.033 –0.039	~90%(La KX-rays)
Cerium-141	β–	0.436	70%	0.145	48%
(32.5 d)		0.581	30%	0.035 –0.042	~17%(Pr KX-rays)
Cerium-144	β–	0.182	19.1%	0.034	0.1%
(284.3 d)		0.216	0.2%	0.040	0.4%
		0.236	4.4%	0.053	0.1%
		0.316	76.3%	0.080	1.5%
				0.100	0.03%
				0.134	10.8%
		via 7.2m^{144}Pr			
	β–	1.534	0.05%	0.059	0%
	i.t.		1.15%	0.696	0.05%
				0.814	0.05%
		via 7.2m^{144}Pr			
	β–	0.808	1.0%	0.696	1.53%
		2.298	1.2%	1.489	0.28%
		2.994	97.75%	2.186	0.72%
Cesium-134	β–	0.09	26%	0.475	1.5%
(2.06 y)		0.42	2.5%	0.563	8.1%
		0.66	71.5%	0.569	14.0%
				0.605	97.5%
				0.796	85.4%

(continued)

Nuclide and Half-Life	Type of Decay	Particle Energies and Transition Probabilities		Electromagnetic Transitions	
		Energy MeV	Transition Probability	Photon Energy MeV	Photons Emitted
				0.802	8.6%
				1.038	1.0%
				1.168	2.0%
				1.365	3.3%
Cesium-137	β–	0.512	94.6%		
(30.17 y)		1.174	5.4%		
	via ^{137m}Ba(2.6 m):			0.662	85.1%
				Ba KX-rays	~7%
				(0.032 –0.038)	
Chlorine-36	β–	0.709	98.1%		
(3.01×10^5 y)	e.c.		1.9%		
Chromium-51	e.c.		100%	0.320	9.83%
(27.7 d)				0.005 –0.006	~22%(V KX-rays)
Cobalt-56	β+	0.4	1%	0.511	from β+
(78.0 d)		1.5	18%	0.847	99.97%
	e.c.		81%	0.977	1.4%
				1.038	14.0%
				1.175	2.3%
				1.238	67.6%
				1.360	4.3%
				1.771	15.7%
				2.015	3.1%
				2.035	7.9%
				2.599	16.9%
				3.010	1.0%

Nuclide and Half-Life	Type of Decay	Particle Energies and Transition Probabilities		Electromagnetic Transitions	
		Energy MeV	Transition Probability	Photon Energy MeV	Photons Emitted
				3.202	3.0%
				3.254	7.4%
				3.273	1.8%
				3.452	0.9%
				others	<1% each
Cobalt-57	e.c.		100%	0.014	9.5%
(271.7 d)				0.122	85.5%
				0.136	10.8%
				0.570	0.01%
				0.692	0.16%
				others	low
				Fe KX-rays	~55%
				(0.006 –0.007)	
Cobalt-58	β+	0.475	15.0%	0.511	from β+
(70.8 d)	e.c.		85.0%	0.811	99.4%
				0.864	0.7%
				1.675	0.5%
				0.006 –0.007	~26%(Fe KX-rays)
Cobalt-60	β–	0.318	99.9%	1.173	99.86%
(5.27 y)		1.491	0.1%	1.333	99.98%
				others	<0.01%
Curium-244	α	5.763	23.6%	0.043	0.02%
(17.8 y)		5.806	76.4%	0.099	0.0013%
		others	low	0.152	0.0014%
				others	low

(continued)

Nuclide and Half-Life	Type of Decay	Particle Energies and Transition Probabilities		Electromagnetic Transitions	
		Energy MeV	Transition Probability	Photon Energy MeV	Photons Emitted
				(up to ~0.8)	
				Pu LX-rays	~8%
				(0.012 –0.023)	
Europium-152	β–	0.185	1.8%	0.122	28.2%
(13.3 y)		0.394	2.4%	0.245	7.4%
		0.705	13.8%	0.344	26.3%
		1.484	8.0%	0.411	2.2%
		others	1.7%	0.444	3.1%
	β+		~0.02%	0.779	12.8%
	e.c.		72.3%	0.867	4.1%
				0.964	14.4%
				01.086	10.0%
				1.090	1.7%
				1.112	13.6%
				1.213	1.4%
				1.299	1.6%
				1.408	20.6%
				~75 others	(<1% each)
Gallium-67	e.c.		100%	0.091	3.6%
(78.26 h)				0.185	23.5%
				0.209	2.6%
				0.300	16.7%
				0.394	4.4%
				0.494	0.1%
				0.704	0.02%
				0.795	0.06%

Nuclide and Half-Life	Type of Decay	Particle Energies and Transition Probabilities		Electromagnetic Transitions	
		Energy MeV	Transition Probability	Photon Energy MeV	Photons Emitted
				0.888	0.17%
				0.008 –0.010	43%(Zn KX-rays)
				via 9.2μs ^{67m}Zn	
				0.093	37.6%
				0.008 –0.010	13%(Zn KX-rays)
Gold-198	β–	0.285	1.32%	0.412	95.45%
(2.696 d)		0.961	98.66%	0.676	1.06%
		1.373	0.02%	1.088	0.23%
Gold-199	β–	0.25	21%	0.050	0.3%
(3.13 d)		0.29	72%	0.158	39.6%
		0.45	7%	0.208	8.8%
				0.069 –0.083	~18%(Hg KX-rays)
Hydrogen-3	β–	0.0186	100%		
(12.43 y)					
Indium-111	e.c.		100%	0.171	90.9%
(2.804 d)				0.245	94.2%
				0.023 –0.027	~84%(Cd KX-rays)
Iodine-125	e.c.		100%	0.035	7%
(60.0 d)				Te KX-rays	138%
				(0.027 –0.032)	

(continued)

Nuclide and Half-Life	Type of Decay	Particle Energies and Transition Probabilities		Electromagnetic Transitions	
		Energy MeV	Transition Probability	Photon Energy MeV	Photons Emitted
Iodine-129	β–	0.150	100%	0.040	7.5%
(1.57 × 10^7 y)				Xe KX-rays (0.030 –0.035)	~69%
Iodine-131	β–	0.247	1.8%	0.080	2.4%
(8.04 d)		0.304	0.6%	0.284	5.9%
		0.334	7.2%	0.364	81.8%
		0.606	89.7%	0.637	7.2%
		0.806	0.7%	0.723	1.8%
	1.3% of ^{131}I decays via ^{131m}Xe (12 d)				
	i.t.		100%	0.164	2%
			(percentages related to disintegrations of ^{131m}Xe)		
Iron-55	e.c.		100%	Mn KX-rays	~28%
(2.69 y)				(0.0059 –0.0065)	
Iron-59	β–	0.084	0.1%	0.143	0.8%
(44.6 d)		0.132	1.1%	0.192	2.8%
		0.274	45.8%	0.335	0.3%
		0.467	52.7%	0.383	0.02%
		1.566	0.3%	1.099	55.8%
				1.292	43.8%
				1.482	0.06%
Krypton-85	β–	0.158	0.43%		
(10.73 y)		0.672	99.57%		

Nuclide and Half-Life	Type of Decay	Particle Energies and Transition Probabilities		Electromagnetic Transitions	
		Energy MeV	Transition Probability	Photon Energy MeV	Photons Emitted
		via ^{85m}Rb(0.96 μs)			
				0.514	0.43%
Lanthanum-140	β−	1.247	11%	0.131	0.8%
(40.27 h)		1.253	6%	0.242	0.6%
		1.288	1%	0.266	0.7%
		1.305	5%	0.329	21%
		1.357	45%	0.432	3.3%
		1.421	5%	0.487	45%
		1.685	18%	0.752	4.4%
		2.172	7%	0.816	23%
		others	low	0.868	5.5%
				0.920	2.5%
				0.925	6.9%
				0.950	0.6%
				1.597	95.6%
				2.348	0.9%
				2.522	3.3%
				others	<0.5% each
Lead-210	α		2×10^{-6}%	0.046	~4%
(22.3 y)	β−	0.015	~80%	0.009 -	
		0.061	~20%	0.017	~21% (Bi LX-rays)
	Daughter ^{210}Bi (5.01 d)				

(continued)

Nuclide and Half-Life	Type of Decay	Particle Energies and Transition Probabilities		Electromagnetic Transitions	
		Energy MeV	Transition Probability	Photon Energy MeV	Photons Emitted
	α	~4.67	~1.3 × 10^{-4}%		
	β–	1.161	~100%		
	Daughter ^{210}Po (138.38 d)				
	α	5.305	100%		
Manganese-54	e.c.		100%	0.835	100%
(312.5 d)				0.0055	~25% (Cr KX-rays)
Mercury-203	β–	0.212	100%	0.279	81.5%
(46.6 d)				0.071 –0.085	12.8% (Tl KX-rays)
Molybdenum-99	β–	0.454	18.3%	0.041	1.2%
(66.2 h)		0.866	1.4%	0.141	5.4%
		1.232	80%	0.181	6.6%
		others	0.03%	0.366	1.4%
				0.412	0.02%
				0.529	0.05%
				0.621	0.02%
				0.740	13.6%
				0.778	4.7%
				0.823	0.13%
				0.961	0.1%

Nuclide and Half-Life	Type of Decay	Particle Energies and Transition Probabilities		Electromagnetic Transitions	
		Energy MeV	Transition Probability	Photon Energy MeV	Photons Emitted
				via 6.02h 99mTc in equilibrium	
				0.002	~0%
				0.141	83.9%
				0.143	0.03%
Neptunium-237	α	4.638	6%	0.029	12%
(2.14×10^6 y)		4.663	3.3%	0.087	13%
		4.765	8%	0.106	0.08%
		4.770	25%	0.118	0.18%
		4.787	47%	0.131	0.09%
		4.802	~3%	0.134	0.07%
		4.816	2.5%	0.143	0.44%
		4.872	2.6%	0.151	0.25%
		others	<2%	0.155	0.10%
				0.169	0.08%
				0.193	0.06%
				0.195	0.21%
				0.202	0.05%
				0.212	0.17%
				0.214	0.05%
				0.238	0.07%
				others	low intensity
	Daughter ^{233}Pa				
		(27.0 d)			
		0.15	27%	0.075	1.3%
		0.17	15%	0.087	2.0%
		0.23	36%	0.104	0.7%

(continued)

Nuclide and Half-Life	Type of Decay	Particle Energies and Transition Probabilities		Electromagnetic Transitions	
		Energy MeV	Transition Probability	Photon Energy MeV	Photons Emitted
		0.26	17%	0.300	6.5%
		0.53	2%	0.312	38%
		0.57	3%	0.340	4.3%
				0.375	0.7%
				0.398	1.3%
				0.416	1.7%
Nickel-63 (100 y)	β–	0.066	100%		
Niobium-95 (35.0 d)	β–	0.160	>99.9%	0.766	99.8%
Phosphorus-32 (14.3 d)	β–	1.709	100%		
Plutonium-238	α	5.445	28.7%	0.043	IC
(87.7 y)		5.499	71.1%	U LX-rays	~13%
		others	0.2%	(0.011 –0.022)	
				U KX-rays	~2.1 × 10^{-4}%
				(0.094 –0.115)	
Polonium-210	α	4.5	0.001%	0.802	0.0012%
(138.38 d)		5.305	100%		
Potassium-42	β–	1.683	0.3%	0.312	0.3%
(12.36 h)		1.995	17.6%	0.900	0.05%
		3.520	82%	1.021	0.02%
		others	0.1%	1.525	17.9%
				1.921	0.04%

Nuclide and Half-Life	Type of Decay	Particle Energies and Transition Probabilities		Electromagnetic Transitions	
		Energy MeV	Transition Probability	Photon Energy MeV	Photons Emitted
Promethium-147	β−	0.103	low	0.121	2.85 × 10^{-3}%
				2.424	0.02%
				others	<0.01% each
(2.623 y)		0.225	~100%		
Radium-226	α	4.598	5.5%	0.186	3.4%
(1600 y)		4.781	94.5%		
via daughters in equilibrium:					
^{222}Rn(3.824 d)	α	5.486	100%		
^{218}Po(3.05 m)	α	6.000	~100%		
	β−	0.277	~0.02%		
^{218}At(~2 s)	α,β−		very low		
^{218}Rn(3.0 × 10^{-2} s)	α		very low		
^{214}Pb(26.8 m)	β−	0.21	0.5%	0.053	IC
		0.51	15.5%	0.242	6.7%
		0.69	42%	0.295	16.9%
		0.74	36%	0.352	32.0%
		1.03	6%		
^{214}Bi(19.8 m)	α	4.9–5.5	0.02%	0.273	5.3%
	β−	0.42	11%	0.609	41.7%
		1.02	23%	0.769	5.3%
		1.51	18%	1.120	14.3%
		1.55	15%	1.238	5.0%
		1.88	9%	1.378	4.8%
		2.6	4%	1.764	15.9%
		3.27	20%	2.204	5.3%
^{214}Po(1.62 × 10^{-4} s)	α	7.688	100%		
^{210}Tl(1.30 m)	β−		very low		

(continued)

Nuclide and Half-Life	Type of Decay	Particle Energies and Transition Probabilities		Electromagnetic Transitions	
		Energy MeV	Transition Probability	Photon Energy MeV	Photons Emitted
^{210}Pb and daughters (not necessarily in equilibrium)					
^{210}Pb(22.3 y)	α		2 × 10^{-6}%	0.046	~4%
	β−	0.015	~80%	Bi LX-rays	~21%
		0.61	~20%	(0.009–0.017)	
daughters:†					
^{210}Bi(5.01 d)	α	4.67	1.3 × 10^{-4}%		
	β−	1.161	100%		
^{210}Po(138.38 d)	α	4.5	0.001%	0.802	0.0012%
		5.305	100%		
^{206}Tl(4.20 m)	β−	present in very low abundance			
		† percentages relate to disintegration of each daughter			
Rubidium-86	β−	0.69	8.8%	1.077	8.8%
(18.7 d)		1.77	91.2%		
Ruthenium-103	β−	0.101	6.3%	0.053	0.4%
(39.26 d)		0.214	89.0%	0.113	~0.01%
		0.456	0.3%	0.242	~0.01%
		0.711	4.4%	0.295	0.3%
				0.444	0.4%
				0.497	88.2%
				0.557	0.8%
				0.610	5.5%
				0.020 –0.023	~0.9(Rh KX-rays)

Nuclide and Half-Life	Type of Decay	Particle Energies and Transition Probabilities		Electromagnetic Transitions	
		Energy MeV	Transition Probability	Photon Energy MeV	Photons Emitted
				via ^{103m}Rh (56 m)	
				0.040	0.1%
				0.020 –0.023	~8%(Rh KX-rays)
Ruthenium-106	β–	0.039	100%		
(369 d)					
	via ^{106}Rh(30.4 s)				
	β–	1.98	1.7%	0.512	20.6%
		2.41	10.5%	0.616	0.7%
		3.03	8.4%	0.622	9.9%
		3.54	78.9%	0.874	0.4%
		others	0.5%	1.050	1.5%
				1.128	0.4%
				1.562	0.2%
				others	<0.1% each
Scandium-46	β–	0.357	~100%	0.889	100%
(83.3 d)		1.48	0.004%	1.121	100%
Selenium-75	e.c.		100%	0.066	1.1%
(119.8 d)				0.097	3.5%
				0.121	17.3%
				0.136	59.0%
				0.199	1.5%
				0.265	59.1%
				0.280	25.2%
				0.401	11.6%

(continued)

Nuclide and Half-Life	Type of Decay	Particle Energies and Transition Probabilities		Electromagnetic Transitions	
		Energy MeV	Transition Probability	Photon Energy MeV	Photons Emitted
				others	<0.05% each
				As KX-rays	~50%
				(0.010 –0.012)	
	via ^{75m}As(16.4 ms):				
				0.024	0.03%
				0.280	5.4%
				0.304	1.2%
				As KX-rays	~2.6%
				(0.010 –0.012)	
Silver-110m	β–	0.084	67.6%	0.116	0%
(249.8 d)		0.531	31%	0.447	3.4%
		others	low	0.620	2.7%
	i.t.		1.4%	0.658	94.2%
				0.678	11.1%
				0,687	6.9%
				0.707	16.3%
				0.744	4.5%
				0.764	22.5%
				0.818	7.2%
				0.885	71.7%
				0.937	34.4%
				1.384	25.7%
				1.476	4.1%
				1.505	13.7%
				1.562	1.2%

Nuclide and Half-Life	Type of Decay	Particle Energies and Transition Probabilities		Electromagnetic Transitions	
		Energy MeV	Transition Probability	Photon Energy MeV	Photons Emitted
	via ^{110}Ag(24.5 s)				
	β–	2.23	0.1%	0.658	0.1%
		2.89	1.3%	others	low intensity
Sodium-22	β+	0.546	90.49%	0.511	from β+
(2.60 y)		1.820	0.05%	1.275	99.95%
	e.c.		9.46%		
Sodium-24	β–	0.284	0.08%	1.369	100%
(15.02 h)		1.392	99.92%	2.754	99.85%
				3.861	0.08%
Strontium-85	e.c.		100%	0.36	0.002%
(64.8 d)				0.88	0.01%
				0.013 –0.015	~60%
					(Rb KX-rays)
				via ^{85m}Rb	
				(0.96 μs)	
				0.514	99.2%
Strontium-89	β–	0.554	~0.01%		
(50.5 d)		1.463	~100%		
				via ^{89m}Y (16 s)	
				0.909	~0.01%
Strontium-90	β–	0.546	100%		
(28.6 y)					
	via ^{90}Y(64.1 h):				

(continued)

Nuclide and Half-Life	Type of Decay	Particle Energies and Transition Probabilities		Electromagnetic Transitions	
		Energy MeV	Transition Probability	Photon Energy MeV	Photons Emitted
		0.513	~0.02%	1.761	IC
		2.274	~99.98%		
Sulfur-35	β–	0.167	100%		
(87.4 d)					
Technetium-99	β–	0.204	low	0.089	6 × 10^{-4}%
(2.13 × 10^5 y)		0.293	~100%		
Technetium-99m	i.t.		100%	0.002	~0%
(6.02 h)				0.141	88.5%
				0.143	0.03%
	Daughter ^{99}Tc				
Terbium-160	β–	0.441	4.4%	0.087	13.8%
(72.3 d)		0.481	10.0%	0.197	5.2%
		0.553	3.3%	0.216	3.9%
		0.575	46.4%	0.299	26.9%
		0.791	6.8%	0.765	2.0%
		0.874	26.8%	0.879	29.8%
		others	2.3%	0.962	9.9%
				0.966	24.9%
				1.178	15.1%
				1.200	2.3%
				1.272	7.6%
				1.312	2.9%
				others	<2% each
Thallium-201	e.c.		100%	0.031	0.22%
(73.1 h)				0.032	0.22%
				0.135	2.65%

Nuclide and Half-Life	Type of Decay	Particle Energies and Transition Probabilities		Electromagnetic Transitions	
		Energy MeV	Transition Probability	Photon Energy MeV	Photons Emitted
				0.166	0.16%
				0.167	10.0%
				0.068 –0.082	~95%(Hg KX-rays)
Thallium-204	β–	0.763	97.4%	0.069 –0.083	~1.5%
(3.78 y)	e.c.		2.6%		(Hg KX-rays)
Thorium-228	α	5.140	0.03%	0.085	1.6%
(1.913 y)		5.176	0.2%	0.132	0.19%
		5.211	0.4%	0.167	0.12%
		5.341	28%	0.216	0.29%
		5.424	71%		
		others	low		
	via daughters in equilibrium (percentages refer to disintegrations of each daughter)				
Radium-224	α	5.447	5.2%	0.241	4.2%
(3.64 d)		5.684	94.8%		
Radon-220	α	5.747	0.07%	0.542	0.07%
(55.3 s)		6.288	99.93%		
Polonium-216	α	5.984	~0.002%	0.808	0.002%
(0.15 s)		6.777	~100%		
Lead-212	β–	0.155	5%	0.115	0.6%
(10.6 h)		0.332	82%	0.239	44.8%
		0.571	13%	0.300	3.4%
Bismuth-212	α	5.607	0.4%	0.040	1%
(60.6 m)		5.768	0.6%	0.288	0.3%
		6.051	25.2%	0.328	0.1%

(continued)

Nuclide and Half-Life	Type of Decay	Particle Energies and Transition Probabilities		Electromagnetic Transitions	
		Energy MeV	Transition Probability	Photon Energy MeV	Photons Emitted
		6.090	9.6%	0.453	0.4%
		others	low	0.727	6.6%
				0.785	1.1%
	β–	0.445	0.7%	0.893	0.4%
		0.572	0.3%	0.952	0.2%
		0.630	1.9%	1.079	0.5%
		0.738	1.5%	1.513	0.3%
		1.524	4.5%	1.621	1.5%
		2.251	55.2%	1.680	0.1%
		others	low	1.806	0.1%
Polonium-212					
(3.05 × 10^{–7} s)	α	8.785	100%		
Thulium-170	β–	0.884	22.8%	0.084	3.4%
(128 d)		0.968	77%		
	e.c.		0.2%		
Tin-113	e.c.		100%	0.255	2.1%
(115.1 d)				0.024 –0.028	73% (in KX-rays)
	Daughter ^{113m}In		99.5%	0.392	64.9%
	i.t.		100%	0.024 –0.028	24%(in KX-rays)
Tritium		see Hydrogen-3			
Tungsten-185	β–	0.304	low	0.125	~0.005%
(75.1 d)		0.429	>99.9%		

Nuclide and Half-Life	Type of Decay	Particle Energies and Transition Probabilities		Electromagnetic Transitions	
		Energy MeV	Transition Probability	Photon Energy MeV	Photons Emitted
Uranium-238	α	4.145	23%	0.048	IC
(4.49 × 10^9 y)		4.195	77%		
daughters in equilibrium:					
^{234}Th	β−	0.100	12%	0.030	IC
(24.1 d)		0.101	21%	0.063	5.7%
		0.193	67%	0.092	3.2%
				0.093	3.6%
^{234m}Pa	β−	2.29	98%	0.043	IC
(1.17 m)		others	0.13%	0.767	0.2%
	i.t.		low	0.810	0.5%
^{234}Pa	β−			1.001	0.6%
(6.70 h)			very low		
^{234}U	α	4.723	27.5%	0.053	0.1%
(2.48 × 10^5 y)		4.773	72.5%		
Xenon-133	β−	0.266	0.9%	0.080	0.4%
(5.25 d)		0.346	99.1%	0.081	36.6%
				0.160	0.05%
				0.030 –0.036	~46%(Cs KX-rays)
Yttrium-88	β+	0.763	~0.2%	0.511	from β+
(106.6 d)	e.c.		99.8%	0.898	93.2%
				1.383	0.04%
				1.836	99.36%
				2.734	0.6%
				3.219	0.009%
				3.52	0.007%
				Sr KX-rays (0.014 –0.016)	~60%

(continued)

Nuclide and Half-Life	Type of Decay	Particle Energies and Transition Probabilities		Electromagnetic Transitions	
		Energy MeV	Transition Probability	Photon Energy MeV	Photons Emitted
Yttrium-90	β–	0.513	~0.02%	1.761	0%
(64.1 h)		2.274	~99.98%		
Yttrium-91	β–	0.340	0.3%	1.205	0.3%
(58.5 d)		1.545	99.7%		
Zinc-65	β+	0.325	1.46%	0.345	~0.003%
(243.8 d)	e.c.		98.54%	0.511	from β+
				0.770	~0.003%
				1.115	50.7%
				0.008 –0.009	~38%(Cu KX-rays)
Zirconium-95	β–	0.365	54.7%	0.724	44.5%
(64.0 d)		0.398	44.6%	0.757	54.6%
		0.887	0.7%		
		1.12	low		
				via 86.6 h ^{95m}Nb in equilibrium	
				0.235	0.2%
	Daughter ^{95}Nb				

Index

A

absorbed dose, 484
absorber, 248, 261–264, 346
  heavy, 252
  thin, 250–251
absorption, 16, 153, 254, 262, 267
absorption coefficients, 167, 197–198, 258, 264
acceptors, 19, 37–41, 130
activity, 483–484
Ag, 6, 221, 236, 438
AgBr, 60, 62, 442
AgCl, 62, 442
AlAs, 11, 25, 100, 237, 439
AlGaAs, 60, 403
alloy semiconductors, 61, 448
alloys, binary, 59, 437, 444
AlN, 11, 52, 67
AlP, 11, 25, 100
α-GaN, 52, 237
alpha particles, 254, 342, 484
AlSb, 11, 25, 52, 340
aluminum antimonide. *See* AlSb
aluminum arsenide. *See* AlAs
aluminum gallium arsenide. *See* AlGaAs
aluminum phosphide. *See* AlP
amorphous material, 67, 428
annealing, 120, 122, 236
anode, 269, 373, 378, 382
  electrodes, 377, 389
  strip, 388–389
antimony. *See* Sb
applications, optoelectronic, 10, 101, 106, 310, 443
Aquadag, 209, 240
atomic, 248
  number, 249, 257, 261, 430
  weights, 253, 419, 448
atoms, 3, 50, 57, 59, 69, 432
  acceptor, 40, 238
  self-interstitial, 71
Au, 6, 52, 221, 240
  contacts, 220, 233
Auger
  electron spectroscopy, 109
  electrons, 256
  recombination, 9
    nonradiative band-to-band, 8

B

B, 38–39, 86, 347–352, 354, 439
ballistic deficiency effects, 371, 375
band
  edges, 31, 43, 219
  morphology, 16
  structure, 3–4, 12–14, 16, 28, 99
bandgap, 5–6, 9–10, 42, 101, 435
  direct, 17, 310
  energy, 5–6, 10, 61, 435
    direct, 452–453
    effective, 61
    function of, 221, 274, 398, 450
  engineering, 101
  indirect, 9
  materials
    direct, 16, 19, 436, 450
    indirect, 13, 16, 19, 40, 163
  narrow, 274
  wide, 232, 274, 443
bands, 5, 13, 17, 23, 28, 32
  heavy-hole, 17, 23
  impurity, 42
  light-hole, 17, 24, 32
barrier, 143, 211–212, 222
  height, 215–219, 228, 231, 241
  potential, 77, 141, 211, 216–217, 228

shape, 216
width, 215, 223, 225, 236, 238
width reduction, 236, 238
BAs, 348
bias, 142, 156, 160, 165, 277
reverse, 142–143, 172, 215–216, 224–225
BiI_3, 335–336, 338
binary
alloys and ternary compounds, 476, 478–480
compounds, 60, 100, 428, 444
bismuth triiodide. *See* BiI_3
BN, 67, 255, 349–350, 351, 354
bonds, 59, 241, 292, 432, 438, 441
boron. *See* B
boron arsenide. *See* BAs
boron nitride. *See* BN
boron phosphide. *See* BP
BP, 255, 289, 348–349
Bragg reflection, 133–134, 190
Bravais, 51, 53, 55, 430
lattice, 51, 54–55, 428–430
Bravias lattice, designations, 51, 428
Bridgman, 90–91, 93, 272
high pressure, 91
methods, 90, 324, 335
horizontal, 90
modified vertical, 338–339
vertical, 309, 321, 336–337
Bridgman–Stockbarger method, 90, 95
Bridgman technique, modified, 317–318
brilliance, 177–178, 180–181
Brinell hardness scale, 434
built-in potential, 150–151, 212, 214–215, 241
bulk
crystals, 67, 327
defects, 69–70, 77, 125
leakage, 277

C

cadmium. *See* Cd
cadmium antimonide. *See* CdSb
cadmium manganese telluride. *See* CdMnTe
cadmium selenide. *See* CdSe
cadmium sulfide. *See* CdS
cadmium telluride. *See* CdTe
cadmium telluride selenide. *See* $Cd_xTe_{1-x}Se$
cadmium zinc selenide. *See* $Cd_xZn_{1-x}Se$
cadmium zinc telluride. *See* CdZnTe
carrier
concentration, 41, 106, 157, 232, 265
generation, 271
lifetimes, 167, 169, 267, 370
mobilities, 42, 68, 159, 373
trapping, 167, 271, 288
velocity, 26–27, 269
carriers, 9, 19, 40, 222, 226, 229
free, 6–7, 36, 40, 209, 341
minority, 39, 277
single, 381, 391, 393, 399, 401–402
thermal generation of, 274, 277
cathode, 164–166, 269–270
signals, 162
cation, 72, 297, 310, 325, 439–440, 442
CCD (charge-coupled device), 197, 250
CCE (charge collection efficiency), 163, 169–170, 270
Cd, 236, 255, 310, 440–441
CdMnTe, 255, 317–318
CdS, 11–12, 25, 67
CdSb, 474
CdSe, 11, 36, 319, 398
CdTe, 93, 102, 107, 311–314
$Cd_xTe_{1-x}Se$, 321–322
$Cd_xZn_{1-x}Se$, 320–321
CdZnTe, 77, 314, 317, 370
detectors, 317, 370, 379, 395
cesium chloride. *See* CsCl
chalcogenides, 323, 338, 442–443
chalcopyrites, 443–444. *See also* $CuFeS_2$
charge, 163, 169, 212, 248–249, 269
carrier mobilities, 43, 99, 292, 438
carriers, 19, 41, 235, 242, 264
charge collection, incomplete, 271, 280, 282
charge collection efficiency. *See* CCE

charge-coupled device. *See* CCD
charge induction, 160, 242, 269, 380–383, 385–386
charged particles, 7, 248–250
chemical vapor deposition. *See* CVD
circular transfer length method. *See* CTLM
collecting electrode, 269, 374, 376, 379, 391
compounds, 56, 59, 64–65, 67, 69
  boron, 255, 348–350
  covalent, 234, 332
  I–VII, 120, 448, 450
  II–VI, 65, 101, 167
  III–V, 439
  III–VI chalcogenides, 323
  layered, 325, 443
  organic, 110, 112, 340
  quaternary, 438, 440
  ternary, 60–61, 348, 438, 440–441
Compton, 251, 262–263
  effect, 256, 259, 263
  scattering, 258
concentrations, dopant , 22, 43
conduction, 28, 38, 40, 43, 146, 226
  band, 5, 15, 23–24, 27–32, 34
  process, 5, 17–18, 28–29, 227
conductivity, 2, 29, 39–41, 43, 45, 155
contact barrier height, 212
contact resistance, 146–149, 209, 230, 238–239
contacts, 127, 130, 146, 227–228, 230
  rectifying, 209, 211, 309
contamination, 86, 88, 94, 96, 108, 113
coplanar grid detector, 242, 389–390, 396
copper. *See* Cu
copper aluminum sulfide. *See* $CuAlS_2$
copper bromide. *See* CuBr
copper chloride. *See* CuCl
copper gallium sulfide. *See* $CuGaS_2$
copper indium selenide. *See* $CuInSe_2$
copper iron sulphide (chalcopyrite). *See* $CuFeS_2$
covalent bonding, 72, 326, 432
crystal
  defects, 69, 87–88, 170
  growth, 68–69, 78, 80
    epitaxial, 105
  imperfections, 70, 391
  lattice, 7, 13, 22, 38, 50, 137
    constant, 61
    planes, 134
  momentum, 15, 22
  planes, 55, 138
  polymorphs, 69
  seed, 82, 88, 90, 97, 107
  structure, 50–51, 54, 133, 427, 429
    wurtzite, 65,67
    zinc blende, 65, 67, 428, 431, 439, 445
  surface, 73, 86–87, 122, 125
  symmetries, 32
  systems, 51, 55, 430
    basic, 51–52, 54–55, 429–430
  types, 51, 53, 428–429
crystalline, 14, 56
  domains, 137–138
  materials, 68, 72, 98, 133, 428
crystallize, 64, 69, 292, 310
CsCl, 428
CTLM (circular transfer length method), 147–148
Cu, 2, 6, 180, 221, 236, 438, 444
$CuAlS_2$, 443
Cubic, 51–53, 64, 428–429
CuBr, 442
CuCl, 60, 442
$CuFeS_2$, 52, 443
$CuGaS_2$, 443
$CuInSe_2$, 443
curie, definition, 416, 483–484
current density, 44, 229, 241, 266
current-voltage characteristics, 141–142, 208–209, 232, 307, 337
CVD (chemical vapor deposition), 103, 105, 272
Czochralski method, 82, 95
  encapsulated, 85
  vapor pressure controlled, 87, 96

D

Debye–Waller factor, 196
deep level transient spectroscopy. *See* DLTS

defects, 43, 69, 72, 122, 124, 170
active, 149, 171
antisite, 70, 72
crystallographic, 125, 434
density, 32, 419, 430
surface state, 219
depletion layer, 215, 224, 229
depletion region, 143, 149, 214, 222, 228–229
detector
applications, 90, 161, 329, 437
capacitance, 151, 276, 278–279, 295
design, 258, 260, 277
fabrication, 80, 242, 333
gas, 346, 381–382
material, 240, 397
performances, 140, 192–193, 370
diamond, 11, 64, 237, 296
structure, 53, 64, 428–429
dielectric, 275, 431
constant, 40, 192, 214, 450, 453
static, 214, 431, 450
direct bandgap semiconductors, 16, 31, 36, 163, 437
dislocation, 9, 43, 70, 72–74, 99, 125
densities, 125
edge, 74–75
line, 73–75
screw, 74–75
DLTS (deep level transient spectroscopy), 171–173
measurements, 174–175
donors, 37–41, 130, 238
dopants, 37, 40, 43, 130, 238
doped semiconductors, 9, 42, 222, 227, 238, 241
doping, 39, 42–43, 57, 225, 402
concentration, 151, 222, 226, 235–236
densities, 9, 42
drift-strip detectors, 388
drift velocities, 18–19, 26, 28, 160

E

effective masses, 18, 22–23, 31–32, 153
effects, photoelectric, 256, 260, 263
electrode design, 377, 380–381
electrodes
drift-strip, 388–389
ring, 391–392
electron, 14–15, 251, 255, 411, 484
acceptors, 38, 442
conduction, 28, 38, 153
drift velocity, 27
effective mass, 25, 225
energy, 14–15, 30
mass, 249, 410
free, 23
mobilities, 19–21, 437, 450
mobility-lifetime products, 437
momentum, 14–15
electron-hole pairs, 5, 7, 163, 248, 268
electronegativities, 57, 220
electronic
noise, 271, 275–276
component, 280–281
properties, 99, 130, 133, 437
electrons, free, 7–9, 18, 29, 32, 256
elemental semiconductors, 25, 27, 59–60, 82, 432
ENC (equivalent noise charge), 276, 279
energy, 12, 73, 414, 435, 483
activation, 40, 72, 171–172
bands, 3–4, 24, 42, 99
binding, 40, 256–258, 321
critical, 180, 186
density of electrons, 33
deposition, 248, 269
Fermi, 30–31, 35, 43, 215, 224
gaps, 7, 100, 238, 442, 453
forbidden, 9, 40, 307
levels, 3–6, 10, 39, 43, 171
Fermi, 212, 303
potential, 3, 14, 68, 213–214, 226
energy loss, 248–252
energy-momentum dispersion, 13
energy of defect formation, 102, 314
energy range, 180, 182, 186, 189, 270
energy resolution, 271, 291, 307, 379, 393, 396
energy states, 9, 15, 43
EPD (etch pit density), 74, 86, 96, 125
epitaxy, 80, 97, 108, 113, 238

equivalent noise charge. *See* ENC
etch pit density. *See* EPD
etching, 73–74, 86, 121, 124–125
EXAFS (extended X-ray absorption fine structure), 195
extended X-ray absorption fine structure. *See* EXAFS

F

Fano
  factor, 271, 273–274, 295
  noise, 102, 271, 273, 398
FE (field emission), 222–224, 226–228
FEL (free-electron laser), 178
Fermi
  energy, 30
  level, 6, 43, 210, 215, 218–219
    pinning, 43, 219–220
Fermi–Dirac probability function, 29, 33
FET (field effect transistor), 278, 298, 392
field effect transistor. *See* FET
field emission. *See* FE
float zone. *See* FZ
free-electron laser. *See* FEL
Frenkel defect, 71–72
Frisch, 382
  grid, 381, 393
FZ (float zone), 88–89, 272

G

Ga, 52, 236, 297, 438–439
GaAs, 2, 16–17, 23, 36, 110, 297
gallium. *See* Ga
gallium arsenide. *See* GaAs
gallium nitride. *See* GaN
gallium phosphide. *See* GaP
gallium selenide. *See* GaSe
gallium telluride. *See* GaTe
gamma ray detectors, 304, 326
gamma rays, 253, 255, 483–484
Γ-valley, 17
GaN, 11, 25, 60, 66–67, 302
GaP, 2, 25, 36, 237, 301, 439
GaSe, 323–324, 441
GaTe, 323, 325, 441
GDMS (glow-discharge mass spectrometry), 131–132
glow-discharge mass spectrometry. *See* GDMS
gold. *See* Au
grain boundaries, 68, 76–78, 170, 312
grid electrodes, 389
growth, 77, 97, 103, 108, 121
  bulk, 82
  epitaxial, 97, 102
growth process, 69, 72, 110, 342

H

Hall
  effect, 153–154, 201
    measurements, 153, 156, 163, 437
  mobilities, 155, 158, 437
harmonics, 185–186
Hecht equation, 163–165, 167–169, 270
hemispherical detector, 377, 379, 393
heteroepitaxy, 97–98, 110
heterostructures, 104–105, 354, 403
hexagonal, 51, 67, 436–437, 445
Hg, 236, 310, 337, 440, 445
HgBrI, 328–329
HgCdTe, 60, 255, 307–308
HgI_2, 270, 288, 326, 347
  detectors,140, 335, 347
HgTe, 106, 310, 440–441
high pressure Bridgman (HPB), 91. *See also* Bridgman
holes, 7–8, 19, 23, 32, 163
  free, 7–8, 13, 32
  heavy, 17, 23–24
  light, 13, 17, 23–24
  mobilities, 19–21, 327, 437, 450
  mobility-lifetime products, 166–167, 379, 437
homoepitaxy, 97, 105

hopping, 43
horizontal Bridgman, 92
Hume-Rothery rules, 57–59

I

ICP (inductively coupled plasma), 131–132
ICP-MS, 131–132
ICP-OES, 131–132
ideality factor, 142–144, 223–224, 241
image charges, 212–213, 218
image force lowering, 212, 228
impurities, 9, 21, 29, 40, 131
  acceptor, 39
  amphoteric, 40
  donor, 38, 318
  ionized, 41, 215
  substitutional, 71
impurity
  atoms, 18, 43, 70, 77
  scattering, 19–21
InAs, 237, 307–308
indirect bandgaps, 12, 16, 435
indium antimonide. *See* InSb
indium arsenide. *See* InAs
indium iodide. *See* InI
indium phosphide. *See* InP
induced charge, 163, 269–270, 380–382, 386
induced signal, 164, 372–373, 385–386
inductively coupled plasma. *See* ICP
ingots, 79, 81, 121
InI, 306–307
InP, 110, 270, 303–304, 435
InSb, 307–308
insertion devices, 177, 179, 186
insulators, 1–3, 5–7, 131, 171
interaction, depth of, 164, 270, 374
interface
  metal-semiconductor, 223, 230
  states, 219, 234
interfacial layers, 196, 218–219, 241
interstitial impurity atoms, 71
intrinsic carrier
  concentrations, 29, 36–37, 229
  density, 35, 37
ionic, 220, 331, 432
iconicity, 297, 432
  scale, 432
ionization, 248–249
  energies, 40, 195
ionizing radiation, 169, 268

J

junction, 150, 172, 213, 223, 241
  metal-semiconductor, 149, 212

K

K-shell, 256–257
KH (Knoop hardness), 120, 433–434, 435, 445
kinetic energy, 14, 249, 253, 256, 258–259
Klein–Nishina formula, 260–261
Knoop hardness. *See* KH

L

L-valley, 17
lapping, 120–124
lattice, 5,9–10, 51, 53
  centering, 52–54, 429
  constants, 100–101, 430
  mismatch, 98–99
  parameter, 50, 100, 430
    function of, 447–449, 451
  point, 50, 65, 427
  scattering, 19, 21
  site, 40, 70, 72
  sites, empty, 71–72
  spacing, 5, 9, 98
  vibrations, 19–20, 271
layers, 98, 106, 342, 403, 443
lead. *See* Pb
lead iodide. *See* PbI_2
lead selenide. *See* PbSe
LEC (liquid encapsulated Czochralski), 85, 87–88, 96, 272

Li, 342, 353
line defects, 70, 72–75
liquid encapsulated Czochralski. *See* LEC
liquid-phase epitaxy. *See* LPE
lithium. *See* Li
lithography, 120, 127–128
LPE (liquid-phase epitaxy), 102–104, 298, 304

M

magnesium sulfide. *See* MgS
magnets, bending, 179–180, 186
majority carriers, 39, 152, 222, 277, 399
mass-action law, 35, 37, 41
mass attenuation coefficients, 264
material properties, 192, 225, 396, 399
material purification, 80–81
materials, 3, 249–250, 253, 343–344, 427
  bulk, 141, 193, 277, 293, 304–305, 323
  contacting, 234, 240
  detector-grade, 85, 294
  direct gap, 10, 17, 31, 339
  group I–VII, 120, 437–438, 445
  group II–VI, 43, 62, 82, 310–311, 437, 440–441, 466, 468
  group IV–IV, 461–462
  group IV–VI, 25, 445–446
  group n–VII, 471–472, 474
  indirect gap, 24, 31
  insulating, 2, 160, 173, 207
  intrinsic, 32, 42, 295
  p-type, 41, 95, 152, 311, 322, 441
  semi-insulating, 298, 321
  soft, 122, 124, 434
MBE (molecular beam epitaxy), 102–103, 108
MCA (multi-channel analyzer), 170, 268
measuring contact resistance, 147
melt, 82, 85, 88, 90, 103–104
melting point, 69, 431, 441–442, 446
mercury. *See* Hg
mercuric bromoiodide. *See* HgBrI
mercuric iodide. *See* HgI_2
mercury cadmium telluride. *See* HgCdTe
mercury telluride. *See* HgTe
metal contact, 145, 149
metal-induced gap states. *See* MIGS
metal organic chemical vapor deposition. *See* MOCVD
metal organic chemical vapor epitaxy, 103
metal organic vapor phase epitaxy. *See* MOVPE
metal-semiconductor
  contacts, 141, 211, 230–231, 236, 241
  interface, 141, 146, 217–219, 222, 239
  junctions, 222, 268
metal work function, 218–219, 221
metallization, 120, 128, 147, 208, 438
  contacting, 233
metals, 1–2, 5–6, 210
method, traveling heater. *See* THM
MgS, 310, 440
MIGS (metal-induced gap states), 218, 221
Miller indices, 54–56
mobility, 18, 44, 153–154, 160, 163, 437
mobility temperature dependence, 20–21
MOCVD (metal organic chemical vapor deposition), 103, 110, 112, 445
Mohs scale, 434–435
molecular beam epitaxy. *See* MBE
molecular weight. *See* MW
momentum, 12–13, 15, 18, 268, 434
monochromator, 182–183, 185, 187–188
mosaicity, 138–140
MOVPE (metal organic vapor phase epitaxy), 103, 308
MTPVT (multi-tube physical vapor transport), 106–107, 113
mu-tau products, 163–165, 167, 288, 399, 453
multi-channel analyzer. *See* MCA
multi-tube physical vapor transport. *See* MTPVT

MW (molecular weight), 430, 447–450, 454

N

n-type, 34, 38, 40, 152
semiconductors, 39, 41, 45, 219, 222, 232, 241
n-type material, 39, 41, 152, 402
neutron
capture, 254, 342, 349
detection, 253, 255, 294, 342, 345, 349, 354
neutrons, 253, 342, 345–347, 350, 352
noise, 271, 275–280, 282
1/f, 278–279
components, 271, 281
current, 278–279
parallel, 278–279, 371
series, 278–279
shot, 102, 275–276, 278, 315
nuclei, 3, 249, 253, 255–256, 261, 283, 430
atomic, 253, 439, 441

O

OES (optical emission spectroscopy), 131–132
ohmic contacts, 141, 230–231, 234, 236, 239
optical emission spectroscopy. *See* OES
optoelectronic, 302–303
organic semiconductors, 340
othorhombic, 52, 429

P

p-type, 34, 39–40, 45, 152
pair, electron-hole, 7, 248, 268
pair creation energy, electron-hole, 273, 398, 435
pair production, 256, 261, 263–264
Pb, 6, 236, 438
PbI_2, 52, 122, 270, 334
PbSe, 25, 307, 442
Pearson
notation, 51, 428
symbols, 53, 55, 428–430
periodic table, 56, 59, 64, 417, 438
phonons, 8–9, 16, 271, 434–436
phosphorus, 38–39, 87
photo-induced current transient spectroscopy. *See* PICTS
photoconductor, 264–265, 334
photoelectric, 263
photoelectric absorption, 188, 257, 262
photoelectron, 251, 256
photons, 8, 16, 255–258, 260–261
scattered, 258–260
photoresist, 127–130
physical properties, 37, 69, 146, 444
PICTS (photo-induced current transient spectroscopy), 173–174
pixel detectors, 18, 384, 386
planar detectors, 152, 375–376, 379, 387
plane defects, 69–70, 76
planes, 54–56, 74, 135, 258
point defects, 69–73, 77, 239
point groups, crystallographic, 54–55, 429–430
Poisson's equation, 214–215, 431
polarization, 176, 192, 195, 258, 260
effects, 192–193, 195, 399, 431
polishing, 119–120, 122–123
polycrystalline materials, 68, 76
polytypes, 293–294, 334, 354
powder diffraction, 135–136
powder XRD, 136
preamplifier, charge sensitive, 276
precipitates, 43, 69, 74, 77, 79, 196

Q

Q-value, 254–255
quartz, 3, 50, 52, 83, 87
quaternary, compounds, 60–61, 100, 103, 440

R

radiation, 132, 175, 248, 268, 483–484
  detection, 249
  detectors, 26, 42, 78, 142, 175, 331
    applications, 101, 293, 318, 322
Rayleigh scattering, 257–258
recombination, 7–9, 167, 169, 228–229, 239, 266
  centers, 171, 239
reflection high energy electron diffraction. *See* RHEED
resistance, 45, 147, 158, 266, 276, 436
resistivity measurements, 147, 158
RHEED (reflection high energy electron diffraction), 109
rhombohedral, 51–53, 293–294, 428–429
ring-drift detectors, 391–392
rise time discrimination. *See* RTD
rocking curve, 137–138, 181, 185, 190
RTD (rise time discrimination), 372–374, 394

S

Si, 2, 10, 27, 37, 64
saturated carrier velocities, 26
Sb, 2, 38, 101, 439
SBH (Schottky barrier height), 219, 235, 402
scattering, 21, 26, 253, 255, 260, 264
  angle, 135, 138, 253, 258, 260–261
  elastic, 255–256, 258
Schottky, 141, 171, 211, 234, 402
  barrier, 149, 172, 211–212, 222, 229, 231–232, 277
  barrier height (*see* SBH)
  contacts, 142, 209, 230, 277, 297, 309
Se, 310–311, 321, 338, 438, 440–442
secondary electrons, 251
selenium. *See* Se
semiconducting materials, 54, 62, 151, 160, 427
semiconductor
  detectors, 195, 248, 250, 264
  groups, 64, 438, 445, 452
  materials, 26, 31, 36, 80, 146, 172, 303, 312
  parameters, 225, 453–454
  radiation detectors, 264–265, 267
  surface, 127, 152, 196, 209, 238–239
semiconductors, 1–3, 6, 10, 29, 210, 434
  covalent, 220–221
  extrinsic, 21, 34, 37, 41, 45
  group IV, 27, 292, 438
  indirect bandgap, 12, 16, 31, 437
  intrinsic, 29, 33–35, 38
  ionic, 220–221, 432
  p-type, 39, 41, 155, 211, 219, 232–233, 237
  wide bandgap, 274
semimetals, 6–7, 16
shaping time, 278–279, 371, 375
Shockley–Ramo theorem, 160, 213, 269, 373, 380
Si and Ge, 8, 24, 26, 28, 31, 270
SiC, 67, 121, 292–294
SiGe, 60, 292, 439
silicon. *See* Si
silicon carbide. *See* SiC
silver. *See* Ag
silver bromide. *See* AgBr
silver chloride. *See* AgCl
simple planar detectors, 79, 160, 343–344, 372, 378
single carrier sensing, 370
single-crystal XRD, 133, 135–136
single crystals, 82, 84, 108, 125, 293
small-pixel effect detectors, 384
Sn
  α-Sn, 292, 438
solid solutions, 50, 56–7, 59, 67–68, 427
solids, 1, 14, 68, 219, 419
space charge, 172, 212, 268–269, 380
space groups, 54–55, 428–430
sphalerite, 65
states, 5, 16, 33–34, 43, 218
  density of, 29, 31–34, 219
  effective density of, 31, 34

strain, 57, 98–99, 138–140
substitutional impurity atoms, 71
substrate, 79, 97, 101, 105
surface, 69, 72, 74, 210, 218, 241
 states, 218–219, 241, 303
  localized electronic, 218
surface leakage current, 228–229, 277
synchrotron
 light sources, 175–176
 radiation, 175–176, 178–179

T

Te, 52, 192, 310, 438, 440–442
 inclusions, 92
techniques
 bi-parametric, 374–375, 393
 float-zone growth, 88–89
 hot probe, 152–153
tellurium. *See* Te
TEM (transmission electron microscopy), 73, 125
ternary compounds, 336–337, 437, 444–445
TFE (thermionic field emission), 222–224, 226–228, 231
thallium bromide. *See* TlBr
thallium bromoiodide. *See* $TlBr_xI_{1-x}$
thallium chalcohalides, 337. *See also* $TlGaSe_2$, Tl_6SeI_4
thallium lead iodide. *See* TlPbI
thermal annealing, 119, 121–122, 130
thermal leakage currents, 102, 315
thermal neutrons, 253–254, 348, 484
thermally stimulated current. *See* TSC
thermionic emission, 222–225, 227–229, 232
thermionic field emission. *See* TFE
THM (traveling heater method), 93–94, 272, 311
Thomson cross section, 257–258, 260–261
Ti, 221, 236, 305–306
tin, gray (α-Sn). *See under* Sn
titanium. *See* Ti
Tl_6SeI_4, 338–339
TlBr, 330–331, 398, 481
$TlBr_xI_{1-x}$, 332–334
$TlGaSe_2$, 337–338
TlPbI, 52, 337
TMG (trimethylgallium), 110–111
TMI (trimethylindium), 110–111
TMZ (traveling molten zone), 93
transfer length method (TLM), 147
 circular, 147–148
transmission electron microscopy. *See* TEM
trapping, 170, 269, 373, 387
 centers, 74, 171, 271, 399
 lengths, 161, 270, 376, 387
 noise, 188, 273, 280–282
traps, deep-level, 8–9
traveling heater method. *See* THM
traveling molten zone. *See* TMZ
trigonal, 51–52, 428
trimethylgallium. *See* TMG
trimethylindium. *See* TMI
triple-axis rocking curve, 139. *See also* XRD
TSC (thermally stimulated current), 171, 202
Tubbs ionicity scale, 432
tunneling, 143, 224–225, 227
 mechanical, 222, 225, 231
twins, 76–78, 91, 318

V

vacancies, 70–72, 256, 271
valence band, 5, 8, 15, 23, 32, 34
valence electrons, 5, 33, 59, 445
 bound, 38
van der Pauw method, 157, 159
van der Waals forces, 325, 442–443
vapor-phase epitaxy. *See* VPE
vapor pressure controlled Czochralski method. *See* VCz method
VCz (vapor pressure controlled Czochralski) method, 87, 96
Vegard's law, 61, 196
velocities, 18, 27, 249, 266, 282, 398
 electron surface recombination, 169

vertical Bridgman, 92, 94
vertical gradient freeze. *See* VGF
VGF (vertical gradient freeze), 94–95
Vickers hardness, 435
VPE (vapor-phase epitaxy), 102–103, 105, 113, 298

W

wafers, 79, 98, 121, 126–127
weight, molecular, 430, 447–450, 454
weighting
  field, 380–381, 383
  potential, 381, 385–386
Wentzel–Kramers–Brillouin. *See* WKB
wigglers, 179–180, 186
WKB (Wentzel–Kramers–Brillouin), 225, 228
work functions, 141, 212, 220, 232–233, 241
wurtzite, 65, 67, 431

X

X-ray absorption fine structure. *See* XAFS
X-ray absorption near edge structure. *See* XANES
X-ray diffraction. *See* XRD
X-rays, 134, 255–256, 483–484
X-valley, 17
XAFS (X-ray absorption fine structure), 183, 195, 198
  measurements, 195–196
XANES (X-ray absorption near edge structure), 195, 197, 199
XRD (X-ray diffraction), 133, 135, 138

Z

zinc. *See* Zn
zinc blende, 52, 65, 297, 428, 439–440
zinc fraction, 101–102, 196, 315, 320
zinc oxide. *See* ZnO
zinc selenide. *See* ZnSe
zinc sulfide. *See* ZnS
zinc telluride. *See* ZnTe
Zn, 39, 52, 65, 116, 236, 310, 314, 441
ZnO, 25, 52, 67, 237, 468
ZnS, 11, 25, 66, 221, 429
ZnSe, 11, 25, 237, 311, 322
ZnTe, 11, 25, 101, 237, 314, 323
zone
  first Brillouin, 14, 31–32
  refining, 80

Printed and bound by CPI Group (UK) Ltd, Croydon, CR0 4YY
21/10/2024
01777112-0011